NUTRITIONAL PATHOPHYSIOLOGY OF OBESITY AND ITS COMORBIDITIES

NUTRITIONAL PATHOPHYSIOLOGY OF OBESITY AND ITS COMORBIDITIES

A Case-Study Approach

SUSAN ETTINGER

Past Associate Professor and Chair, Clinical Nutrition Department, New York Institute of Technology;
Adjunct Associate Professor, Nutrition and Food Science, Hunter College, City University of New York, New York, NY, USA;
Research Associate, Division of Endocrinology Diabetes & Nutrition, Mount Sinai St. Luke's-Roosevelt Hospital, New York, NY, USA

AMSTERDAM • BOSTON • HEIDELBERG • LONDON
NEW YORK • OXFORD • PARIS • SAN DIEGO
SAN FRANCISCO • SINGAPORE • SYDNEY • TOKYO
Academic Press is an imprint of Elsevier

Academic Press is an imprint of Elsevier
125 London Wall, London EC2Y 5AS, United Kingdom
525 B Street, Suite 1800, San Diego, CA 92101-4495, United States
50 Hampshire Street, 5th Floor, Cambridge, MA 02139, United States
The Boulevard, Langford Lane, Kidlington, Oxford OX5 1GB, United Kingdom

Medical Disclaimer
Medicine is an ever-changing field. Standard safety precautions must be followed, but as new research and clinical experience broaden our knowledge, changes in treatment and drug therapy may become necessary or appropriate. Readers are advised to check the most current product information provided by the manufacturer of each drug to be administered to verify the recommended dose, the method and duration of administrations, and contraindications. It is the responsibility of the treating physician, relying on experience and knowledge of the patient, to determine dosages and the best treatment for each individual patient. Neither the publisher nor the authors assume any liability for any injury and/or damage to persons or property arising from this publication.

British Library Cataloguing-in-Publication Data
A catalogue record for this book is available from the British Library

Library of Congress Cataloging-in-Publication Data
A catalog record for this book is available from the Library of Congress

ISBN: 978-0-12-803013-4

For Information on all Academic Press publications
visit our website at https://www.elsevier.com

Working together
to grow libraries in
developing countries

www.elsevier.com • www.bookaid.org

Publisher: Mica Haley
Acquisition Editor: Tari Broderick
Editorial Project Manager: Fenton Coulthurst
Production Project Manager: Edward Taylor
Designer: Victoria Pearson

Typeset by MPS Limited, Chennai, India

Contents

Foreword

This unique text by Dr. Susan Ettinger, entitled *Nutritional Pathophysiology of Obesity and Its Comorbidities* provides current information on the metabolic dysregulations that are common to a variety of diseases associated with obesity. In this presentation she provides an integrated approach to the known molecular actions of dietary components, and cites evidence suggesting the role of specific dietary components in preventing and/or ameliorating disease progression. The author notes methodologic and design problems that flaw prior and current research and offers future research avenues. Each chapter in this book is extensively researched to identify relationships between pathways of disease and nutrient functions. These relationships can be tested in bench and clinical protocols.

In addition to the nine chapters, the text is supplemented by five Essentials. These are appendices that describe nutrient relationships common to many of the diseases and comorbidities under discussion. The first three Essentials discuss nutrients grouped by common functions: (1) nutrients that modulate oxidative damage; (2) nutrients required for bone calcification and structure; and (3) nutrients consisting of trace elements and vitamins required for life but toxic if not carefully handled. Two additional Essentials review current understanding of the influence of diet on the microbiota and intestinal permeability, with recommendations for patient feeding.

The mechanistic approach to the molecular actions of nutrients is enhanced by the following: references to population studies that identify individuals and groups at risk for depletion due to the inadequacy of their diet; concomitant diseases; and medical and pharmaceutical therapy and stages of the life cycle. Each chapter provides current published recommendations for medical nutrition therapy. By juxtaposing disease processes and molecular functions of nutrients the text suggests an array of biological plausibilities in need of future investigation.

The mechanisms and resources presented in this text are invaluable to biomedical researchers and clinicians alike, who seek new ways to utilize the full potential of diet as adjuvant to medical therapy. This will facilitate the achievement of the ultimate objective, which is to prevent and control obesity and its comorbid disease states.

Dr. Sami Hashim

Dr. Hashim has 40 years of distinguished experience in the field of metabolism and nutrition at St Luke's and Roosevelt Hospitals. He is the senior attending physician in the Division of Endocrinology and Metabolism and Professor of Nutritional Medicine at Columbia University College of Physicians and Surgeons.

S. Hashim

Foreword

As a nutrition educator, I am delighted with the creation of *Nutritional Pathophysiology of Obesity and Its Comorbidities*. In this text, Dr. Ettinger has delineated current understanding of the interlocking themes that potentiate the metabolic dysregulations and characterize obesity and metabolic syndrome. She follows the pathology of these deviant pathways as they exert tissue damage and lead to comorbid conditions associated with obesity, such as diabetes, cardiovascular disease, and involvement of the bone, lung and immune defense.

Most importantly for students, she links these metabolic dysregulations to current understanding of the molecular functions of dietary components. She cites experimental evidence suggesting that aberrant nutrient availability, whether from dietary inadequacy, distribution, or excretion, can potentiate these pathways, worsening the comorbid state. I found the chapter on chronic renal failure especially intriguing. While published medical nutrition therapy, also delineated in the chapter, focuses largely on supporting the patient with nutritional therapy, Dr. Ettinger points out mechanisms through which targeted modification of dietary components can minimize kidney damage and potentially slow the progression to end-stage disease. She also suggests strategies for specific dietary changes to achieve these treatment goals.

Finally, the level of science presented in this text is at a level ideal for training advanced masters and doctoral students. I believe students will find the chapters, together with the five Essentials appendices very informative as they work through the case scenarios. Because the text has a single author, the information it provides is highly integrated, with frequent reference to discussions in other sections. This contextual strategy allows the student to solve case problems, calling upon a range of insights from multiple disciplines. Since I expect that advanced nutrition students will enter careers in nutrition research and practice, the links presented throughout this text will provide a rich source of unresolved questions that can be developed into testable study designs for clinical research protocols.

Dr. Jennifer Nasser

Dr. Nasser is an Associate Professor of the College of Nursing and Health Professions at Drexel University, being Director of the PhD Program in Nutrition Sciences. Her clinical research focuses on dopamine-mediated mechanisms of food intake regulation in humans and its impact on metabolic homeostasis, particularly as it applies to aging, eating disorders, and obesity.

J. Nasser

Preface

Two roads diverged in a wood and I –
I took the one less traveled by
And that has made all the difference **Robert Frost (1920)**

This work derives from my abiding interest in the ways everyday food is used by our bodies to build structure, make repairs and provide the means to think and work.

As a child growing up on a small farm in Iowa, I was immersed in all aspects of home food production and preparation and gained perspective on the close links between food and life. I came to New York to study classical music, but soon discovered that my interests had changed and I reverted to my abiding passion, the study of food and its relationship to health. As a novice in science, I sought advice on course prerequisites from nutrition programs in the greater New York area. Serendipitously, my first call was to the Institute of Human Nutrition (IHN) at Columbia University. Ms. Ada Joyner, administrator of the program was eminently patient as she answered all my questions and guided me through the premedical program given at Columbia University School of General Studies. She was single-handedly responsible for my obtaining a rigorous grounding in the basic hard sciences that underlie nutrition science. I know now that had I contacted other nutrition and dietetic programs, the prerequisites required would have been less rigorous and my path would have been different.

The heavy science courses intensified my desire to know more. I was fascinated by the brilliant reds and greens elicited as electrons interacted with metals in heme and chlorophyll. Even at this early stage in my career, I attempted to integrate nutrition and basic science. The 12 credit organic chemistry course allowed mentored time to construct a protocol for the synthesis of vitamin E and to build the molecule bond by bond in my instructor's laboratory. It was great fun.

Armed with my science prerequisites, I crossed the street and applied to the Master of Science in Nutrition at Teachers College (TC), Columbia University. In one of the first classes, the assignment was to envision the scope of nutrition science as a healthcare profession. As a newcomer to the field and fresh from my premedical courses, it is perhaps understandable that I viewed nutrition science as an integral element of medical sciences. I presented this concept in my student paper and was firmly struck down by the professor. She informed me that the scope of nutrition practice was limited to patient feeding and counseling and did not include in-depth mechanistic study of nutrition in biological processes. I was shocked. Where was the integration with medicine? Who was trained to translate bench research into clinical practice and to integrate these two inextricably linked bodies of knowledge? I aired my views with the program director. She agreed with my vision, but gently referred to me as a "maverick."

I could not believe that there were no programs that trained nutrition scientists in medicine. However, an exhaustive search of the doctoral programs listed by the American Society for Nutrition Sciences revealed, in fact, no well-trodden educational pathway that provided expert knowledge and skill at this interface. As no program existed, I reasoned that it would have to be created. I recruited Dale Green, the other "maverick" at TC, and we used the skills in curriculum development gained in our Master of Education in Nutrition program to design our vision of an ideal curriculum for a PhD in Clinical Nutrition.

I met with medical course directors at Columbia College of Physicians and Surgeons. They liked the idea and assured me that they would welcome nutrition students into their courses. I also gained permission to crossregister in the MS program at the IHN and take their advanced nutrition science courses. Dale and I avidly sought out seminars both at Columbia at other institutions throughout the greater New York area that featured national and international nutrition experts. We gained a wide knowledge of the expanding body of knowledge that comprised nutrition science and the current research themes in the area.

Fortified with our newly created curriculum, we approached these experts. Some dismissed the idea out of hand. "If you want to gain this knowledge, go to medical school." Well-respected dietitians told us that experience, not advanced education, was the way in which a dietitian obtained knowledge. Dr. Jules Hirsch at the Rockefeller University was more sympathetic and offered to discuss the curriculum with other nutrition experts at an upcoming conference. He returned saying that his colleagues all thought the idea was interesting but unworkable and that we should "go to medical school." In retrospect, I believe he must also have put in a good word with the admissions committee at the IHN, because I was accepted to the PhD program as a single subject ($n = 1$) of our proposed experimental curriculum. Unfortunately, since the curriculum was a radical departure from the usual doctoral nutrition curriculum, the door was firmly closed after I was admitted and Dale's application was rejected.

With my MS and MEd degrees in hand I was hired to work on a project to teach nutrition to medical students at the New York College of Medicine. My position on the project was primarily supportive, as teaching was done by physicians and biochemists. The nutrition content provided to the medical students consisted primarily of diagnosis and treatment of nutritional anemias and a review of published dietetic guidelines for specific patients. Since my masters level status did not qualify me to do any actual teaching, I was permitted to spend my time shadowing a gastroenterologist who patiently explained the pathogenesis of primary biliary cirrhosis, inflammatory bowel disease, and other maladies suffered by his patients. I was also allowed to follow clinicians in neonatal and adult settings. This experience was invaluable and reinforced my understanding that I needed to obtain more medical knowledge and doctoral training.

At the time I entered the PhD program at Columbia University, medical courses were offered through the Graduate School of Arts and Sciences and open to graduate students. I eagerly registered for all the basic science courses including Histology, Physiology, Neural Sciences, and Pharmacology. My favorite course was the 10 credit pathology course. I was fascinated by the detective work involved in understanding disease processes and especially loved the case studies and the laboratory called "man in the pan." Students were presented with a tray of organs from a deceased patient and asked to identify the processes that produced the gross and microscopic changes in the organs. The final clinical medical course, "Abnormal Human Biology," was not open to graduate students, but I appealed to the course director and was allowed to participate on a noncredit basis. This comprehensive course, covering all of internal medicine, was immense. In addition to lectures, it included case discussions with attending physicians and hands-on laboratories that illustrated physiologic concepts. I was especially delighted because, in contrast to my medical school colleagues, I clearly saw the many nutritional implications contained in the integrated medical course and its case discussions.

Dale and I had envisioned that at the end of coursework, our doctoral candidate would undertake clinical rotations and conduct independent clinical research under the guidance of a mentor. As the IHN faculty were primarily bench researchers, I searched elsewhere for my doctoral mentor. My final choice was problematic since, while my mentor was a hematologist and I did learn to draw blood on my subjects, I was unable to negotiate clinical experience. Thus my doctoral studies were largely focused at the bench where I studied folate, copper, and zinc status in a geriatric population. I did learn a great deal about blood, and the data I generated shed light on possible changes in nutrient distribution in aging and inflammation. In my enthusiasm to understand underlying processes, however, I interpreted my data to suggest that aging was accompanied by increased inflammation. This concept, although generally now believed to be the case, was not well accepted at the time and my thesis was called "science fiction."

Ultimately, after much blood, sweat, and tears, I graduated one glorious spring morning and celebrated with champagne and strawberries.

I pursued my interest in immune function, inflammation, and aging during postdoctoral studies at Cornell University Medical Center in the Division of Geriatrics and gained knowledge and skill in cell culture and molecular biology. During this time, I was contacted by Dr. William Heston, Dean of Natural Sciences at New York Institute of Technology (NYIT). Dr. Heston wanted to begin a new Master of Science in Clinical Nutrition at NYIT. He had engaged as a consultant one of the nutrition experts who had been a major critic of our curriculum. As Dr. Heston outlined his plans at our first interview, I was stunned to discover that he was describing my own curriculum. I felt like the mother of Moses being asked to nurse her own child. I had no choice but to accept the challenge. This was not the PhD program I had envisioned, but it was a start.

I modeled the new curriculum on my own educational process. Because at the masters level we could not provide medical courses, I introduced key concepts within a new course I called Nutritional Pathophysiology. I used Robbins, Pathologic Basis of Disease as a text and inserted mechanistic nutrient functions from the literature. Students loved the course. To integrate the program with medicine, I worked with Dr. Schiowitz, dean of our osteopathic school, to develop a dual MS/DO degree program. To provide credentialing for our students, I wrote self-studies for the American Dietetic Association and New York State Education Department. I still hoped to create a doctoral program at NYIT and developed multiple curricula models combining nutrition courses with courses from the osteopathic school. Sadly, repeated efforts to bring this curriculum to the doctoral level were met with steadfast administrative resistance. This roadblock prompted several surreptitious excursions to other institutions in the attempt to persuade these outside nutrition administrators to work with me to implement my curriculum. Again, the response was deemed "interesting but not workable." One dean stated that clinical nutrition "was not pure science" and that a PhD in Clinical Nutrition was not in the interest of his institution.

Refusing to give up, I applied for extramural funds to test elements of my curriculum at the postmasters level. I was funded by the National Cancer Institute (NCI) to develop cancer nutrition training. Although participating oncologists and researchers readily agreed to mentor our students in cancer rotations, the students, even with enrichment courses and tutoring, were unprepared to meet the challenge. Evaluation of the program confirmed my initial diagnosis: without a rigorous science background and doctoral-level training in medical sciences and research methodology, it was not possible to produce the medical nutrition scientist envisioned those many years ago. As there was no way that this training could be implemented at NYIT, I decided it was time to escape and try another path.

Returning to my roots within the Columbia University community, I became a member of the New York Obesity Research Center (NYORC) and embarked on bench research to understand the behavior of adipocytes—but I did not give up on my desire to see the creation of a doctoral-level medical nutrition scientist. While the NYORC could not grant academic degrees, it did have a postdoctoral program. Dr. Xavier Pi-Sunyer, Director of the NYORC, agreed to act as principal investigator on a program funded by the National Institute of Diabetes, Digestive and Kidney Diseases (NIDDK) to provide postdoctoral training in the language and practice of medicine and clinical research to candidates with a PhD in nutrition. All members of the steering committee reasoned that having obtained a doctoral degree in nutrition, candidates should have acquired in-depth knowledge of nutrition and skill in conducting independent research. This turned out to be an erroneous assumption. Careful evaluation of the program revealed that the background knowledge of the participants did not include medical sciences and the molecular nutrition knowledge of some participants was limited. Although enrichment and tutoring strategies were provided, they could not overcome the prior knowledge deficit. Thus, while all of our graduates gained from the experience, none achieved the medical nutrition competencies we had envisioned.

Moving forward more years than I care to count, I find that in the intervening years the educational landscape at the interface of nutrition and medicine has not substantively improved. Despite over 50 years of attempts, nutrition education for physicians remains variable. Many programs do not offer nutrition; some offer a few hours of training that consists largely of applied dietetics. Some programs teach the medical students to cook. At the same time, there is no educational standard for doctoral nutrition programs. Thus, while some doctoral nutrition curricula focus on molecular bench research, others train students in public health and community research methodology. The Registered Dietitian is still credentialed at the baccalaureate level and practices applied nutrition in health and wellness. I repeat, where is the integration with medicine?

Failure to train medical nutrition scientists stands in stark contrast to the sophisticated research technology that permits us to examine cellular behavior in the living animal and to explore molecular functions of nutrients in the diverse and fundamental processes that control life. Mechanistic studies clearly demonstrate that food and the bioactive components it contains are essential for optimal health and disease prevention. These elegant studies have shown that nutrients can be in short supply due to reduced intake, increased losses, or dysregulated distribution and that altered nutrient status increases risk for morbidity and mortality. We now know a great deal about the complexities in assessing nutrient status and are well aware that serum levels may provide inaccurate and misleading information. Yet this information is poorly translated from the bench to the bedside.

The purpose of this work is not to present a manual for nutrition support, nor is it a scholarly paper describing the body of my own work. Since we have not yet created a cadre of medical nutrition scientists competent to consult with the physician on the use of targeted nutrition strategies as adjuvant medical therapy, I have extracted major mechanistic themes from the biomedical literature relating to the role of nutrients in obesity and its comorbidities. Where possible I have built upon published cases to describe a typical patient and have incorporated typical dietary and lifestyle characteristics that modulate risk for the condition under discussion. When suitable cases were not available, I used information from the published literature to invent a hypothetical patient as a point of reference for the subsequent discussion. Within the framework of our current understanding of disease pathogenesis, I have layered evidence from the literature on the functions of nutrients in these pathways. The abundance or research in the literature has also revealed a multitude of testable hypotheses with potential for major implications for human health. Thus, this work is a first attempt to provide in one venue, an overview of the interface of nutrition and medicine. It is meant to encourage clinicians and researchers to consider the totality of the diet, micro-nutrients and non-nutrient components as well as macronutrients, and to develop protocols to test these hypotheses. It is hoped that the data generated by these studies will provide support for development of guidelines and recommendations for both patients and the wider community at large.

In addition to awards from NCI and NIDDK that allowed me to test this curriculum, I wish to thank the many mentors, teachers, students, friends, family, and coworkers who encouraged me throughout my career. I am grateful to my friends and colleagues who critiqued sections of this work, especially Dr. Sami Hashim for his insightful reviews of these chapters. More importantly, I want to thank those who sought to dissuade me from this endeavor. Their comments and critiques helped shape my vision of the scope of nutrition science and the role of the medical nutrition scientist in the preservation of health and prevention of disease. If this text sparks interest and inspires innovative clinical research and practice, it will have served its purpose. It will in small part have encouraged "the end of the beginning."

Now this is not the end. It is not even the beginning of the end

It is, perhaps, the end of the beginning *Winston Churchill (1942)*

Susan Ettinger

1

Obesity and Metabolic Syndrome

CHIEF COMPLAINT (TYPICAL PATIENT)

A 41-year-old African-American woman requested consultation about losing weight. She was somewhat overweight as a child, has a history of dieting throughout adolescence and adulthood, and has gotten successively heavier, especially after the birth of her four children ages 8, 9, 12, and 15.

MEDICAL HISTORY

- Prediabetes with mild hyperglycemia (fasting blood glucose levels 115–125 mg/dL).
- Workups for other endocrine abnormalities have been negative in three previous examinations.
- The patient has been hospitalized four times for vaginal childbirth; no gestational diabetes.
- She has no known history of gallbladder disease, allergy, or hyperlipidemia.
- Although her husband noted that she snores at night, she does not wake from snoring, nor does she fall asleep at odd times during the day.
  - *Skin*: no history of rashes or unusual skin pigmentation, stretch marks, or skin tags;
  - *Neurologic*: no headaches, tremors, seizures, or depression;
  - *Endocrine*: normal menstrual cycle, no abnormal heat or cold intolerances;
  - *Cardiovascular*: no hypertension, heart disease, orthopnea, or dyspnea;
  - *Joints*: no swelling, heat, or redness.

FAMILY HISTORY

The patient lives with her husband (44 years) and children (8, 9, 12, and 15 years). She works as a social worker; her husband is a plumber. The patient reports no history for diabetes, although many of her first-degree relatives have been diagnosed with hypertension and hyperlipidemia. Her father died at 68 years of a heart attack. Her mother (65 years) has hypertension and hyperlipidemia controlled with medication.

PHYSICAL EXAMINATION

- *General appearance*: obese woman in no acute distress; no cushingoid features.
- *Exam*: nonpalpable thyroid; no hirsutism or striae; no dorsal, cervical, or supraclavicular fat; no acanthosis nigricans; no edema.
- *Vital signs*: temperature: 98.4°F; HR 88 bpm; BP 145/90 mmHg.
- *Anthropometric data*:
  - Height 5′4″ (163 cm); weight 265 lb (120.2 kg); BMI 45.5 kg/m^2; adjusted IBW 81.2 kg (calculation from http://www.globalrph.com/ibw_calc.cgi/);
  - Waist circumference 46 inches (106.7 cm);
  - Triceps skinfold 8 cm (80 mm);
  - Mid arm muscle circumference 28 cm (280 mm).

LABORATORY DATA

- *Blood chemistries*: within normal limits.
- *Complete blood count*: within normal limits.
- *Thyroid function test*: normal.

	Patient's value	Normal values
Fasting serum glucose	115 mg/dL	65–100 mg/dL
Hgb A1c	5.9%	<5.5%
Potassium	3.8 mEq/L	2.8–6.2 mEq/L
Total serum cholesterol	230 mg/dL	<200 mg/dL
Serum triglycerides	200 mg/dL	<150 mg/dL
Serum iron	52 µg/dL	50–170 µg/dL

1

DIET HISTORY

- The patient does not smoke and drinks 1—2 glasses of wine or beer with dinner about every 2 days.
- The patient drinks two or three cups of coffee a day.
- Throughout her life, the patient has tried a number of weight-loss diets with 20—40 pounds of weight loss on each occasion. Each time the weight returned within a year, and she typically leveled off at 10—15 pounds more than before.
- The patient walks 2—3 blocks to and from the subway each day, but does not have regular planned exercise.
- The patient's highest lifetime weight was 285 pounds, she is now at 265 pounds. Her low adult weight was 140 pounds when she lost ~60 pounds for her wedding 20 years ago.

A 24-HOUR DIETARY RECALL

Breakfast

- Coffee (2 cups) with skim milk (1/2 cup);
- Cereal (3/4 cup) with a banana and (1/2 cup) skim milk.

Lunch

- Grilled chicken fillet (3 oz) sandwich with lettuce and tomato on a roll (no mayo);
- Diet coke.

Afternoon Snack

- Large coffee with Splenda and coffee creamer.

Dinner

- Eggplant Parmesan with low-fat mozzarella;
- Steamed broccoli rabe with lemon;
- Green salad with low-fat ranch dressing (~1/4 cup);
- Red wine (1 glass).

After Dinner

- Fat-free sorbet (1 cup);
- 5 low-fat chocolate chip cookies (1.5 inches each).

Intake	kcal 1602	Protein 47 gm
Requirements	kcal 1935[a]	Protein 64.96 (Adj. IBW × 0.8 × kg)

[a]Harris Benedict calculations using 1.2 activity factor. http://www-users.med.cornell.edu/~spon/picu/calc/beecalc.htm/.

RESOURCES

Obesity is not a personal failing. In trying to lose weight, the obese are fighting a difficult battle. It is a battle against biology, a battle that only the intrepid take on and one in which only a few prevail (Friedman, 2003).

1.1 FOOD AND ORGANIC LIFE

The laws of thermodynamics apply in living organisms as in the physical environment. To survive, an organism must regulate its supplies of energy and substrate to repair and replace body systems that continually undergo entropy and degradation. Food intake is the only way in which substrate can be provided, thus in a very real sense, we are what we eat. In order to free up time to go about the activities of daily living, food intake is, necessarily, episodic, thus most organisms consume meals at intervals throughout the day. To assure that the food taken in provides an adequate supply of substrate to maintain life, organisms have elaborated complex systems of central and peripheral sensors that monitor intake of food, passage of its metabolites through the intestinal tract, absorption, transport in the circulation, utilization, and storage for later use. Mediators such as leptin and gut hormones trigger an integrated web of neural, endocrine, paracrine, and other signals that regulate energy expenditure and body composition to maintain homeostasis as illustrated in Figs. 1.1—1.3 below.

As the organism grows, escapes predation, reproduces, and ages, the signals are modulated and body composition altered in response to substrate need (Wells, 2006). The vast literature exploring this integrated system has been reviewed extensively (Mayer and Thomas, 1967; Weigle, 1994; Rosenbaum et al., 1997; Simpson and Raubenheimer, 2005; Friedman, 2014; van der Klaauw and Farooqi, 2015) and will not be discussed in detail here.

While the ability to store substrate to meet physiological needs is protective, and indeed essential, in recent years humans across the globe have begun to store fuel in excess of these needs; as a result, the prevalence of obesity with its comorbid complications has risen dramatically. In 1980, 28.8% of men and 29.8% of women worldwide had a body mass index (BMI = weight in kg/height in m^2) of 25 or greater. In 2013, the prevalence of overweight and obesity (BMI >30) had increased to 36.9% and 38.0%, respectively (Ng et al., 2014). The increase in obesity has been linked to increased healthcare costs; estimates of the cost of obesity vary by the analytical method used. Recent estimates for the annual US medical care costs of obesity-related illness in adults have ranged from $85.7 billion (Finkelstein et al., 2009) to $209.7 billion (Cawleya and Meyerhoeferd, 2011), depending on the patient's age and underlying medical conditions. These statistics might lead one to

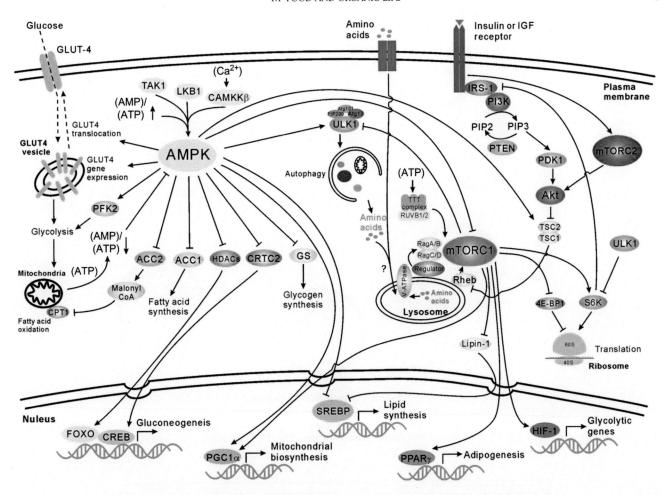

FIGURE 1.1 Nutrient sensing and downstream actions. AMPK activation controls metabolism of glucose and lipids to increase catabolic pathways that provide substrate for ATP generation and to decrease biosynthetic pathways that utilize energy. The mammalian target of rapamycin complex 1 (mTORC1) is integrated with AMPK. It promotes growth and anabolic processes in response to amino acid availability. Note that although two mTOR complexes exist, only mTOR 1 is sensitive to amino acids. Stimulatory interactions are indicated with ↓, and inhibitory interactions are indicated with ⊥.

GLU4, glucose transporter type 4; *CPT1*, carnitine palmitoyltransferase-1; *TAK1*, TGF-β-activated kinase 1; *LKB1*, liver kinase B1 (a key AMPK activator tumor suppressor); *CAMKKβ*, calmodulin-dependent protein kinase kinase β; *ACC*, acetyl CoA carboxylase; *PFK2*, phosphofructokinase 2; *GS*, glycogen synthase; *HDACs*, histone deacetylases; *CRTC2*, CREB-regulated transcription coactivator 2; *FOXO*, forkhead box protein O; *CREB*, cAMP response element-binding protein; *ULK*, UNC-51-like kinase; *FIP200*, 200 kDa FAK family kinase-interacting protein; *ATG*, autophagy-related; *PI3K*, phosphoinositide 3-kinase; *IRS1*, insulin receptor substrate 1; *PTEN*, phosphatase and tensin homolog; *PDK1*, 3-phosphoinositide-dependent protein kinase 1; *TSC1/2*, tuberous sclerosis 1/2; *Rheb*, Ras homolog enriched in brain; *v-ATPase*, vacuolar H⁺-adenosine triphosphatase; *4E-BP1*, eukaryotic initiation factor 4E-binding protein 1; *S6K*, ribosomal S6 kinase; *Ragulator*, a protein complex responsible for lysosomal recruitment and activation of Rag GTPases; *PGC1α*, peroxisome proliferator-activated receptor-γ coactivator 1α; *PPARγ*, peroxisome proliferator-activated receptor-γ; *SREBP*, sterol regulatory element-binding protein; *HIF*, hypoxia-inducible factors. *Reproduced from Yuan, H.-X., Xiong, Y., et al., 2013. Nutrient sensing, metabolism, and cell growth control. Mol. Cell 49 (3), 379–387—Science Direct.*

ask: "If body mass (weight) is maintained in an ideal range by a complex network of sensors and signals, why then is obesity rapidly becoming endemic in the US and globally?" To address this question, it is necessary to consider the human genetic background, the form and composition of food eaten, as well as environmental factors such as toxic exposures and food insecurity. In addition, several methodologic considerations must be addressed, including whether BMI, waist circumference, or other indices most specifically address the cardiometabolic risk for obesity (Visscher et al., 2015).

The role of genetics in susceptibility to obesity is indisputable (Friedman, 2014). Evolutionary biologists have proposed that populations are becoming obese because of accumulated polymorphisms in the genes regulating energy homeostasis. Because failure to provide adequate substrate results in death of the organism before reproduction can take place, genes regulating adequate intake must have been under selective pressure to remain operational since prehistoric times. In contrast, genes preventing excess fat storage would be under less evolutionary pressure and

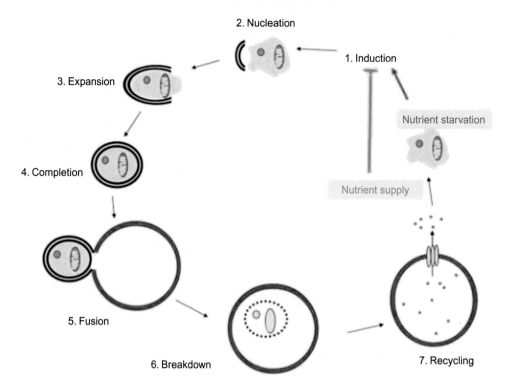

FIGURE 1.2 **Autophagy.** Cellular homeostasis requires that biosynthetic and degradation processes be tightly coordinated to respond to changing environmental conditions. Autophagy adjusts cellular content to the changing physiological needs. Note that mTORC1 localizes to the lysosome (red circle) to detect newly released amino acids. *Reproduced from Todde, V., Veenhuis, M., et al., 2009. Autophagy: principles and significance in health and disease. Biochimica et Biophysica Acta (BBA)—Mol. Basis Dis. 1792 (1), 3—13 (Todde et al., 2009)—Science Direct.*

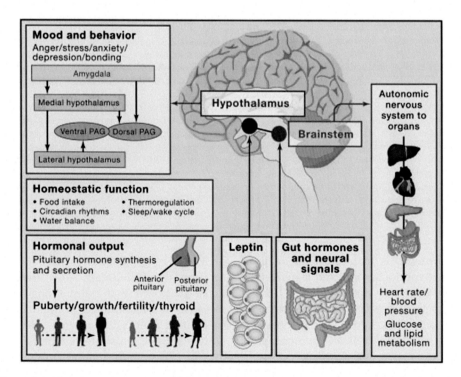

FIGURE 1.3 **Leptin and gut hormones integrate multiple systems to maintain energy homeostasis.** Leptin released from adipocytes, mediators from the gut, and other neuronal and hormonal signals are integrated through the brain to regulate food intake, energy expenditure, mood, and behavior as well as other systems that maintain homeostatic functions in response to changing environmental and physiological settings. *Reproduced from van der Klaauw, A.A., Farooqi, I.S., 2015. The hunger genes: pathways to obesity. Cell 161 (1), 119—132—Science Direct.*

could have "drifted" such that polymorphisms are more common, making more individuals at risk of becoming obese (Simpson and Raubenheimer, 2005). Conversely, the "thrifty gene" hypothesis suggests that gene polymorphisms that allowed our primitive ancestors to store more fuel allowed survival during periodic famines and successful reproduction (Neel, 1999). More recently, evidence has been presented suggesting that the genes of prehistoric peoples have not remained static, but were modified by their ethnic backgrounds and by the environmental conditions they encountered in their migrations across the globe (Sellayah et al., 2014). Epigenetic modification of gene expression, occurring from conception through adulthood can also alter the plasticity of the phenotype related to body size. The impacts of nutrients and dietary exposures on epigenetic patterns and their relationships to obesity and its comorbidities have been recently reviewed (Milagro et al., 2013). Nonetheless, while polymorphisms and epigenetic modifications can alter energy balance in the individual, genetic change occurs slowly in a population. Thus, genetic changes are unlikely to be the sole cause of the dramatic global rise in obesity in recent years.

Several possibilities have been advanced for the "obesity epidemic," including the reduction in physical activity in a technological age and increased intake of high-calorie, processed fast foods (Heber, 2010). Low calcium intake, inadequate, or disrupted sleep duration (Chaput et al., 2010), and even excessive dieting with reduction in lean body mass (Dulloo et al., 2015) have been linked to obesity. Whatever the cause, obesity and weight cycling have been associated with significant health risks (Montani et al., 2015). Unfortunately, once an individual's mass of adipose tissue (AT) has expanded, long-term efforts to maintain weight loss have been largely ineffective. The estimate made almost two decades ago, that 90−95% of persons who achieve weight loss subsequently regain the weight (Rosenbaum et al., 1997), remains largely true today, despite increasing efforts by healthcare professionals, fashion impresarios, diet food vendors, and the media to encourage the population to achieve the elusive slim ideal. Thus, a corollary question is raised: "What causes adipose tissue dysregulation, and what can the obese individual do to reduce risk for its comorbid complications?"

1.2 NUTRIENT SENSORS AND DOWNSTREAM TARGETS

Systems that permit biological organisms to sense their cellular energy balance and maintain energy homeostasis are highly conserved in evolution, from unicellular prokaryotes to multicellular eukaryotes, including humans. A master sensor of energy availability to the body is adenosine monophosphate-activated kinase (AMPK), activated through a complex signaling cascade. When intracellular adenosine triphosphate (ATP) falls and the concentrations of adenosine diphosphate (ADP) and/or adenosine monophosphate (AMP) increase, AMPK is activated and coordinates energy utilization in liver, muscle, and fat. AMPK stimulates catabolic processes that generate ATP and inhibits anabolic pathways that utilize ATP, thereby conserving energy. These actions are integrated with the mammalian target of rapamycin (mTOR), a sensor that detects amino acids and growth-stimulating signals and regulates cell growth accordingly. Other sensors are potential regulators as well. For example, members of the dioxygenase family (prolylhydroxylase, lysine demethylase, and DNA demethylase) have been proposed as corollary sensors of oxygen availability and metabolic status (Yuan et al., 2013).

As shown in Fig. 1.1, AMPK consists of a heterotrimeric complex that is regulated by multiple signals and directly binds AMP, ADP, or ATP. ATP concentrations are normally several-fold higher than ADP, but because the complex binds ADP more tightly than ATP, it detects rising ADP concentrations. Binding ADP results in AMPK phosphorylation and creates a conformational change that activates the sensor. Intracellular concentrations of ADP are usually higher than AMP, suggesting that in conditions of moderate stress, ADP is the relevant activator. Severe hypoxic stress causes AMP levels to rise and bind AMPK. AMP binding not only phosphorylates the enzyme, but protects the enzyme from dephosphorylation, thus increasing AMPK activity by at least 10-fold. These actions make AMP the essential regulator under severe energy stress (Xiao et al., 2011). The role of AMPK in regulating these processes has been recently reviewed (Krishan et al., 2015).

Intracellular ATP is generated by mitochondrial oxidation of acetyl coenzyme A (acetyl CoA) derived from ingested and stored macronutrients, predominately glucose and lipids as described elsewhere (refer to Chapter 5: Cardiomyopathy and Congestive Heart Failure) High-energy electrons are transferred to nicotine adenine dinucleotide (NADH) and flavin adenine dinucleotide (FADH2) and passed to the electron transport chain located on the inner mitochondrial membrane. (NADH and FADH2 are redox cofactors that shuttle high-energy electrons from catabolic reactions to the mitochondrial oxidative phosphorylation chain for processing.) The resultant proton motive force ($\Delta\mu_H$) drives ATP synthesis (Mitchell, 1961).

AMPK increases the availability of glucose as an energy source for ATP generation by increasing the expression and translocation of the glucose transporter 4 (GLUT4) to the plasma membrane. It also activates the rate-limiting enzyme for glycolysis, phosphofructo-kinase, thereby stimulating the formation of pyruvate, substrate for acetyl CoA formation in the mitochondria. AMPK expedites long-chain fatty acid entry into the mitochondria as substrate for β-oxidation and conversion to acetyl CoA as described below. Mitochondrial biogenesis, necessary for efficient ATP production from these substrates, is also stimulated by AMPK. Together these pathways are tightly regulated to maintain an ATP concentration far in excess of ADP at an optimal ATP:ADP ratio of approximately 10:1.

To conserve energy in times of stress, AMPK also inhibits anabolic pathways that utilize ATP. Energy-requiring pathways for lipid biosynthesis are reduced by phosphorylating and inhibiting acetyl CoA carboxylase 1 (ACC1) and hydroxymethylglutaryl CoA reductase (HMGCR), the rate-limiting steps in the synthesis of fatty acids and cholesterol, respectively. Inhibition of ACC1 prevents cytoplasmic acetyl CoA units from polymerizing to form the six-carbon intermediate in fatty acid biosynthesis, malonyl CoA. Malonyl CoA also exerts feedback inhibition on carnitine palmitoyltransferase-1, the enzyme that facilitates long-chain fatty acid entry to the mitochondrion for metabolism via the β-oxidation pathway. Thus, by reducing the amount of malonyl CoA formed, AMPK both inhibits ATP-requiring fatty acid biosynthesis and facilitates entry of long-chain fatty acids into the mitochondria for ATP synthesis. AMPK phosphorylates and inactivates HMGCR, limiting the ATP-utilizing pathway for cholesterol biosynthesis. AMPK suppresses sterol regulatory element-binding protein 1, a transcription factor that upregulates both ACC1 and fatty acid synthase. Thus AMPK exerts a multifaceted effect on lipid metabolism to both increase ATP synthesis and decrease ATP utilization, as pictured in Fig. 1.1. Glucose storage as glycogen is also inhibited by AMPK.

1.2.1 Growth and Amino Acid Utilization

To ensure that cells and tissues grow only when conditions are favorable, AMPK interacts with another nutrient sensor that integrates information from the environment. Amino acids and glucose, growth factors as well as stress, are sensed through the highly conserved Ser/Thr protein kinase target of rapamycin (TOR) (Jewell et al., 2013). In organisms as diverse as yeasts, fruit flies, and mammals, nutrient availability activates the TOR complex 1 (TORC1) and downstream signals promote cell growth by stimulating protein synthesis and other biosynthetic pathways and by inhibiting cellular catabolism. In the fed state, amino acids are provided by food. If the amino acid pool is

limited, as in fasting, organisms can liberate amino acids by activating autophagy, the catabolic process that removes amino acids and nutrients from cells and tissues to maintain essential cell functions (Dann and Thomas, 2006; Jewell et al., 2013; Demetriades et al., 2014) as described below.

Mammalian TORC1 (mTORC1) senses the amino acid pool by two distinct small GTPases, Rag GTPase and Rheb. Rag GTPase controls mTORC1 localization to the lysosome where it detects amino acids released from degraded cellular proteins. Rheb controls mTORC1 kinase activity initiated by growth factors and nutrients. Input from both signals is required for full activation of mTORC1 (Chantranupong et al., 2015). Fig. 1.1 depicts the crosstalk between AMPK and mTORC1. In mammals, AMPK inhibits mTORC1 by two mechanisms. First, it phosphorylates and activates the tuberous sclerosis 2 GTPase-activating protein, a major inhibitor of the pathway. Second, it phosphorylates raptor, a subunit of mTORC1, also resulting in inhibition of mTORC1 kinase activity. Under conditions of starvation, mTORC1 is inactivated by AMPK and by limited amino acid signaling. These controls limit excessive utilization of amino acids for growth during starvation.

Autophagy is essential for cell survival, especially during fasting. As illustrated in Fig. 1.1, AMPK signals autophagy when protein and energy supplies are low. Autophagy is inhibited in the fed state by insulin stimulation of mTORC1, thereby preventing activation of UNC-51-like kinase, the primary signal for autophagy. In fasting, induced autophagy begins with formation of a double-membrane structure within the cell (Fig. 1.2). This structure envelops cytoplasmic nutrients and organelles creating an autophagosome. Because it can surround and take up portions of the cytoplasm, the autophagosome facilitates mitochondrial degradation (mitophagy), as well as glycogen and lipid droplet (LD) breakdown to provide nutrient substrate during starvation. The autophagosome fuses with the lysosome—endosome system and the captured cellular material undergoes acid proteolysis by lysosomal enzymes. Released metabolites can be utilized for essential cell functions (Lamb et al., 2013).

All cells have the capacity to generate autophagosomes for general housekeeping use, to destroy worn-out organelles, and remove cell debris. While the process has been observed for over 50 years, the origin of the autophagosome membrane, the precise signals for formation of the autophagosome and the signals for amino acid transport out of the autolysosome remain unclear (Jewell et al., 2013). Since dysregulated autophagy plays a role in several diseases, including inflammatory bowel disease, cancer, and neurodegeneration, this area is under intense investigation as recently reviewed (Shibutani and Yoshimori, 2014).

1.3 AT FORM AND FUNCTION

Animals, plants, and other organisms including fungi store lipids in cytoplasmic inclusions called lipid droplets surrounded by a monolayer of phospholipids and associated proteins (Londos et al., 1999; Martin and Parton, 2006). The presence of fuel-storing LDs in cells is critical for normal function. For example, LDs in heart muscle are closely associated with mitochondria, providing easy access to lipid substrate (Wang et al., 2013). However, if the number or size of myocardial LD exceeds a threshold, excess LDs have been associated with cardiac dysfunction (Kienesberger et al., 2013). Although the precise steps in LD genesis and regulation are still unclear, current understanding is that lipids are delivered to cells in triglyceride (TG)-rich lipoproteins (chylomicrons and very-low-density lipoproteins (VLDL)). Lipoprotein lipase, attached to the luminal surface of the capillary endothelium by heparin sulfated proteoglycans, hydrolyzes neutral TG to free fatty acids and monoglycerides. Nonesterified free fatty acids (NEFAs) are transferred into the cells primarily via a saturable, protein-facilitated process. The best described NEFA translocase is the scavenger receptor CD36 (Goldberg et al., 2009), although other transporters are known to exist. Cells take up NEFAs and free cholesterol and convert them to neutral TG and cholesterol esters in the endoplasmic reticulum (ER). LDs form in discrete regions of the ER that appear to contain proteins with the ability to form a hydrophobic domain between the phospholipid leaflets, creating a new cytosolic organelle that contains neutral lipids surrounded by a single monolayer of phospholipids. This is the nascent LD (Martin and Parton, 2005).

As stimulated preadipocytes progress through differentiation, they form multiple tiny LDs in their cytoplasm. To become mature adipocytes adapted for storing lipids, these multilocular preadipocytes must fuse the tiny LDs into a single large LD, orders of magnitude larger than the LDs observed in other tissues. To facilitate micro-LD clustering and fusion, fat-specific protein 27 (FSP27) is expressed on the LD membrane (Jambunathan et al., 2011). Fusion also requires autophagy (Maixner et al., 2012); inhibition of autophagy results in failure of LD to fuse into the single large LD with capacity to store TG.

A major function of mature adipocytes is to control uptake and release of potentially toxic NEFAs. Stored neutral lipids (triglycerides, TG) are hydrolyzed in response to counterregulatory hormones (cortisol, epinephrine, growth hormone, and glucagon). Hydrolysis is accomplished through the activity of proteins associated with the LD membrane, including perilipin and hormone-sensitive lipase. Hydrolyzed NEFAs and free cholesterol required for metabolism, membrane biosynthesis, and steroid synthesis are released. NEFAs can also act as powerful signaling molecules to control their own metabolism. For example, they act as ligands for peroxisome proliferator-activated receptors (PPARs) in the liver for metabolic adaptation to starvation, in the skeletal muscle for fatty acid oxidation during fasting and starvation, and for mitochondrial biogenesis. In the preadipocyte, NEFAs activate PPARγ, inducing differentiation to mature adipocytes competent to store fatty acids as neutral TG (Nakamura et al., 2014; Papackova and Cahova, 2015).

Adipocytes make up a large portion of AT and interact with the extensive collagen matrix containing fibroblast-like mesenchymal stem cells, a capillary network, macrophages, and adaptive immune cells. As a tissue, AT is metabolically very active, secreting an array of bioactive hormones and mediators (Sethi and Vidal-Puig, 2007). Table 1.1 lists some of the complex molecular and biochemical signals secreted by AT that enable rapid system responses to changes in nutritional cues and requirements; leptin is considered a master regulator, controlling not only food intake and energy expenditure, but mood, immune, and other physiologic functions as pictured in Fig. 1.3 (van der Klaauw and Farooqi, 2015). Adiponectin secreted by the adipocyte is a multifunctional target of PPARγ and may also act as a spacer between adipocytes, allowing individual adipocytes greater access to extracellular fluid, thereby facilitating insulin sensitivity (Nakamura et al., 2014). Adipocytes also produce insulin. Recently, mRNA for insulin and insulin protein have been demonstrated in adipocytes, but not preadipocyes. Insulin production in adipocytes is distinct from, but coordinated with, insulin production in the pancreas. Thus, adipocytes from patients with type II diabetes produced significantly less insulin than cells from control subjects (Sanjabi et al., 2015).

1.3.1 Visceral Versus Subcutaneous Fat: Brown Versus White Fat

Depending on its location, AT responds differently to stimulation as has been recently reviewed (Lee et al., 2013). In animal models, AT in and around the visceral organs (visceral adipose tissue, VAT) takes up glucose rapidly, is subject to apoptosis, remodels and enlarges in response to excessive energy intake, especially high-fat diets (Strissel et al., 2007). In contrast, subcutaneous adipose tissue (SAT) in the abdominal and gluteal region is much less metabolically active and appears to act as a "metabolic sink," sequestering potentially toxic NEFAs as neutral TGs. A possible explanation for this functional difference is the production of insulin by mature adipocytes (Sanjabi et al., 2015). While mature human adipocytes from both VAT and SAT produced insulin, insulin production by VAT cells was approximately sixfold

TABLE 1.1   Factors Secreted by Adipose Tissue

General function	Factors currently identified
Lipid metabolism	Lipoprotein lipase (LPL), free fatty acids, glycerol, apolipoprotein E
Steroid hormones	Estrone, estradiol, testosterone
Growth factors and cytokines	IGF-1, nerve growth factor (NGF), vascular endothelial growth factor (VEGF), leptin, tumor necrosis factor-α (TNF-α), interleukin-1β (IL-1β), interleukin-6 (IL-6)
Vasoactive factors	Monobutyrin, angiotensinogen angiotensin II, atrial natriuretic peptide, eicosanoids, prostaglandins E2 (PGE2), prostaglandins F2a (PGF2a), prostacyclin (prostaglandin I2/PGI2)
Complement system	Factor B, factor C, C3, C1q, factor D (adipsin/acylation-stimulating protein (ASP/C3desARg))
Binding proteins	Retinol BP, IGF-BPs, sTNFRs
Extracellular matrix proteins	Monocyte chemotactic protein-1 (MCP-1), matrix metalloproteinases
Others	Adiponectin (Acrp30/AdipoQ), cholesteryl ester transfer protein, plasminogen activator inhibitor 1, haptoglobin, LPA, lysophosphatidic acid, resistin, visfatin/PBEF, omentin, fasting-induced adipose factor, metallothionen, apelin

BP, binding protein; IGF-BPs, insulin-like growth factor binding proteins; LPA, lysophosphatidic acid; PBEF, pre-B-cell colony-enhancing factor; sTNFRs, soluble tumor necrosis factor receptors.

Adapted from Sethi, J.K., Vidal-Puig, A.J., 2007. Thematic review series: adipocyte biology. Adipose tissue function and plasticity orchestrate nutritional adaptation. J. Lipid Res. 48 (6), 1253–1262.

higher than by SAT cells. The authors determined that the magnitude of insulin production was related to LD size, and fell dramatically ex vivo, in adipocytes exposed to high media glucose or lipopolysaccharide (LPS); both agents induce an inflammatory state and stimulate lipolysis.

In human studies, enlargement of VAT has been strongly linked to insulin resistance, hepatic steatosis, and other metabolic dysregulations (Despres et al., 2008). It is possible that hyperinsulinemia produced, in part, by the mass of hypertrophic VAT adipocytes, accounts for the initial rapid uptake of glucose by adipocytes, followed by downregulation of GLUT4 transporters and adipocyte insulin resistance (Gonzalez et al., 2011). Recent work suggests that hyperinsulinemia also drives hepatocyte lipogenesis while failing to downregulate gluconeogenesis (GNG) (Cook et al., 2015). This situation results in the dual threats of hyperglycemia and hypertriglyceridemia and leads to hepatic steatosis associated with VAT hypertrophy. SAT adipocytes, in comparison, retain the ability to undergo hyperplasia in response to excess energy availability. If this capacity is compromised or exceeded, especially in the obese patient, high circulating lipids can increase the size and number of LD in parenchymal cells at ectopic sites such as the liver, heart, muscle, and pancreas with serious health consequences (Kienesberger et al., 2013).

Mammals store lipids in both white adipose tissue (WAT) and brown adipose tissue (BAT). BAT and WAT differ developmentally, functionally, as well as structurally (Gesta et al., 2007). BAT contains abundant mitochondria and multiple small LDs. The iron content in the mitochondrial respiratory chain accounts for its darker brown color, hence the name, "brown fat." In contrast, WAT mature adipocytes are unilocular, with one large LD and few mitochondria. BAT develops prenatally, whereas WAT matures postnatally. Finally, BAT is innervated by the sympathetic nervous system and highly vascularized, facilitating energy dissipation through adaptive thermogenesis. It is of interest that autophagy was shown to be necessary for differentiation of the WAT phenotype; inhibition of autophagy induced a "browning" of the AT. A comprehensive examination of human BAT and WAT development has revealed a transient expression of BAT markers during adipocyte differentiation, suggesting that BAT and WAT reflect different stages along a differentiation continuum (Guennoun et al., 2015). The ability of BAT to release energy as heat is due to the presence of uncoupling protein 1 (UCP1) in the inner mitochondrial membrane. This facultative proton transporter bleeds off the chemiosmotic proton gradient created by the respiratory chain. Without a strong gradient, ATP synthase does not produce ATP, and can actually waste energy as heat. Adaptive thermogenesis is vital to maintain body heat in neonates with a high surface area to volume ratio. BAT is less prominent in the adult human with greater volume to surface area ratio and better capacity to maintain stable body temperature. Since it is possible that increased BAT in obese adults could dissipate excess calories, regulation of molecular pathways in the adipocyte to favor BAT has been suggested as a therapeutic target for weight maintenance (Hansen and Kristiansen, 2006).

1.4 AT DYSREGULATION AND METABOLIC SYNDROME

Metabolic Syndrome (MS) refers to the term "syndrome X" originally coined (Reaven, 1988) to describe a cluster of metabolic abnormalities associated with insulin resistance and increasing risk for type II diabetes. Reaven proposed that resistance to insulin-stimulated glucose uptake can be compensated by hyperinsulinemia. Unfortunately, this compensation is not benign. The comorbidities of MS include hypertension not reduced with antihypertensive treatment. Hypertriglyceridemia with decreased high-density lipoprotein (HDL) concentrations and other risk factors for cardiovascular disease are also clustered in this syndrome. Insulin resistance was identified as a fundamental "disorder" that occurred prior to diabetes. Since he identified MS in a subpopulation of lean, but insulin-resistant individuals, Reaven included obesity as an "associated" comorbidity and not necessarily the cause of the insulin resistance syndrome.

In the years that followed, abundant biochemical and clinical evidence has demonstrated that insulin resistance is indeed a key factor associated with MS abnormalities. These include a typical atherogenic state with dyslipidemia, specifically with high TG and apolipoprotein B concentrations, an increased proportion of small dense LDL particles and a reduced concentration of smaller HDL-cholesterol particles. Abnormalities also include a prothrombotic profile, elevated blood pressure, rising blood glucose, and a state of chronic, low-grade inflammation (Despres et al., 2008).

Central to this cluster of potentially life-threatening disease states is AT dysfunction, now thought to play a major role in MS pathogenesis (Lafontan, 2014). Hypoxia is prominent in dysfunctional AT with closely packed hypertrophic and adiponectin-depleted adipocytes; hypoxia leads to generation of oxidative and ER stress. Activated adipocytes and infiltrating immune cells in dysfunctional AT secrete inflammatory mediators as outlined in Fig. 1.4 (McArdle et al., 2013). Mediators stimulate infiltrating macrophages to change character from M2, characterized as secreting antiinflammatory mediators, to M1 forms that highly express genes for proinflammatory mediators such as tumor necrosis factor alpha (TNFα), interleukin-6 (IL-6), monocyte chemoattractive protein 1 (MCP1), and inflammatory nitric oxide synthase (Lumeng et al., 2007).

Cellular stress in dysfunctional AT is known to induce autophagy (see above) resulting in either cell survival, or, if the cell is too severely damaged, cell death. Adipocyte death and fallout reduces lipid storage capacity and potentiates NEFA lipotoxicity. When

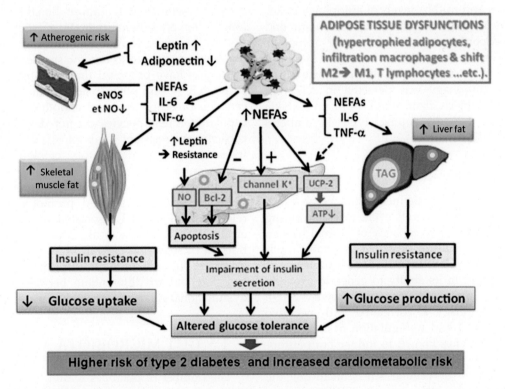

FIGURE 1.4 **Adipose tissue dysfunction.** Risk for metabolic syndrome due to obesity can be tied to adipose tissue dysfunction. Altered secretory capacity of the cells results in multiple effects in the blood vessels, muscles, liver, and pancreas. UCP-2, uncoupling protein-2; ATP, adenosine triphosphate; TNF, tumor necrosis factor; IL, interleukin. *Reproduced from Lafontan, M., 2014. Adipose tissue and adipocyte dysregulation. Diabetes Metab. 40 (1), 16–28—Science Direct.*

the obese patient with dysfunctional AT also consumes nutrients in excess of needs, hyperinsulinemia with associated hyperglycemia can ensue. In the liver, hyperinsulinemia increases hepatic lipid synthesis, increasing the lipid load and increased deposition of the fat in ectopic sites such as the liver, muscle, and other tissues. Reduced glucose uptake in the muscle and fat results in altered glucose tolerance with increased risk for type II diabetes and cardiovascular disease. Causes and consequences of AT dysfunction have been well reviewed elsewhere (Blüher, 2013; Sam and Mazzone, 2014).

1.4.1 AT Dysfunction: Autophagy and Senescence

As described above (Fig. 1.3), autophagy is required in periods of nutrient starvation to recycle metabolic substrate, prevent cell damage, and maintain metabolism through lysosomal turnover of cellular components. Autophagy has also been shown to modulate lipid storage capacity by regulating LD biogenesis and breakdown (Kovsan et al., 2010). Since autophagy is inhibited by insulin and mTORC1, adipocytes in the fed state would be expected to display few autophagic markers. Unexpectedly, AT from human obese subjects exhibited more autophagic markers than AT from lean controls. Protein and mRNA levels of autophagy genes were increased in VAT compared with SAT tissue and levels were proportional to the degree of adipocyte hypertrophy (Kovsan et al., 2011). The degree of autophagy appeared to be related to insulin resistance in the AT, since less autophagy was detected in AT from "healthy obese" subjects than in AT from prediabetic patients at risk for developing type II diabetes (Kosacka et al., 2015).

1.4.2 Adipocyte Senescence

Fifty years ago, Hayflick defined cellular senescence as a state of "permanent" cell cycle arrest resulting from the limited replicative capacity of normal human diploid fibroblasts in culture; he called this state "replicative senescence" (Hayflick, 1965). Autophagy is linked to senescence in its ability to generate irreversible cell cycle arrest (White and Lowe, 2009). Activation of autophagy is correlated with negative feedback of mTORC1 and upregulation of a subset of autophagy-related genes known to induce both autophagy and senescence (Young et al., 2009). Senescence can also be triggered by cellular stress, including the DNA damage response (Wang et al., 2009) and aberrant hyperproliferative stimuli (d'Adda di Fagagna, 2008). Senescent cells are not quiescent; they are metabolically active, but dysfunctional. They increase AT

dysfunction by secreting a spectrum of proinflammatory cytokines and destructive secretions; this secretory cocktail has been characterized as a senescence-associated secretory phenotype (SASP). When exposed to SASP adjacent cells exhibit ER stress (Campisi and d'Adda di Fagagna, 2007; Rodier et al., 2009; Rodier and Campisi, 2011; Salama et al., 2014).

The senescent phenotype is associated with larger size and flattened cell shape. In order to undergo senescence, cells would have to engaged in targeted degradation of specific cellular components, perhaps of the cytoskeleton, through autophagy (Young et al., 2009). To date, no single biomarker has been identified as specific to senescent cells, however they can be detected by high concentrations of lysosomal β-galactosidase as well as biomarkers p16INK4a and p21; the latter are tumor suppressor proteins that inhibit the cell cycle. Despite their flattened phenotype and two- or more fold larger size than normal cells, senescent cells do not have more DNA than normal cells, suggesting the arrest occurs in the initial G_1 phase of the cell cycle. Multiple foci of nuclear DNA damage and excessive accumulation of heterochromatin are also characteristic of senescent cells (Rodier and Campisi, 2011). Most importantly, senescent cells do not readily die. Cell types are variably susceptible to apoptotic death, however senescent fibroblasts were shown to withstand the apoptotic stimulus, ceramide, longer than endothelial cells (Hampel et al., 2004).

Preadipocytes, derived from mesenchymal fibroblast-like cells, are known to become senescent. Kirkland and colleagues have proposed that repeated replication, ER stress, NEFA, and other toxic stimuli can trigger senescence in preadipocytes and endothelial cells lining AT capillaries. Preadipocytes do not develop lipid-storing capacity, are resistant to apoptosis, develop SASP with its secretion of inflammatory cytokines, extracellular matrix modulators, and other mediators. SASP mediators can reprogram M2 macrophages to M1 cells, further impeding adipogenesis and promoting lipolysis with release of NEFA and associated lipotoxicity (Kirkland et al., 2002; Tchkonia et al., 2010). Thus, an increased fraction of senescent cells in human AT not only takes up space, but reduces the AT capacity to store lipids and has potential to differentiate between "healthy" and dysfunctional AT. More research is required in this area.

1.5 GUT MICROBIOTA AND OBESITY

The number of microbes within the human gastrointestinal tract has been estimated as more than 10-fold that of the cells in the human host; in short, we are less than 10% human. Although gastric acid, bile, and pancreatic secretions suppress bacterial colonization of the stomach and small intestine, unique bacterial species can occupy their own specialized ecological niches along the full length of the gastrointestinal tract (Gibson and Roberfroid, 1995) as discussed in Essentials IV: Diet, Microbial Diversity and Gut Integrity (Fig. 1.5).

Although individual humans appear to maintain a relatively constant microbial pattern over time, relative overgrowth of Firmicutes versus Bacteroidetes has been observed in obese animals and humans. This altered ratio (dysbiosis) was associated with changes in gene expression and increased capacity to extract short-chain fatty acids (SCFA) and other nutrients from the diet (Turnbaugh et al., 2006). The diet consumed by the obese and weight-reduced human could mediate this change, since improvement in dysbiosis was seen in humans following weight loss on diets limiting either fat or carbohydrate (Ley et al., 2006). While mechanisms linking gut microbiota, obesity, and the MS are complex, two key hypotheses are under investigation (Harris et al., 2012). The first is that gut microbes and/or their toxic metabolites (eg, lipopolysaccharide, LPS) translocate the intestinal epithelial cell (IEC) barrier, activate the innate immune response, and induce chronic systemic inflammation and subsequent

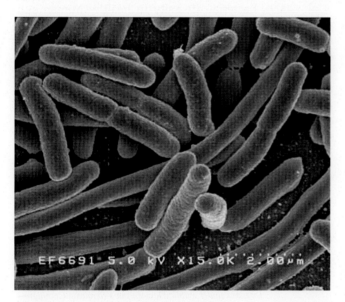

FIGURE 1.5 Scanning electron micrograph of *Escherichia coli*, grown in culture and adhered to a cover slip. *E. coli* is a Gram-negative, facultative anaerobic bacterium of the family Enterobacteriaceae, commonly found in the colon of humans. Most strains are harmless, and in fact, benefit their host by producing metabolites such as butyrate, niacin, and vitamin K_2 (Bentley and Meganathan, 1982) and by preventing intestinal colonization with pathogenic bacteria. *Figure in public domain. "Escherichia Coli NIAID" by Credit: Rocky Mountain Laboratories, NIAID, NIH—NIAID: for the public domain images, there is no copyright, no permission required, and no charge for their use. Licensed under Public Domain via Wikimedia Commons—https://commons.wikimedia.org/wiki/File:EscherichiaColi_NIAID. jpg#/media/File:EscherichiaColi_NIAID.jpg/.*

metabolic dysfunction. The second is that food metabolites derived from bacterial action, for example, SCFA, signal biological responses both at the IEC barrier and systemically in the host. These lines of investigation suggest that dysbiosis, altered microbial diversity, abnormal permeability of the intestinal epithelial barrier, or altered sensitivity of the intestinal cells to diet-derived compounds or their bacterial metabolites, can predispose to obesity and its comorbities (Shen et al., 2013) as discussed elsewhere (refer to Essentials IV: Diet, Microbial Diversity and Gut Integrity).

1.5.1 Microbiota and the Immune System

A complex crosstalk exists among the microbiota, IEC, and the gut-associated immune system. Microbes secrete signals that direct development of key lymphocyte subsets. Microbial signals direct class switching in human intestinal B cells, govern differentiation of intestinal Th17 effector T cells and regulatory T cells, and determine the ratio of Th1 and Th2 effector cells (refer to Chapter 3: Type I Diabetes and Celiac Disease). Microbes also produce metabolites that support the IEC growth and maintain barrier integrity (O'Hara and Shanahan, 2006). The microbiota communicates with the gut-associated innate and adaptive immune responses via immunosensory cells including dendritic cells, antigen-presenting cells, and intestinal M (microfold) cells associated with Peyer's patches. Intestinal enterocytes secrete afferent signals (chemokines and cytokines) that direct innate and adaptive immune cells to the infected site, while M cells transport luminal antigens to subadjacent dendritic cells and other antigen-presenting cells. Dendritic cells and enterocytes detect and discriminate commensal and pathogenic bacterial antigens by means of pattern recognition receptors (PRRs) that recognize specific microbial molecular patterns. PRRs include the toll-like receptor (TLR) family that signals activation of inflammatory cascades (eg, NFκB). In the healthy gut, optimal TLR expression and activity contribute to homeostasis. Pathogenic microbes and their toxic metabolites can also gain access to the bloodstream through paracellular transport when intestinal permeability is increased (refer to Essentials IV: Diet, Microbial Diversity and Gut Integrity; see discussion in Chapter 2: The Obese Gunshot Patient: Injury and Septic Shock).

1.5.2 Gut Microbe-Obesity Hypothesis 1: Endotoxemia, Obesity, and MS

Obesity and MS are characterized both by chronic systemic inflammation and by the presence of metabolic endotoxemia detected as increased circulating LPS concentrations (Manco et al., 2010;

Moreno-Navarrete et al., 2012). LPS bound to LPS-binding protein (LPB) has also been measured as an index of endotoxemia. LBP concentrations were higher in presumably healthy obese Chinese subjects compared with lean controls and strongly associated with MS and type II diabetes (Sun et al., 2010). These results are supported by animal studies that produced metabolic endotoxemia and associated obesity, insulin resistance, and diabetes by feeding mice a high-fat (72%) diet. Recent studies have revealed that the endotoxins activate TLR4, stimulating the innate immune system and inducing insulin resistance (Jialal et al., 2014). Adipocytes also express TLR4. Recently, LPS binding to TLR4 on human visceral adipocytes was associated with a reduction in adiponectin secretion (Taira et al., 2015); insulin production by adipocytes was also decreased by LPS binding (Sanjabi et al., 2015). Taken together, current evidence suggests that LPS, together with its coreceptor proteins (LBP; CD14) bind TLR4 and signal downstream inflammatory cascades. One of these cascades, nuclear factor kappa beta (NFκβ), triggers release of inflammatory mediators such as IL-6, IL-1β, TNF-α, and monocyte chemotactic protein-1. It must be noted that species differ dramatically (>10,000-fold) in their sensitivity to LPS. The LPS dose that produces shock in humans is at the low end of the spectrum, while the lethal dose (LD$_{50}$) for mice is at the high end (Warren et al., 2010). Thus, studies in animals may not be directly applicable to humans.

Links between endotoxemia and dysregulation of energy homeostasis have been observed for many years. Over 20 years ago, an acute bolus of LPS produced hyperglycemia, hypoaminoacidemia, and

hyperlipidemia, together with increased splanchnic glucose output in healthy human subjects. The changes could not be attributed to hormonal changes, but were associated with increased TNFα and IL-6 at 90 minutes (Fong et al., 1990). Subsequent trials with low-dose LPS produced hypertriglyceridemia in rodents with increased hepatic lipoprotein production. Since inactivation of proinflammatory cytokines (TNFα, IL-1) did not block the increase in serum TG, these results suggested that LPS, alone or with unknown mediators, induced dyslipidemia (Feingold et al., 1992) (Fig. 1.6)

1.5.3 Gut Microbe-Obesity Hypothesis 2: Microbiota Metabolites and Energy Regulation

The second line of inquiry is that ingested food is metabolized by the microbiota and that these metabolites modify energy balance and lipid distribution. Observations of dysbiosis in the microbial diversity patterns in obese, compared with lean, humans and experimental animals, have revealed a reversal in the relative density of the two major bacterial phyla in obese subjects; energy-harvesting Firmicutes are increased and Gram-negative Bacteroidetes are decreased (Turnbaugh et al., 2009). Bacterial diversity in healthy human subjects was strongly associated with consumption of a diverse, fiber-rich diet (Claesson et al., 2012) as discussed below.

Bacteria, principally in the colon, produce short chain fatty acids (SCFAs) from resistant starch and undigested proteins. The most physiologically important SCFAs, defined as carboxylic acids with aliphatic tails less than six carbons, are acetate (C2), propionate (C3), and

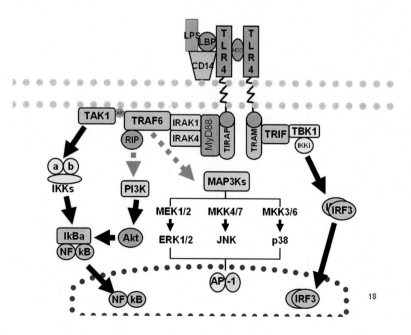

FIGURE 1.6 **LPS stimulates innate immune response.** LPS bound to its binding protein (LBP) and CD14 stimulates the pattern recognition protein, TLR4 and sets the innate immune response in motion. The inflammatory NFκB pathway stimulates secretion of inflammatory mediators, TNF-α, and IL-1. *Figure is in public domain. "Toll-like receptor pathways revised" by the original uploader was Subclavian at English Wikipedia—Transferred from en.wikipedia to commons. Licensed under Public Domain via Wikimedia Commons— https://commons.wikimedia.org/wiki/File:Toll-like_receptor_ pathways_revised.jpg#/media/File:Toll-like_receptor_pathways_ revised.jpg/.*

butyrate (C4). As recently reviewed (Layden et al., 2013), once formed, SCFA are rapidly absorbed by colonocytes and some butyrate is diverted for use as the primary colonic energy source. Most of the SCFA are released into the portal vein for transport to the liver where butyrate and propionate, but not acetate, are extracted by hepatocytes. Acetate along with residual butyrate and propionate leave the liver by the hepatic vein and enter the peripheral circulation where SCFA bind to cell surface receptors of the G-protein-coupled receptor family. These receptors, designated FFAR2 (grp43) and FFAR3 (gpr41), have been detected in cells of the human immune system, especially monocytes and neutrophils as well as in human AT, muscle, liver, pancreatic islets, and other tissues.

SCFA bound to these receptors modulate several pathways involved in energy utilization. For example, SCFA binding by intestinal L cells stimulates synthesis of glucagon-like peptide 1 (GLP1) (Tolhurst et al., 2012), thereby increasing glucose-dependent insulin secretion by the pancreatic β cells (refer to Chapter 4: Type II Diabetes, Peripheral Neuropathy and Gout). Recently, a transgenic mouse model was used to demonstrate that gpr43 binding suppresses insulin signaling in white AT, thereby inhibiting TG accumulation, while gpr43 binding at the liver and muscle promotes NEFA oxidation. Thus SCFA-gpr43 binding appears to act as a sensor for postprandial energy, enhancing its utilization by skeletal muscle, reducing adipocyte storage, and maintaining metabolic homeostasis (Marques et al., 2015). SCFAs are also important epigenetic modulators, for example by providing the acetate moiety for histone acetylation.

SCFA are also active participants in regulation of intestinal gluconeogenesis (GNG) and subsequent neuronal signaling to brain targets. This gut—brain axis can provide an explanation for earlier observations that reported improved glucose tolerance in mice fed a high-fat diet supplemented with butyric acid (Gao et al., 2009). In contrast to hepatic GNG which potentiates hyperglycemia and glucose intolerance, intestinal GNG is upregulated during the postabsorptive phase and improves glucose control by signaling brain targets.

Intestinal GNG has been estimated to provide 20—25% of total endogenous glucose during fasting. This glucose production not only supports fasting blood glucose, but it improves insulin sensitivity by stimulating peripheral and liver glucose uptake and proportionally inhibiting hepatic GNG (Gautier-Stein et al., 2006). As indicated in Fig. 1.7, SCFA, especially butyrate and propionate, upregulate intestinal GNG genes by different mechanisms (Gautier-Stein et al., 2006; De Vadder et al., 2014). Glucose in the portal vein appears to be sensed by the glucose receptor SGLT3, and signals the brain through the vagus and

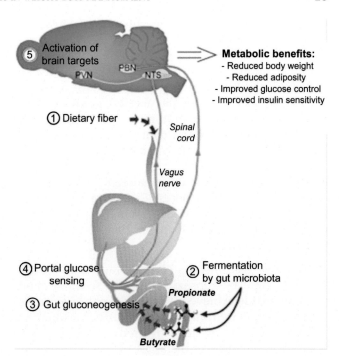

FIGURE 1.7 **Role of SCFA in the gut—brain axis.** Propionate and butyrate differentially activate intestinal gluconeogenesis (IGN). Butyrate does not bind FFAR3, but acts by an unknown mechanism associated with intracellular rise in AMP, AMPK phosphorylation and expression of genes required for IGN. Propionate is both a substrate for IGN and potentiates neural signaling by a process that involves FFAR3 binding. Glucose produced by IGN is detected by a portal vein glucose sensor that transmits the signal to the brain along neuronal tracts. *From De Vadder, F., Kovatcheva-Datchary, P., et al., 2014. Microbiota-generated metabolites promote metabolic benefits via gut-brain neural circuits. Cell 156 (1—2), 84—96—Science Direct.*

spinal neuronal circuits to modulate hunger and support glucose homeostasis. Additionally, as discussed below, peptides derived from dietary protein are released in the portal vein and act as antagonists of the μ-opioid receptors on neurons in portal vein walls and both promote intestinal GNG and signal through neural circuits to the brain to inhibit food intake (Duraffourd et al., 2012).

1.6 GOALS AND PRACTICAL ISSUES IN WEIGHT LOSS MANAGEMENT

Guidelines for weight loss and management have been devised in the United States by the National Heart, Lung and Blood Institute (NHLBI), in cooperation with the National Institute of Diabetes and Digestive and Kidney Diseases (NIDDK). In 1998 "Clinical Guidelines on the Identification, Evaluation, and Treatment of Overweight and Obesity in Adults" was published (Expert Panel on the Identification et al., 1998). This report was a systematic review (SR) of the published scientific literature found in MEDLINE from Jan. 1980 to Sep. 1997. The Expert Panel reviewed this literature to determine appropriate treatment strategies for overweight and obesity. In 2005, the NHLBI initiated development of a review and update of overweight/obesity guidelines. To assure improved quality and impact of the guidelines, policies to assure rigor and minimize bias were put in place and the most recent recommendations published (Jensen et al., 2014).

Nonsurgical weight loss treatment is considered successful if the patient loses 10–20% of high weight and maintains this weight loss through changes in lifestyle that can be sustained over time. Specific goals have been established:

- The *initial goal* of weight loss therapy is to reduce body weight by approximately 10% from baseline. If this goal is achieved, further weight loss can be attempted, if indicated through further evaluation.
- A *reasonable time line* for a 10% reduction in body weight is 6 months of therapy.
  - For overweight patients with BMIs in the typical range of 27–35, a decrease of 1255–2092 kJ/day (300–500 kcal/day) will result in weight losses of about 0.23–0.45 kg/week and a 10% loss in 6 months.

    For more severely obese patients (BMIs >35), deficits of up to 2092–4184 kJ/day (500–1000 kcal/day) will lead to weight losses of about 0.45–0.90 kg/week and a 10% weight loss in 6 months (Fig. 1.8).
- It should be noted that severely obese individuals (BMI > 40) have a higher basal energy expenditure and may safely lose ~10% of their weight in a 6-month period with a smaller calorie deficit (Blackburn et al., 2010).
- Weight loss at the rate of 0.45–0.90 kg/week (energy deficit of 2092–4184 kJ/day) occurs safely for up to 6 months. After 6 months, the rate of weight loss usually declines and weight plateaus because of a lesser energy expenditure at the lower weight.
- Experience reveals that lost weight usually will be regained unless an intensive weight maintenance program consisting of dietary therapy, physical activity, and behavior therapy is continued indefinitely.
- After 6 months of weight loss treatment, efforts to maintain weight loss should be put into place.
- If more weight loss is needed, another attempt at weight reduction can be made. This will require further adjustment of the diet and physical activity prescriptions.

Very-low-calorie diets (VLCD ≤ 800 kcal/day) are not recommended. Although they may produce a rapid initial weight loss, upon return to actual food intake, the rate of weight regain is also rapid. VLCD are too low in calories to provide an adequate nutrient intake (including dietary fiber), and because they are often provided in liquid form, they do not require the patient to make long-lasting behavioral changes. Finally, very-low-fat diets have been shown to increase the risk for gallstones (Hoy et al., 1994). For patients unable to achieve significant weight reduction, prevention of further weight gain is an important goal; such patients may also need to participate in a weight management program. Bariatric surgery as a tool for weight loss is also discussed.

1.6.1 Clinical Diagnosis of MS

Since insulin sensitivity is difficult to measure in clinical practice, several agencies, including the National Cholesterol Education Program-Adult Treatment Panel III (NCEP-ATP III) and International Diabetes Federation (IDF) have devised simple clinical parameters listed in Table 1.2. The cut-off values identify individuals likely to be insulin-resistant and who exhibit the atherogenic and diabetogenic abnormalities related to an impaired insulin action. It should be noted that no direct marker of insulin resistance is contained in the NCEP-ATP III or IDF clinical criteria and that the prevalence of insulin resistance varies depending on the MS clinical criteria used (Grundy et al., 2005).

With the use of these simple criteria, investigators found that a clinical diagnosis of the MS (either by NCEP-ATP III or IDF criteria) was associated with an increased *relative* risk of cardiovascular disease. However not all subgroups of patients with three of the five criteria will have similar clinical prognoses. For example, type II diabetic patients who are hyperglycemic, obese, and hypertensive can be metabolically quite distinct from nondiabetic patients with high TG and low HDL-cholesterol levels and abdominal obesity. It should also be noted that the commonly used BMI is relatively insensitive to body composition and metabolic risk. The guidelines in Table 1.2 use waist circumference to assess risk related to obesity. Indices such as waist:hip ratio and waist:height ratio are also used to arrive at a more accurate predictor of metabolic risk. Note also that the well-described failure of obese subjects to achieve and maintain dramatic weight loss on a diet and exercise program may mask the improved metabolic profile and VAT:SAT ratio and improved risk prognosis that accompanies diet and lifestyle changes.

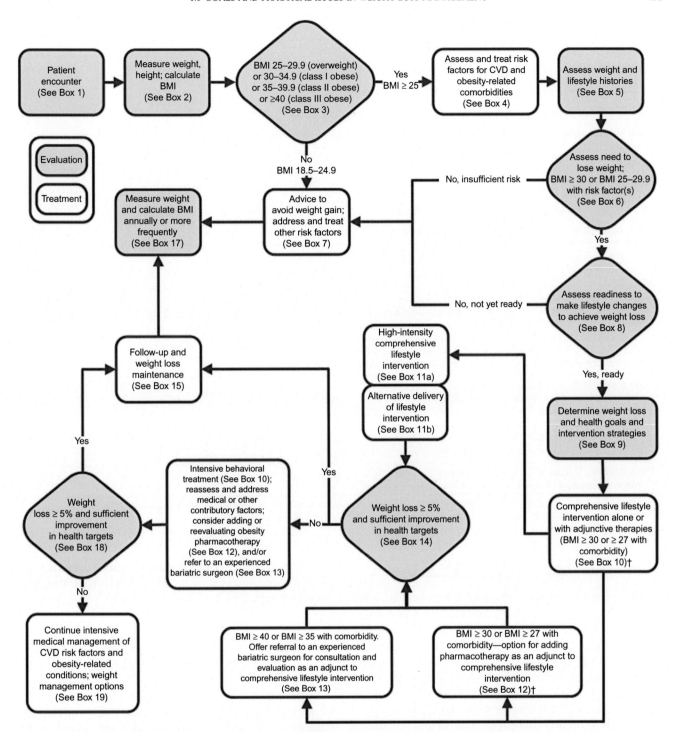

FIGURE 1.8 **Treatment Algorithm for Chronic Disease Management for Primary Care of Patients With Overweight and Obesity***. *This algorithm applies to the assessment of overweight and obesity and decisions based on that assessment. †BMI cutpoint determined by the FDA and listed on the package inserts of FDA-approved obesity medications. BMI indicates body mass index; CVD, cardiovascular disease; and FDA, U.S. Food and Drug Administration. The Expert Panel addressed five critical questions (CQ). CQ1 addresses the expected health benefits of weight loss as a function of the amount and duration of weight loss. CQ2 addresses the health risks of overweight and obesity and seeks to determine if the current waist circumference cutpoints and the widely accepted BMI cutpoints defining persons as overweight (BMI 25–29.9 kg/m²) and obese (BMI ≥30 kg/m²) are appropriate for population subgroups. CQ3 asks which dietary intervention strategies are effective for weight loss efforts while CQ4 addresses the efficacy and effectiveness of a comprehensive lifestyle approach (diet, physical activity, and behavior therapy) to achieve and maintain weight loss. CQ5 seeks to determine the efficacy and safety of bariatric surgical procedures, including benefits and risks (Jensen et al., 2014). *Reproduced from Jensen, M.D., Ryan, D.H., et al., 2014. 2013 AHA/ACC/TOS guideline for the management of overweight and obesity in adults: a report of the American college of cardiology/American heart association task force on practice guidelines and the obesity society. J. Am. Coll. Cardiol. 63 (25 PART B), 2985–3023—Science Direct.*

TABLE 1.2   Guidelines for Diagnosing Metabolic Syndrome: Three of the Following Five Conditions (Grundy et al., 2005)

- Fasting glucose ≥100 mg/dL (or receiving drug therapy for hyperglycemia)

- Blood pressure ≥130/85 mmHg (or receiving drug therapy for hypertension)

- Triglycerides ≥150 mg/dL (or receiving drug therapy for hypertriglyceridemia)

- HDL-C <40 mg/dL in men or <50 mg/dL in women (or receiving drug therapy for reduced HDL-C)

- Waist circumference ≥102 cm (40 in) in men or ≥88 cm (35 in) in women; if Asian-American, ≥90 cm (35 in) in men or ≥80 cm (32 in) in women

Complaints of chest pain, dyspnea, or claudication (symptoms of possible complications) may warrant more detailed studies such as:

- Electrocardiography (rest/stress ECG)

- Ultrasonography (vascular or rest/stress echocardiography)

- Stress single-photon emission computed tomography (SPECT); cardiac positron emission tomography (PET)

TABLE 1.3   Lifestyle Recommendations to Reduce Symptoms of Metabolic Syndrome

- *Lower LDL-c: evidence is strong (A)*
  - Consume vegetables, fruits, and whole grains, low-fat dairy, poultry, fish, legumes, nontropical oils and nuts
  - Avoid sweets, sugar-sweetened beverages, and red meats
    - Adapt dietary pattern to appropriate calorie requirements; modify for cultural and medical needs
    - DASH diet, USDA food pattern and AHA diet patterns are recommended
    - Aim for a diet with 5–6% calories from saturated fat; reduce calories from *trans* fats

- *Lower blood pressure: evidence is strong (A)*
  - Follow diet for LDL-cholesterol lowering (above)
  - Reduce sodium intake
    - Reduce sodium intake to 2400 mg sodium/day
    - *Evidence moderate (B)*
      - Greater BP lowering achieved by reducing sodium to 1500 mg/day

- *Physical activity to lower LDL-c and BP: evidence is moderate (B)*
  - Aerobic exercise 3–4 times per week at ~40 min per session

Adapted from Eckel, R.H., Jakicic, J.M., et al., 2014. 2013 AHA/ACC guideline on lifestyle management to reduce cardiovascular risk: a report of the American college of cardiology/American heart association task force on practice guidelines. Circulation 129 (25 Suppl. 2), S76–S99.

1.7 MANAGEMENT OF MS

Hippocrates was one of the first physicians to write about obesity. His prescription, less food and more exercise, still remains the mainstay of treatment (Precope, 1952). In 2005, with revisions published in 2013, the American Heart Association/National Heart, Lung and Blood Institute issued a scientific statement on the diagnosis and management of MS (Grundy et al., 2005; Eckel et al., 2014). The prime emphasis in management of the MS per se is to modify underlying risk factors for clinical atherosclerotic disease. Closely related is the goal to reduce risk for type II diabetes mellitus. Diet modification and physical activity have been shown to reduce all of the metabolic risk factors. If absolute risk is still too high, consideration can be given to incorporating drug therapy into the regimen. Priorities for drug therapy include elevations of low density lipoprotein C (LDL-C), blood pressure, and glucose; current guidelines for their management should be followed. It is, of course, important that efforts should be made to bring about smoking

cessation and reduction of excess alcohol consumption. Lifestyle recommendations abstracted from the most current report are found in Table 1.3.

1.8 DIET PATTERNS THAT MODIFY RISK FOR OBESITY AND MS

Although there is sound evidence that weight loss can be achieved by reducing calories, there is continued dispute as to the best mix of macronutrient calories in the diet. A low-calorie/low-fat diet is effective in achieving successful weight loss, however there is a growing body of data to suggest that when energy intake is restricted, the macronutrient composition of the diet produces minimal differences in weight loss and maintenance. This question was addressed with a meta-analysis of 5 trials including 447 individuals with baseline BMI of at least 25. While low-carbohydrate diets produced greater weight loss at 6 months, the difference was no longer significant at 12 months. At 12 months, TG and HDL levels were more favorable in the low-carbohydrate group, but total cholesterol and LDL values were more favorable in the low-fat group (Nordmann et al., 2006). Compliance was greater for both diets when combined with a comprehensive behavioral support program (Foster et al., 2010). In addition to weight loss, both diets appear to favorably influence clinical indicators of the MS, and to downregulate inflammation associated with adipose tissue (AT) dysfunction (McArdle et al., 2013).

It is widely misperceived that American adults consume excess protein. This is based in part on the recommended dietary allowance (RDA) of 0.8 g/kg body weight per day. Analysis of NHANES data from 2001 to 2010 ($n = 23,876$ adults) revealed a range of intake from 0.69 to 1.51 g/kg body weight (Pasiakos et al., 2015). However, this analysis also revealed that usual protein intake was inversely associated with BMI and waist circumference, and positively associated with HDL cholesterol and reduced risk for cardiometabolic disease. The investigators concluded that a higher protein intake improves MS risk. These findings are supported by abundant reports suggesting beneficial effects, but no negative consequences of consuming protein at approximately twice the current RDA. These studies prompted the Institute of Medicine to establish an acceptable macronutrient distribution range for protein of 10–35% of energy for adults; this range is associated with reduced risk for chronic diseases, while providing adequate intakes of essential nutrients (Medicine, 2005). In 2013, Protein Summit 2.0 brought together experts to examine available data on protein requirements in the maintenance of optimal health (Rodriguez, 2015). A review of the role of protein in weight management suggested that protein (1.2–1.6 g/kg body weight) has modest beneficial effects on thermogenesis and resting energy expenditure. This level of protein intake also improved satiety, and adherence to an energy-reduced diet (Leidy et al., 2015). Supportive data from a recent European multicenter trial, the Diet Obesity Gene (DioGenes), was recently published. Investigators followed 932 obese families for 1 year to test the effectiveness of diet patterns with a slight increase in dietary protein (HP) and/or a lowered glycemic index (LGI). Following an initial weight loss program for adult participants, compared with the HP and LGI groups, the HP/LGI group showed no weight regain, a reduced drop-out rate, and reduced body fatness, with positive effects on blood pressure, lipids, and inflammation (Astrup et al., 2015).

Data from recent clinical trials also suggest that the Mediterranean diet, with a higher total-fat content made up largely of monounsaturated fatty acids found in olive and canola oils, nuts, and seeds, normalized indices of MS in some populations (Jiménez-Gómez et al., 2010), while diets high in saturated fatty acids (SFA) induced a proinflammatory gene expression profile in AT (van Dijk et al., 2009). In an attempt to clarify the impact of macronutrients on obesity and MS, an extensive search of human dietary trials within the PubMed database was conducted. Most studies included nondiabetic subjects, however a smaller number of studies included patients with type II diabetes (Deer et al., 2015). Positive associations with MS were found with diets enriched in SFAs, and possibly with fructose. The authors noted that other variants known to modify metabolic effects, such as subject fitness, health status, intake of micronutrients, and nonnutrient bioactive components and environmental exposures were not measured and commented that the design of most trials is insufficient to conduct a comprehensive mechanistic analysis of specific diet components on the prevention and treatment of overweight/obesity.

1.9 SPECIFIC DIET COMPONENTS THAT TARGET OBESITY AND MT

As discussed below and throughout this text, the human diet is comprised of individual components with potential to modulate food intake and energy homeostasis. Unfortunately, the design of many clinical trials have failed to examine the diet pattern with respect to the content of specific micronutrients and bioactive food components, nor have they evaluated mechanistic biological plausibility of targeted diet modification. Readers are directed to recent comprehensive reviews of the mechanisms through which dietary components can prevent and/or treat obesity and MS (Blackburn et al., 2010; Navarro et al., 2015; Torres-Fuentes et al., 2015). Selected mechanisms are discussed below.

1.9.1 Dietary Fiber and Microbiome Diversity

Dietary fiber is composed predominantly of carbohydrate polymers with glycosidic linkages that cannot be hydrolyzed by digestive enzymes. Fibers can be separated into insoluble (cellulose and lignins) and soluble (galacto-oligosaccharides and fructo-oligosaccharides) fractions based on their fermentation by gut bacteria. Resistant starch is a digestible glucose polymer (in α 1–4 linkage) that reaches the colonic microbiota because its form (encased in a seed or grain) or processing (retrograde starch formed by moist heat and subsequent cooling) renders it resistant to digestion in the small intestine. While insoluble fiber provides fecal bulk, soluble plant fiber is preferentially metabolized by unique microbial species and facilitates their growth and expansion (Turnbaugh et al., 2009). A large majority of the microbes that reside in gastrointestinal tract belong either to the Firmicutes or Bacteroidetes phyla; these two phyla account for greater than 98% of the 16S rRNA sequences detected in the gut microbiota of mammals. Bacteroidetes ($\sim$50% of the microbiota) are Gram-negative anaerobes that colonize the entire gastrointestinal tract, including the oral cavity and stomach, despite the presence of gastric acid and digestive enzymes. This phylum is specialized to degrade resistant dietary polymers, such as plant wall compounds (β-glucans, pectin, and xylan) and can survive on host-derived carbohydrates such as N-glycans in mucins or chondroitin sulfates in a fasted animal. In contrast, the other major intestinal phylum, Firmicutes, includes Gram-positive organisms such as the *Clostridium*, *Lactobacillus*, and Peptostreptococcaceae that thrive on resistant starch. Thus, a diet rich in plant wall compounds facilitates Bacteroidetes abundance, while a high-resistant starch diet favors Firmicutes proliferation.

A connection between dietary fiber and MT symptoms was described over 40 years ago with the observation that rural African peoples eating a diet high in unrefined foods had a low prevalence of diabetes, cardiovascular disease, gastrointestinal diseases, and cancer compared to Africans eating highly refined Western-style diets (Burkitt and Trowell, 1977). Subsequent observations that microbial diversity and function are different in obese compared with lean subjects have led to the proposal that the microbiome can exert an "obesogenic" influence by harvesting calories from nondigested food in the colon. Evidence suggests that abnormal microbial diversity also contributes to gut barrier dysfunction mediatd by dietary components discussed in Essentials IV: Diet, Microbial Diversity and Gut Integrity, as well as by enteroendocrine cells that produce glucagon-like peptides 1 and 2, peptide YY and other peptides that control food intake; the endocannabinoid system also controls gut permeability (Geurts et al., 2014). The pathogenesis of the dysfunctional microbiota associated with obesity is as yet unclear. It is likely that a weight-reduction diet containing rich sources of soluble fiber, resistant starch, as well as other nondigestible food items including inulin-type fructans, arabinoxylans, chitin, glucans, and polyphenols can produce beneficial changes in microbial diversity and function (He et al., 2015).

1.9.2 Simple Sugars and Starch as Sources of Glucose

Foods and beverages that taste sweet, whether or not they contain calories, are detected by sweet receptors found, not only in the mouth, but throughout the gastrointestinal tract. Intestinal glucose absorption is mediated by sodium-dependent glucose transporter 1 (SGLT1) and glucose transporter 2 (GLUT2). Activation of the intestinal sweet taste receptors, heterodimers of the G-protein-coupled receptors T1R2 and T1R3, promote increased availability of brush-border GLUTs and facilitate increased glucose absorption. A recent study has shed light on the plasticity of intestinal glucose transport in lean and massively obese human subjects (Ait-Omar et al., 2011). In normal individuals under fasting conditions, apical SGLT1 captures trace glucose from the lumen, while basolateral GLUT2 transfers glucose from the blood to the rapidly dividing enterocyte to meet its energy needs. After healthy humans consumed a sugar-rich meal, GLUT2 was transiently moved from the basal to the apical enterocyte membrane, tripling the initial rate of sugar uptake. The spike of glucose in the portal vein stimulated insulin secretion by the pancreas; insulin signaling in the enterocyte contributed to the internalization of apical GLUT2 and a return to basal location and function. In contrast, in 73% of massively obese patients, GLUT2 remained at the apical membrane despite a rise in insulin. The authors postulated that permanent apical GLUT2 placement would not only facilitate high glucose absorption from a sugar-rich diet, but would facilitate glucose efflux to the lumen as soon as the glucose gradient between blood and lumen was reversed. The efflux of glucose could fuel intestinal bacteria, contributing to observed changes in gut microbiota in obesity (Zhang et al., 2009). Recently, accelerated proximal intestinal glucose absorption was reported in morbidly obese patients (Nguyen et al., 2015). The authors postulated that increased glucose absorption was related to upregulation of SGLT1 expression and increased incretin promotion, resulting in hyperinsulinemia and hyperglycemia. Although not measured, it is possible that the hyperinsulinemia disrupted the insulin signal in the enterocyte. Reduced insulin signaling would result in GLUT2 retention in the apical membrane, with subsequent increased glucose absorption. These observations suggest that diet strategies that reduce the abundance of free and rapidly digested glucose have potential to reduce the postprandial glucose spike and minimize subsequent insulin overshoot following a meal.

1.9.3 Dietary Fructose

Our early ancestors consumed fructose primarily in the form of fruits and honey. Large quantities of fructose were virtually absent in our diet until a few hundred years ago, when sucrose from cane and beets became a commercial commodity. Sucrose, a disaccharide containing one molecule each of fructose and glucose, is hydrolyzed by *sucrase* at the enterocyte brush border. Glucose is absorbed as described above and enters the highly regulated glycolytic pathway for conversion to pyruvate. In contrast, although fructose has the same chemical formula as glucose ($C_6H_{12}O_6$), its metabolism differs markedly from that of glucose (Mayes, 1993; Johnson et al., 2009; Dekker et al., 2010). Following sucrose hydrolysis, fructose is absorbed on a concentration gradient by the energy-independent GLUT5 transporter. Absorbed fructose is almost completely extracted by the liver where it is metabolized by substrate-regulated, fructose-specific enzymes that shuttle fructose into the triose-phosphate pool. Thus increased dietary fructose availability leads to increased production of pyruvate and lactate. Pyruvate is converted to acetyl CoA in the mitochondria by thiamin-requiring pyruvate dehydrogenase. Acetyl CoA produced in excess of need is shifted back to the cytosol for fatty acid synthesis and esterification to TGs. Newly formed TGs are secreted from the liver as VLDL. Through these pathways, chronic high-fructose consumption can functionally deplete thiamin, augment VLDL output, and be manifest as hypertriglyceridemia.

High-fructose consumption reliably produces obesity and MS in animal models (Zhou et al., 2014). Chronic dietary fructose administration has been shown to upregulate fructose transporters (primarily GLUT5) and fructokinase (ketohexokinase), a key enzyme in the fructose pathway (Patel et al., 2015). These adaptations increase the fraction of fructose absorbed from the diet, resulting in excess fructose absorption on a normal fructose intake. Furthermore, fructokinase requires ATP to phosphorylate fructose; increased fructokinase activity thereby depletes ATP and raises AMP in many cells, including vascular endothelial and proximal tubular cells in the kidney. Increased AMP signals low cellular energy, stimulates AMPK phosphorylation, and leads to transient arrest of protein synthesis, production of inflammatory proteins, endothelial dysfunction, and oxidative stress. Additionally, AMP is converted into xanthine, which is metabolized by xanthine oxidoreductase to uric acid. In animal models, a transient rise in serum uric acid is seen after acute ingestion of fructose; chronic fructose ingestion (several weeks) can even raise fasting serum uric acid levels. Uric acid is known to inhibit nitric oxide bioavailability. Because insulin requires nitric oxide to stimulate cellular glucose uptake, it has been proposed that fructose-induced hyperuricemia may account, in part, for fructose-induced increased risk for MS (Cirillo et al., 2006).

Chronic fructose consumption has been implicated in other aspects of obesity and the MS (Johnson et al., 2013). Fructose consumption upregulates not only its own transporter, GLUT5, but also sodium and chloride transporters, resulting in a state of salt overload that can increase blood pressure. Mechanisms through which excess fructose can activate vasoconstrictors such as endothelin 1, inactivate vasodilators such as NO, and overstimulate the sympathetic nervous system have been recently reviewed (Klein and Kiat, 2015). Fructose consumption has also been linked to risk for type II diabetes (DiNicolantonio et al., 2015). The authors cite evidence from both human and animal studies that fructose increases hepatic de novo lipogenesis and reduces fatty acid oxidation, leading to hepatic insulin resistance, adaptive increased insulin secretion, and progressive β-cell dysfunction. In addition, fructose can increase hepatic gluconeogenesis, stimulating even greater insulin release. This is especially true when fructose is consumed with excess calories (Faeh et al., 2005).

Arguments have been presented that, in humans, the impact of usual dietary fructose intake as sucrose or high-fructose corn syrup is less clear (Tappy and Le, 2010; van Buul et al., 2014). Despite evidence from epidemiological studies suggesting that consumption of sweetened beverages (containing either sucrose or a mixture of glucose and fructose) is associated with a high-energy intake, increased body weight, and the occurrence of metabolic and cardiovascular disorders, meta-analyses are inconclusive. In 2012, NIH and USDA sponsored a workshop to determine whether there is an obesogenic role for fructose. Participants discussed threshold doses, especially in susceptible populations, beyond which these sugars promote insulin resistance and diabetes. A major consideration was the physiologic relevance of studying the metabolism of the pure fructose monosaccharide, since this form is seldom consumed by humans and the participants called for well-controlled long-term studies (Laughlin et al., 2014). It should be noted that a serious bias appears to exist between systematic reviews (SRs) sponsored by industry and those not industry-supported (Bes-Rastrollo et al., 2013). Of SRs without any conflict of interest, 83% found sugar-sweetened beverages (SSBs) a potential risk factor for weight gain; for SRs with conflict of interest, 83% reported that evidence was insufficient to support a positive association between SSB consumption and weight gain or obesity.

At this time there is no unequivocal evidence that fructose intake at moderate doses, especially consumed in fruit and honey is directly related to the adverse metabolic effects seen in animals following a fructose load. Some studies have even reported a beneficial effect of fructose in lowering glycemic excursions, especially if energy intake is not in excess. While there had been concern that consumption of free fructose, as provided in high-fructose corn syrup, could cause adverse effects, no direct evidence was presented, suggesting that consumption of fructose as high-fructose corn syrup is not physiologically different from fructose consumed as sucrose.

1.9.4 Nonnutritive Artificial Sweeteners

Excess consumption of readily absorbable sugars and starches is clearly associated with risk for obesity and

MS. Thus the identification and approval by the FDA (US Food and Drug Association) of seven nonnutritive artificial sweeteners (NNS): acesulfame K, aspartame, luo han guo fruit extract, neotame, saccharin, stevia, and sucralose has been welcomed by the public health community. NNS have been recommended as alternatives to allow individuals at risk for obesity to enjoy pleasurable sweet foods without consuming excess energy, as has been recently reviewed (Fitch and Keim, 2012; Swithers, 2013). However, while use of NNS to reduce sugar and carbohydrate consumption should be expected to reduce the risk for obesity and its complications, studies in humans are limited, conflicting, and fraught with confounding variables. The American Heart Association and American Diabetes Association have recently reviewed available human evidence and developed a scientific statement regarding use of NNS (Gardner et al., 2012). They concluded that "when used judiciously, NNS could facilitate reductions in added sugars intake, thereby resulting in decreased total energy and weight loss/weight control, and promoting beneficial effects on related metabolic parameters."

The chemical constituents and metabolism of each NNS have been recently reviewed (Shankar et al., 2013). Their unique interactions with signaling pathways and with the microbiome have been proposed as the basis for the contradictory results of human trials using NNS. For example, it is possible that by providing a sweet taste signal without the attendant caloric spike, NNS alters neuronal signaling pathways, impairing the individual's ability to regulate food intake (Swithers et al., 2013; Swithers, 2015). On the other hand, some NNS are poorly absorbed and come in direct contact with the gut microbiome. Humans and mice given saccharin at the maximal FDA recommended level for 12 weeks showed significantly increased weight, waist–hip ratio, fasting blood glucose, glycosylated hemoglobin (HbA1C), and impaired glucose tolerance test. These changes were associated with changes in microbial patterns similar to those reported in obese humans and animals (Suez et al., 2014). It should be noted that saccharin impaired glucose tolerance to a greater extent than sucralose and aspartame in the mouse model, suggesting that each NNS interacts with gut bacteria in a unique fashion.

1.9.5 Metabolic Endotoxemia and Inflammation in Obesity

Several agents in the human diet have the potential to increase intestinal permeability by causing IEC injury, modulating transcellular transporter distribution, and by disrupting tight junction complexes and increasing paracellular transport from the gut lumen (refer to Essentials IV: Diet Impact on Microbiota Diversity and Gut Integrity). Increased intestinal permeability is associated with transport of the Gram-negative bacterial endotoxin, lipopolysaccharide (LPS), across the enterocyte barrier and into the bloodstream. While it has no enzymatic or intrinsic activity, the "toxicity" of LPS is conferred entirely by the organism that senses its presence and reacts to it in injurious ways. As indicated above, (and in Fig. 1.6) LPS with its binding protein (BP) and CD14 are bound to membrane-bound TLR4 that transmits proinflammatory signals to the cell. The response of an individual to LPS translocation depends on several variables including the ability of LPS to stimulate mediator production and the intensity of the cell response (Munford, 2005). The plausibility that LPS translocation and subsequent inflammation can potentiate obesity and MT is suggested by several observations. LPS is found in the serum of normal individuals, but LPS-BP was elevated in obesity and associated with insulin resistance (Moreno-Navarrete et al., 2012). Cell culture studies using human adipocytes have revealed that LPS bound to TLR4 on the adipocyte triggered the innate immune response with increased TNFα and IL-6 secretion (Creely et al., 2007).

Few studies have examined the interactions of LPS on other adipose tissue cell components. Recent work (Velmurugan et al., 2015) suggests that LPS can reduce adipose tissue macrophage (ATM) concentrations of hydrogen sulfide (H_2S) (refer to Essentials I: Life in an Aerobic World: Nutrients That Modify Transcriptional Control of Antioxidant Protection). Steady-state intracellular H_2S were reduced both in ATM from mice with diet-induced obesity and in macrophage RAW264.7 cells exposed to LPS. Reduced intracellular H_2S concentration led to increased Ca^{2+} influx and enhanced production of inflammatory cytokines; treatment with the exogenous H_2S donor, GYY4137, prevented Ca^{2+} influx. LPS also reduced adiponectin and leptin production in murine VAT cells (Taira et al., 2015), and reduced insulin production and secretion from human VAT and SAT cells (Sanjabi et al., 2015). Finally, circulating LPS is known to interact directly with adipocytes, the liver, and other systems to produce many of the symptoms of MS as recently reviewed (Konrad and Wueest, 2014). LPS produced hypertriglyceridemia in rodents at doses orders of magnitude lower than those which produce septic shock in humans. At these low doses, LPS produced hypertriglyceridemia by increasing hepatic lipoprotein production; at higher doses, hypertriglyceridemia was due to reduced lipoprotein catabolism (Feingold et al., 1992).

1.9.6 Protein

As discussed above, dietary protein is not only critical for supporting endogenous protein synthesis and maintaining lean body tissue, but it also influences targets of weight regulation, for example, satiety, thermogenesis, and energy efficiency (Stock, 1999; Layman et al., 2009; Westerterp-Plantenga et al., 2009). A prime mechanism for these satiety effects appears to be indirect via intestinal gluconeogenesis (GNG). A protein-enriched diet (PED) induced the expression of genes regulating intestinal GNG (glucose-6 phosphatase catalytic subunit (G6PC), phosphoenolpyruvate carboxykinase-cytosolic form (PEPCK-C) and glutaminase), resulting in secretion of glucose into the portal vein in the postabsorptive period (Mithieux and Gautier-Stein, 2014). Studies in mice revealed that peptides and protein digests are also released into the portal vein where they behave as antagonists of μ-opioid receptors present in neurons of the vessel walls. Signals are transmitted via ascending vagus and spinal cord tracts and signal the brain (Fig. 1.7). These signals promote gut GNG with subsequent inhibition of food intake (Duraffourd et al., 2012). Gut glucose release accounts for 5–10% of total endogenous glucose production after a standard carbohydrate-enriched meal, compared with 20–25% of glucose produced after PED. The ability of protein intake to induce satiety and maintain organ mass appears to depend on the consumption of complete proteins that contain the proportions of amino acids necessary for body needs (Layman and Walker, 2006).

1.9.7 Fat-Soluble Vitamins and Precursors

Preliminary work has shown that lipid-soluble vitamins (LSV; Vitamins A, D, E, and K) and precursor carotenoids (refer to Essentials II: Heavy Metals, Retinoids and Precursors: Carotenoids and to Essentials III: Nutrients in Bone Structure and Calcification: Vitamin D) regulate gene expression in multiple tissues and can modulate key processes relating to adipocyte differentiation, lipid storage capacity, and survival as recently reviewed (Landrier et al., 2012). Several studies have demonstrated that LSV are sequestered in LDs, and suggest that low serum LSV levels in obesity reflect this sequestration. More investigations into the mechanisms of intracellular trafficking, mobilization, and activation are required to obtain a clear picture of the effects of LSV intake and distribution on adipocyte biology. For example, LSV in adipocyte LDs have the potential to modulate several key processes via the regulation of adipocyte gene expression. How this modulation relates to the pathogenesis of obesity and its associated pathologies requires further investigation.

1.9.8 Polyphenols and Flavonoids

A wide variety of nonnutritive phenolic compounds are found in the food supply, as described elsewhere (refer to Essentials IV: Diet, Microbial Diversity and Gut Integrity: Diet Exposures That Reduce Gut Permeability).

At this time, the dearth of clinical trials has prevented development of guidelines for safe and effective use of polyphenols as pharmaceutical alternatives for treatment of obesity and MT. A review of available information on use of green tea extract (catechins) and soy (genestein) reveals inconsistent results. Given their potential for cytotoxic and genotoxic (Ross, 2008) actions, it is not surprising that adverse results, including elevated liver enzymes and even death, were reported in subjects taking plant extracts and supplements (Hurt and Wilson, 2012). On the other hand, dietary sources of these components include fruits and vegetables as well as coffee, tea, spices and chocolate. Intake of these foods is highly recommended for inclusion in diet regimens to reduce obesity and modulate symptoms of MS.

1.9.9 Resveratrol

Resveratrol (3,5,4′ trihydroxystilbene) is a polyphenol naturally enriched in the skins of red grape, purple berries (eg, blueberries, blackberries), peanut butter, and dark chocolate. Like other polyphenols it increases the activity of AMP-activated protein kinase (AMPK), thus reducing ATP utilization and increasing substrate availability as recently reviewed (Fullerton and Steinberg, 2010; de Ligt et al., 2015). Resveratrol also activates sirtuin 1 (SIRT1), a histone deacetylase that requires NAD^+ for activation. Interactions between AMPK and SIRT1 are complex and not yet clarified, and may explain the beneficial effects of resveratrol on energy homeostasis. Resveratrol activation of AMPK also phosphorylates and activates PGC1α, resulting in the upregulation of mitochondrial biogenesis (Nogueiras et al., 2012).

Studies in rodents and experimental models have revealed beneficial effects of resveratrol supplementation on preventing obesity and its comorbidities. Improved mitochondrial function, insulin sensitivity, and reduced liver fat accumulation have been demonstrated in supplemented rodents; these changes have been attributed to activation of the AMPK, SIRT1–PGC1a axis. Human trials have not been as consistent, likely due to ineffective dosage and/or short duration of the clinical trials. A meta-analysis of 11 clinical trials with 388 subjects revealed that resveratrol significantly improved glucose control and insulin sensitivity in persons with diabetes but did not affect glycemic measures in nondiabetic persons (Liu et al., 2014). Studies are also underway to investigate the possibility that resveratrol, through enhancing mitochondrial biogenesis and inducing UCP1 can promote preadipocyte differentiation to BAT, thereby stimulating adaptive thermogenesis. Given its potential to improve mitochondrial function and limit fat storage, future research should explore the utility of including food sources of resveratrol in diet patterns used to achieve and maintain healthy weight.

1.9.10 Ketogenic Diets

Low-carbohydrate diets and ketogenic diets have been proposed as an effective means to improve satiety, initiate weight loss, and increase utilization of stored fat for energy (Sumithran and Proietto, 2008), at least in the short term (Johnstone et al., 2008). The ketone body, β-hydroxybutyrate (βOHB) (refer to Chapter 3: Type I Diabetes and Celiac Disease), has been viewed as a carrier of energy from hepatic fat oxidation to peripheral tissues for utilization. Recently, βOHB has been described as a signaling molecule and histone deacetylase inhibitor (Newman and Verdin, 2014), with implications to the biology of aging. This body of work suggests the intriguing possibility that low-carbohydrate ketogenic diets that increase βOHB production could be an endogenous avenue linking the diet to gene expression and would enhance not only weight loss, but some of the benefits of lifespan extension as seen in model organisms. Clearly, translation of these possibilities to human obesity requires a great deal of investigation.

1.9.11 Sulfur-Donating Vegetables

The discovery that hydrogen sulfide gas (H_2S), although toxic in high concentrations, is one of three physiologically important "gasotransmitters" along with nitric oxide and carbon monoxide, has led to extensive work on H_2S functions and sources (Wang, 2012; Kolluru et al., 2013). Diet modification strategies that include nutrients required for endogenous H_2S production and dietary components that donate sulfur have been described elsewhere (refer to Essentials I: Life in an Aerobic World: Nutrients, Transcriptional Control of Antioxidant Protection). A recent report (Tsai et al., 2015) demonstrated that enzymes for endogenous H_2S production are upregulated during preadipocyte differentiation and that both endogenous and exogenous sources of H_2S regulate LD formation and lipolysis in mature adipocytes. Since obesity is characterized by dysfunctional AT with limited ability to safely store lipid substrates in mature adipocytes and/or effectively differentiate preadipocytes into fat-storing cells (Jankovic et al., 2015),

these preliminary research reports suggest a role for dietary H_2S sources in modification of obesity and prevention of its comorbidities. Potential therapeutic strategies might include potentiation of endogenous H_2S biosynthesis (eg, vitamin B_6) and provision of dietary H_2S donors (eg, cabbage, garlic, and egg yolk). Additionally, the influence of diet components on microbial H_2S production and subsequent beneficial or deleterious effects on biological systems must be clarified (Motta et al., 2015).

1.10 FUTURE RESEARCH NEEDS

While several scientific societies have published recommendations for the prevention and treatment of obesity and MS, these guidelines provide only general diet outlines and do not incorporate targeted diet modification strategies. Focused clinical trials are needed to explore the safety and effectiveness of structuring the individual diet to incorporate these components. For example, a weight-loss diet adequate in protein and low in free sugars (especially fructose) that contains adequate dietary fiber to maintain a robust microbiota and provide substrate for SCFAs is supported by the literature but requires further testing. Additionally, increased consumption of nutrient-dense vegetables, including sulfur donors, root vegetables, and mushrooms as a source of soluble fiber and possibly limiting grains known to increase intestinal permeability and facilitate metabolic endotoxemia in susceptible individuals should be tested for efficacy and subject compliance. Finally, given the difficulty of weight-loss maintenance and potential for regain, targeted dietary strategies that promote "healthy" adipose tissue should be explored.

References

Ait-Omar, A., Monteiro-Sepulveda, M., et al., 2011. GLUT2 accumulation in enterocyte apical and intracellular membranes: a study in morbidly obese human subjects and ob/ob and high fat−fed mice. Diabetes. 60 (10), 2598−2607.

Astrup, A., Raben, A., et al., 2015. The role of higher protein diets in weight control and obesity-related comorbidities. Int. J. Obes. 39 (5), 721−726 (2005).

Bentley, R., Meganathan, R., 1982. Biosynthesis of vitamin K (menaquinone) in bacteria. Microbiol. Rev. 46 (3), 241−280.

Bes-Rastrollo, M., Schulze, M.B., et al., 2013. Financial conflicts of interest and reporting bias regarding the association between sugar-sweetened beverages and weight gain: A systematic review of systematic reviews. PLoS Med. 10 (12), e1001578.

Blackburn, G.L., Wollner, S., et al., 2010. Lifestyle interventions for the treatment of class III obesity: a primary target for nutrition medicine in the obesity epidemic. Am. J. Clin. Nutr. 91 (1), 289S−292S.

Blüher, M., 2013. Adipose tissue dysfunction contributes to obesity related metabolic diseases. Best Pract. Res. Clin. Endocrinol. Metab. 27 (2), 163–177.

Burkitt, D.P., Trowell, H.C., 1977. Dietary fibre and western diseases. Irish Med. J. 70 (9), 272–277.

Campisi, J., d'Adda di Fagagna, F., 2007. Cellular senescence: when bad things happen to good cells. Nat. Rev. Mol. Cell Biol. 8 (9), 729–740.

Cawleya, J., Meyerhoeferd, C., 2011. The medical care costs of obesity: an instrumental variables approach. J. Health Econ. 31 (1), 219–230.

Chantranupong, L., Wolfson, R.L., et al., 2015. Nutrient-sensing mechanisms across evolution. Cell. 161 (1), 67–83.

Chaput, J.P., Sjödin, A.M., et al., 2010. Risk factors for adult overweight and obesity: the importance of looking beyond the 'big two'. Obes. Facts. 3 (5), 320–327.

Cirillo, P., Sato, W., et al., 2006. Uric acid, the metabolic syndrome, and renal disease. J. Am. Soc. Nephrol. 17 (Suppl. 3), S165–S168.

Claesson, M.J., Jeffery, I.B., et al., 2012. Gut microbiota composition correlates with diet and health in the elderly. Nature. 488 (7410), 178–184.

Cook, J.R., Langlet, F., et al., 2015. Pathogenesis of selective insulin resistance in isolated hepatocytes. J. Biol. Chem. 290 (22), 13972–13980.

Creely, S.J., McTernan, P.G., et al., 2007. Lipopolysaccharide activates an innate immune system response in human adipose tissue in obesity and type 2 diabetes. Am. J. Physiol.—Endocrinol. Metab. 292 (3), E740–E747.

d'Adda di Fagagna, F., 2008. Living on a break: cellular senescence as a DNA-damage response. Nat. Rev. Cancer. 8 (7), 512–522.

Dann, S.G., Thomas, G., 2006. The amino acid sensitive TOR pathway from yeast to mammals. FEBS Let. 580 (12), 2821–2829.

Deer, J., Koska, J., et al., 2015. Dietary models of insulin resistance. Metabolism. 64 (2), 163–171.

Dekker, M.J., Su, Q., et al., 2010. Fructose: a highly lipogenic nutrient implicated in insulin resistance, hepatic steatosis and metabolic syndrome. Am. J. Physiol. Endocrinol. Metab. 299 (5), E685–E694.

de Ligt, M., Timmers, S., et al., 2015. Resveratrol and obesity: can resveratrol relieve metabolic disturbances? Biochimica et Biophysica Acta—Mol. Basis Dis. 1852 (6), 1137–1144.

Demetriades, C., Doumpas, N., et al., 2014. Regulation of TORC1 in response to amino acid starvation via lysosomal recruitment of TSC2. Cell. 156 (4), 786–799.

Despres, J.-P., Lemieux, I., et al., 2008. Abdominal obesity and the metabolic syndrome: contribution to global cardiometabolic risk. Arterioscler. Thromb. Vasc. Biol. 28 (6), 1039–1049.

De Vadder, F., Kovatcheva-Datchary, P., et al., 2014. Microbiota-generated metabolites promote metabolic benefits via gut-brain neural circuits. Cell. 156 (1–2), 84–96.

DiNicolantonio, J.J., O'Keefe, J.H., et al., 2015. Added fructose: a principal driver of type 2 diabetes mellitus and its consequences. Mayo Clin. Proc. 90 (3), 372–381.

Dulloo, A.G., Jacquet, J., et al., 2015. How dieting makes the lean fatter: from a perspective of body composition autoregulation through adipostats and proteinstats awaiting discovery. Obes. Rev. 16, 25–35.

Duraffourd, C., De Vadder, F., et al., 2012. Mu-opioid receptors and dietary protein stimulate a gut-brain neural circuitry limiting food intake. Cell. 150 (2), 377–388.

Eckel, R.H., Jakicic, J.M., et al., 2014. 2013 AHA/ACC guideline on lifestyle management to reduce cardiovascular risk: a report of the American college of cardiology/American heart association task force on practice guidelines. Circulation. 129 (25 Suppl. 2), S76–S99.

Expert Panel on the Identification, E., T. o. Overweight, et al., 1998. Executive summary of the clinical guidelines on the identification, evaluation, and treatment of overweight and obesity in adults. Arch. Intern. Med. 158 (17), 1855–1867.

Faeh, D., Minehira, K., et al., 2005. Effect of fructose overfeeding and fish oil administration on hepatic de novo lipogenesis and insulin sensitivity in healthy men. Diabetes. 54 (7), 1907–1913.

Feingold, K.R., Staprans, I., et al., 1992. Endotoxin rapidly induces changes in lipid metabolism that produce hypertriglyceridemia: low doses stimulate hepatic triglyceride production while high doses inhibit clearance. J. Lipid Res. 33 (12), 1765–1776.

Finkelstein, E.A., Trogdon, J.G., et al., 2009. Annual medical spending attributable to obesity: payer-and service-specific estimates. Health Aff. 28 (5), w822–w831.

Fitch, C., Keim, K.S., 2012. Position of the academy of nutrition and dietetics: use of nutritive and nonnutritive sweeteners. J. Acad. Nutr. Diet. 112 (5), 739–758.

Fong, Y.M., Marano, M.A., et al., 1990. The acute splanchnic and peripheral tissue metabolic response to endotoxin in humans. J. Clin. Invest. 85 (6), 1896–1904.

Foster, G.D., Wyatt, H.R., et al., 2010. Weight and metabolic outcomes after 2 years on a low-carbohydrate versus low-fat diet. Ann. Intern. Med. 153 (3), 147–157.

Friedman, J., 2014. 20 years of leptin: leptin at 20: an overview. J. Endocrinol. 223 (1), T1–T8.

Friedman, J.M., 2003. A war on obesity, not the obese. Science. 299 (5608), 856–858.

Fullerton, M.D., Steinberg, G.R., 2010. SIRT1 takes a backseat to AMPK in the regulation of insulin sensitivity by resveratrol. Diabetes. 59 (3), 551–553.

Gao, Z., Yin, J., et al., 2009. Butyrate improves insulin sensitivity and increases energy expenditure in mice. Diabetes. 58 (7), 1509–1517.

Gardner, C., Wylie-Rosett, J., et al., 2012. Nonnutritive sweeteners: current use and health perspectives: a scientific statement from the American Heart Association and the American Diabetes Association. Diabetes Care. 35 (8), 1798–1808.

Gautier-Stein, A., Zitoun, C., et al., 2006. Transcriptional regulation of the glucose-6-phosphatase gene by cAMP/vasoactive intestinal peptide in the intestine: ROLE OF HNF4α, CREM, HNF1α, and C/EBPα. J. Biol. Chem. 281 (42), 31268–31278.

Gesta, S., Tseng, Y.-H., et al., 2007. Developmental origin of fat: tracking obesity to its source. Cell 131 (2), 242–256.

Geurts, L., Neyrinck, A.M., et al., 2014. Gut microbiota controls adipose tissue expansion, gut barrier and glucose metabolism: novel insights into molecular targets and interventions using prebiotics. Benef. Microbes. 5 (1), 3–17.

Gibson, G.R., Roberfroid, M.B., 1995. Dietary modulation of the human colonic microbiota: introducing the concept of prebiotics. J. Nutr. 125 (6), 1401–1412.

Goldberg, I.J., Eckel, R.H., et al., 2009. Regulation of fatty acid uptake into tissues: lipoprotein lipase- and CD36-mediated pathways. J. Lipid Res. 50 (Suppl.), S86–S90.

Gonzalez, E., Flier, E., et al., 2011. Hyperinsulinemia leads to uncoupled insulin regulation of the GLUT4 glucose transporter and the FoxO1 transcription factor. Proc. Nat. Acad. Sci. 108 (25), 10162–10167.

Grundy, S.M., Cleeman, J.I., et al., 2005. Diagnosis and management of the metabolic syndrome: an American Heart Association/National Heart, Lung, and Blood Institute Scientific Statement. Circulation. 112 (17), 2735–2752.

Guennoun, A., Kazantzis, M., et al., 2015. Comprehensive molecular characterization of human adipocytes reveals a transient brown phenotype. J. Trans. Med. 13 (1), 135.

Hampel, B., Malisan, F., et al., 2004. Differential regulation of apoptotic cell death in senescent human cells. Exp. Gerontol. 39 (11–12), 1713–1721.

Hansen, J.B., Kristiansen, K., 2006. Regulatory circuits controlling white versus brown adipocyte differentiation. Biochem. J. 398 (2), 153–168.

Harris, K., Kassis, A., et al., 2012. Is the gut microbiota a new factor contributing to obesity and its metabolic disorders? J. Obes. 2012, 879151.

Hayflick, L., 1965. The limited in vitro lifetime of human diploid cell strains. Exp. Cell Res. 37 (3), 614–636.

He, B., Nohara, K., et al., 2015. Transmissible microbial and metabolomic remodeling by soluble dietary fiber improves metabolic homeostasis. Sci. Rep. 5, 10604.

Heber, D., 2010. An integrative view of obesity. Am. J. Clin. Nutr. 91 (1), 280S–293S.

Hoy, M.K., Heshka, S., et al., 1994. Reduced risk of liver-function-test abnormalities and new gallstone formation with weight loss on 3350-kJ (800-kcal) formula diets. Am. J. Clin. Nutr. 60 (2), 249–254.

Hurt, R.T., Wilson, T., 2012. Geriatric obesity: evaluating the evidence for the use of flavonoids to promote weight loss. J. Nutr. Gerontol. Geriatr. 31 (3), 269–289.

Jambunathan, S., Yin, J., et al., 2011. FSP27 promotes lipid droplet clustering and then fusion to regulate triglyceride accumulation. PLoS ONE. 6 (12), e28614.

Jankovic, A., Korac, A., et al., 2015. Redox implications in adipose tissue (dys)function—a new look at old acquaintances. Redox Biol. 6, 19–32.

Jensen, M.D., Ryan, D.H., et al., 2014. 2013 AHA/ACC/TOS guideline for the management of overweight and obesity in adults: a report of the American college of cardiology/American heart association task force on practice guidelines and the obesity society. J. Am. Coll. Cardiol. 63 (25 PART B), 2985–3023.

Jewell, J.L., Russell, R.C., et al., 2013. Amino acid signalling upstream of mTOR. Nat. Rev. Mol. Cell Biol. 14 (3), 133–139.

Jialal, I., Kaur, H., et al., 2014. Toll-like receptor status in obesity and metabolic syndrome: a translational perspective. J. Clin. Endocrinol. Metab. 99 (1), 39–48.

Jiménez-Gómez, Y., Marín, C., et al., 2010. A low-fat, high-complex carbohydrate diet supplemented with long-chain (n-3) fatty acids alters the postprandial lipoprotein profile in patients with metabolic syndrome. J. Nutr. 140 (9), 1595–1601.

Johnson, R.J., Nakagawa, T., et al., 2013. Sugar, uric acid, and the etiology of diabetes and obesity. Diabetes. 62 (10), 3307–3315.

Johnson, R.J., Perez-Pozo, S.E., et al., 2009. Hypothesis: could excessive fructose intake and uric acid cause type 2 diabetes? Endocr. Rev. 30 (1), 96–116.

Johnstone, A.M., Horgan, G.W., et al., 2008. Effects of a high-protein ketogenic diet on hunger, appetite, and weight loss in obese men feeding ad libitum. Am. J. Clin. Nutr. 87 (1), 44–55.

Kienesberger, P.C., Pulinilkunnil, T., et al., 2013. Myocardial triacylglycerol metabolism. J. Mol. Cell. Cardiol. 55 (1), 101–110.

Kirkland, J.L., Tchkonia, T., et al., 2002. Adipogenesis and aging: does aging make fat go MAD? Exp. Gerontol. 37 (6), 757–767.

Klein, A.V., Kiat, H., 2015. The mechanisms underlying fructose-induced hypertension: a review. J. Hypert. 33 (5), 912–920.

Kolluru, G.K., Shen, X., et al., 2013. Hydrogen sulfide chemical biology: pathophysiological roles and detection. Nitric Oxide. 35, 5–20.

Konrad, D., Wueest, S., 2014. The gut-adipose-liver axis in the metabolic syndrome. Physiology. 29 (5), 304–313.

Kosacka, J., Kern, M., et al., 2015. Autophagy in adipose tissue of patients with obesity and type 2 diabetes. Mol. Cell. Endocrinol. 409, 21–32.

Kovsan, J., Bashan, N., et al., 2010. Potential role of autophagy in modulation of lipid metabolism. Am. J. Physiol.—Endocrinol. Metab. 298 (1), E1–E7.

Kovsan, J., Blüher, M., et al., 2011. Altered autophagy in human adipose tissues in obesity. J. Clin. Endocrinol. Metab. 96 (2), E268–E277.

Krishan, S., Richardson, D.R., et al., 2015. Adenosine monophosphate–activated kinase and its key role in catabolism: structure, regulation, biological activity, and pharmacological activation. Mol. Pharmacol. 87 (3), 363–377.

Lafontan, M., 2014. Adipose tissue and adipocyte dysregulation. Diabetes Metabol. 40 (1), 16–28.

Lamb, C.A., Yoshimori, T., et al., 2013. The autophagosome: origins unknown, biogenesis complex. Nat. Rev. Mol. Cell Biol. 14 (12), 759–774.

Landrier, J.-F., Marcotorchino, J., et al., 2012. Lipophilic micronutrients and adipose tissue biology. Nutrients. 4 (11), 1622.

Laughlin, M.R., Bantle, J.P., et al., 2014. Clinical research strategies for fructose metabolism. Adv. Nutr. 5 (3), 248–259.

Layden, B.T., Angueira, A.R., et al., 2013. Short chain fatty acids and their receptors: new metabolic targets. Trans. Res. 161 (3), 131–140.

Layman, D.K., Evans, E.M., et al., 2009. A moderate-protein diet produces sustained weight loss and long-term changes in body composition and blood lipids in obese adults. J. Nutr. 139 (3), 514–521.

Layman, D.K., Walker, D.A., 2006. Potential importance of leucine in treatment of obesity and the metabolic syndrome 1–3. J. Nutr. 136 (1), 319S–323S.

Lee, M.-J., Wu, Y., et al., 2013. Adipose tissue heterogeneity: implication of depot differences in adipose tissue for obesity complications. Mol. Aspects Med. 34 (1), 1–11.

Leidy, H.J., Clifton, P.M., et al., 2015. The role of protein in weight loss and maintenance. Am. J. Clin. Nutr. 101 (6), 1320S–1329S.

Ley, R.E., Turnbaugh, P.J., et al., 2006. Microbial ecology: human gut microbes associated with obesity. Nature. 444 (7122), 1022–1023.

Liu, K., Zhou, R., et al., 2014. Effect of resveratrol on glucose control and insulin sensitivity: a meta-analysis of 11 randomized controlled trials. Am. J. Clin. Nutr. 99 (6), 1510–1519.

Londos, C., Brasaemle, D.L., et al., 1999. Perilipins, ADRP, and other proteins that associate with intracellular neutral lipid droplets in animal cells. Semin. Cell Dev. Biol. 10 (1), 51–58.

Lumeng, C.N., Bodzin, J.L., et al., 2007. Obesity induces a phenotypic switch in adipose tissue macrophage polarization. J. Clin. Inves. 117 (1), 175–184.

Maixner, N., Kovsan, J., et al., 2012. Autophagy in adipose tissue. Obes. Facts. 5 (5), 710–721.

Manco, M., Putignani, L., et al., 2010. Gut microbiota, lipopolysaccharides, and innate immunity in the pathogenesis of obesity and cardiovascular risk. Endocr. Rev. 31 (6), 817–844.

Marques, T.M., Wall, R., et al., 2015. Dietary trans-10, cis-12-conjugated linoleic acid alters fatty acid metabolism and microbiota composition in mice. Br. J. Nutr. 113 (5), 728–738.

Martin, S., Parton, R.G., 2005. Caveolin, cholesterol, and lipid bodies. Semin. Cell Dev. Biol. 16 (2), 163–174.

Martin, S., Parton, R.G., 2006. Lipid droplets: a unified view of a dynamic organelle. Nat. Rev. Mol. Cell Biol. 7 (5), 373–378.

Mayer, J., Thomas, D.W., 1967. Regulation of food intake and obesity. Science. 156 (3773), 328–337.

Mayes, P.A., 1993. Intermediary metabolism of fructose. Am. J. Clin. Nutr. 58 (5), 754S–765S.

McArdle, M.A., Finucane, O.M., et al., 2013. Mechanisms of obesity induced inflammation and insulin resistance: insights into the emerging role of nutritional strategies. Front. Endocrinol. 4, 52.

Medicine, I.O., 2005. Dietary Reference Intakes for Energy, Carbohydrate, Fiber, Fat, Fatty Acids, Cholesterol, Protein, and Amino Acids (Macronutrients). The National Academies Press, Washington, DC.

Milagro, F.I., Mansego, M.L., et al., 2013. Dietary factors, epigenetic modifications and obesity outcomes: Progresses and perspectives. Mol. Aspects Med. 34 (4), 782–812.

Mitchell, P., 1961. Coupling of phosphorylation to electron and hydrogen transfer by a chemi-osmotic type of mechanism. Nature. 191 (4784), 144–148.

Mithieux, G., Gautier-Stein, A., 2014. Intestinal glucose metabolism revisited. Diabetes Res. Clin. Pract. 105, 295–301.

Montani, J.P., Schutz, Y., et al., 2015. Dieting and weight cycling as risk factors for cardiometabolic diseases: who is really at risk? Obes. Rev. 16, 7–18.

Moreno-Navarrete, J.M., Ortega, F., et al., 2012. Circulating lipopolysaccharide-binding protein (LBP) as a marker of obesity-related insulin resistance. Int. J. Obes. 36 (11), 1442–1449.

Motta, J.-P., Flannigan, K.L., et al., 2015. Hydrogen sulfide protects from colitis and restores intestinal microbiota biofilm and mucus production. Inflamm. Bowel Dis. 21 (5), 1006–1017.

Munford, R.S., 2005. Invited review: detoxifying endotoxin: time, place and person. J. Endotoxin Res. 11 (2), 69–84.

Nakamura, M.T., Yudell, B.E., et al., 2014. Regulation of energy metabolism by long-chain fatty acids. Prog. Lipid Res. 53, 124–144.

Navarro, E., Funtikova, A.N., et al., 2015. Can metabolically healthy obesity be explained by diet, genetics, and inflammation? Mol. Nutr. Food Res. 59 (1), 75–93.

Neel, J.V., 1999. Diabetes mellitus: a "thrifty" genotype rendered detrimental by "progress"? 1962. Bull. World Health Org. 77 (8), 694–703.

Newman, J.C., Verdin, E., 2014. Ketone bodies as signaling metabolites. Trends Endocrinol. Metab. 25 (1), 42–52.

Ng, M., Fleming, T., et al., 2014. Global, regional, and national prevalence of overweight and obesity in children and adults during 1980?2013: a systematic analysis for the Global Burden of Disease Study 2013. Lancet. 384 (9945), 766–781.

Nguyen, N.Q., Debreceni, T.L., et al., 2015. Accelerated intestinal glucose absorption in morbidly obese humans: relationship to glucose transporters, incretin hormones, and glycemia. J. Clin. Endocrinol. Metab. 100 (3), 968–976.

Nogueiras, R., Habegger, K.M., et al., 2012. Sirtuin 1 and sirtuin 3: physiological modulators of metabolism. Physiol. Rev. 92 (3), 1479–1514.

Nordmann, A.J., Nordmann, A., et al., 2006. Effects of low-carbohydrate vs low-fat diets on weight loss and cardiovascular risk factors: a meta-analysis of randomized controlled trials. Arch. Intern. Med. 166 (3), 285–293.

O'Hara, A.M., Shanahan, F., 2006. The gut flora as a forgotten organ. EMBO Rep. 7 (7), 688–693.

Papackova, Z., Cahova, M., 2015. Fatty acid signaling: the new function of intracellular lipases. Int. J. Mol. Sci. 16 (2), 3831–3855.

Pasiakos, S.M., Lieberman, H.R., et al., 2015. Higher-protein diets are associated with higher HDL cholesterol and lower BMI and waist circumference in US adults. J. Nutr. 145 (3), 605–614.

Patel, C., Douard, V., et al., 2015. Transport, metabolism, and endosomal trafficking-dependent regulation of intestinal fructose absorption. FASEB journal: official publication of the Federation of American Societies for Experimental Biology 29 (9), 4046–4058.

Precope, J., 1952. Hippocrates on Diet and Hygiene. Zeno, London.

Reaven, G., 1988. Banting lecture 1988. Role of insulin resistance in human disease. Diabetes. 37 (12), 1597–1607.

Rodier, F., Campisi, J., 2011. Four faces of cellular senescence. J. Cell Biol. 192 (4), 547–556.

Rodier, F., Coppe, J.-P., et al., 2009. Persistent DNA damage signalling triggers senescence-associated inflammatory cytokine secretion. Nat. Cell Biol. 11 (8), 973–979.

Rodriguez, N.R., 2015. Introduction to protein summit 2.0: continued exploration of the impact of high-quality protein on optimal health. Am. J. Clin. Nutr. 101 (6), 1317S–1319S.

Rosenbaum, M., Leibel, R.L., et al., 1997. Obesity. N. Engl. J. Med. 337, 396–407.

Ross, J.A., 2008. Environmental and genetic susceptibility to MLL-defined infant leukemia. JNCI Monogr. 2008 (39), 83–86.

Salama, R., Sadaie, M., et al., 2014. Cellular senescence and its effector programs. Genes Dev. 28 (2), 99–114.

Sam, S., Mazzone, T., 2014. Adipose tissue changes in obesity and the impact on metabolic function. Trans. Res. 164 (4), 284–292.

Sanjabi, B., Dashty, M., et al., 2015. Lipid droplets hypertrophy: a crucial determining factor in insulin regulation by adipocytes. Sci. Rep. 5, 8816.

Sellayah, D., Cagampang, F.R., et al., 2014. On the evolutionary origins of obesity: a new hypothesis. Endocrinology. 155 (5), 1573–1588.

Sethi, J.K., Vidal-Puig, A.J., 2007. Thematic review series: adipocyte biology. Adipose tissue function and plasticity orchestrate nutritional adaptation. J. Lipid Res. 48 (6), 1253–1262.

Shankar, P., Ahuja, S., et al., 2013. Non-nutritive sweeteners: review and update. Nutrition. 29 (11–12), 1293–1299.

Shen, J., Obin, M.S., et al., 2013. The gut microbiota, obesity and insulin resistance. Mol. Aspects Med. 34 (1), 39–58.

Shibutani, S.T., Yoshimori, T., 2014. A current perspective of autophagosome biogenesis. Cell Res. 24 (1), 58–68.

Simpson, S.J., Raubenheimer, D., 2005. Obesity: the protein leverage hypothesis. Obes. Rev. 6 (2), 133–142.

Stock, M.J., 1999. Gluttony and thermogenesis revisited. Int. J. Obes. 23 (11), 1105–1117.

Strissel, K.J., Stancheva, Z., et al., 2007. Adipocyte death, adipose tissue remodeling, and obesity complications. Diabetes. 56 (12), 2910–2918.

Suez, J., Korem, T., et al., 2014. Artificial sweeteners induce glucose intolerance by altering the gut microbiota. Nature. 514 (7521), 181–186, Advance online publication.

Sumithran, P., Proietto, J., 2008. Ketogenic diets for weight loss: a review of their principles, safety and efficacy. Obes. Res. Clin. Pract. 2 (1), 1–13.

Sun, L., Yu, Z., et al., 2010. A marker of endotoxemia is associated with obesity and related metabolic disorders in apparently healthy chinese. Diabetes Care. 33 (9), 1925–1932.

Swithers, S.E., 2013. Artificial sweeteners produce the counterintuitive effect of inducing metabolic derangements. Trends Endocrinol. Metab. 24 (9), 431–441.

Swithers, S.E., 2015. Not so sweet revenge: unanticipated consequences of high-intensity sweeteners. Behav. Anal. 38 (1), 1–17.

Swithers, S.E., Sample, C.H., et al., 2013. Adverse effects of high-intensity sweeteners on energy intake and weight control in male and obesity-prone female rats. Behav. Neurosci. 127 (2), 262–274.

Taira, R., Yamaguchi, S., et al., 2015. Bacterial cell wall components regulate adipokine secretion from visceral adipocytes. J. Clin. Biochem. Nutr. 56 (2), 149–154.

Tappy, L., Le, K.-A., 2010. Metabolic effects of fructose and the worldwide increase in obesity. Physiol. Rev. 90 (1), 23–46.

Tchkonia, T., Morbeck, D.E., et al., 2010. Fat tissue, aging, and cellular senescence. Aging Cell. 9 (5), 667–684.

Todde, V., Veenhuis, M., et al., 2009. Autophagy: principles and significance in health and disease. Biochimica et Biophysica Acta (BBA)—Mol. Basis Dis. 1792 (1), 3–13.

Tolhurst, G., Heffron, H., et al., 2012. Short-chain fatty acids stimulate glucagon-like peptide-1 secretion via the G-protein—coupled receptor FFAR2. Diabetes. 61 (2), 364–371.

Torres-Fuentes, C., Schellekens, H., et al., 2015. A natural solution for obesity: bioactives for the prevention and treatment of weight gain. A review. Nutr. Neurosci. 18 (2), 49–65.

Tsai, C.-Y., Peh, M.T., et al., 2015. Hydrogen sulfide promotes adipogenesis in 3T3L1 cells. PLoS ONE. 10 (3), e0119511.

Turnbaugh, P.J., Hamady, M., et al., 2009. A core gut microbiome in obese and lean twins. Nature. 457 (7228), 480–484.

Turnbaugh, P.J., Ley, R.E., et al., 2006. An obesity-associated gut microbiome with increased capacity for energy harvest. Nature. 444 (7122), 1027–1131.

van Buul, V.J., Tappy, L., et al., 2014. Misconceptions about fructose-containing sugars and their role in the obesity epidemic. Nutr. Res. Rev. 27 (1), 119–130.

van der Klaauw, Agatha A., Farooqi, I.S., 2015. The hunger genes: pathways to obesity. Cell. 161 (1), 119–132.

van Dijk, S.J., Feskens, E.J., et al., 2009. A saturated fatty acid—rich diet induces an obesity-linked proinflammatory gene expression profile in adipose tissue of subjects at risk of metabolic syndrome. Am. J. Clin. Nutr. 90 (6), 1656–1664.

Velmurugan, G.V., Huang, H., et al., 2015. Depletion of H2S during obesity enhances store-operated Ca^{2+} entry in adipose tissue macrophages to increase cytokine production. Sci. Signal. 8 (407), ra128-ra128.

Visscher, T.L.S., Heitmann, B.L., et al., 2015. A break in the obesity epidemic[quest] Explained by biases or misinterpretation of the data[quest]. Int. J. Obes. 39 (2), 189–198.

Wang, C., Jurk, D., et al., 2009. DNA damage response and cellular senescence in tissues of aging mice. Aging Cell. 8 (3), 311–323.

Wang, H., Lei, M., et al., 2013. Analysis of lipid droplets in cardiac muscle. Methods Cell Biol. 116, 129–149.

Wang, R., 2012. Physiological implications of hydrogen sulfide: a whiff exploration that blossomed. Physiol. Rev. 92 (2), 791–896.

Warren, H.S., Fitting, C., et al., 2010. Resilience to bacterial infection: difference between species could be due to proteins in serum. J. Infect. Dis. 201 (2), 223–232.

Weigle, D.S., 1994. Appetite and the regulation of body composition. FASEB J. 8 (3), 302–310.

Wells, J.C.K., 2006. The evolution of human fatness and susceptibility to obesity: an ethological approach. Biol. Rev. 81 (2), 183–205.

Westerterp-Plantenga, M.S., Nieuwenhuizen, A., et al., 2009. Dietary protein, weight loss, and weight maintenance. Ann. Rev. Nutr. 29 (1), 21–41.

White, E., Lowe, S.W., 2009. Eating to exit: autophagy-enabled senescence revealed. Genes Dev. 23 (7), 784–787.

Xiao, B., Sanders, M.J., et al., 2011. Structure of mammalian AMPK and its regulation by ADP. Nature. 472 (7342), 230–233.

Young, A.R.J., Narita, M., et al., 2009. Autophagy mediates the mitotic senescence transition. Genes Dev. 23 (7), 798–803.

Yuan, H.-X., Xiong, Y., et al., 2013. Nutrient sensing, metabolism, and cell growth control. Mol. Cell. 49 (3), 379–387.

Zhang, H., DiBaise, J.K., et al., 2009. Human gut microbiota in obesity and after gastric bypass. Proc. Nat. Acad. Sci. 106 (7), 2365–2370.

Zhou, X., Han, D., et al., 2014. A model of metabolic syndrome and related diseases with intestinal endotoxemia in rats fed a high fat and high sucrose diet. PLoS ONE. 9 (12), e115148.

Further Reading

Kumar, V., et al., 2014. Robbins and Cotran Pathologic Basis of Disease, ninth ed. Elsevier Saunders, pp. 444–448., Chapter 9. Environmental and Nutritional Diseases: Obesity.

2

The Obese Gunshot Patient: Injury and Septic Shock

CHIEF COMPLAINT (TYPICAL PATIENT)

Mr. R was brought to the emergency room by a friend after he had been shot twice in the abdomen during a robbery at his store. He was vomiting blood and complained of severe back and "stomach" pain. He responded poorly to questions, stating the pain was "too bad for me to think."

PATIENT HISTORY

Patient's friend provided basic information. Patient is a 19-year-old Hispanic male who works as a clerk in a convenience store. He lives with his brother, his brother's wife, and their two children, ages 2 and 5. Patient is attending community college part-time.

PHYSICAL EXAMINATION

- *Anthropometrics*: height 5′7″; weight 246 lb; BMI 38.5.
- *General appearance*: obese patient in severe pain and distress.
- *Vital signs*: temp. 102.6°F; BP 115/65 mmHg; HR 135 bpm.
- *HEENT*: NG tube in place for decompression.
- *Neurologic*: sedated.
- *Extremities*: 4+ bilateral pedal edema.
- *Skin*: warm, moist.
- *Chest/lungs*: lungs clear to auscultation and percussion; maintained on mechanical ventilation.
- *Peripheral vascular*: pulses full, no bruits.
- *Abdomen*: distension, wound vacuum-assisted closure (VAC) in place, three tubes draining peritoneal fluid, hypoactive bowel sounds present in all regions. Liver percusses approximately 8 cm

at midclavicular line, one fingerbreadth below right costal margin.

PERTINENT LABORATORY DATA

Value	Patient's values	Normal values
• Albumin	2.4 gm/dL	3.5–5 gm/dL
• Total protein	5.2 gm/dL	6–8 gm/dL
• Prealbumin	3.0 mg/dL	16–35 mg/dL
• Transferrin	190 mg/dL	215–365 mg/dL
• Sodium	146 mEq/L	136–145 mEq/L
• Potassium	4.0 mEq/L	3.5–5.5 mEq/L
• PO$_4$	2.2 mg/dL	2.3–4.7 mEq/L
• Osmolality	317 mmol/kg/H$_2$O	285–295 mM/kg/H$_2$O
• Glucose	164 mg/dL	70–110 mg/dL
• BUN	23 mg/dL	8–18 mg/dL
• Creatinine	1.4 mg/dL	0.6–1.2 mg/dL
• Calcium	7.1 mg/dL	9–11 mg/dL
• ALT	435 U/L	3–36 U/L
• AST	190 U/L	0–35 U/L
• Alk phos	540 U/L	30–120 U/L
• CPK	167 U/L	55–170 U/L
• CRP	245 mg/dL	<1.0 mg/dL
• LDH	750 U/L	207–378 U/L
• Chol	180 mg/dL	120–199 mg/dL
• HDL-C	40 mg/dL	>45 mg/dL
• TG	274 mg/dL	40–160 mg/dL

(Continued)

Nutritional Pathophysiology of Obesity and its Comorbidities
DOI: http://dx.doi.org/10.1016/B978-0-12-803013-4.00002-8

(Continued)

Value	Patient's values	Normal values
• HbA$_{1C}$	7%	3.9–5.2%
• WBC	$15.2 \times 10^3/mm^3$	$4.8–11.8 \times 10^3/mm^3$
• RBC	$3.2 \times 10^3/mm^3$	$4.5–6.2 \times 10^3/mm^3$
• Hct	35%	40–54%
• MCV	89 μm^3	82 μm^3
• PT	9 s	11–16 s

URINE SPECIMEN

• Appearance	dark yellow—cloudy	pale yellow—clear
• Sp grv	1.078	1.003–1.030
• Protein	+1	neg
• Glucose	+1	neg
• Bacteria	5	0
• Mucus	5	0
• Yeast	2	0

MEDICAL AND NUTRITION HISTORY

General: information obtained from patient's brother. Patient has not gained or lost a significant amount of weight recently, nor has he followed a special diet. Patient has no known allergies and has apparently been in good health. Brother noted that patient was fond of jelly donuts and went out with his friends frequently for evenings of beers and bar food.

EMERGENCY TREATMENT

Patient was taken to surgery where he underwent an exploratory damage-control laparotomy, gastric repair, control of liver hemorrhage and resection of proximal jejunum, leaving his gastrointestinal (GI) tract in discontinuity.

HOSPITAL COURSE

Patient was transferred to the Trauma Intensive Care Unit and maintained on mechanical ventilation.

• Day 2: he was taken to surgery to remove packs and to reestablish bowel continuity. An abdominal VAC was placed (Negative-pressure wound therapy (NPWT) uses a vacuum dressing to promote healing in acute or chronic wounds. The subatmospheric pressure facilitates wound healing by drawing fluid from the wound and increasing blood flow to the area.

The VAC is a widely available commercial product. A systematic review of trials with this device has demonstrated its safety and effectiveness (Xie et al., 2010)). Three Jackson—Pratt drains were left in place.

• Day 3: patient returned to surgery where an anastomotic leak was detected. A gastrojejunostomy tube was inserted through the patient's stomach, with the jejunal limb shortened to provide antegrade intraluminal drainage as well as retrograde jejunostomy drainage.

• Day 7: patient returned to surgery for an abdominal washout, insertion of a distally placed J-tube for feeding and a VAC change.

• Patient subsequently returned to the OR for multiple washouts and reapplication of a wound VAC.

Nutritional consult was ordered by the trauma surgeon after initial surgery on Day 1.

RESOURCES

2.1 BIOLOGICAL RESPONSE TO CELL INJURY

Unintentional injury remains the fifth leading cause of death in the United States. Damage to the organism, from whatever cause (knife slash, gunshot wound, blunt force trauma, and the like) kills or severely compromises cells adjacent to the injury to variable degrees. Cells so severely stressed that they are no longer able to adapt will ultimately die. Cells less severely damaged can recover, allowing cellular adaptations due to the injury to be normalized if the damaging stimulus is removed. The host responds to trauma by engaging an ancient, redundant system of metabolic cascades aimed at ridding the organism of noxious damaging agents, removing necrotic debris, and providing adequate nutrients for cell adaptation and recovery. While this innate host defense is ultimately protective, if inappropriately triggered or poorly controlled, it can be highly destructive. In the critically injured patient, these metabolic cascades can be dysregulated, leading to further tissue injury and nutrient deprivation, multiorgan dysfunction, prolonged hospitalization, increased morbidity and often, death (Fig. 2.1).

Necrosis is defined as cell death that leaves behind cellular debris. Significant necrosis results from mechanical injury and other pathological exposures such as ischemia, toxins infections, and trauma. The organism has several ways to remove dangerous material, including recruitment of phagocytes that ingest and degrade necrotic debris and by detoxification enzymes that facilitate xenobiotic excretion. If necrosis is severe, as in extensive burns, surgical debridement of the dead tissue is necessary to eliminate its potential for stimulating inflammation.

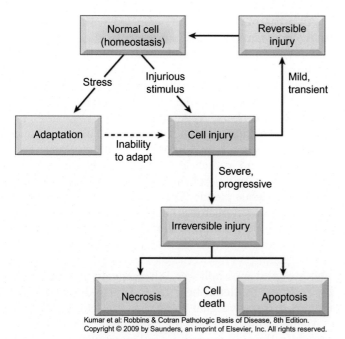

Kumar et al: Robbins & Cotran Pathologic Basis of Disease, 8th Edition.
Copyright © 2009 by Saunders, an imprint of Elsevier, Inc. All rights reserved.

FIGURE 2.1 **Possible sequelae following cell injury.** When the organism is stimulated by environmental stressors, a new functional steady state is rapidly achieved. Injured cells can adapt and maintain the intracellular milieu within a narrow physiologic range. Four general cellular adaptions include: *hypertrophy, hyperplasia, atrophy,* and *metaplasia.* If the injurious stressor is removed, the adapted cell can return to a stable baseline homeostasis. If stress exceeds the cell's ability to adapt and injury becomes irreversible, the cell dies by apoptosis or necrosis. *Reproduced with permission from Kumar, V., et al., 2015. Robbins and Cotran Pathologic Basis of Disease, 9th Ed. Elsevier Saunders.*

If the injury compromises the cell membranes, cellular contents can leak into the extracellular space and damage adjacent cells, eliciting an inflammatory attack and widening the area of injury. Damage to lysosomal membranes releases acidic and proteolytic lysosomal enzymes into the cytoplasm, killing the cell. An alternate form of cell death, *apoptosis*, results when the cell signals its own demise. The signal can emanate from deprivation of growth factors, DNA damage, or other internal stimulus. For example, physiologic apoptosis occurs during tissue remodeling and embryonic development. During apoptosis, DNA is cleaved by endonucleases without loss of membrane integrity and apoptotic fragments are removed by phagocytes. Cells that die by apoptosis do not leave debris and do not elicit an inflammatory response.

2.2 TRAUMATIC INJURY: GUNSHOT WOUNDS

Mechanical trauma inflicts injury depending on the shape of the colliding object, the amount of energy discharged at impact, and the tissues or organs that receive the impact. Several definitions of mechanical trauma include:

- *Abrasion*: a wound caused by scraping the skin surface. Abrasions usually remove only the epidermal skin layer.
- *Contusion*: a bruise, usually caused by blunt trauma. Contusions usually involve vascular damage and extravasation of blood into tissue.
- *Laceration*: a tear in the skin due to excessive stretching, as caused by blunt force trauma. Most lacerations have rough, jagged edges. Intact blood vessels may bridge the laceration.
- *Incised wound*: a wound inflicted by a sharp instrument such as a knife. Blood vessels are severed.
- *Puncture wound*: a wound caused by an instrument that penetrates the tissue and creates an exit wound.

The biological consequences of gunshot wounds have been reviewed (Maiden, 2009). A bullet discharged from a gun causes injury by crushing the tissue as it enters the body. The type and severity of injury are dependent on the bullet placement (wound location), the type of bullet, and the elasticity of the bullet target; for example, tissues with high elasticity (eg, muscle) can absorb impact forces, while solid organs such as the liver, are likely to rupture on impact. The force of the projectile creates a permanent cavity in the tissue. Entry velocity is also transmitted to tissues adjacent to the bullet path boundary, creating a temporary "blast effect" expansion. If the bullet strikes a bone, fragments can ricochet, striking tissues and organs not in the original bullet path; bone fragments can also act as projectiles and exert further tissue damage. Injury is less severe if the bullet passes through the body and leaves by an exit wound. In this case, some fraction of the bullet's kinetic energy remains in the bullet and is transferred to an external target. Hollow-point bullets are especially damaging because they increase in diameter once within the target, maximizing tissue injury, blood loss, and shock. The hollow-point bullet usually remains inside the body, thereby transferring all of its kinetic energy to that target. Shotgun shells contain variable numbers of round metal pellets that spread as the shot is fired. The size of the entrance wound increases with distance of the target from the firearm. Thus, shotgun blasts can result in massive surface tissue damage but may not result in puncture wounds. Bullets of any type carry with them residue of the combusted powder that generated their velocity. The powder penetrates the body with the bullet, and can also be seen on the skin adjacent to the entrance wound, especially if the bullet is fired at close range.

2.3 INNATE HOST DEFENSE

2.3.1 The Skin Ecology

Entrance of a projectile, such as a bullet, into the body carries with it any resident microbial agents on the bullet and its powder, as well as microbes present on the skin at the point of entry. Normally, the skin forms a thick, waterproof permeability barrier that separates the body from a hostile environment. Epithelial cells destined to form the epidermis (keratinocytes) migrate from the underlying basal layer. Keratinocytes differentiate as they travel to the surface, developing lamellar bodies filled with lipids such as fatty acids (FAs), cholesterol, sphingomyelin, and enzymes that modify the lipid structures. At the uppermost skin layer (the stratum corneum), the lamellar bodies are extruded to form an extracellular barrier enriched in neutral lipids consisting largely of ceramides (50%), cholesterol (25%), and FAs (15%). Having lost their lamellar bodies, keratinocytes shed their organelles and become corneocytes encased in an envelope made up of extensively crosslinked proteins such as involucrin and loricin. These components form a hydrophobic barrier that controls the movement of water and electrolytes across the skin barrier (Feingold, 2007).

All areas of the skin are colonized with commensal (indigenous) microbes. While much of the skin is cool and dry, some areas are moist, warm, and conducive to microbial growth. Specific microbial species have evolved to live in specialized niches of the human skin; in health, these populations are relatively stable. Microbial numbers and species are regulated by metabolite secretions from microbes (eg, oleic acid and porphyrins), by antimicrobial peptides including β-defensins and cathelicidins secreted by the corneocytes, and by antimicrobial lipids and proteins secreted by the sebaceous glands (Drake et al., 2008; Belkaid and Segre, 2014). Skin microbes engage in extensive crosstalk with cells of the immune system. Microbial products and secretions can enter the accessory glands and hair follicles, signaling secretion of cytokines from dermal lymphocytes and macrophages. Alternatively, microbes can be captured by dendritic cells and their digested fragments can directly signal the immune system (Fig. 2.2).

While commensal skin microbes protect intact skin, a breach in the skin made by a bullet or other wounding agent allows these microbes to enter the body where they can act as pathogens, contributing to inflammation and tissue damage. Microbial entry into the body is especially dangerous in patients with nutrient-rich blood, as in hyperglycemia or hyperlipidemia, or during acute inflammation as described below.

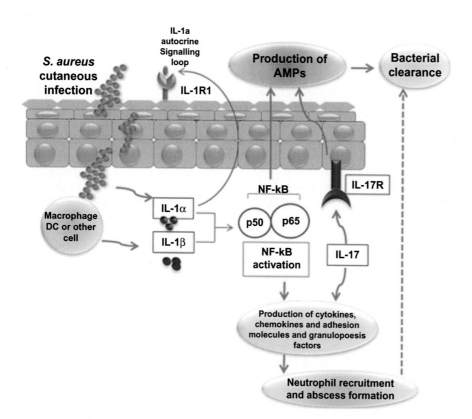

FIGURE 2.2 Commensal skin microbes, including bacteria, fungi, and viruses produce antimicrobial peptides (AMPs) that interact with antimicrobial peptides and lipids produced by keratinocytes. Cutaneous infection triggers macrophage/dendritic cell cytokine production, activation of nuclear factor kappa beta (NFκB) and production of cationic antimicrobial peptides (CAMPs) such as human β-defensins and cathelicidin that amplify the immune response and promote microbial killing. *Reproduced from Ryu, S., Song, P., et al., 2014. Colonization and infection of the skin by S. aureus: immune system evasion and the response to cationic antimicrobial peptides. Int. J. Mol. Sci. 15 (5), 8753 (Ryu et al., 2014). Creative Commons Attribution License.*

2.3.2 The Gut Microbiota

It is now well understood that the entire length of the gastrointestinal tract is colonized with microbes. While microbial growth in the proximal gastrointestinal tract is reduced somewhat by stomach acid, some microbial species occupy an ecological niche even in this harsh environment (Yang et al., 2013). Commensal microbes increase in number and diversity in the distal GI tract, especially in the colon. Components of the normal intestinal flora, now characterized as the *microbiota*, exert a major protection against environmental pathogens. As with skin commensal microbes, the gut microbiota competes with pathogens for nutrients and produces antibacterial substances that inhibit the growth of pathogens. Perhaps more importantly, the microbiota engages in extensive crosstalk with body systems, especially with the gut epithelial cells, altering the quantity and composition of mucin secretion, and with the cells of the immune system to create a modulated immune defense (Yang et al., 2013).

Commensal intestinal microbes are not entirely benign; if an injury penetrates the intestinal wall, microbial commensals can act as pathogens. Even if the wall is not injured, mediators of trauma and stress are known to loosen tight junctions between mucosal cells, allowing bacteria and endotoxins to pass between cells and reach extraintestinal sites such as mesenteric lymph nodes, spleen, and bloodstream (bacterial translocation). Some endotoxins such as lipopolysaccharides (LPS) found in the outer membrane of Gram-negative bacteria, are potent stimuli of the immune response as described in Chapter 1, Obesity and Metabolic Syndrome. The diet contains elements that also loosen tight junctions and could predispose to bacterial translocation following trauma (refer to Essentials IV: Diet, Microbial Diversity, and Gut Integrity).

Intestinal epithelial cells (IECs) are covered with a thick, viscous mucus barrier consisting largely of mucin glycoproteins (MUC2), other goblet cell products, as well as antimicrobial peptides (β-defensin and lysozymes) from Paneth cells and secretory IgA secreted by enterocytes. Microbes colonize the outer, "loose" mucus layer but are not generally present in the inner layer, adherent to the IECs. If the mucus layers are degraded due to reduced mucin biosynthesis or destroyed by exogenous agents, microbes can penetrate the protective mucous barrier and adhere to the surface epithelium, increasing intestinal permeability, and enhancing susceptibility to injury (Kim and Ho, 2010). Additionally, mucin regulates the function of intestinal dendritic antigen-presenting cells and IECs, allowing the host to maintain tolerance toward food and commensal antigens (Belkaid and Grainger, 2013). Disruption of microflora diversity, bacterial

overgrowth, reduction in the thickness and function of the mucus layer, increased permeability of the mucosal barrier (Ulluwishewa et al., 2011), and deficiencies of host immune defense can act synergistically to increase risk for bacteremia and endotoxemia.

2.3.3 The IEC Barrier

The IEC barrier consists of a monolayer of cells reinforced by desmosomes attached to keratin filaments in the cytoplasm. The monolayer is continually replenished by multipotent stem cells at the base of the crypts of Lieberkühn that give rise to four cell lineages: enterocytes that form the absorptive monolayer, mucus-secreting goblet cells, enteroendocrine cells, and Paneth cells. Over a 4–5-day life cycle, IECs proliferate, migrate up the intestinal villi, differentiate, and die. Enterocytes are linked by several types of junctions. Adherens junctions provide strong mechanical attachment between cells, while gap junctions facilitate communication between adjacent cells by allowing passage of ions and small molecules. Tight junctions fuse adjacent outer leaflets, limiting paracellular transport of molecules from the lumen into the submucosal layer. At the apical region, the IEC monolayer expresses a carbohydrate-rich glycocalyx containing receptors and signaling molecules, overlaid with a thick coat of mucus. The IEC monolayer, especially the Paneth cells, produces antimicrobial agents that provide a chemical barrier against pathogens and control transit of luminal contents into the bloodstream (Zasloff, 2002). Only molecules capable of binding to apical surface receptors and transporters can transit across the epithelial barrier (transcytosis). Transport between cells (paracellular transit) is limited by tight junctional complexes (TJ) as shown in Fig. 2.3. Transmembrane proteins (occludin and claudin) seal the paracellular space, while zonulin-1 (ZO-1) links TJ proteins to F-actin, a component of the apical perijunctional actomyosin ring. Contraction of the perijunctional cytoskeleton is regulated by phosphorylation of myosin light chain (MLC) by its kinase (MLCK) (Turner et al., 1997) and RhoA/Rho-kinase; both regulate cytoskeleton shape and movement. Severe ischemic injury resulting in decreased perfusion of the IEC monolayer is associated with disruption of the intestinal barrier function and increased paracellular permeability (Ulluwishewa et al., 2011).

Abundant data on the type of agents that disrupt TJ and their mechanisms of action have been obtained in vitro using Caco-2 cell monolayers. (The Caco-2 cell line is derived from a human colon carcinoma. These cells retain characteristics similar to IEC cells, in that they form a polarized monolayer and have a

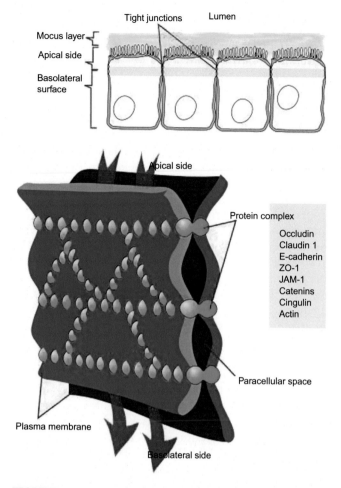

FIGURE 2.3 Tight junction components. The transmembrane proteins, occludin and claudin, together with junctional adhesion molecule 1 (JAM-1) and other proteins form a tight junction barrier and seal the paracellular space. These and accessory junctional components are attached to cytoskeletal filaments and other elements to provide stability to the cell. *Reproduced from Mariana Ruiz LadyofHats (Public domain), via Wikimedia Commons.*

well-defined brush border on the apical surface and retain intercellular tight junctions. It must be remembered that transformation changes some aspects of the cellular metabolism, and that colonic cells are different from cells in the proximal intestine.) Tight junction integrity requires that serine and threonine residues on occludin are phosphorylated. Bacterial LPS induces expression of a serine/threonine protein phosphatase 2A (PP2A) that dephosphorylates occludin and redistributes it from the junctional surface to the cytoplasm. Inflammatory stimuli also increase production of oxygen radicals capable of combining with nitric oxide (NO) to form peroxynitrite radicals that nitrate tyrosine residues in the PP2A catalytic subunit, thereby increasing its activity and moving occludin to the cytoplasm. Without occludin, the tight junction becomes "leaky" and the membrane permeable.

Disassembly of tight junctions in the intestinal barrier causes paracellular channels to open, allowing access of microbes and their products (eg, LPS) to the lymphatics and the bloodstream (Ulluwishewa et al., 2011). Even if the intestine is not directly damaged (eg, by gunshot injury) reduced splanchnic blood flow decreases IEC perfusion. (*Splanch* comes from a Greek word, *splēn*, meaning organ. It is used to describe nerves and vessels that serve the internal organs. The splanchnic circulation of the gastrointestinal tract originates at the celiac trunk, the superior and the inferior mesenteric arteries.) Hypoxia shifts enterocyte metabolism to anaerobic glycolysis with production of lactic acid. Mucosal acidosis and ischemia injure the IEC monolayer and allow translocation of microbes and their products.

2.3.4 The Innate Immune System

A diverse array of evolutionarily ancient hematopoietic cell types, including dendritic cells, monocytes, macrophages, and granulocytes comprise the innate immune system. These cell populations collaborate with each other, with the adaptive immune system and with nonhematopoietic cells to promote immunity, inflammation, and tissue repair. Innate lymphoid cells (ILCs) are a subset of the innate immune system characterized by lymphoid morphology but lacking expression of cell surface molecules and defined as cell lineage marker negative cells as recently reviewed (Artis and Spits, 2015; Sonnenberg and Artis, 2015). Three subsets of ILC are known, subsets are specifically activated by diverse stimuli including microbes, allergens, neuropeptides, hormones, eicosanoids, cytokines, and other mediators, and play major roles in immunity and maintenance of tissue repair and homeostasis. At the intestinal barrier, ILCs regulate interactions between the host and the microbiota, chiefly by limiting inappropriate immune responses to bacteria. Experimental loss of murine ILCs was associated with lower levels of antimicrobial peptides expressed by intestinal cells. The ILC3 subset regulates interactions between dendritic cells and the microbiota and maintains separation between inflammatory microbial products and the host immune system.

Injury increases microbial crosstalk with immune cells that stimulates a spectrum of events leading to further TJ disruption. Injury-stimulated ILCs secrete cytokines (interferon gamma, IFN-γ and tumor necrosis alpha, TNF-α) that dysregulate occludin expression and inhibit MLCK-mediated phosphorylation of MLC. As a result, TJ junctions are disrupted and the intestinal barrier becomes more permeable. IFN-γ also promotes the redistribution of TJ proteins by a micropinocytosis process. Interleukins (IL-4 and IL-13) induce an increase in intestinal permeability through induction of epithelial apoptosis and expression of the pore-forming TJ protein claudin-2. Mast cells produce TNF-α, mast cell protease 1 (mcpt1), and lipid mediators, including histamine, platelet-aggregating factor, and prostaglandins. Mcpt1 degrades the TJ protein occludin-1, altering barrier function. Eosinophil-derived major basic protein downregulates occludin-1 expression in colonic epithelial cells (Groschwitz and Hogan, 2009).

Activated intestinal macrophages secrete inflammatory nitric oxide synthase (iNOS) that generates excessive amounts of NO. (Inducible iNOS differs from endothelial NOS (eNOS) in that it is activated by inflammatory mediators and produces large amounts of NO as a part of host defense against microbial and other attacks.) While the small amounts of NO normally produced by eNOS protect the cells from vasoconstrictive injury, overproduction of NO during inflammation upregulates expression of the adhesion molecule, intracellular adhesion molecule-1 (ICAM-1), necessary for polymorphonuclear leukocyte (PMN) infiltration into the mucosal barrier. NO also increases the expression of inflammatory transcription factors including nuclear factor kappa beta (NFκB) and Signal transducer and activator of transcription 3 (STAT 3). These factors bind DNA and increase expression of inflammatory cytokines (IL-6) and granulocyte-colony stimulating factor (G-CSF), stimulating PMN production in the bone marrow. It is unclear whether the dysfunction of intestinal smooth muscle seen with iNOS induction is a direct result of NO excess, or an indirect alteration mediated by associated reactive oxygen radicals. In any case, NO has been implicated as a major causal factor in reduced cellular viability with villous damage and loss of intestinal barrier function following endotoxin exposure.

2.3.5 Innate Response to Injury

If a microorganism breaches the epithelial barriers, the innate immune system recognizes the pathogen using several types of receptors. At least 10 toll-like receptors (TLRs) are present on the surface of human cells, especially on phagocytic and dendritic cells and on the epithelial cells lining the respiratory and gastrointestinal tracts. LPS expressed on Gram-negative bacteria, as well as lipoteichoic acid and peptidoglycan (PG) on Gram-positive bacteria, activate different members of the TLR family (Akira et al., 2006). Other carbohydrates (mannan, glucan, and chitin) on fungi and parasites as well as lipid mediators, chemokines, and bacteria-specific N-formyl-methionyl peptides are also detected. While these pathogen-associated molecular pattern (PAMP) molecules are also expressed by resident commensal microbes, the innate immune system has evolved the ability to detect and respond to sudden changes in concentration or the appearance of these molecules in normally sterile parts of the body.

Tissue injury can also cause sterile inflammation without infection (Rock et al., 2010). This type of inflammation is driven by a precursor of IL-1β (Pro- IL-1β) produced by macrophages following recognition of products of the injury. Sterile irritants, including uric acid and cholesterol crystals, asbestos, products of dead and dying cells, extracellular adenosine triphosphate (ATP) and other proinflammatory particles are recognized as damage-associated molecular patterns (DAMPs). DAMPs bind to pattern recognition receptors (PRRs) on the macrophage and initiate formation of a large multiprotein complex, the inflammasome. Pro-IL-1β remains inactive and intracellular until cytosolic PRRs are further activated. At this point, Pro-IL-1β is cleaved by caspase-1, mature IL-1β is then secreted and participates in the integrated response to injury. One of the most well-studied inflammasomes is NLRP3 (NLRP3 refers to components of the inflammasome (NACHT, LRR, and PYD domains-containing protein 3 components). When NLRP3 recognizes DAMPs, it recruits caspase-1 and initiates sequential steps ultimately leading to host-cell death (Mariathasan and Monack, 2007; Summersgill et al., 2014) as illustrated in Fig. 2.4.

2.4 MEDIATORS OF THE ACUTE-PHASE RESPONSE

In response to injury, secreted mediators coordinate a whole-body acute-phase response (APR) involving hepatocytes, muscles, fat, and bone marrow, to mount an effective inflammatory response. Symptoms of APR include an increase in body temperature, anorexia, drowsiness, and the hemodynamic responses to shock. Vascular changes include expression of endothelial adhesion molecules, leukocyte proliferation, adherence, and translocation into the area of damage. Phospholipids in the plasma membranes of participating cells are used as substrate for lipid mediators that increase procoagulant and decrease anticoagulant activity facilitating formation of a hemostatic plug. Secondary mediators including IL-8, IL-6, and platelet-derived growth factor are secreted and vascular fibroblasts proliferate and increase production of collagen, collagenase, protease, and eicosanoid biosynthesis in response to these mediators.

The bone marrow responds to mediator stimuli by increasing circulating leukocyte availability. The leukocyte count can rise to >20,000 cells/mL in acute

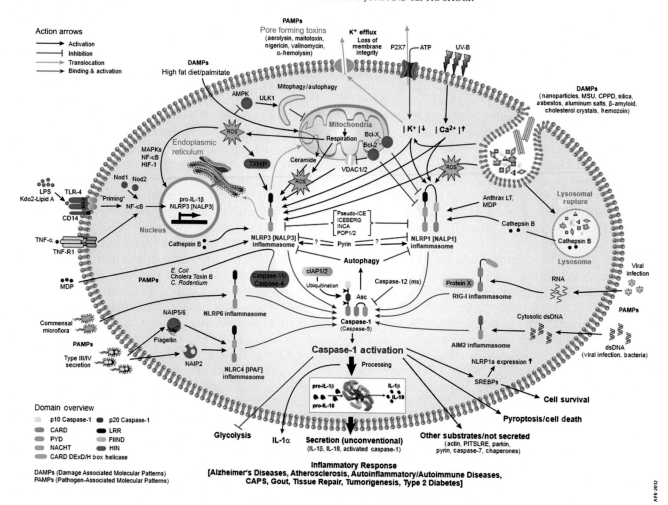

FIGURE 2.4 **Inflammasome formation and action.** The NLRP3 inflammasome complex is formed when membrane receptors on the macrophage recognize PAMPs, DAMPs, and other signals and express pro-IL-1β. This inactive mediator remains intracellular until cytosolic PPRs are further activated to form a large, multiprotein cytoplasmic complex, the inflammasome. The cysteine protease caspase-1 is activated, leading to the cleavage and activation of pro-IL-1β to IL-1β and other cytokines such as IL-18, that orchestrate the inflammatory response and ultimately, cause host-cell death. *Reproduced from A. Traina, Adipogen International, 2012. CC BY-SA 3.0 <http://creativecommons.org/licenses/by-sa/3.0>, via Wikimedia Commons.*

inflammation; bacterial infections raise the neutrophil count (*neutrophilia*) while viral infections increase lymphocytes (*lymphocytosis*) and allergies and parasitic infestations raise the absolute number of eosinophils (*eosinophilia*). Initially following injury, immature cells are released from the postmitotic pool in the bone marrow. The morphology of the immature neutrophil is distinct from the lobed nucleus in the mature neutrophil. The nucleus of immature neutrophils is a solid band, thus an increase in the number of "band cells" (also called a *left shift*) suggests an acute-phase process. Prolonged infection increases production of colony stimulating factors (CSFs) that increase leukocyte production in the bone marrow. If the infection is extreme, the number of newly formed leukocytes can climb as high as 40,000–100,000 cells/mL; these numbers indicate a *leukemoid reaction* of a magnitude resembling true leukemia.

It should be noted that some viruses and other microbial infections are associated with decreased leukocyte production (*leukopenia*).

2.4.1 Dietary Inflammasome Suppressors

2.4.1.1 Resveratrol (3,4',5-Trihydroxystilbene)

Resveratrol is a natural phenolic phytoalexin found in plant species, such as red grapes, peanuts, and mulberries (Frémont, 2000). The *trans*-isomer in grapes is naturally converted to the *cis*-isomer during fermentation by yeast isomerases. The *trans*-isomer has been shown to exhibit antiinflammatory, antiaging, immunomodulatory, and other protective effects. Other isomers may also be protective. A recent study demonstrated that pretreatment of macrophages with *cis*-resveratrol not only reduced pro-IL-1β production and IL-1β secretion, but also suppressed ATP-induced transcription and activation of caspase-1 and caspase-4, endoplasmic reticulum stress, and reactive oxygen species production. Furthermore, *cis*-resveratrol attenuated cyclooxygenase-2 expression and prostaglandin E2 production (Huang et al., 2014).

2.4.1.2 Curcumin (Diferuloylmethane)

The rhizome from the perennial herb, turmeric (*Curcuma longa*) contains a phenolic product, curcumin, that also has been prized for its antiinflammatory, immunomodulatory, and protective effects (Basnet and Skalko-Basnet, 2011). A mechanistic study (Gong et al., 2015) demonstrated that curcumin inhibited mature IL-1β in LPS-primed murine peritoneal and bone marrow macrophages. It also reduced the level of cleaved caspase-1 (measured by western blot and ELISA) and prevented K^+ efflux, the common trigger for NLRP3 inflammasome activation. Lysosome disruption and reactive oxygen species (ROS) were also suppressed. Data from this study suggested that the curcumin-induced K^+ efflux inhibition downregulation of extracellular signal regulated kinase signaling inhibited NLRP3 inflammasome activation. The authors also reported that curcumin administration in vivo improved survival of mice suffering from lethal endotoxic shock.

2.4.2 The Complement System and Related Cascades

The complement cascade includes about 20 soluble proteins that participate in immune defense. Complement precursors are made in the liver and circulate in the plasma as inactive forms. Once activated by pathogen sensing or antibody binding, complement proteins interact to form complexes that stimulate the adaptive immune response, recruit phagocytes and inflammatory cells, or directly lyse microbial and foreign cells. Plasma contains other inactive proteins including kinen, fibrinogen, and coagulation cascades that are also activated and interact to facilitate inflammation, repair, and healing. It must be remembered that vitamin K is necessary for γ carboxylation of specific inactive coagulation factors; carboxylation allows these proteins to bind calcium and become activated as discussed elsewhere (refer to Essentials III: Nutrients for Bone Structure and Calcification: Vitamin K) (Fig. 2.5).

2.4.3 Lipid Mediators

Inflammatory mediators stimulate cleavage of polyunsaturated FAs (PUFAs) from membrane

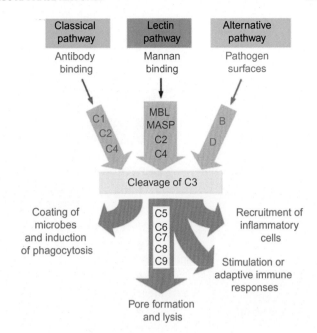

FIGURE 2.5 **The complement cascade.** Following activation, complement components form complexes that potentiate the inflammatory response. *Reproduced with permission from Molecular Biology of the Cell 6th Edition. Garland Science.*

phospholipids. The released FAs are enzymatically oxidized by cyclooxygenase and 5-lipoxygenase to form a variety of paracrine hormones, especially prostaglandins and leukotrienes. These hormones and their metabolites mediate vasoactive and inflammatory functions as indicated in Fig. 2.6. The relative concentration of FAs in membrane phospholipids reflects dietary intake. In general, FAs from grains and grain-fed animals are predominately omega 6 (n-6) FA, while oily fish and some algae produce n-3 FA. The ratio of dietary n-6:n-3 FA is reflected in membrane FA composition. While n-6 and n-3 FA are metabolized by the same enzymes, the eicosanoid products have different activities. n-6 FAs (C20:4; arachidonic acid) are substrates for proinflammatory mediators and n-3 FAs substrates: C20:5 and C22:6; eicosapentaenoic (EPA) and docosahexaenoic acids (DHA) respectively, have antiinflammatory characteristics. The n-3FAs enter the eicosanoid pathways and produce prostaglandins and leukotrienes of the "3" series that elicit less vasoconstrictive and inflammatory actions. (Membrane FA are cyclized by cyclooxygenase in the formation of prostaglandins. This process requires two double bonds. Thus FA with four double bonds (arachidonic acid) will produce PG of the "2" series, while FA with five double bones (EPA) produce PG of the "3" series. Lipoxygenase does not cyclize the FA, hence leukotrienes (LT 4 and 5, respectively), are formed.G-protein receptors (GPR120) that specifically bind n-3 FA have been reported to improve insulin sensitivity in animal

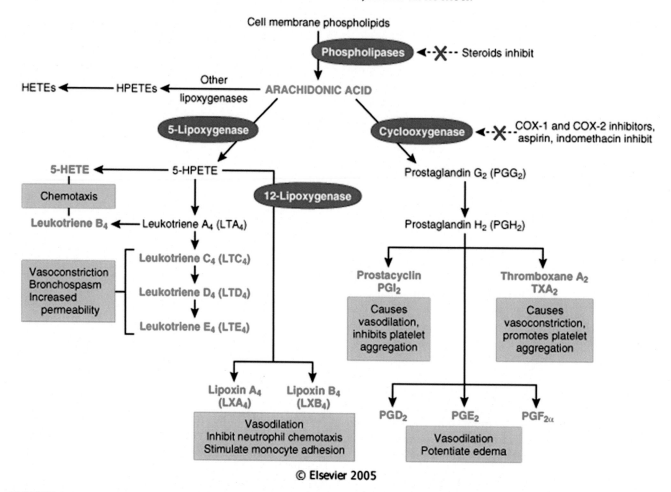

FIGURE 2.6   **The eicosanoid pathway.** In response to inflammatory signals, arachidonic acid in the phospholipid bilayer is enzymatically modified to produce proinflammatory and vasoactive prostaglandins and leukotrienes. If the membrane contains n-3 fatty acids, EPA and DHA are cleaved and enter the same pathways to generate prostaglandins of the 3 series and leukotrienes and lipoxins of the 5 series; paracrine hormones generated with n-3 FA substrate exert less deleterious consequences. Steroids and aspirin inhibit key enzymes in this pathway (red X). *Reproduced with permission from Robbins—Elsevier.*

models fed a high-fat diet (Oh et al., 2010), as discussed below.

Recently, other antiinflammatory, counter-regulatory lipid mediators derived from membrane PUFA have been identified. These mediators include lipoxins, resolvins, protectins, cyclopentenones, and presqualene diphosphate (Haworth and Levy, 2007). Some products (resolvins and protectins) are associated with wound healing and are being intensively studied as expediting factors in the resolution of hyperinflammation following trauma (Ariel and Timor, 2013).

In an alternate pathway, highly unsaturated FAs, arachidonic acid as well as EPA and DHA can be nonenzymatically oxidized by free radicals to prostaglandin-like isoprostanes. These molecules have been used as markers for oxidative stress and are elevated with obesity and aging, as well as with trauma and tissue damage. Saturated FAs (SFAs), released

from adipose tissue in response to counter-regulatory hormones (cortisol, glucagon, epinephrine, and growth hormone) also stimulate the inflammatory response (Glass and Olefsky, 2012). SFAs can bind to TLRs, stimulating transforming growth factor activating kinase 1 (TAK1) and activating downstream pathways such as NFκB, thereby enhancing inflammation and increasing cytokine secretion (Fig. 2.7).

2.4.4 Epigenetic Regulation of the APR

The distinct, stereotypic clinical phases of acute inflammation appear to be modulated by epigenetic changes. Following injury or insult, an acute, proinflammatory phase evolves. As the patient recovers, hyperinflammation is superseded by an adaptive, antiinflammatory phase and finally, resolution. Depending on the type of injury, the nutritional status of

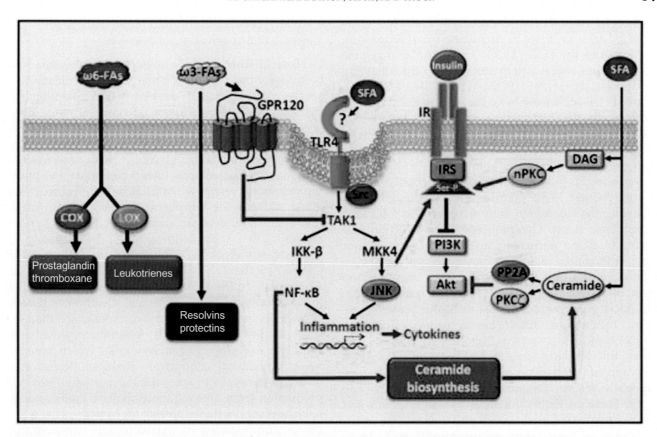

FIGURE 2.7 **Coordinated actions of lipid mediators.** SFA signals may interact with TLRs or independently to stimulate inflammatory pathways. SFA can also reduce insulin signaling and augment inflammation-dependent ceramide biosynthesis to inhibit AKT activation, reducing anabolic signals. These effects can be inhibited by n-3 fatty acids that, in addition to providing substrate for antiinflammatory resolvins and protectins, bind to G-protein receptor 120 (GPR120) and inhibit transforming growth factor activating kinase 1 (TAK1), thereby reducing the NFκB and JNK inflammatory pathways. *Reproduced with permission from Glass, C.K., Olefsky, J.M., 2012. Inflammation and lipid signaling in the etiology of insulin resistance. Cell Metab. 15 (5), 635–645—Science Direct.*

the host, and the postinjury care, these phases can be swift or extended, with significant impact on the outcome. Chromatin remodeling and gene reprogramming control phases of acute systemic inflammation by epigenetic shifting of the expression of specific sets of genes. Resolution of inflammation and host survival occurs when epigenetic reprogramming is terminated by, as yet, unknown mechanisms (McCall et al., 2011). Initially, the insult is recognized by sensors that upregulate inflammatory mediators. NFκB is recognized as being a master regulator that incites transcription of proinflammatory genes for interleukin I (IL-1β) and tumor necrosis factor α (TNFα). The hyperinflammatory phase is followed by events that deactivate NFκB and facilitate its proteasome-dependent protein degradation. Reduction in NFκB stimulated inflammatory mediator generation and formation of silent heterochromatin signals an antiinflammatory phase. If the antiinflammatory phase does not occur promptly, the clinical outcome is poor. Resolution of inflammation occurs when epigenetic remodeling is terminated and homeostasis returns (McCall et al., 2011).

2.5 HYPERINFLAMMATION, SEPSIS, AND SHOCK

Severe inflammatory response to injury or microbial activity is called *sepsis*, from a Greek word meaning putrefaction and decay. It refers to the potentially fatal whole-body inflammation (also called systemic inflammatory response syndrome or SIRS). If the injurious stimulus is severe and prolonged, SIRS can overwhelm the entire body; SIRS is an ominous development in critical care and accounts for >50% of all patient deaths. The lethal potential of SIRS is caused largely by the damage done to multiple organs (multiple organ dysfunction syndrome or MODS) due to the widespread host response, complicated by septic shock and reduced organ perfusion. Clinical signs of sepsis include fever or hypothermia, tachypnea (rapid breathing), confusion, and edema, as well as tachycardia (rapid heart rate), decreased urination, elevated blood sugar, and metabolic acidosis. Rapid breathing results when the body adapts to metabolic acidosis by blowing off carbon dioxide. (Metabolic acidosis is characterized by a decrease in circulating bicarbonate ions. This elevates the PCO_2/HCO_3^- ratio and changes the hydrogen ion concentration according to the following equation: $[H+] = 24 \times (PCO_2/[HCO_3^-])$. PCO_2 is reduced by increasing alveolar ventilation, thereby reducing the PCO_2/HCO_3- ratio toward, but not to, normal. For acid—base tutorials, see http://fitsweb.uchc.edu/student/selectives/TimurGraham/Compensatory_responses_metabolic_acidosis.html/.)

Shock results when a dramatic reduction of blood volume reduces tissue perfusion. It is a common event following severe injury with massive loss of blood (*hemorrhagic shock*). Compromised heart function that decreases tissue perfusion can precipitate *cardiogenic shock*. If the initial injury has introduced microbial contamination into the wound, the patient may develop *septic shock*. (Medscape has a very good review of septic shock. This resource is free to health professionals upon registration. http://emedicine.medscape.com/article/168402-overview#aw2aab6b2b2/.) In septic shock, microbial products and necrotic debris activate macrophages and other cells to produce mediators, especially the proinflammatory cytokines, TNFα and IL-1β. These mediators initiate a cascade of events that dilate peripheral blood vessels and activate the hypothalamus to raise body temperature (fever). These changes produce warm, flushed skin, rather than the cool, clammy skin of hypovolemic shock. Symptoms of septic shock can be nonspecific and include fever, chills or rigors, confusion and anxiety, difficulty breathing, fatigue and malaise, or nausea and vomiting. Intracellular volume and tissue perfusion are further compromised by vasodilation and increased vascular permeability, activation of thrombosis, and/or hemorrhage. Again, the result is cellular hypoxia, lactic acidosis, and death. Whatever its cause, hypovolemia of shock reduces tissue perfusion, worsening ischemic tissue damage, triggering activation of the clotting cascades and increasing the risk for formation of thrombin and life-threatening blood clots. Although it is potentially reversible, prolonged tissue hypoxia can cause irreversible cellular damage and cell death.

2.5.1 Hypovolemia and Shock Secondary to a Gunshot Wound

A gunshot wound to the abdomen causes massive blood loss that precipitates hypovolemic shock and can have the following lethal consequences:

- Initially, vasomotor compensatory mechanisms are activated to perfuse vital organs.
- As shock progresses, circulatory and metabolic imbalances worsen, producing acidosis.

- Finally, cellular, tissue, and organ damage are so severe that survival is not possible.

These life-threatening events following trauma have been described as the "lethal triad." Patients with severe blood loss can develop acute coagulopathy, hypothermia, and acidosis (Jansen et al., 2009). Hypoperfusion results in tissue hypoxia with reduced capacity for aerobic respiration. To compensate, the tissues switch to anaerobic glycolysis, with production of lactic acid. Lactic acidosis lowers tissue pH and blunts the vasomotor response. Arterioles fail to constrict and blood pools in the peripheral tissues, causing further tissue anoxia and organ damage. As cellular hypoxia worsens, the mitochondria lose the capacity to generate sufficient ATP to maintain sodium efflux (Na^+-K^+ATPase) as illustrated in Fig. 2.8. Accumulated intracellular sodium exerts osmotic fluid influx and cells swell. Lysosomal membranes become permeable and proteolytic enzymes leak into the cytosol to further injure the cells. Inflammatory mediators stimulate excessive NO biosynthesis which compromises myocardial contraction. Reduced perfusion of the kidneys results in acute tubular necrosis. Poor ATP production from anaerobic metabolism limits endogenous heat production with risk for hypothermia, which can be compounded by exposure and administration of cold resuscitation fluids, including blood. A core temperature below 35°C is a strong predictor of mortality following trauma (Martin et al., 2005).

2.5.2 Vascular Permeability Exacerbates Hypoperfusion

Injury induces a transient (3—5 seconds) vasoconstriction followed by vasodilation. These changes allow the capillary beds in the affected area to fill with blood, explaining the redness (erythema or *rubor*) and swelling (*tumor*) characteristic of inflammation. Normally, the vascular endothelial monolayer is linked by tight junctional complexes to prevent fluid and plasma proteins from leaking from the blood into the tissue. As in the gastrointestinal (GI) tract, inflammatory mediators disassemble the endothelial junctions, allowing protein-rich fluid (exudate) to enter the extracellular spaces. As protein is lost from the blood, remaining blood has less oncotic force to return fluid to the vasculature, resulting in the edema observed with inflammation. The exact composition of the extracellular exudate depends on the severity of the injury and the presence of blood or infection in the wound. Partially dehydrated blood remaining in the vessels is viscous and moves sluggishly through the capillaries (*stasis*). Microscopically, packed red blood cells can be seen in dilated capillary vessels.

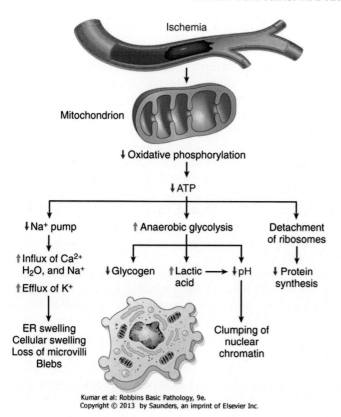

Kumar et al: Robbins Basic Pathology, 9e.
Copyright © 2013 by Saunders, an imprint of Elsevier Inc.

FIGURE 2.8 **Ischemia and its sequelae.** The cell adapts to cellular hypoxia and reduced ATP production by increasing anaerobic glycolysis. This process not only fails to maintain glycogen stores, but it also produces lactic acid, thereby lowering intracellular pH. Low intracellular pH denatures proteins and causes nuclear chromatin to "clump." The injured cell pumps H^+ ion into the extracellular fluid and blood, lowering blood pH, resulting in systemic acidosis. Reduced ATP compromises protein synthesis; ribosomes detach from the endoplasmic reticulum. ATP-dependent channel function is dysregulated, allowing influx of Ca^{2+}, H_2O, and Na^+, disrupting cell signaling and causing swelling of the cell and its organelles with loss of function. *Reproduced with permission from Kumar, V., et al., 2015. Robbins and Cotran Pathologic Basis of Disease, 9th Ed. Elsevier Saunders.*

2.5.3 Acute Coagulopathy of Trauma (ACT)

Coagulation is intimately related to inflammatory processes. Widespread activation of the coagulation cascade deposits tiny fibrin clots in the microcirculation, causing ischemia in organs and hemolysis of red cells as they negotiate the narrowed vessels. Disseminated intravascular coagulation is a serious, potentially fatal complication of shock. As platelets and clotting factors are used up in clot formation, a bleeding diathesis (propensity) and hemostatic failure ensues. Acute Coagulopathy of Trauma (ACT) is multifactorial and results in increased transfusion requirements with increased incidence of organ dysfunction, critical care stay, and mortality. At least six key initiators of coagulopathy are seen in trauma patients. These include tissue injury, shock, hemodilution, hypothermia, acidemia, and

inflammation. All hemostatic components, including platelets, endothelium, and coagulation cascades play essential roles in a relative inhibition of stable clot formation by anticoagulant and fibrinolytic pathways; the degree of participation varies with type of injury and effects of subsequent medical therapies. Coagulation proteases may be lost or inhibited due to widespread activation and consumption of the proteases or because their concentrations are diluted, due to hypothermia and acidosis, or through the activation of anticoagulant and fibrinolytic pathways.

2.6 BODY COMPOSITION AND MACRONUTRIENT CHANGES IN APR

Trauma compromises nutritional status due to poor appetite, anorexia from disease, treatment or pain, gastrointestinal lesions that impair absorption, and/or metabolic alterations that dysregulate nutrient utilization or excretion. For example, fever increases resting energy expenditure while APR mediators redistribute nutrients as described below. The degree to which the APR compromises nutritional status is strongly influenced by the nutritional intake prior to the trauma, stage of life cycle (infant, pregnancy, elderly), therapeutic modalities (drugs, tests, vaccines), the nature of the trauma, and/or the injurious agent and nutritional intervention post trauma. Evidence for these interactions and future research priorities has been reviewed (Raiten et al., 2015).

2.6.1 Alterations in Macronutrient Metabolism and Body Composition with APR

Host response and changes in body composition secondary to critical injury were described over 40 years ago (Beisel, 1975), although mechanisms that drive this response are only now being elucidated. Trauma triggers an alteration in energy utilization by modifying substrate availability and activating alternative pathways to meet increased needs. Prominent changes include tissue wasting (cachexia) with reductions in muscle, fat, and bone. Increased macronutrient availability is reflected as hyperglycemia and hyperlipidemia with insulin resistance mediated predominantly by counter-regulatory hormones (Watters et al., 1986). While in health, reduced protein intake is accompanied by reduced nitrogen excretion to maintain nitrogen balance, this adaptation is not observed in the critically ill patient. In febrile illness, nitrogen balance becomes negative as muscle protein is redistributed to provide precursor materials to meet both catabolic and anabolic processes. Body fat and glycogen stores are mobilized to provide for energy demands. Mediators include neurohormonal events such as activation of the sympathetic nervous system, stimulation of the hypothalamic–pituitary–adrenal axis with increased cortisol production, and an increase in the secretion of glucagon relative to that of insulin (Wilmore, 1991). Skeletal muscle is the major reservoir for release of amino acids and gluconeogenic substrate during sepsis.

The spectrum of wasting observed with sepsis and inflammation was originally attributed to the protein, cachectin, more recently defined as the primary inflammatory mediators, TNFα and IL-1β (Fong et al., 1989). These mediators simultaneously inhibit insulin release and increase glucagon secretion from the pancreas, promoting insulin resistance and reducing anabolism in fat, muscle, and other tissues. Glycogenolysis and gluconeogenesis are stimulated both by inflammatory mediators and by release of stress hormones, resulting in the hyperglycemia observed in sepsis. Hyperglycemia impairs host response by reducing bactericidal activity of neutrophils and also by increasing expression of adhesion molecules on the endothelium. Elevated circulating triglyceride (TG) levels with inflammation are due, in part, to the actions of inflammatory mediators that downregulate lipoprotein lipase (LPL) and prevent TG clearance. Endotoxemia also increases circulating TG levels. Rats treated with LPS had a threefold increase in plasma TG concentrations; LPS-induced TG increase was not due to increased TG output by the liver, but to the endotoxin-induced suppression of LPL activity (Bagby et al., 1987).

Muscle wasting occurs early and rapidly in critical illness and is more dramatic in patients with multiorgan failure (MOF) (Puthucheary et al., 2013). APR-induced changes in at least five proteolytic pathways have previously been identified, including the lysosomal, Ca^{2+}-dependent, and ubiquitin–proteasome-dependent pathways, as well as the caspase systems and matrix metalloproteinases (MMPs) (Ventadour and Attaix, 2006). Protein wasting in a longitudinal cohort of critically ill ICU patients was assessed by muscle loss, quantitated by serial ultrasound measures of the rectus femoris cross-sectional area. Histological analysis of the cross-sectional area of the vastus lateralis muscle, the ratio of DNA to protein, as well as biochemical measures of muscle protein breakdown rates and signaling pathways were assessed in a subset of patients. All patients were given enteral feeding normalized to ideal body weight. Unexpectedly, early high-protein feeding actually increased protein loss. Data indicated that muscle wasting in these patients could be explained by a combination of reduced protein synthesis and increased muscle breakdown.

2.6.2 Protein and Energy Needs in the Trauma Patient

APR-induced altered nutrient distribution complicates patient-feeding strategies. While use of enteral nutrition to improve clinical outcomes is well supported, recommendations for calories and proteins remain uncertain. A recent meta-analysis of clinical trials was conducted using mortality as the primary endpoint and infection, gastrointestinal intolerance, hospital and intensive care unit (ICU) lengths of stay, and mechanical ventilation days as secondary indicators. Data indicated that in patients who were not malnourished, high-energy intake did not improve outcomes and actually increased complications in critically ill patients. A moderate nutrient intake (33.3–66.6% of goal energy) may reduce mortality, and a higher protein intake (≥0.85 g/kg per day) may decrease the infection rate (Tian et al., 2015). These observations are supported by a recently published metabolomic analysis in a fed and fasted porcine model. Prior feeding status significantly altered metabolic response to hemorrhagic shock. Whereas fasted animals depended on proteolysis to provide both substrate and energy, fed animals continued anabolic activities in muscle and liver with a lag time to switch from anabolism to catabolism. Fed animals relied mainly on glucose, and less on amino acids for energy production. It is of interest that fed animals also exhibited a higher death rate (47% vs 28%), especially in the postresuscitation period (Witowski et al., 2015).

2.6.3 Metabolic Complications Secondary to Obesity and Hyperglycemia

Obesity limits the physiologic response to acute illness in a number of ways (Choban and Dickerson, 2005). The morbidly obese patient (BMI >40 or >35 with severe comorbidity) retains fluid. Increased total body water can worsen respiratory insufficiency, compromise cardiac function, and reduce tolerance to adequate IV fluids. Sleep apnea, common in the morbidly obese patient can cause chronic hypoventilation and present problems with mechanical ventilation. Fatty liver, also frequently seen in this population, exacerbates hepatic dysfunction manifested as lipid intolerance and greater plasma TG elevations. Mediators of critical illness increase muscle proteolysis, releasing amino acid and glucose substrate for host response and repair processes. While some studies suggested that increased adipose tissue mass paradoxically improves outcome, others reported increased risk for mortality in obese patients (Neville et al., 2004).

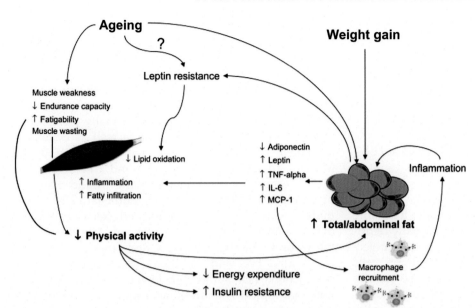

FIGURE 2.9 **Pathologic connections: weight gain and sarcopenia.** *Reproduced with permission from Zamboni, M., Mazzali, G., et al., 2008. Sarcopenic obesity: a new category of obesity in the elderly. Nutr. Metab. Cardiovas. Dis. 18 (5), 388–395 (Zamboni et al., 2008)—Science Direct.*

The term, sarcopenia, is from the Greek words "sarx" (flesh) and "penia" (loss), and was initially coined to describe reduced muscle mass with aging (Rosenberg, 1989, 1997; Narici and Maffulli, 2010). This gradual age-associated loss of lean body mass has been attributed to an imbalance between protein synthesis and breakdown, possibly related to blunted anabolic responses to meals, insulin, and other hormones. However, other aspects of aging, including inflammation, reduced mitochondria function and sedentary lifestyle have also been implicated (Ribeiro and Kehayias, 2014). The impact of obesity and diabetes on muscle mass and function is not well studied, but reductions in muscle strength and power have been suggested by animal studies and a few clinical reports (Hilton et al., 2008; Buendía et al., 2015; Zhang et al., 2015). It is likely that even young obese patients may suffer from prior decrements in muscle mass and function that are compounded by the known muscle catabolism that occurs during trauma (Fong et al., 1989).

Thus, although an obese patient may look "well nourished," muscle mass to total weight may be compromised. As indicated in Fig. 2.9, adipose mass, especially visceral adipose tissue increases inflammatory mediators that can compromise muscle mass and function and induce muscle lipotoxicity from intramuscular lipid deposition (Zhang et al., 2015). Care must be taken to assure that protein requirements are met by exogenous administration to prevent protein malnutrition in the obese patient as described below.

The diabetic patient already suffers from metabolic stress secondary to hyperglycemia and MS. Hyperglycemia increases production and/or expression of proinflammatory mediators, leukocyte adhesion factors, ROS, and oxidative damage. It also compromises the innate immune response, especially chemotaxis and phagocytic capacity (Weekers et al., 2003). Since leukocytes have no capacity to store fuel, they require adequate, but not excessive glucose concentrations for optimal function. Inflammatory stimuli increase glucose uptake in these cells, facilitated by insulin and its receptors. Ex vivo studies have revealed that both hypo- and hyperglycemia impair immune-cell functions and promote inflammatory responses, while clamp studies have revealed proinflammatory effects of hyperglycemia as well as immune-promoting effects of insulin (Calder et al., 2007). Immune dysfunction in the diabetic patient has been associated with increased tissue damage, slowed repair, disability, and death in critical care patients (Clement et al., 2004). In obese and diabetic patients, counter-regulatory stress hormones and mediators accelerate catabolism, hepatic gluconeogenesis, and lipolysis, raising blood glucose, FFAs, ketones, and lactate levels. Glucose toxicity blunts insulin secretion, resulting in further hyperglycemia. The vicious cycle of stress-induced hyperglycemia and hypoinsulinemia subsequently causes maladaptive responses in immune function, fuel production, and synthesis of mediators that cause further tissue and organ dysfunction. Given the compromised immune defense associated with hyperglycemia and inflammation, aggressive management of blood glucose with insulin was proposed (Clement et al., 2004). The authors cited clinical trials that reported improved recovery and decreased mortality when blood glucose averaged <110 mg/dL (6.1 mmol/L). More recently however, the American College of Physicians published Best Practice Advice recommending a target blood glucose level of

7.8–11.1 mmol/L (140–200 mg/dL) if insulin therapy is used in SICU/MICU patients. They cautioned that targets less than 7.8 mmol/L (<140 mg/dL) should be avoided to reduce risk for hypoglycemia and its deleterious sequelae (Qaseem et al., 2014).

2.7 TARGETED NUTRIENT STRATEGIES FOR WOUND HEALING

Trauma is commonly associated with delayed wound healing (Blass et al., 2013). Normally, following injury, cells adjacent to the injured area partially dedifferentiate and migrate, extending lamellipodial projections into the wound and exerting a pull on cells behind them (Jacinto et al., 2001). The precise balance between migration, proliferation, and differentiation of epithelial cells is accompanied by biosynthesis of collagen and other connective tissue elements to replace damaged tissue, angiogenesis, and neural ingrowth to perfuse and innervate the newly formed tissue. Comorbidities of obesity, especially hyperglycemia, as well as poor perfusion and infection, can delay healing. Since nutrients play key roles in processes that enhance healing, it is likely that trauma-induced nutrient depletion and redistribution are also implicated in delayed healing with trauma. However, despite the mechanistic roles played by several key nutrients described below, a recent review of intervention trials did not find conclusive evidence that supplementation with nutrient cocktails or individual nutrients improved healing of pressure ulcers (Langer and Fink, 2014). The authors did note that most studies were small and had methodological flaws that prevented more robust outcomes. It is also possible that the processes associated with pressure ulcers are not the same as those encountered in the trauma patient.

As described elsewhere (refer to Essentials I: Life in an Aerobic World. Ascorbate and Chapter 9: Osteoporosis and Fracture Risk), ascorbate and copper are required for crosslinking of newly formed collagen. Zinc, especially topical zinc, improves wound healing through a number of mechanisms as reviewed (Lansdown et al., 2007). Zinc bound to metallothionein (MT) is donated for DNA polymerases during the mitotic epithelial burst, zinc addition mimics the action of growth factors and has also been shown to upregulate growth factors such as insulin-like growth factor. Protection against oxidative stress and bacterial toxins has been attributed to zinc-induced upregulation of cysteine-rich MT. Zinc facilitates fibroblast infiltration and integrin expression, necessary for keratinization and keratinocyte migration.

The zinc-metalloenzyme, alkaline phosphatase is a marker for dermal blood vessels and is upregulated during early stages of angiogenesis and connective tissue proliferation. The metal is also a critical component of matrix metalloproteinases (MMPs) that contain one catalytic zinc ion and one structural zinc ion. MMPs are capable of degrading virtually all components of the extracellular matrix as well as cytokines and their receptors, adhesion molecules, and latent MMPs. All cell types in the wound can synthesize MMPs, especially keratinocytes along wound margins, macrophages, fibroblasts, and endothelial cells. Thus, under regulation by mediators and cell extracellular matrix contacts, MMPs carry out autodebridement of the wound, cleaving collagens, and other substrates in the wound bed and facilitating healing and wound contraction. The authors reviewed clinical trials and note that topical zinc dressings improved wound healing even in patients with normal zinc status.

2.7.1 Vitamin A and Precursors

Retinoic acid (vitamin A) sources and utilization have been discussed elsewhere (see Essentials II: Heavy Metals, Retinoids, and Precursors: Vitamin A and Carotenoids). Vitamin A has been used topically to treat damaged skin. Postulated mechanisms include increasing procollagen synthesis, inhibiting MMPs, and reducing the tortuosity of elastic fibers. Retinoic acid also increases epidermal hyaluronic acid (HA), a high-molecular-weight glycosaminoglycan that provides a reservoir for growth factors; its hydrating capacity allows it to maintain high pericellular concentrations of growth factors and essential nutrients. Thus HA is essential for keratinocyte proliferation and differentiation as reviewed (Sayo et al., 2013). Carotenoids, lutein and zeaxanthin, while not retinoic acid precursors, have structural similarities and also upregulated hyaluronic acid synthase 3 (HAS3), one of three isoforms that catalyzed the terminal step in hyaluronan synthesis (Li et al., 2015) in a three-dimensional keratinocyte model of human skin. It is of interest that HA incorporated into a fibrin containing topical dressing reduced bleeding, increased the rate of reepithelialization, neovascularization, and collagen deposition in experimental wounds. It also encouraged the reestablishment of sebaceous glands and hair follicles at some locations (Anilkumar et al., 2011).

2.7.2 Honey

Honey, although not studied as a dietary component for wound healing, has been applied topically for millennia by ancient peoples. The Edwin Smith papyrus (1500 BC) was the first written record that mentioned the treatment of burns with honey and grease (Pećanac et al., 2013). The antimicrobial actions of honey have been reviewed (Vandamme et al., 2013). Honey contains glucose oxidase that, when diluted by wound exudate, produces hydrogen peroxide (H_2O_2) that triggers NFκB, activates neutrophil phagocytosis, and upregulates genes for inflammatory mediators that recruit and amplify the response. Depending on the nectar source, honey also contains diverse antibacterial phytochemicals. Manuka honey (from the *Leptospermum scoparium* tree, New Zealand) is especially interesting because it has antimicrobial actions independent of H_2O_2 activity. Honey is hygroscopic. Its low water activity reduces bacterial growth and its high sugar content draws fluid from the wound by osmosis. Finally, honey has a pH between 3.2 and 4.5, which further inhibits microbial growth.

The utility of honey-impregnated dressings for diabetic foot ulcers and especially for wounds contaminated with antibiotic-resistant bacteria has been reviewed (Molan, 2006; Jenkins et al., 2011; Guthrie et al., 2014; Tsang et al., 2015). However, heterogeneous patient populations and potential for bias and imprecision in the evidence have not, as yet, allowed a conclusion as to the superiority of honey-impregnated dressings to standard therapy (Jull Andrew et al., 2015).

2.8 MICRONUTRIENTS, PATHOGENS, AND RESPONSE TO INJURY

Trauma, burns, and critical illness also alter the distribution of micronutrients. Vitamins, minerals, and bioactive components in the food supply are vital to all living organisms including pathogens. Micronutrient availability to humans is highly dependent on dietary components (Hunt, 2003) and distribution is dramatically altered by obesity (Via, 2012). Thus, many obese trauma patients may have preexisting nutrient deficits. Iatrogenic malnutrition secondary to the actions of inflammatory mediators, an oxidative internal environment, as well as adverse effects of therapeutic drugs, medical procedures, and interventions can exacerbate preexisting nutrient inadequacies. The risk for malnutrition in the hospitalized patient was documented over 40 years ago (Butterworth, 1974) and is still a topic of controversy (Preiser et al., 2015). Thus, despite a dramatically increased body mass, the obese patient must be assessed for preexisting micronutrient malnutrition which can become life-threatening secondary to nutrient deficits induced by trauma.

2.8.1 Iron Metabolism in the Trauma Patient

Obesity is often accompanied by microcytic anemia in the context of high serum ferritin (SF) concentrations and increased iron stores in the adipocytes and other tissues. Iron metabolism and regulation have been discussed elsewhere (refer to Essentials II: Heavy Metals, Retinoids, and Precursors: Iron and Chapter 5: Diabetic Cardiomyopathy and Congestive Heart Failure). It has been proposed that obesity-associated dysregulation of iron homeostasis is due to the chronic inflammation seen with obesity. Inflammation triggers hepcidin upregulation with subsequent inhibition of iron release from enterocytes and macrophages (Bekri et al., 2006). It is possible that other hormones and mediators associated with obesity, in addition to hepcidin, also modify iron distribution. For example, paracrine interactions between adipose tissue macrophages and adipocytes, and during lymphocyte and red cell development, have been implicated in controlling iron distribution (Orr et al., 2014). It should also be noted that obesity-associated iron retention in the cells in the presence of high serum ferritin could be an indication of iron overload with associated proinflammatory tissue damage as discussed elsewhere (refer to Essentials II: Heavy Metals, Retinoids, and Precursors: Iron).

Experimentally, reduced serum iron levels secondary to the rapid upregulation of hepcidin with infection and inflammation has a direct antimicrobial action in vitro, but methods used to achieve these changes exceeded physiological range (Ganz and Nemeth, 2015). These observations counter the long held assumption that the fall in circulating iron observed with infection is a defense mechanism to restrict iron uptake by pathogens. It has now been recognized that the siderophores elaborated by microbes can take up iron even in severely depleted environments. Thus, hypoferremia in the infected patient does not appear to be the mechanism for host defense. Ganz and colleagues propose that because iron tightly bound to transferrin is inaccessible to microbial ferric transporters, the hypoferremic response mediated by hepcidin actually frees up binding sites on transferrin, allowing transferrin to bind iron released from injured cells during trauma and infection. This limits the availability of nontransferrin-bound iron (NTBI) that would otherwise be readily available for utilization by microorganisms (Ganz and Nemeth, 2015).

In this context, the wisdom of iron supplementation to treat microcytic anemia in the injured and/or infected patient is controversial. Treatment of anemia with inorganic iron salts or complexes has been reported to favor pathogen growth. An in vitro model using enteric bacteria demonstrated that in response to increasing concentrations of iron, pathogens grew rapidly and adhered to monolayers of Caco-2 cells, whereas non-pathogenic bacteria did not (Kortman et al., 2012). Additionally, administration of iron supplementation to animals in vivo caused intestinal inflammation and damage. The probable cause of the injury is due to NTBI participation in the Fenton reaction leading to oxidative/nitrosative stress and tissue damage (Koskenkorva-Frank et al., 2013).

Adverse effects of iron supplementation have also been observed in humans. Elevated intestinal permeability and small intestinal injury, measured as an increased lactulose:mannitol (L:M) ratio, were reported in Zambian schoolchildren given iron supplements (Nchito et al., 2006). Two double-blind randomized control trials in 6-month-old Kenyan infants ($n = 115$) consuming maize porridge daily for 4 months demonstrated an iron-dependent change in fecal bacterial species in favor of enteric pathogens as measured by 16S pyrosequencing and targeted real-time PCR (qPCR) (Jaeggi et al., 2014). Additionally, iron supplements increased fecal calprotectin, a marker of intestinal inflammation, and increased the incidence of diarrhea in infants (Sazawal et al., 2006). In malaria-endemic regions, iron supplements have been shown to increase malaria severity and mortality by mechanisms as yet unclear. It has been proposed that NTBI resulting from supplements increases sequestration of malaria-infected cells in the capillaries of the brain and intestine, further increasing intestinal permeability and worsening cerebral malaria (Hurrell, 2011).

2.8.2 Microcytic Anemia in the Trauma Patient: To Treat or Not to Treat

Microcytic anemia that develops in patients with cancer, chronic kidney disease, and congestive heart failure is associated with a relatively poor prognosis. However, before providing iron supplements to an anemic trauma patient, it is important to distinguish between the anemia of chronic disease (ACD) and iron-deficiency anemia (IDA). In ACD, inflammatory mediators induce changes in iron homeostasis, reducing erythropoietin production, erythroid progenitor cell proliferation, and red cell life span, resulting in reduction in red cell mass. Hepcidin levels are increased, degrading ferroportin and preventing iron release from enterocytes and macrophage stores. At the same time, DMT1 expression on the macrophage membrane is induced by hepcidin, increasing iron uptake (Weiss and Goodnough, 2005). ACD patients usually present with a mild to moderate normochromic, normocytic anemia (hemoglobin level, 8–9.5 g per

deciliter) and a low reticulocyte count, indicating underproduction of red cells.

In contrast, IDA is caused by an absolute depletion in total body iron and is associated with a low hepcidin concentration. The major cause of IDA is acute blood loss through injury, as in the trauma patient, or chronic loss, as in the patient with gastrointestinal lesions. IDA is characterized by increased concentrations of the soluble transferrin receptor (TfR), a truncated fragment of the membrane TfR, total iron-binding capacity, and reduced SF. In ACD, levels of TfR are negatively affected by inflammatory mediators while SF is elevated. Several biomarkers have been proposed to differentiate these two types of anemia, especially in the trauma patient at risk for infection (Brugnara, 2003; Engle-Stone et al., 2013).

2.8.3 Zinc Requirements in the Trauma Patient

Zinc is essential to maintain barrier function, the first line of defense against environmental agents as discussed above and elsewhere (refer to Essentials II: Heavy Metals, Retinoids, and Precursors: Zinc). The role of zinc in intestinal barrier function was studied in vitro in Caco-2 monolayers cultured in zinc-deficient medium. A reduction in transepithelial electrical resistance and changes in tight and adherens junctions were observed, with delocalization of ZO-1, occludin, β-catenin, and E-cadherin to the cytoplasm. F-actin and β-tubulin were disorganized and ZO-1, occludin, and β-tubulin concentrations reduced as well. Dephosphorylation of occludin and hyperphosphorylation of β-catenin and ZO-1 were also noted in this study (Finamore et al., 2008).

Zinc also plays a major role both in innate and adaptive immunity as illustrated in Fig. 2.10. Traumatic injury, even without infection, can lead to hyperinflammation and with subsequent sepsis, multiple organ system failure and death. A recent study has suggested a role for zinc in limiting hyperinflammation (Summersgill et al., 2014). Early studies identified zinc as an intrinsic regulator of inflammasome function, including caspase-1 activation and mature IL-1β secretion from LPS-primed macrophages. IL-1β and other mediators stimulate the stereospecific events of the acute-phase immune defense response (Brough et al., 2009). The importance of zinc in limiting hyperinflammation was supported by a recent study demonstrating that sustained Zn^{2+} depletion acted as a stimulus to NLRP3 inflammasome formation and contributed to hyperinflammation, possibly by damaging the lysosome and releasing factors that induce cell death (Aits and Jäättelä, 2013).

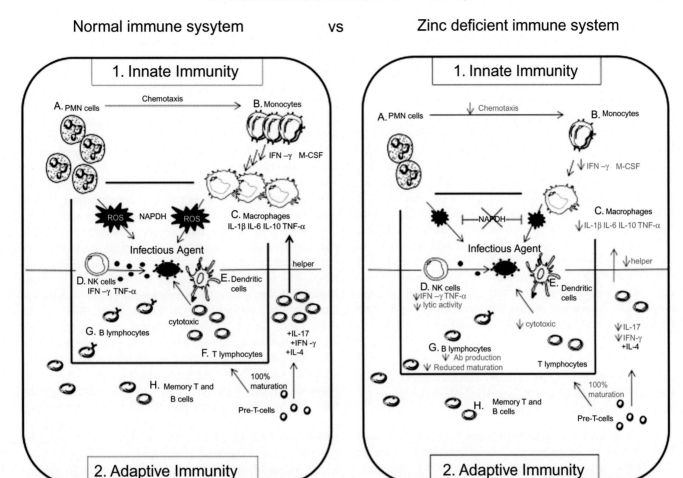

FIGURE 2.10 **Zinc impact on immunity.** Zinc deficiency compromises both the innate and adaptive immune systems. It decreases chemotaxis, reduces the number, maturation and activity of immune cells (PMNs, monocytes, NK cells, dendritic cells, and lymphocytes). It also reduces phagocytic NADPH, limiting oxygen-dependent phagocyte killing of microbial agents. *Reproduced with permission from Bonaventura, P., Benedetti, G., et al., 2015. Zinc and its role in immunity and inflammation. Autoimm. Rev. 14 (4), 277–285 (Bonaventura et al., 2015)—Science Direct.*

2.8.4 Zinc Availability in Obese Patients

Lower zinc concentrations were found in the tissues of genetically obese mice (Kennedy et al., 1986) and in human obese subjects (Ozata et al., 2002; Wiechuła et al., 2012). While an increase in fat mass appears to dysregulate zinc availability, the mechanism that underlies this dysregulation remains unknown. It is possible that obesity-induced zinc dysregulation reduces zinc availability and thus contributes to inflammation and oxidative damage observed in obesity. To address this question, 223 overweight/obese otherwise healthy individuals were placed on low-zinc and normal-zinc diets (Costarelli et al., 2010). Subjects on a low-zinc diet were less able to respond to oxidative stress and experienced a deeper inflammatory status (higher proinflammatory markers), altered lipid profile (high LDL and low HDL cholesterol), and increased insulin production compared to obese

subjects with normal zinc intake. These results suggest that despite the apparent zinc dysregulation associated with obesity, a zinc-adequate diet can partially compensate and reduce the risk for inflammation and other comorbidities of obesity. It should be remembered that urinary zinc was higher and intracellular zinc lower in patients with type II diabetes (SjÖGren et al., 1988).

2.8.5 Zinc Supplementation in the Trauma Patient

A systematic review of the literature was conducted on zinc supplementation to patients who were critically ill following a life-threatening event. Seven articles were identified, but three were excluded because zinc was given as a cocktail with other nutrients such as n-3 FA, arginine, and glutamine. When data from the four remaining randomized control trials (RCTs) were aggregated, zinc supplementation was associated with a trend toward reduction of mortality. While a larger mortality effect was suggested by the data, the studies were too small to confirm this effect. The authors review the relative safety, risks, and benefits of parenteral zinc (100% available) and enteral zinc with its variable bioavailability (refer to Essentials II: Heavy Metals, Retinoids, and Precursors) and conclude that an enteral dose of 20 mg is probably insufficient to restore zinc status in the critically ill patient, and that the optimal dose of parental zinc is not known. Injury and infection are known to stimulate hypozincemia, with zinc sequestration in the liver and other organs. At this time, it is not known whether the low zinc indices in the critically ill patient represent compensatory hypozincemia that improves with recovery, or whether they are an indication of sepsis-associated increased utilization and loss and zinc repletion should be provided (Heyland et al., 2008).

2.8.6 Vitamin C (Ascorbate) Requirements in Critical Care (Refer to Essentials I: Life in an Aerobic World. Ascorbate)

Ascorbic acid deficiency (in scurvy) is characterized by hemorrhage, impaired secretory functions, vasomotor instability, hematologic alterations, impaired wound healing, depressed immune response, and psychological disturbances. Since the critical care patient displays many of these symptoms, the ascorbate requirements of these patients have recently been reviewed (Berger and Oudemans-Van Straaten, 2015). A major factor in the trauma patient is oxidative damage, derived from complications of the initial injury including septic shock and ischemia/reperfusion injury, as well as from emergency treatment such as oxygen therapy (Cornet et al., 2013). The arsenal of defenses against oxidative damage includes the enzyme cascades, superoxide dismutase, catalase and glutathione peroxidase, as well as proteins (albumin), bilirubin, uric acid, and other lipid- and water-soluble metabolites. Vitamin C was identified as the primary plasma antioxidant defense during sepsis; ascorbate also reduces vitamin E radicals and protects biomembranes from lipid peroxidation (Stocker et al., 1987). Ascorbate levels in the serum fall precipitously following trauma; the fall is greater if the patient develops septic shock (Oudemans-van Straaten et al., 2014).

Under normal conditions, vascular smooth muscle tone is regulated by NO derived from eNOS; the complex regulation of eNOS has been recently reviewed (Shu et al., 2015). NO activates guanylyl cyclase in the vascular smooth muscle cell, raising cGMP and cAMP and relaxing the vessel. Superoxides produced during sepsis react with inflammatory (iNOS derived) NO to produce peroxynitrites that inactivate endothelial prostacyclin (PGI_2) synthase and limit PGI_2 production from membrane arachidonic acid. Ascorbate reduces superoxide and peroxy radical production by inhibiting inducible iNOS, and prevents activation of the Jak2/Stat1/IRF1 signaling pathway, blocking superoxide production via NADPH oxidase (Wu et al., 2007). Ascorbate can also reduce physiologically relevant circulating radicals, including urate and glutathione radicals (Lykkesfeldt et al., 2014) restoring their activity.

Vascular changes that accompany sepsis include reduced density of perfused capillaries, increased vascular permeability, and reduced arteriolar responsiveness. While these changes, if localized, facilitate healing, if they occur system-wide, they can induce tissue hypoxia, mitochondrial dysfunction, ATP depletion, and organ failure, despite adequate fluid resuscitation and blood oxygen. Persuasive evidence (Wilson, 2009) suggests that ascorbate protects against vascular changes by multiple mechanisms. Ascorbate reduces endothelial dysfunction by inhibiting oxidation of tetrahydrobiopterin (BH4), thus preventing eNOS uncoupling and reducing superoxide formation (refer to Chapter 6: Atherosclerosis and Arterial Calcification). Ascorbate maintains vascular tight junctions by inhibiting PP2A activation, which dephosphorylates occludin (refer to Essentials IV: Diet, Microbial Diversity, and Gut Integrity) (Han et al., 2010; Zhou et al., 2012). Ascorbate also inhibits microcirculatory flow impairment by inhibiting TNFα-induced ICAM expression, which triggers leukocyte stickiness and sludging. It should be noted that many of these studies were done in animal models, only some of which had the l-gulonolactone oxidase gene inactivated to model ascorbate-deficient human physiology (Fig. 2.11).

A recent study tested whether vitamin C restores endotoxemia-compromised blood flow in humans. LPS was administered to healthy human subjects to produce acute endotoxemia and forearm blood flow (FBF) was measured following provision of placebo and two levels of intravenous vitamin C. As expected, LPS decreased plasma vitamin C concentrations and reduced the acetylcholine (ACh)-dependent increase in FBF by up to 76%. Vitamin C supplementation raised the mean plasma ascorbate concentration to 3.2 or 4.9 mmol/L in the two treatment groups and restored the FBF to baseline (Aschauer et al., 2014).

Host defense requires a competent innate and adaptive immune response. Inflammatory cells concentrate ascorbate at levels as great as 100-fold across the

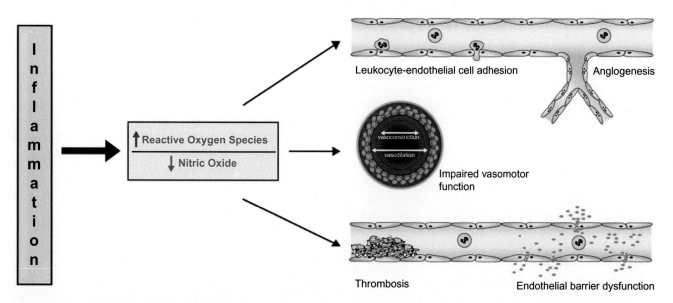

FIGURE 2.11 **Vascular changes in inflammation.** *Reproduced with permission from Kvietys, P.R., Granger, D.N., 2012. Role of reactive oxygen and nitrogen species in the vascular responses to inflammation. Free Radic. Biol. Med. 52 (3), 556–592 (Kvietys and Granger, 2012)—Science Direct.*

membrane (Bergsten et al., 1990), suggesting that the nutrient is essential for function. Upon mediator stimulation, PMNs are recruited on a chemotactic gradient and peak at the infected site within 24–48 hours. PMNs have short half-lives and rapidly undergo apoptosis. PMNs from scorbutic guinea pigs contained 16-fold less ascorbate than normal, were morphologically distorted, killed only 12% of phagocytosed bacteria, and displayed no chemotactic responses in vitro (Goldschmidt, 1991). PMN from ascorbate-deficient mice lacking the functional gene for L-gulonolactone oxidase (Gulo), while morphologically normal, had limited phagocytic capacity and failed to undergo normal apoptosis, dying in a few days in necrosis, the debris providing chemotactic inflammatory signals. Ascorbate-deficient PMNs from these mice also had elevated levels of hypoxia-inducible factor α (HIFα) (refer to Essentials I: Life in an Aerobic World. Ascorbate), previously shown by the authors to inhibit PMN apoptosis under hypoxic conditions (Vissers and Wilkie, 2007). These findings were confirmed ex vivo in macrophages from Gulo-deficient mice subjected to sterile inflammation. Increased expression of inflammatory mediators (IL-1β, TNFα, and MCP-1) and decreased antiinflammatory mediator expression were observed in macrophages from ascorbate-deficient mice. Gulo-deficient mice exposed to an intraperitoneal infusion of a fecal stream solution (fecal stream solution was prepared by rehydrating mouse fecal pellets; supernate was used as to induce peritonitis) suffered multiple organ damage and coagulation abnormalities; ascorbic acid infusion (200 mg/kg) attenuated the damage (Fisher et al., 2013). Of great

interest was the ability of ascorbate to regulate neutrophil extracellular trap (NET) formation. Neutrophils have the ability to extrude antimicrobial material including DNA, trapping and killing pathogens; excessive NET in sepsis can damage organs and tissues (Mohammed et al., 2013). Additionally, ascorbate enhanced facilitation of *efferocytosis*, removal of dead/dying cells, allowing suppression of inflammation with resolution (Ravichandran and Lorenz, 2007; Divangahi, 2015), thus dampening proinflammatory responses.

2.8.7 Recommendations for Ascorbate Supplementation

Several studies performed in septic patients have found that plasma ascorbate levels correlate inversely with the incidence of organ failure and directly with survival. Plasma ascorbate concentration is a function of the distribution volume and cellular uptake, absorption, and renal excretion. It is likely that increased cellular uptake and leukocyte turnover are primary causes of low ascorbate levels in these patients (Oudemans-van Straaten et al., 2014). Extremely low plasma ascorbate levels following trauma were not normalized by 300 or even 1000 mg/day of parenteral ascorbate supplements (Long et al., 2003). The authors suggested that increased ascorbate supplements (~3000 mg/day) should be given at least for the first 3 days following trauma (Berger and Oudemans-Van Straaten, 2015). Adjuvant supplements (2–6 g/day) given to burn patients during the first 24 hours reduced resuscitation fluid volume requirements and wound edema, as well as severity of respiratory dysfunction (Tanaka et al., 2000). Data from a randomized, double-blind, placebo controlled, phase I trial support these recommendations. Patients with severe sepsis received 50 or 200 mg/kg ascorbic acid infusions every 6 hours for 4 days and were monitored for arterial hypotension, tachycardia, hypernatremia, and nausea or vomiting. Sequential Organ Failure Assessment (SOFA) was made and plasma levels of ascorbic acid, C-reactive protein, procalcitonin, and thrombomodulin were measured. Ascorbic acid infusions rapidly raised plasma ascorbate levels from the low baseline (17.9 μM; normal range = 50−70 μM). No adverse safety events were observed and SOFA scores promptly declined. Proinflammatory biomarkers were reduced and thrombomodulin did not rise, as it did in control subjects (Fowler et al., 2014).

2.9 CURRENT RECOMMENDATIONS FOR PATIENT FEEDING

Except on few occasions, the patient appears to die from the body's response to infection rather than from it. (Sir William Osler, 1904)

Strategies for feeding the trauma patient have undergone multiple iterations over the years as has been recently reviewed (Rosenthal et al., 2015). The authors summarized current practice guidelines from the Society of Critical Care Medicine and American Society for Parenteral and Enteral Nutrition (Table 2.1). Additional, and overlapping recommendations were published by the North American Surgical Nutrition Summit (McClave et al., 2013) and Canadian Guidelines 2015. Optimal therapy for obese critical

TABLE 2.1 Current Evidence-based Goals/recommendations for Enteral or Parenteral Nutrition McClave et al. (2009)

Enteral preferred when gut is functional	Start enteral nutrition early (24−48 h) after adequate resuscitation
Immune-enhancing diets in patients with increased risk of nosocomial infection	Hold small bowel feeds if patient is on an increasing dose of catecholamine
Gastric or postpyloric feeds are equivalent for aspiration risk, but prefer postpyloric feeds.	Enteral glutamine for burns and trauma patients
Provide antioxidants and trace minerals	Soluble fiber for management of "tube-feed" diarrhea
Start parenteral nutrition after 7 day if not meeting at least 60% of goal with enteral nutrition	Use nurse-driven enteral protocol with emphasis on management of intolerance

care patients has not been conclusively established; challenges and considerations for patient feeding in adult (Martinez et al., 2015) and pediatric (Bechard et al., 2013) obese patients have been reviewed. In addition to the nutrients described in detail above, several other nutrients have been implicated in patient management (Pierre et al., 2013). A detailed composite of nutritional management of the patient is found elsewhere (refer to Essentials V: Nutritional Support in Critically Ill Patients).

2.9.1 Glutamine

Glutamine is a nonessential amino acid produced primarily in the muscle. Branched chain amino acid breakdown releases ammonia (NH_3). Two NH_3 are transaminated sequentially to a citric acid cycle intermediate, α ketoglutarate, forming glutamine. Glutamine is readily taken up by cells and deaminidated to glutamate and NH_3. Catabolic insults increase glutamine demand to provide intermediates for the citric acid cycle and as substrate for gluconeogenesis. Intermediates branch off at several points to provide purines, pyrimidines, and other substrates (proline, aspartic acid, and other nonessential amino acids) for cell proliferation and wound repair. Glutamine breakdown also produces NADPH, thereby maintaining redox status of the cell (Soeters, 2015). Glutamine transports NH_3 safely to the kidney to maintain acid−base status and between organs for biosynthesis of nucleotides, amino sugars, arginine, glutathione, and glucosamine as recently reviewed (Rosenthal et al., 2015). Glutamine supplementation has also been shown as effective in reducing peripheral insulin resistance in stressed humans and in animal models.

Glutamine biosynthesis is not impaired during critical illness, but its levels fall in plasma and tissues because of increased demand. For this reason, glutamine supplementation has been recommended for critically ill patients (Al Balushi et al., 2013). Several difficulties with glutamine supplementation have been demonstrated. Glutamine may not be efficiently metabolized in patients with liver failure, leading to potentially toxic levels in the plasma. Glutamine administered parenterally does not improve bowel integrity, especially in long-term inflammatory states. If the glutamine is enterally administered, it is metabolized primarily by splanchnic organs and its bioavailability is poor. A recent randomized and blinded two-by-two factorial trial of 1223 critically ill adults with MOF in 40 ICUs in Canada, the United States, and Europe were studied. The patients were receiving mechanical ventilation and provided with glutamine, antioxidants, both, or placebo. Increased mortality at

6 months was reported for the glutamine groups (Heyland et al., 2013). Several mechanisms have been implicated in these findings. The most likely cause, seen only in patients with multiorgan system failure, could relate to the high dose of glutamine relative to other amino acids and total protein. Despite the large volume of human trials attesting to its benefit, glutamine supplementation remains controversial, and should probably not be given to patients with renal failure or MOF until these safety questions are clarified (Rosenthal et al., 2015).

2.9.2 Vitamin D

It has been known at least since Dr. Niels Ryberg Finsen won a Nobel Prize in 1903 for discovering that sun rays can kill *Mycobacterium tuberculosis* (http://www.nobelprize.org/nobel_prizes/medicine/laureates/1903/finsen-bio.html) that vitamin D plays a role in immune defense. A molecular explanation for this phenomenon has recently been provided with the demonstration that toll like receptor (TLR) stimulation of cells participating in innate defense catalyzes conversion of $25(OH)D_3$ (25D) to active $1,25(OH)_2D_3$ (1,25D) as well as upregulation of the vitamin D receptor (VDR) and downstream targets of VDR such as the antimicrobial peptide, cathelicidin (Liu et al., 2006). Additionally, the VDR and its ligand are required for dermal responses to injury and the absence of VDR leads to defects in development of granulation tissue. The molecular basis appears to involve VDR interaction with the TGF-β signaling pathway (Luderer et al., 2013) as well as with β-catenin and calcium signaling (Oda et al., 2015).

Clinical vitamin D deficiency has been defined by the Institute of Medicine as 25D level of less than $20 \, ng/mL$ ($50 \, nM$) (Norman et al., 2007; Food and Nutrition Board, 2010). As described elsewhere (refer to Essentials III: Nutrients for Bone Structure and Calcification: Vitamin D), 25D levels are low in about 50% of the North American elderly population and in about two-thirds of the populations in the rest of the world. Younger individuals are also below the recommended level, especially obese patients and those with darker skin. Several surveys of critical care patients have revealed that low vitamin D status prior to admission (Braun et al., 2011) was associated with increased all-cause mortality and sepsis (Upala et al., 2015). A recent meta-analysis concluded hypovitaminosis D is associated with adverse outcomes after diverse surgical procedures (Iglar and Hogan, 2015). These observations suggest that 25D status should be assessed in all trauma patients, and especially in obese trauma patients.

2.9.3 Carnitine

Carnitine (β-hydroxy-γ-*N*-trimethylaminobutyric acid) (refer to Chapter 5: Cardiomyopathy and Congestive Heart Failure) is widely distributed in the food supply and is a conditionally essential nutrient required for the efficient transport of long-chain FAs and branched chain amino acids into the mitochondria for oxidation. Compared with control patients, septic patients had increased urinary excretion of free carnitine and lower plasma levels of short-chain acylcarnitines associated with the level of hypermetabolism (Nanni et al., 1985); carnitine excretion in sepsis has also been related to cell injury and tissue wasting (Iapichino et al., 1988). Reviews of the literature have revealed that carnitine action in septic shock downregulated spontaneous and LPS-triggered TNFα production, reduced severity of the illness, and accelerated recovery (Famularo et al., 1995; Flanagan et al., 2010).

Despite its potential benefit in sepsis, carnitine use has not been evaluated for safety or efficacy. A recent RCT studied 31 patients in vasopressor septic shock with a SOFA score of >5 (Puskarich et al., 2014). L-carnitine solution was administered intravenously as a 4-g bolus followed by an 8-g infusion over a 12-hour period. No significant difference in serious adverse effects was observed in L-carnitine-treated patients versus patients treated with placebo. Although no statistical difference was observed in the SOFA score at 24 hours, patients in the L-carnitine arm had a significant reduction in mortality at 28 days and a nonsignificant reduction at 1 year. Although this study was small, confined to a single institution, and apparently did not assess the prior nutritional (carnitine) status or the impact of nutritional intervention, it did demonstrate the safety and possible efficacy of carnitine supplementation in this high-risk population. Larger intervention trials are needed.

2.9.4 Trace Elements including Selenium (Se)

Several trace elements and other nutrients have been identified as important in inflammation and repair. These include copper, chromium, manganese, vitamin A, vitamin D, calcium, phosphorus, and vitamin K. The impact of trace elements in critical care has been reviewed (Rech et al., 2014).

2.9.5 Selenium Regulation During Stress

The concentration of Se secreted by the liver as selenoprotein P (Sepp1) is ~$65 \, \mu g$ Se/L plasma, while that of glutathione peroxidase (Gpx3) is ~$17 \, \mu g$ Se/L plasma. Together these components comprise the regulated selenium pool (refer to Essentials I: Life in an Aerobic World: Single Unit Nutrients: Selenium). Plasma Se concentrations were low in ICU patients on admission and fell further with infection, organ failure, and longer duration of illness. Low-plasma Se was associated with more tissue damage and increased mortality (Sakr et al., 2007). Several explanations have been advanced to explain the association of sepsis with low circulating selenoproteins, including reduced intake, hemodilution due to intravenous fluids, escape from interstitial compartments via capillary permeability, losses of body fluids in surgical drains and exudation, phlebotomy, malnutrition, and renal replacement therapy (Sakr et al., 2014). Other possible factors include drugs (statins, corticosteroids, diuretics), prior smoking, alcoholism, and other environmental factors (Hardy et al., 2012). On the other hand, measured Se losses did not appear sufficient to account for the fall in selenoproteins in burns patients (Berger et al., 1992). To test whether Se fall with sepsis represents a negative acute phase response (APR), mice were injected with LPS and selenium indices were measured (Renko et al., 2009). The authors determined that the rapid fall in circulating selenoproteins could be explained by specific impairment of hepatic translation of Sepp1 mRNA during LPS-induced APR. Several inhibitors were identified that disrupted Se metabolism, transport, and supply to peripheral tissues. Clinical observations support this plausibility (Hawker et al., 1990).

If the fall in Se with APR is metabolically determined, should Se be supplemented? Experimental data suggest that Se plays major roles in immune defense as reviewed by Aribi et al. (2015). Se modifies interactions between macrophages and lymphocytes, regulates NO availability, suppresses proinflammatory mediators, and modulates TLR signaling, thus inhibiting the NFκB pathway and the inflammatory genes it controls. Ex vivo studies showed that sodium selenite acted on macrophage activation, phagocytosis, and bacterial killing in a dose-dependent manner, suggesting that inadequate Se could impair macrophage killing activities, by mechanisms not yet clearly defined (Aribi et al., 2015). Se also plays a key role in defense against oxidative damage during sepsis due to its requirement as a catalytic center for GPxs. Sepp1, produced by the liver and transported to the tissues, provides Se for GPx activation. Cells produce selenoproteins based on their essentiality (selenium hierarchy). Thus, if Sepp1 availability is low, GPx1, GPx3, and other "nonessential" selenoprotein concentrations rapidly fall, preserving Se for "essential" enzymes (McCann and Ames, 2011).

The literature is replete with clinical trials that have examined Se supplementation on a variety of outcomes, including mortality. Results have been inconclusive as

recently reviewed (Manzanares et al., 2015). Some trials suggested that Se supplementation might contribute to mortality, while others reported improved response. The current Cochrane review concluded that, despite multiple clinical trials, current evidence was not sufficient to recommend supplementation of critically ill patients with selenium. Methodological inadequacies, including sample size, design, and outcomes were cited (Allingstrup and Afshari, 2015).

This lack of conclusive data could result because low plasma selenium does not reflect selenium deficiency, but rather a fall or redistribution of selenium with inflammation as well described for iron and zinc. Furthermore, selenium has been shown to exert a spectrum of toxic effects at unexpectedly low concentrations. Thus, upregulation of antioxidant enzyme levels with Se supplementation may actually reflect self-induced oxidative stress (Jablonska and Vinceti, 2015). Clearly, well-designed clinical trials are needed to resolve this question.

References

Aits, S., Jäättelä, M., 2013. Lysosomal cell death at a glance. J. Cell Sci. 126 (9), 1905–1912.

Akira, S., Uematsu, S., et al., 2006. Pathogen recognition and innate immunity. Cell. 124 (4), 783–801.

Al Balushi, R.M., Cohen, J., et al., 2013. The clinical role of glutamine supplementation in patients with multiple trauma: a narrative review. Anaesth. Intensive Care. 41 (1), 24–34.

Allingstrup, M., Afshari, A., 2015. Selenium supplementation for critically ill adults. Cochrane Database Syst. Rev. Available from: http://dx.doi.org/10.1002/14651858.CD003703.pub3.

Anilkumar, T.V., Muhamed, J., et al., 2011. Advantages of hyaluronic acid as a component of fibrin sheet for care of acute wound. Biologicals. 39 (2), 81–88.

Aribi, M., Meziane, W., et al., 2015. Macrophage bactericidal activities against staphylococcus aureus are enhanced in vivo by selenium supplementation in a dose-dependent manner. PLoS ONE. 10 (9), e0135515.

Ariel, A., Timor, O., 2013. Hanging in the balance: endogenous anti-inflammatory mechanisms in tissue repair and fibrosis. J. Pathol. 229 (2), 250–263.

Artis, D., Spits, H., 2015. The biology of innate lymphoid cells. Nature. 517 (7534), 293–301.

Aschauer, S., Gouya, G., et al., 2014. Effect of systemic high dose vitamin C therapy on forearm blood flow reactivity during endotoxemia in healthy human subjects. Vascul. Pharmacol. 61 (1), 25–29.

Bagby, G.J., Corll, C.B., et al., 1987. Triacylglycerol kinetics in endotoxic rats with suppressed lipoprotein lipase activity. Am. J. Physiol. 253 (1 Pt 1), E59–E64.

Basnet, P., Skalko-Basnet, N., 2011. Curcumin: an anti-inflammatory molecule from a curry spice on the path to cancer treatment. Molecules. 16 (6), 4567–4598.

Bechard, L.J., Rothpletz-Puglia, P., et al., 2013. Influence of obesity on clinical outcomes in hospitalized children: a systematic review. JAMA Pediatr. 167 (5), 476–482.

Beisel, W.R., 1975. Metabolic Response to Infection. Ann. Rev. Med. 26, 9–20.

Bekri, S., Gual, P., et al., 2006. Increased adipose tissue expression of hepcidin in severe obesity is independent from diabetes and NASH. Gastroenterology. 131 (3), 788–796.

Belkaid, Y., Grainger, J., 2013. Mucus coat, a dress code for tolerance. Science. 342 (6157), 432–433.

Belkaid, Y., Segre, J.A., 2014. Dialogue between skin microbiota and immunity. Science. 346 (6212), 954–959.

Berger, M.M., Cavadini, C., et al., 1992. Selenium losses in 10 burned patients. Clin. Nutr. 11 (2), 75–82.

Berger, M.M., Oudemans-Van Straaten, H.M., 2015. Vitamin C supplementation in the critically ill patient. Curr. Opin. Clin. Nutr. Metab. Care. 18 (2), 193–201.

Bergsten, P., Amitai, G., et al., 1990. Millimolar concentrations of ascorbic acid in purified human mononuclear leukocytes. Depletion and reaccumulation. J. Biol. Chem. 265 (5), 2584–2587.

Blass, S.C., Goost, H., et al., 2013. Extracellular micronutrient levels and pro-/antioxidant status in trauma patients with wound healing disorders: results of a cross-sectional study. Nutr. J. 12, 157.

Bonaventura, P., Benedetti, G., et al., 2015. Zinc and its role in immunity and inflammation. Autoimm. Rev. 14 (4), 277–285.

Braun, A., Chang, D., et al., 2011. Association of low serum 25-hydroxyvitamin D levels and mortality in the critically ill. Crit. Care Med. 39 (4), 671–677.

Brough, D., Pelegrin, P., et al., 2009. Pannexin-1-dependent caspase-1 activation and secretion of IL-1β is regulated by zinc. Eur. J. Immunol. 39 (2), 352–358.

Brugnara, C., 2003. Iron deficiency and erythropoiesis: new diagnostic approaches. Clin. Chem. 49 (10), 1573–1578.

Buendía, R.G., Zambrano, M.E., et al., 2015. ¿Existe sarcopenia en pacientes menores de 30 años por criterio de bioimpedanciometría? Acta Medica Colombiana. 40, 132–137.

Butterworth Jr., C.E., 1974. The skeleton in the hospital closet. Nutr. Today. 9 (2), 4–8.

Calder, P., Dimitriadis, G., et al., 2007. Glucose metabolism in lymphoid and inflammatory cells and tissues. Curr. Opin. Clin. Nutr. Metab. Care. 10 (4), 531–540.

Choban, P.S., Dickerson, R.N., 2005. Morbid obesity and nutrition support: is bigger different? Nutr. Clin. Pract. 20 (4), 480–487.

Clement, S., Braithwaite, S.S., et al., 2004. Management of diabetes and hyperglycemia in hospitals. Diabetes Care. 27 (2), 553–591.

Cornet, A.D., Kooter, A.J., et al., 2013. The potential harm of oxygen therapy in medical emergencies. Crit. Care. 17 (2), 313.

Costarelli, L., Muti, E., et al., 2010. Distinctive modulation of inflammatory and metabolic parameters in relation to zinc nutritional status in adult overweight/obese subjects. J. Nutr. Biochem. 21 (5), 432–437.

Divangahi, M., 2015. Efferocytosis: burying cell corpses to regulate tolerance and immunity. Oncotarget. 6 (17), 14721–14722.

Drake, D.R., Brogden, K.A., et al., 2008. Thematic review series: skin lipids. Antimicrobial lipids at the skin surface. J. Lipid Res. 49 (1), 4–11.

Engle-Stone, R., Nankap, M., et al., 2013. Plasma ferritin and soluble transferrin receptor concentrations and body iron stores identify similar risk factors for iron deficiency but result in different estimates of the national prevalence of iron deficiency and iron-deficiency anemia among women and children in Cameroon. J. Nutr. 143 (3), 369–377.

Famularo, G., De Simone, C., et al., 1995. Carnitine and septic shock: a review. J. Endotoxin Res. 2 (2), 141–147.

Feingold, K.R., 2007. Thematic review series: skin lipids. The role of epidermal lipids in cutaneous permeability barrier homeostasis. J. Lipid Res. 48 (12), 2531–2546.

Finamore, A., Massimi, M., et al., 2008. Zinc deficiency induces membrane barrier damage and increases neutrophil transmigration in caco-2 cells. J. Nutr. 138 (9), 1664–1670.

Fisher, B.J., Kraskauskas, D., et al., 2013. Attenuation of sepsis-induced organ injury in mice by vitamin C. J. Parenteral Enteral Nutr. 38 (7), 825–839.

Flanagan, J.L., Simmons, P.A., et al., 2010. Role of carnitine in disease. Nutr. Metab. 7 (1), 1–14.

Food and Nutrition Board, Dietary Reference Intakes for Calcium and Vitamin D, 2010, National Academy Press; Washington, DC, I. o. M. Food and Nutrition Board.

Fong, Y., Moldawer, L.L., et al., 1989. Cachectin/TNF or IL-1 alpha induces cachexia with redistribution of body proteins. Am. J. Physiol. 256 (3 Pt 2), R659–R665.

Fowler, A., Syed, A., et al., 2014. Phase I safety trial of intravenous ascorbic acid in patients with severe sepsis. J. Trans. Med. 12 (1), 32.

Frémont, L., 2000. Biological effects of resveratrol. Life Sci. 66 (8), 663–673.

Ganz, T., Nemeth, E., 2015. Iron homeostasis in host defence and inflammation. Nat. Rev. Immunol. 15 (8), 500–510.

Glass, C.K., Olefsky, J.M., 2012. Inflammation and lipid signaling in the etiology of insulin resistance. Cell Metab. 15 (5), 635–645.

Goldschmidt, M.C., 1991. Reduced bactericidal activity in neutrophils from scorbutic animals and the effect of ascorbic acid on these target bacteria in vivo and in vitro. Am. J. Clin. Nutr. 54 (6), 1214S–1220S.

Gong, Z., Zhou, J., et al., 2015. Curcumin suppresses NLRP3 inflammasome activation and protects against LPS-induced septic shock. Mol. Nutr. Food Res. 59 (11), 2132–2142.

Groschwitz, K.R., Hogan, S.P., 2009. Intestinal barrier function: molecular regulation and disease pathogenesis. J. Allergy Clin. Immunol. 124 (1), 3–20.

Guthrie, H.C., Martin, K.R., et al., 2014. A pre-clinical evaluation of silver, iodine and Manuka honey based dressings in a model of traumatic extremity wounds contaminated with Staphylococcus aureus. Injury. 45 (8), 1171–1178.

Han, M., Pendem, S., et al., 2010. Ascorbate protects endothelial barrier function during septic insult: role of protein phosphatase type 2A. Free Radic. Biol. Med. 48 (1), 128.

Hardy, G., Hardy, I., et al., 2012. Selenium supplementation in the critically ill. Nutr. Clin. Pract. 27 (1), 21–33.

Hawker, F.H., Stewart, P.M., et al., 1990. Effects of acute illness on selenium homeostasis. Crit. Care Med. 18 (4), 442–446.

Haworth, O., Levy, B.D., 2007. Endogenous lipid mediators in the resolution of airway inflammation. Eur. Respir. J. 30 (5), 980–992.

Heyland, D., Muscedere, J., et al., 2013. A randomized trial of glutamine and antioxidants in critically ill patients. N. Engl. J. Med. 368 (16), 1489–1497.

Heyland, D.K., Jones, N., et al., 2008. Zinc supplementation in critically ill patients: a key pharmaconutrient? J. Parenteral Enteral Nutr. 32 (5), 509–519.

Hilton, T.N., Tuttle, L.J., et al., 2008. Excessive adipose tissue infiltration in skeletal muscle in individuals with obesity, diabetes mellitus, and peripheral neuropathy: association with performance and function. Phys. Ther. 88 (11), 1336–1344.

Huang, T.-T., Lai, H.-C., et al., 2014. cis-Resveratrol produces anti-inflammatory effects by inhibiting canonical and non-canonical inflammasomes in macrophages. Innate Immun. 20 (7), 735–750.

Hunt, J.R., 2003. Bioavailability of iron, zinc, and other trace minerals from vegetarian diets. Am. J. Clin. Nutr. 78 (3), 633S–639S.

Hurrell, R.F., 2011. Safety and efficacy of iron supplements in malaria-endemic areas. Ann. Nutr. Metab. 59 (1), 64–66.

Iapichino, G., Radrizzani, D., et al., 1988. Carnitine excretion: a catabolic index of injury. J. Parenteral Enteral Nutr. 12 (1), 35–36.

Iglar, P., Hogan, K., 2015. Vitamin D status and surgical outcomes: a systematic review. Patient Saf. Surg. 9 (1), 14.

Jablonska, E., Vinceti, M., 2015. Selenium and human health: witnessing a Copernican revolution? J. Environ. Sci. Health, Part C. 33 (3), 328–368.

Jacinto, A., Martinez-Arias, A., et al., 2001. Mechanisms of epithelial fusion and repair. Nat. Cell Biol. 3 (5), E117–E123.

Jaeggi, T., Kortman, G.A.M., et al., 2014. Iron fortification adversely affects the gut microbiome, increases pathogen abundance and induces intestinal inflammation in Kenyan infants. Gut. 64 (5), 731–742.

Jansen, J.O., Thomas, R., et al., 2009. Damage control resuscitation for patients with major trauma. BMJ. 338, b1778.

Jenkins, R., Burton, N., et al., 2011. Manuka honey inhibits cell division in methicillin-resistant *Staphylococcus aureus*. J. Antimicrob. Chemother. 66 (11), 2536–2542.

Jull Andrew, B., Cullum, N., et al., 2015. Honey as a topical treatment for wounds. Cochrane Database of Systematic Reviews. Available from: http://dx.doi.org/10.1002/14651858.CD005083.pub4.

Kennedy, M.L., Failla, M.L., et al., 1986. Influence of genetic obesity on tissue concentrations of zinc, copper, manganese and iron in mice. J. Nutr. 116 (8), 1432–1441.

Kim, Y., Ho, S., 2010. Intestinal goblet cells and mucins in health and disease: recent insights and progress. Curr. Gastroenterol. Rep. 12 (5), 319–330.

Kortman, G.A.M., Boleij, A., et al., 2012. Iron availability increases the pathogenic potential of *Salmonella Typhimurium* and other enteric pathogens at the intestinal epithelial interface. PLoS ONE. 7 (1), e29968.

Koskenkorva-Frank, T.S., Weiss, G., et al., 2013. The complex interplay of iron metabolism, reactive oxygen species, and reactive nitrogen species: insights into the potential of various iron therapies to induce oxidative and nitrosative stress. Free Radic. Biol. Med. 65, 1174–1194.

Kvietys, P.R., Granger, D.N., 2012. Role of reactive oxygen and nitrogen species in the vascular responses to inflammation. Free Radic. Biol. Med. 52 (3), 556–592.

Langer, G., Fink, A., 2014. Nutritional interventions for preventing and treating pressure ulcers. Cochrane Database Syst. Rev. Available from: http://dx.doi.org/10.1002/14651858.CD003216.pub2.

Lansdown, A.B.G., Mirastschijski, U., et al., 2007. Zinc in wound healing: theoretical, experimental, and clinical aspects. Wound Repair Regen. 15 (1), 2–16.

Li, R., Turner, S.D., et al., 2015. Xanthophylls lutein and zeaxanthin modify gene expression and induce synthesis of hyaluronan in keratinocyte model of human skin. Biochem. Biophys. Rep. 4, 52–58.

Liu, P.T., Stenger, S., et al., 2006. Toll-like receptor triggering of a vitamin D-mediated human antimicrobial response. Science. 311 (5768), 1770–1773.

Long, C.L., Maull, K.I., et al., 2003. Ascorbic acid dynamics in the seriously ill and injured. J. Surg. Res. 109 (2), 144–148.

Luderer, H.F., Nazarian, R.M., et al., 2013. Ligand-dependent actions of the vitamin D receptor are required for activation of TGF-β signaling during the inflammatory response to cutaneous injury. Endocrinology. 154 (1), 16–24.

Lykkesfeldt, J., Michels, A.J., et al., 2014. Vitamin C. Adv. Nutr. Int. Rev. J. 5 (1), 16–18.

Maiden, N., 2009. Ballistics reviews: mechanisms of bullet wound trauma. Forensic Sci. Med. Pathol. 5 (3), 204–209.

Manzanares, W., Langlois, P.L., et al., 2015. Pharmaconutrition with selenium in critically ill patients: what do we know? Nutr. Clin. Pract. 30 (1), 34–43.

Mariathasan, S., Monack, D.M., 2007. Inflammasome adaptors and sensors: intracellular regulators of infection and inflammation. Nat. Rev. Immunol. 7 (1), 31–40.

Martin, R.S., Kilgo, P.D., et al., 2005. Injury-associated hypothermia; an analysis of the 2004 National Trauma Data Bank. Shock. 24 (2), 114–118.

Martinez, E.E., Ariagno, K., et al., 2015. Challenges to nutrition therapy in the pediatric critically ill obese patient. Nutr. Clin. Pract. 30 (3), 432–439.

McCall, C.E., El Gazzar, M., et al., 2011. Epigenetics, bioenergetics, and microRNA coordinate gene-specific reprogramming during acute systemic inflammation. J. Leukocyte Biol. 90 (3), 439–446.

McCann, J.C., Ames, B.N., 2011. Adaptive dysfunction of selenoproteins from the perspective of the triage theory: why modest selenium deficiency may increase risk of diseases of aging. FASEB J. 25 (6), 1793–1814.

McClave, S.A., Kozar, R., et al., 2013. Summary points and consensus recommendations from the North American surgical nutrition summit. J. Parenteral Enteral Nutr. 37 (Suppl. 5), 99S–105S.

McClave, S.A., Martindale, R.G., et al., 2009. Guidelines for the provision and assessment of nutrition support therapy in the adult critically ill patient: Society of Critical Care Medicine (SCCM) and American Society for Parenteral and Enteral Nutrition (A.S.P.E.N.). J. Parenteral Enteral Nutr. 33 (3), 277–316.

Mohammed, B.M., Fisher, B.J., et al., 2013. Vitamin C: a novel regulator of neutrophil extracellular trap formation. Nutrients. 5 (8), 3131–3150.

Molan, P.C., 2006. The evidence supporting the use of honey as a wound dressing. Int. J. Low. Extrem. Wounds. 5 (1), 40–54.

Nanni, G., Pittiruti, M., et al., 1985. Plasma carnitine levels and urinary carnitine excretion during sepsis. J. Parenteral Enteral Nutr. 9 (4), 483–490.

Narici, M.V., Maffulli, N., 2010. Sarcopenia: characteristics, mechanisms and functional significance. Br. Med. Bull. 95 (1), 139–159.

Nchito, M., Friis, H., et al., 2006. Iron supplementation increases small intestine permeability in primary schoolchildren in Lusaka, Zambia. Trans. R. Soc. Trop. Med. Hygiene. 100 (8), 791–794.

Neville, A.L., Brown, C.R., et al., 2004. Obesity is an independent risk factor of mortality in severely injured blunt trauma patients. Arch. Surg. 139 (9), 983–987.

Norman, A.W., Bouillon, R., et al., 2007. 13(th) workshop consensus for vitamin D nutritional guidelines. J. Steroid Biochem. Mol. Biol. 103 (3–5), 204–205.

Oda, Y., Tu, C.L., et al., 2015. Vitamin D and calcium regulation of epidermal wound healing. J. Steroid Biochem. Mol. Biol. Available from: http://dx.doi.org/10.1016/j.jsbmb.2015.08.011.

Oh, D.Y., Talukdar, S., et al., 2010. GPR120 is an omega-3 fatty acid receptor mediating potent anti-inflammatory and insulin-sensitizing effects. Cell. 142 (5), 687–698.

Orr, J.S., Kennedy, A., et al., 2014. Obesity alters adipose tissue macrophage iron content and tissue iron distribution. Diabetes. 63 (2), 421–432.

Oudemans-van Straaten, H.M., Spoelstra-de Man, A.M.E., et al., 2014. Vitamin C revisited. Crit. Care. 18 (4), 460.

Ozata, M., Mergen, M., et al., 2002. Increased oxidative stress and hypozincemia in male obesity. Clin. Biochem. 35 (8), 627–631.

Pećanac, M., Janjić, Z., et al., 2013. Burns treatment in ancient times. Med. Pregl. 66 (5–6), 263–267.

Pierre, J.F., Heneghan, A.F., et al., 2013. Pharmaconutrition review: physiological mechanisms. J. Parenteral Enteral Nutr. 37 (Suppl. 5), 51S–65S.

Preiser, J.-C., van Zanten, A., et al., 2015. Metabolic and nutritional support of critically ill patients: consensus and controversies. Crit. Care. 19 (1), 35.

Puskarich, M.A., Kline, J.A., et al., 2014. Preliminary safety and efficacy of L-carnitine infusion for the treatment of vasopressor-dependent septic shock: a randomized control trial. J. Parenteral Enteral Nutr. 38 (6), 736–743.

Puthucheary, Z.A., Rawal, J., et al., 2013. Acute skeletal muscle wasting in critical illness. JAMA. 310 (15), 1591—1600.

Qaseem, A., Chou, R., et al., 2014. Inpatient glycemic control: best practice advice from the clinical guidelines committee of the American College of Physicians. Am. J. Med. Qual. 29 (2), 95—98.

Raiten, D.J., Ashour, F.A.S., et al., 2015. Inflammation and nutritional science for programs/policies and interpretation of research evidence (INSPIRE). J. Nutr. 145 (5), 1039S—1108S.

Ravichandran, K.S., Lorenz, U., 2007. Engulfment of apoptotic cells: signals for a good meal. Nat. Rev. Immunol. 7 (12), 964—974.

Rech, M., To, L., et al., 2014. Heavy metal in the intensive care unit: a review of current literature on trace element supplementation in critically ill patients. Nutr. Clin. Pract. 29 (1), 78—89.

Renko, K., Hofmann, P.J., et al., 2009. Down-regulation of the hepatic selenoprotein biosynthesis machinery impairs selenium metabolism during the acute phase response in mice. FASEB J. 23 (6), 1758—1765.

Ribeiro, S.M.L., Kehayias, J.J., 2014. Sarcopenia and the analysis of body composition. Adv. Nutr. Int. Rev. J. 5 (3), 260—267.

Rock, K.L., Latz, E., et al., 2010. The sterile inflammatory response. Ann. Rev. Immunol. 28 (1), 321—342.

Rosenberg, I.H., 1989. Summary comments. Am. J. Clin. Nutr. 50 (5), 1231—1233.

Rosenberg, I.H., 1997. Sarcopenia: origins and clinical relevance. J. Nutr. 127 (5), 990S—991S.

Rosenthal, M.D., Vanzant, E.L., et al., 2015. Evolving paradigms in the nutritional support of critically ill surgical patients. Curr. Probl. Surg. 52 (4), 147—182.

Ryu, S., Song, P., et al., 2014. Colonization and infection of the skin by S. aureus: immune system evasion and the response to cationic antimicrobial peptides. Int. J. Mol. Sci. 15 (5), 8753.

Sakr, Y., Maia, V., et al., 2014. Adjuvant selenium supplementation in the form of sodium selenite in postoperative critically ill patients with severe sepsis. Crit. Care. 18 (2), R68.

Sakr, Y., Reinhart, K., et al., 2007. Time course and relationship between plasma selenium concentrations, systemic inflammatory response, sepsis, and multiorgan failure. Br. J. Anaesth. 98 (6), 775—784.

Sayo, T., Sugiyama, Y., et al., 2013. Lutein, a nonprovitamin A, activates the retinoic acid receptor to induce HAS3-dependent hyaluronan synthesis in keratinocytes. Biosci. Biotechnol. Biochem. 77 (6), 1282—1286.

Sazawal, S., Black, R.E., et al., 2006. Effects of routine prophylactic supplementation with iron and folic acid on admission to hospital and mortality in preschool children in a high malaria transmission setting: community-based, randomised, placebo-controlled trial. Lancet. 367 (9505), 133—143.

Shu, X., Keller, T.C.S.I.V., et al., 2015. Endothelial nitric oxide synthase in the microcirculation. Cell. Mol. Life Sci. 72 (23), 4561—4575.

SjÖGren, A., FlorÉN, C.-H., et al., 1988. Magnesium, potassium and zinc deficiency in subjects with type ii diabetes mellitus. Acta Med. Scand. 224 (5), 461—466.

Soeters, P.B., 2015. Glutamine structure and function: a starter pack. In: Rajendram, R., Preedy, V.R., Patel, V.B. (Eds.), Glutamine in Clinical Nutrition. Springer, New York, pp. 3—20. (Chapter 1).

Sonnenberg, G.F., Artis, D., 2015. Innate lymphoid cells in the initiation, regulation and resolution of inflammation. Nat. Med. 21 (7), 698—708.

Stocker, R., Glazer, A.N., et al., 1987. Antioxidant activity of albumin-bound bilirubin. Proc. Nat. Acad. Sci. 84 (16), 5918—5922.

Summersgill, H., England, H., et al., 2014. Zinc depletion regulates the processing and secretion of IL-1[beta]. Cell Death Dis. 5, e1040.

Tanaka, H., Matsuda, T., et al., 2000. Reduction of resuscitation fluid volumes in severely burned patients using ascorbic acid administration: a randomized, prospective study. Arch. Surg. 135 (3), 326—331.

Tian, F., Wang, X., et al., 2015. Effect of initial calorie intake via enteral nutrition in critical illness: a meta-analysis of randomised controlled trials. Crit. Care. 19 (1), 180.

Tsang, K.-K., Kwong, E.W.-Y., et al., 2015. The anti-inflammatory and antibacterial action of nanocrystalline silver and Manuka honey on the molecular alternation of diabetic foot ulcer: a comprehensive literature review. Evid. Based Complement. Alternat. Med. 2015, 218283.

Turner, J.R., Rill, B.K., et al., 1997. Physiological regulation of epithelial tight junctions is associated with myosin light-chain phosphorylation. Am. J. Physiol. 273 (4 Pt 1), C1378—C1385.

Ulluwishewa, D., Anderson, R.C., et al., 2011. Regulation of tight junction permeability by intestinal bacteria and dietary components. J. Nutr. 141 (5), 769—776.

Upala, S., Sanguankeo, A., et al., 2015. Significant association between vitamin D deficiency and sepsis: a systematic review and meta-analysis. BMC Anesthesiol. 15 (1), 1—11.

Vandamme, L., Heyneman, A., et al., 2013. Honey in modern wound care: a systematic review. Burns. 39 (8), 1514—1525.

Ventadour, S., Attaix, D., 2006. Mechanisms of skeletal muscle atrophy. Curr. Opin. Rheumatol. 18 (6), 631—635.

Via, M., 2012. The malnutrition of obesity: micronutrient deficiencies that promote diabetes. ISRN Endocrinol. 2012, 8.

Vissers, M.C.M., Wilkie, R.P., 2007. Ascorbate deficiency results in impaired neutrophil apoptosis and clearance and is associated with up-regulation of hypoxia-inducible factor 1α. J. Leukocyte Biol. 81 (5), 1236—1244.

Watters, J.M., Bessey, P.Q., et al., 1986. Both inflammatory and endocrine mediators stimulate host responses to sepsis. Arch. Surg. 121 (2), 179—190.

Weekers, F., Giulietti, A.-P., et al., 2003. Metabolic, endocrine, and immune effects of stress hyperglycemia in a rabbit model of prolonged critical illness. Endocrinology. 144 (12), 5329—5338.

Weiss, G., Goodnough, L.T., 2005. Anemia of chronic disease. N. Engl. J. Med. 352 (10), 1011—1023.

Wiechuła, D., Loska, K., et al., 2012. Chromium, zinc and magnesium concentrations in the pubic hair of obese and overweight women. Biological Trace Element Research. 148 (1), 18—24.

Wilmore, D.W., 1991. Catabolic Illness. N. Engl. J. Med. 325 (10), 695—702.

Wilson, J.X., 2009. Mechanism of action of vitamin C in sepsis: ascorbate modulates redox signaling in endothelium. BioFactors. 35 (1), 5—13.

Witowski, N., Lusczek, E., et al., 2015. A four-compartment metabolomics analysis of the liver, muscle, serum, and urine response to polytrauma with hemorrhagic shock following carbohydrate prefeed. PLoS ONE. 10 (4), e0124467.

Wu, F., Schuster, D.P., et al., 2007. Ascorbate inhibits NADPH oxidase subunit p47phox expression in microvascular endothelial cells. Free Radic. Biol. Med. 42 (1), 124—131.

Xie, X., McGregor, M., et al., 2010. The clinical effectiveness of negative pressure wound therapy: a systematic review. J. Wound Care. 19 (11), 490—495.

Yang, I., Nell, S., et al., 2013. Survival in hostile territory: the microbiota of the stomach. FEMS Microbiol. Rev. 37 (5), 736—761.

Zamboni, M., Mazzali, G., et al., 2008. Sarcopenic obesity: a new category of obesity in the elderly. Nutr. Metab. Cardiovascul. Dis. 18 (5), 388—395.

Zasloff, M., 2002. Antimicrobial peptides of multicellular organisms. Nature. 415 (6870), 389—395.

Zhang, P., Peterson, M., et al., 2015. Visceral adiposity is negatively associated with bone density and muscle attenuation. Am. J. Clin. Nutr. 101 (2), 337–343.

Zhou, G., Kamenos, G., et al., 2012. Ascorbate protects against vascular leakage in cecal ligation and puncture-induced septic peritonitis. Am. J. Physiol. 302 (4), R409–R416.

Further Reading

Albert, B., et al., 2014. Molecular Biology of the Cell 6th Ed. Chapter 23. Pathogens and Infection; Chapter 24.The Innate and Adaptive Immune System. Garland Science.

WebPath Firearms Tutorial accessed at http://library.med.utah.edu/WebPath/TUTORIAL/GUNS/GUNINTRO.html.

Kumar, V., et al., 2014. Robbins and Cotran Pathologic Basis of Disease 9th Ed. Chapter 2. Cellular Responses to Stress and Toxic Insults: Adaption, Injury and Death; Chapter 3. Inflammation and Repair. Elsevier Saunders.

Wilmore, D.W., 1991. Catabolic illness—strategies for enhancing recovery. NEJM. 325, 695–702. Available from: http://dx.doi.org/10.1056/NEJM199109053251005.

3

Type I Diabetes and Celiac Disease

CHIEF COMPLAINT (TYPICAL PATIENT)

MS, an 18-year-old white male senior in high school, presented to the health center complaining of fatigue, recent weight loss, blurred vision, intermittent diarrhea and gastric cramping, frequent urination, and increased thirst for the past 4 weeks. Mr. S had originally attributed these symptoms to the stress and high caffeine intake he used to keep his grades up. He came to the center because he has lost ∼10 lbs despite a ravenous appetite and serious attempts to increase his caloric intake.

PAST MEDICAL HISTORY AND MEDICATION

MS has no known family history of diabetes, hypertension, heart disease, or obesity. His father and one sibling have had intermittent gastric problems diagnosed as irritable bowel syndrome. His paternal grandmother (62 years) has chronically suffered from GI symptoms and now has severe osteoporosis.

DIET HISTORY (24-HOUR RECALL)

Snack (7 am Home)	Coffee with sugar	1 cup, 2 packets
	Whole milk	3 tbsp.
Breakfast (10 am Food truck)	Orange juice	16 oz.
	Bagel	1 large
	Cream cheese	1 oz. (2 tbsp.)
Lunch (1:30 pm Home)	Beef bologna	4 oz.
	American cheese	2 oz.
	White bread (enriched)	4 slices
	Mayonnaise	2 tbsp.

(Continued)

(Continued)

	Corn chips	1 oz. bag
	Coca-Cola (regular)	12 oz.
Snack (5 pm Out)	Frozen yogurt/cherries	1 cup
Dinner (8 pm Fast food bar)	Pizza with pepperoni	2 slices (1/4 of 15″ pie)
	Coca-Cola, regular	12 oz.
Evening snack (11 pm Home)	Doughnut, jelly	2
	Whole milk	1 cup

Diet Analysis

Total Calories		3239 kcal/day
Carbohydrate	397 g	49% of total kcal
Fat	137 g	38% of total kcal
Protein	105.3 g	13% of total kcal
Requirements	2383 kcal	(By Harris Benedict using 1.3 activity factor)
	60.4 g protein	(RDA for young adult = kg × 0.8 g/kg)

SOCIAL HISTORY

MS admits that he drinks alcohol on weekends with his friends, but no more than 4–6 12-oz cans of beer per week. He does not smoke. He drinks 3–4 cups of coffee per day with milk and sugar. He does not follow a special diet. Because his study schedule for advanced placement courses to enhance college admission is very heavy, his diet is erratic. He often cannot take time for meals and snacks when time permits. He plays basketball on the weekends, and works out in the gym at least 2–3 times per week.

Review of Systems

General	Fatigue and weight loss
Eyes	Blurred vision
Gastrointestinal	Watery diarrhea, recurrent intestinal cramping and distension
Endocrine	Increased thirst (polydipsia), frequent urination (polyuria)

Physical Examination

Vital signs	Temp. 98.4°F; HR 80 bpm; respiration 12 bpm; BP 102.70 mmHg
Anthropometrics	Height 5'10" (178 cm)
	Current weight 166 lb (75.5 kg)
	BMI 24 kg/m^2
	Usual weight 176 lb (80 kg)
	Usual BMI 25 kg/m^2
General	Thin male resting comfortably
Head/neck	Facial pallor, pale conjunctiva
Cardiac	Regular rate and rhythm, no murmurs or gallops
Abdomen	Mild bloating, hypoactive bowel sounds, no hepatosplenomegaly
Extremities	Subcutaneous muscle wasting
Rectal	Guaiac-positive stool

Laboratory Data		Normal Range
HEMATOLOGY		
Hematocrit	28%	39–49%
Hemoglobin	9.3 g/dL	13.5–17.5 g/dL
MCV	69.9 μm^3 (fl)	80–100 μm^3 (fl)
RDW	20%	11–15%
WBC	8.3 mm^3	4.5–11 mm^3
Differential	Normal	
Platelet count	221 × 10^3/mm^3	150–450 10^3/mm^3
ESR	70 mm/h	1–15 mm/h
Reticulocyte count	2%	0.5–1.5%
BLOOD CHEMISTRY		
Serum glucose	300 mg/dL	65–110 mg/dL
HbA1c	8.8%	4–6%
Serum sodium	133 mg/L	133–145 mg/L

(Continued)

(Continued)

Laboratory Data		Normal Range
Serum potassium	3.3 mEq/L	3.5–5.3 mEq/L
Serum chloride	101 mg/L	97–107 mg/L
Serum bicarbonate	22 mg/L	24–32 mg/L
Plasma ketones	1:8 dilution	None
Serum calcium	6.9 mg/dL	8.6–10 mg/dL
Serum albumin	3.2 g/dL	3.4–4.8 mg/dL
Serum phosphorous	1.8 mg/dL	2.7–4.5 mg/dL
Total protein	6.3 g/dL	6.4–8.3 g/dL
Alk. phosphatase	245 U/L	25–100 U/L
Total bilirubin	0.2 mg/dL	0.3–1.2 mg/dL
URINE CHEMISTRY		
Ketones	4+	Neg.
Glucose	2+	Neg.

Diagnosis

Based on patient's medical history, physical exam, and test results, he is diagnosed with type I diabetes and referred to the diabetes clinic. He is also referred for a GI workup.

RESOURCES

3.1 PATHOGENESIS OF TYPE I DIABETES MELLITUS

Type I Diabetes Mellitus (DM) is characterized by rapid destruction of pancreatic β cells leading to an absolute deficiency of insulin. It often presents in patients younger than 20 years and is estimated to account for 5–10% of all DM cases. The pathogenesis includes a genetic predisposition with a superimposed environmental trigger that initiates rapid immune destruction of pancreatic β cells and severely compromises insulin production (Fig. 3.1).

Although the prevalence of type I DM has increased in the last two decades; the cause of this increase is not yet clear (Dabelea et al., 2014). A variety of putative mechanisms known to increase risk for type I DM have been reviewed (Egro, 2013). These can be divided into protective and potentiating factors. Breast milk appears protective, whereas cow's milk is diabetogenic. Similarly, the hygiene hypothesis suggests that increased microbial exposure in early childhood reduces risk while recent viral infection increases risk. Adequate vitamin D from sunlight, diet, or supplements has also been shown to be protective in modulating immune response to environmental agents as discussed below. As described

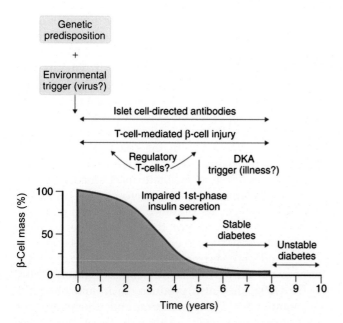

FIGURE 3.1 A summary of the sequence of events that lead to pancreatic β-cell loss and ultimately to the clinical evolution of type 1 diabetes. *DKA*, diabetic ketoacidosis. *Source: Reproduced from Cecil Textbook of Medicine 23rd Ed (Figure 247-2) with permission from Elsevier.*

elsewhere (refer to Essentials III: Nutrients for Bone Structure and Calcification: Vitamin D), vitamin D deficiency is widespread and more common in individuals with dark skin and in obese persons. Our growing understanding of vitamin D function in immunomodulation during pregnancy suggests that maternal vitamin D deficiency may compromise fetal vitamin D status in utero and potentially increase risk for type I DM in the offspring (Tamblyn et al., 2015). Clearly, risk for developing type I DM is multifactorial and no one potentiating factor has yet been identified.

Putative genes implicated in the pathogenesis of type I DM include polymorphisms on the HLA genes, the insulin gene, and genes that regulate T-cell activation. (Human leukocyte antigen (HLA) genes are the human versions of the major histocompatibility complex (MHC) genes found in animals. The proteins, unique to the individual, are expressed on the cell surface and present antigens to T-cells, thereby stimulating an adaptive immune response.) Autoimmune markers to detect type I DM include islet cell autoantibodies (eg, GAD65) as well as autoantibodies to insulin, the tyrosine phosphatases IA-2 and IA-2β, and the zinc transporter, ZnT8 (Howson et al., 2012) (refer to Essentials II: Heavy Metals, Retinoids and Precursors: Zinc). The disease has strong HLA associations, with linkage to the DQA and DQB genes. These HLA-DR/DQ alleles can be either predisposing or protective. Whatever the initiating factor, autoimmune β-cell destruction proceeds at variable

rates; it is often rapid in infants and children, and slower in adults. In addition to multiple genetic polymorphisms, autoimmune destruction of β-cells is related to environmental factors that are still poorly defined.

Some of the environmental factors predisposing to type I DM may relate to nutrient availability, distribution, and/or utilization as discussed below. Although patients are not typically obese when they present with type I DM, obesity should not preclude the diagnosis. Patients may also present with, or subsequently develop, other autoimmune disorders such as Hashimoto thyroiditis, celiac disease, Graves' disease, Addison's disease, vitiligo, autoimmune hepatitis, myasthenia gravis, or pernicious anemia. Some authors have proposed that the pathogeneses of type I DM and type II DM intersect at significant points, since both DM types have been associated with inflammatory-mediated loss of pancreatic β cells, can be exacerbated by obesity, and are characterized by metabolic abnormalities such as hyperglycemia, hyperlipidemia, and other symptoms of metabolic syndrome (Donath and Halban, 2004). Another potential environmental factor is the composition of the gut microbiome. Preliminary research recently reviewed (Knip and Siljander, 2016) indicates that the intestinal microbiota of individuals at risk for type I DM is different from that of healthy individuals. Specifically, preclinical type I DM microbiota has a predominance of *Bacteriodetes* and reduction of butyrate-producing bacteria. The bacterial diversity is also low, as is the community stability. These early studies could not determine whether change in gut bacteria preceded initiation of type I DM or modified its progression.

To clarify discussion of the nutritional pathology of each DM type, the DM chapters are structured as follows.

- Chapter 3. Type I DM focuses on nutrients and immune/inflammatory destruction of pancreatic β cells.
- Chapter 4. Type II DM focuses on nutrients and metabolic dysregulations characteristic of DM.

3.2 TYPE I DM AS DYSREGULATED IMMUNE RESPONSE

Type I DM can be considered as a loss of tolerance to tissue self-antigens. The defect can impair central tolerance in the thymus, or peripheral tolerance that regulates the development of effector T cells in secondary lymphoid tissues such as the gut-associated lymphoid tissue (GALT) (Jeker et al., 2012). Lymphocyte trafficking between primary (thymus) and secondary lymphoid tissues is modulated by concentration gradients of the lipid mediator, sphingosine-1-phosphate (S1P) as reviewed in Hla (2005). While it is not known what initiates the autoimmune attack that destroys the pancreatic β cells, there is abundant evidence from animal models of type I DM such as the nonobese diabetic (NOD) mouse and the BioBreeding (BB) rat that islet β cells are destroyed by an immune attack as reviewed in Rabinovitch and Suarez-Pinzon (2007).

Briefly, the individual's genome responds to specific environmental factor(s) that predispose it to the development of pathogenic autoreactive T cells that target pancreatic β cells (Quintana et al., 2008). Naïve CD4$^+$ helper T cells have the capacity to differentiate into two major types of effector cells. They can become autoreactive T cells with cytotoxic characteristics or they can become more tolerogenic regulatory T cells (T^{regs}). In infancy and childhood, T^{regs} are formed in the thymus and enter the peripheral circulation where they survive for about 2 months and must be continually renewed. In the adult, T^{regs} are formed primarily in secondary lymphoid organs such as the GALT. The conversion is modulated by gut metabolites via the forkhead transcription factor (FOXP3) as reviewed by Wang et al. (2009) and Kinoshita and Takeda (2014) and described below. T^{reg} production requires a local source of IL-2 (eg, naïve T cells) and expression of an IL-2 receptor (CD25). Development of these tolerogenic cells is facilitated by exposure to TGF-β, retinoic acid (RA) and by T cell receptor activation, and is impaired by exposure to IL-6 and IL-21.

Normally, the protective regulatory T cells (T^{regs}) deter attack on islet β cells. If the T^{reg} cells fail to develop adequately, the autoreactive T cells dominate, leading to islet inflammation (insulitis) with infiltration by macrophages, CD4$^+$ helper and CD8$^+$ cytotoxic cells that specifically destroy the β cells. As reviewed in Akimova et al. (2010), the forkhead transcription factor (FOXP3) downregulates T-cell expression of proinflammatory genes (IL-2, IL-4, and IFN-γ), and upregulates expression of CD25 and other T^{reg}-associated genes. Thus FOXP3$^+$ T^{regs} are important in limiting autoimmunity and maintaining peripheral tolerance.

FOXP3 itself can be epigenetically modified by reversible acetylation (HAT) and deacetylation (HDAC). Therapeutic manipulation of FOXP3 acetylation using histone deacetylase inhibitors (HDACIs) is being studied as a means to promote the development, proliferation, and suppressive functions of FOXP3$^+$ T^{regs} in autoimmune states (Wang et al., 2009). In brief, HDACIs exert antiinflammatory effects through suppressing cytokine production by antigen-presenting cells (APCs) including dendritic cells (DCs) and monocytes. Cytokine suppression by HDACI impairs the differentiation of CD4$^+$ T cells into autoreactive T cells but does not impair development of T^{regs} expressing FOXP3. In addition to synthetic HDACIs, several naturally occurring HDACIs (eg, butyrate) are being tested for their ability to exert these antiinflammatory effects.

Butyrate is an HDACI produced by gut microbiota. This four carbon fatty acid binds G-protein-coupled receptors on the enterocyte (GPR43 and GPR109A) as described elsewhere (refer to Chapter 1: Obesity and Metabolic Syndrome: Gut Microbiota and Obesity). GPR43 is activated by all three short-chain fatty acids (SCFA: acetate, propionate, and butyrate); GPR109 is activated only by butyrate and by niacin, also synthesized by gut microbiota. In the colon, butyrate is the endogenous ligand for GPR109, while in immune cells GPR109 is required to facilitate niacin-mediated suppression of inflammation and atherosclerosis (Lukasova et al., 2011). Thus, the presence of microbial commensals competent to produce these metabolites may be decisive and can trigger conversion of CD4$^+$ cells to T^{regs} (Round and Mazmanian, 2010; Smith et al., 2013). Microbes not only produce SCFAs that bind to GPR43 and GPR109a; they also produce niacin and other factors required for T^{reg} production and downregulation of autoreactive T cells. Although these gut-related mechanisms (Fig. 3.2) have been studied most often with respect to other autoimmune conditions and not specifically with type I DM, several working hypotheses specifically targeting type I DM have recently been suggested (Hu et al., 2015; Li and Atkinson, 2015).

3.2.1 Aryl Hydrocarbon Receptor Induction and Type I DM

The ligand-activated transcription factor, aryl hydrocarbon receptor (AhR), has been implicated as another determinant in the conversion of naïve T cells to their effector phenotypes. AhR activation induced adaptive CD4$^+$CD25$^+$ T^{regs} in mice during an acute graft-versus-host (GvH) response and prevented the generation of allospecific cytotoxic T lymphocytes. The synthetic AhR ligand used was a determinant in the suppression of experimental autoimmune encephalitis in association with an expanded population of Foxp3$^+$ T^{regs} (Kerkvliet and Wong, 2009). These observations have sparked research in this area.

AhR is an ancient protein highly conserved in invertebrates as well as vertebrates. It binds promiscuously to a wide array of environmental carcinogens and toxins including polycyclic aromatic hydrocarbons (PAHs), and dioxin-like compounds as well as to heterocyclic amines produced by cooking meat (Sugimura et al., 2004). AhR not only binds toxins, but also binds to endogenous aromatic metabolites including tryptophan derivatives, steroid hormones such as estrogens and natural aromatic plant components including flavonoids, polyphenolics, and indoles (Denison and Nagy, 2003). When activated by ligand binding, the inactive AhR dissociates from its protein complex and is translocated into the nucleus where it binds the dioxin (xenobiotic) responsive element

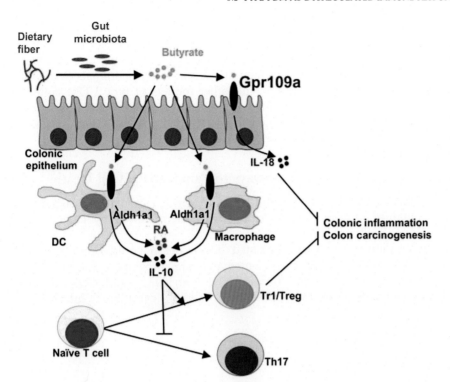

FIGURE 3.2 **Dietary fiber, butyrate, and T^{reg}.** Dietary fiber enhances proliferation of specific commensal microbes that produce butyrate and niacin; these nutrients induce IL-18 that suppresses colonic inflammation via binding Gpr109a. Butyrate and niacin also induce IL-10 and retinol dehydrogenase (Aldh1a1) by macrophages and dendritic cells. Dietary retinal is converted to retinoic acid (RA), a key determinant in TGF-β-mediated conversion of naïve T cells into Foxp3$^+$ T^{reg} cells. Without adequate RA, TGF-β converts cells to autoreactive T cells (Coombes et al., 2007; Mucida et al., 2007). IL-10 stimulates T^{reg} formation and inhibits differentiation into autoreactive T cells (Singh et al., 2014). *Source: Reproduced from Singh, N., Gurav, A., et al. 2014. Activation of Gpr109a, receptor for niacin and the commensal metabolite butyrate, suppresses colonic inflammation and carcinogenesis. Immunity 40 (1), 128–139 with permission from Science Direct.*

(AHRE, DRE, or XRE). Binding promotes expression of multiple protective genes such as phase I and II detoxification enzymes (Okey, 2007). Dietary components also trigger AhR to modulate phase I/phase II enzymes. The research is confusing in that AhR has been shown to respond to the same ligand in different ways in a wide variety of cultured cells and experimental animal models (Conney, 2003).

AhR activation by dietary components has been most often studied using isolated components in vitro or in animal models with respect to cancer. For example, under some experimental conditions, bioactive components such as curcumin and caffeine inhibited carcinogenesis, whereas in others, tumor development was enhanced. Under the same conditions, AhR bound to the same ligand exhibited different and gender-specific carcinogenic potential in different organs and in specific cancer cell types (Safe et al., 2013). Little is known about the human AhR response to chronic exposure from dietary components and a great deal of experimental work will be required to understand the relationship between dietary AhR activation and autoimmune response. Further it will be necessary to determine whether and how this mechanism plays a role in pancreatic β cell death in type I DM.

3.2.2 Vitamin D and Type I DM

Vitamin D bioavailability and activation is discussed elsewhere (refer to Essentials III: Nutrients for Bone Structure and Calcification: Vitamin D). 25D is secreted from liver stores and transported by vitamin D binding protein (DBP) to target tissues. The most well studied target is the renal proximal tubule. Kidney production of vitamin D hormone (1,25D) from 25D is tightly regulated by several factors including parathyroid hormone and fibroblast growth factor 23 (FGF23). Vitamin D is hydroxylated in the 1 position by 1α-hydroxylase enzyme (CYP27B1). Kidney derived 1,25D exerts its action at the bone and enterocyte as a ligand to vitamin D receptor (VDR) with subsequent modification of target gene expression. The strict control of its activation and relatively short half-life (4–6 hours) tightly limits the effect of kidney derived 1,25D on its targets.

It is now known that 1α-hydroxylase is contained in many nonrenal tissues that can also produce 1,25D. In these cells, 25D activation is not controlled by parathyroid hormone and FGF23. In the immune cells, the enzyme is controlled by immune stimuli (Stoffels et al., 2006). Because it does not circulate, intracellular 1,25D formed in immune cells can rise to the supraphysiologic levels required for its immunomodulary actions as reviewed in Mathieu (2015).

DCs appear to be a major target for the immunomodulary effects of 1,25D. DC differentiation is inhibited by 1,25D, locking the DCs into an immature, tolerogenic state. These 1,25D-modified, antigen-presenting cells produce less of the proinflammatory cytokines (IL12, IL23) and more antiinflammatory mediators such as IL10; IL10 can recruit and differentiate T^{reg}

subsets and dampen Th1 and Th17 responses (Fig. 3.3). In addition to its action in modulating transcription of immune-related genes, a recent report suggests that 1,25D may also act by modulating the metabolic activity of the DC, upregulating glycolysis and increasing lactate secretion (Ferreira et al., 2015).

Lymphocytes also convert 25D to 1,25D. While resting T lymphocytes express low VDR, upon activation they upregulate VDR expression; VDR binds 1,25D and stimulates expression of over 100 1,25D-responsive genes. Thus, 1,25D can modulate cytokine secretion and cell function as reviewed by Baeke et al. (2010).

NOD mice treated life-long with oral vitamin D exhibited a decrease in insulitis and type I DM (Takiishi et al., 2014). These mice had decreased autoregulatory effector T cells and increased T^{reg} cells. While additional trials showed a protective immunomodulatory effect on NOD mice, high vitamin D doses that entirely

prevented type I DM also produced hypercalcemia and bone decalcification. A structural analog of vitamin D produced decreased calcemic effects and also inhibited the development of insulitis in these animals (Mathieu, 2015). Work with other models of type I DM animals including VDR knockout mice showed contradictory results, suggesting that other compensatory mechanisms may be involved.

3.2.3 Pathways to Pancreatic β Cell Death

The multiple ways in which T cells can cause pancreatic β cell death have been reviewed by Donath and Halban (2004).

T-cell-mediated β cell death could occur through MHC class 1-mediated signaling by cytotoxic CD8$^+$ T cells. Alternatively, both CD4$^+$ and CD8$^+$ T cells produce potentially lethal cytokines. Interferon-γ

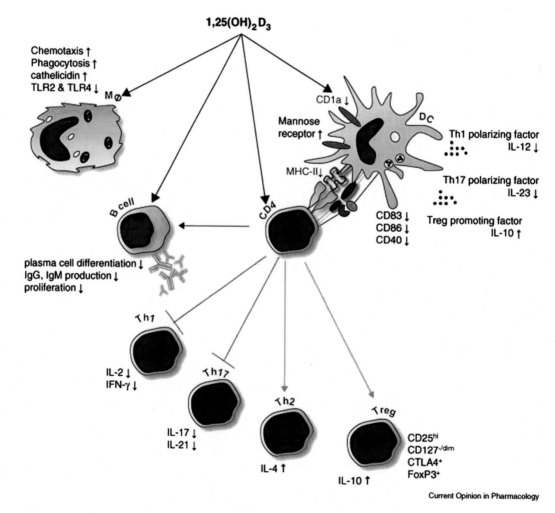

FIGURE 3.3 **Immunomodulary effects of 1,25D.** Intracellular activation of 1,25D modifies the functions of a host of immune cells including macrophages, dendritic cells, B and T lymphocytes. The primary result is reduction in autoreactive effector T cells (Th1, Th17) number and proliferation of T^{reg} cells (Baeke et al., 2010). Maturation of B cells to plasma cells and subsequent antibody secretion is also inhibited. As with DC, treatment with 1,25D renders B cells tolerogenic. *Source: Reproduced from Baeke, F., Takiishi, T., et al. 2010. Vitamin D: modulator of the immune system. Curr. Opin. Pharmacol. 10 (4), 482–496. with permission from Science Direct.*

(IFNγ) induces the FAS death receptor and chemokine production by β cells. FAS-ligand expressing T cells can bind their targets and initiate apoptosis. Chemokines recruit mononuclear cells to the area, increasing inflammation. Further, INFγ can activate macrophages and induce production of proinflammatory cytokines such as IL-1β and TNFα. In comparison to other islet cells, β-cells express receptors for IL-1 and appear to be more sensitive to IL-1β − induced apoptosis. Finally, cytokine production is associated with production of reactive oxygen species (ROS) that can produce irreparable damage in the islet β cell.

3.3 METABOLIC DYSREGULATION, CLINICAL PRESENTATION, AND DIAGNOSIS OF TYPE I DM

3.3.1 Hyperglycemia

Although most tissues can use fatty acids as a fuel source, glucose is essential for selected tissues, especially the brain and central nervous system (Cahill, 1971). Circulating glucose concentration is controlled by three processes: glucose production by the liver, glucose uptake and utilization by the peripheral tissues, and the actions of hormones, including insulin, glucagon, and counter-regulatory hormones on glucose uptake, utilization, and excretion. The signaling pathways that control the metabolism of dietary fuels in the fed and fasted states have been discussed elsewhere (refer to Chapter 1: Obesity and Metabolic Syndrome. Nutrient Sensors and Downstream Targets). In the fasting state, pancreatic insulin production by the β cells is low and glucagon production by the α cells is high. Glucagon facilitates hepatic gluconeogenesis and glycogenolysis and downregulates glycogen synthase, thereby providing sufficient glucose to maintain blood levels. Glucagon also stimulates carnitine acyltransferase, facilitating fatty acid uptake into the mitochondria for ATP biosynthesis. In the fed state, dietary fuels are readily available. Since high blood glucose levels in the blood exert undesirable osmotic effects on fluid balance and are also cytotoxic, glucose levels must be rapidly returned to normal by the action of insulin. This anabolic hormone promotes glucose uptake and utilization by the tissues, predominately muscle and adipose tissues, and glucose storage as glycogen.

In type I DM, compromised insulin production secondary to pancreatic β cell destruction is the primary cause of fasting hyperglycemia. Normally, insulin in the portal vein exerts a restraining effect on hepatic gluconeogenesis. It also suppresses glucagon secretion by pancreatic α cells. If insulin production is severely suppressed, portal glucagon levels rise and stimulate

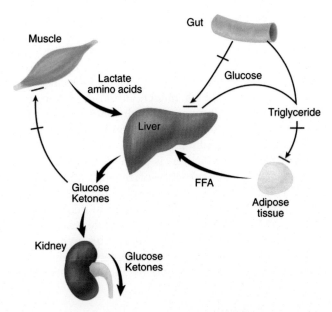

FIGURE 3.4 **The effects of severe insulin deficiency on body fuel metabolism.** Lack of insulin leads to mobilization of substrates for gluconeogenesis and ketogenesis from muscle and adipose tissue, accelerated production of glucose and ketones by the liver, and impaired removal of endogenous and exogenous fuels by insulin-responsive tissues. The net results are severe hyperglycemia and hyperketonemia that overwhelm renal glucose reabsorption and result in glucosuria. *FFA*, free fatty acids. *Source: Reproduced from Cecil Textbook of Medicine 23rd Ed (Figure 247-1) with permission from Elsevier.*

hepatic glucose production as indicated above. Lack of systemic insulin signaling prevents glucose uptake into peripheral tissues through the insulin-dependent GLUT 4 transporters. Because glucagon also increases mitochondrial utilization of fatty acids (refer to Chapter 5: Cardiomyopathy and Congestive Heart Failure. ATP-Dependent Excitation − Contraction), high glucagon output further limits mitochondrial glucose oxidation (Fig. 3.4). Taken together, the consequences that stem from an absolute lack of insulin in type I DM include profound hyperglycemia and glucosuria (or glycosuria). Subcutaneous insulin therapy can improve glucose uptake by peripheral tissues, but cannot effectively increase portal vein insulin levels. Thus the liver does not receive adequate insulin to arrest glucose production or store glucose as glycogen. Peripheral tissues develop insulin resistance from chronic insulin deprivation and exhibit toxic effects from chronic hyperglycemia. Renal glycosuria only partially compensates for the other defects in glucose disposal.

3.3.2 Elevated Free Fatty Acids and Ketones

When insulin is present, adipocytes take up and esterify free fatty acids (FFA) to TG. When insulin is lacking, counter-regulatory hormones such as glucagon and epinephrine stimulate adipocyte lipolysis and FFA release. To reduce the lipotoxic effects of elevated concentrations of FFA, some of the circulating lipids are incorporated into lipid droplets in inappropriate tissues such as the liver and muscles. Elevated FFAs also provide excess substrate for mitochondrial oxidation, providing the liver with energy for gluconeogenesis. FFAs also interfere with insulin production by the pancreatic β cell and impair insulin signaling.

Mitochondrial function is also compromised by insulin lack. In the presence of insulin, both glucose and FA are metabolized to acetyl CoA and enter the citric acid (TCA) cycle. TCA oxidation requires that acetyl CoA (2C) binds to an intermediate, oxaloacetate (4C), to form citric acid. The acetyl group on citric acid is oxidized and the oxaloacetate moiety is released to reenter the TCA cycle bound to another acetyl CoA.

Oxaloacetate is formed de novo from pyruvate obtained only through glycolysis. During fasting or under conditions when FFAs are in excess and intracellular glucose is limiting, oxaloacetate is preferentially utilized for gluconeogenesis and not available to condense with acetyl CoA and facilitate its oxidation through the TCA cycle. Under these conditions, acetyl CoA is formed into ketone bodies (Fig. 3.5).

The regulation of ketone body biosynthesis has been recently reviewed (Newman and Verdin, 2014). Briefly, adipocyte lipolysis increases FA availability, β oxidation, and acetyl CoA production. In type I diabetes, reduced portal insulin levels are not sufficient to stimulate glycolysis, thus oxaloacetate is limiting. Acetyl CoA moieties are therefore condensed to form ketones (acetoacetate, β hydroxybutyrate, acetone); total insulin lack can sharply elevate ketone

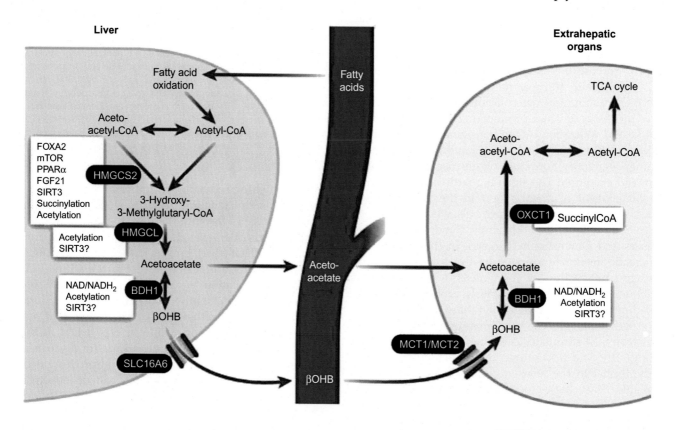

TRENDS in Endocrinology & Metabolism

FIGURE 3.5 **Ketone biosynthesis and utilization.** The rate-limiting step in ketone biogenesis is condensation of acetyl CoA with acetoacetyl Co A to form HMG-CoA (β-hydroxy-β-methylglutaryl-CoA). HMG-CoA is also an intermediate in the synthesis of cholesterol and other complex lipids such as ubiquinone (coenzyme Q-10). Ketogenesis occurs in the mitochondria, whereas cholesterol synthesis occurs in the cytoplasm, thus both processes are independently regulated. Ketone biogenesis is nutritionally regulated by at least two pathways. The forkhead box transcription factor 2 (FOXA2) binds to the *Hmgcs2* gene promoter and activates transcription of the mitochondrial gene, HMG-CoA synthase. FOXA2 is inactivated by insulin signaling via phosphorylation and nuclear export and is activated by glucagon-stimulated acetylation. Sirtuin 1 (SIRT1), a highly conserved NAD + -dependent deacetylase that detects energy availability and modulates metabolic processes, deacetylates FOXA2, and activates ketogenesis. Ketone production is also regulated by mTORC1 (mammalian target of rapamycin complex 1) that exerts control over nutrient distribution during fasting (Newman and Verdin, 2014) as discussed in Chapter 1: Obesity and Metabolic Syndrome: Nutrient Sensors and Downstream Targets. *Source: Reproduced from Newman, J.C., Verdin, E., 2014. Ketone bodies as signaling metabolites. Trends Endocrinol. Metab. 25 (1), 42−52 with permission from Science Direct.*

biosynthesis and precipitate potentially fatal ketoacidosis. Rising blood ketones further inhibit insulin action and stimulate secretion of counter-regulatory hormones. Catecholamines, growth hormone, and cortisone increase lipolysis, delivering ever more FA to the liver, where they saturate the β oxidation pathway, leading to hyperlipidemia and liver deposition (hepatic steatosis). High levels of ketones in type I DM also induce inflammation and oxidative stress as reviewed by Hoffman et al. (2013). Ketones also induce systemic proinflammatory cytokines (IL-1β, IL-6, TNF-α, and IL-8); increased regulatory IL-10, complement active peptides, acute-phase proteins, and T-lymphocyte activation have also been reported.

3.3.3 Diagnostic Criteria for Type I Diabetes

Type I DM patients usually present with symptoms illustrated in Fig. 3.6 that reflect end-organ response to inadequate insulin availability. Children and adolescents may present with ketoacidosis, while other patients may have moderate fasting hyperglycemia that rapidly changes to frank hyperglycemia and/or ketoacidosis with infection or other stress. Type I DM can occur at any age, even in the eighth and ninth decades of life. While adult presenters may retain residual β-cell function that initially prevents ketoacidosis, these individuals eventually become dependent on insulin for survival.

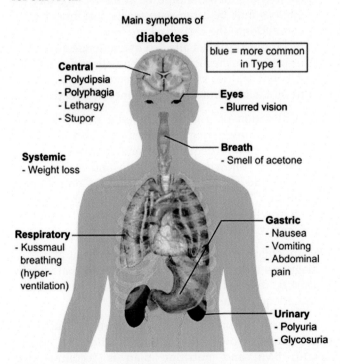

FIGURE 3.6 **Presenting symptoms of DM.** Symptoms listed in blue are common presenting symptoms of type I DM. *Source: By Mikael Häggström (Public domain), via Wikimedia Commons.*

Measurement of blood glucose is essential, not only for diagnosis (Newman and Verdin, 2014) but for management decisions. A random plasma glucose ≥200 mg/dl (11.1 mM/L) meets the definition of DM, however additional measurement of hemoglobin A_1C (A_1C) fraction will provide information on the duration of hyperglycemia. Type I diabetes is defined by one or more autoimmune markers. Measurement of these markers listed above, especially in first-degree relatives and those patients with lower blood glucose who report transient hyperglycemic episodes can identify risk for developing type 1 DM (American Diabetes Association, 2016a,b).

3.4 STANDARDS OF CARE FOR TYPE I DIABETES

3.4.1 General Recommendations

Medical nutrition therapy (MNT) recommendations recently revised by the American Diabetes Association (2016a,b) for children and adolescents with type I diabetes are similar to those from 2010 (American Diabetes Association, 2010, 2016a,b). It should be noted that, despite the recommendations listed below, a recent survey revealed that almost half of the US adults with either type of diabetes did not meet recommended goals for diabetes care (Ali et al., 2013).

In general, MNT should be "patient-centered" and focus on achieving blood glucose goals without excessive hypoglycemia, as well as lipid and blood pressure goals, and goals for normal growth and development. Nutrient recommendations compiled over a decade ago form the basis for MNT recommendations (Franz et al., 2002) and are similar to requirements for healthy children and adolescents.

In general, the diets of US children are inadequate in the following ways:

- Too low in fruits and vegetables;
- Inadequate in dietary fiber;
- Levels of saturated fat well above the National Cholesterol Education Program (NCEP) recommendations.

Consultation with a registered dietitian with experience in pediatric nutrition and diabetes is recommended. Meal plans must be individualized to accommodate food preferences, cultural influences, physical activity patterns, and family eating patterns and schedules. The meal planning approach selected must assist families in learning about the effects of specific foods on blood glucose levels. The system must also be comprehensible and one that can be implemented within the context of the family's lifestyle and eating patterns.

3.4.2 Diet and Exercise

In individuals taking insulin and/or insulin secreta-gogues, physical activity can cause hypoglycemia if medication dose or carbohydrate consumption is not altered. For individuals on these therapies:

- Added carbohydrate should be ingested if preexercise glucose levels are <100 mg/dL (5.6 mmol/L).

In summary, General Guidelines include:

- Consultation with a dietitian to develop/discuss the medical nutrition plan is encouraged;
- Evaluate height, weight, BMI, and nutrition plan annually;
- Calories should be adequate for growth and restricted if child becomes overweight.

The Dietary Guidelines for Americans (2010) (http://health.gov/dietaryguidelines/2015-scientific-report/. Note that 2015 guidelines will be published later in 2016) recommends that all Americans aim for the major goals:

- Balance calories with physical activity to manage weight;
- Consume more of certain foods and nutrients such as fruits, vegetables, whole grains, fat-free and low-fat dairy products, and seafood;
- Consume fewer foods with sodium (salt), saturated fats, trans fats, cholesterol, added sugars, and refined grains;
- With regard to the glycemic effects of carbohydrates, the total amount of carbohydrate in meals and snacks is more important than the source or type.
- Individuals receiving intensive insulin therapy should adjust their premeal insulin dosages based on the carbohydrate content of meals.

There is some evidence for the following statements (Franz et al., 2002):

- Individuals receiving fixed daily insulin dosages should try to be consistent in day-to-day carbohydrate intake.
- Although the use of low-glycemic index food may reduce postprandial hyperglycemia, there is not sufficient evidence of long-term benefit to recommend use of low-glycemic index diets as a primary strategy in food/meal planning for individuals with type I diabetes.
- As for the general public, consumption of fiber is to be encouraged; however, there is no reason to recommend that people with type I DM consume a greater amount of fiber than other Americans.
- Percentages of carbohydrate should be based on individual nutrition assessment.

Additional recommendations for type I DM patients (American Diabetes Association, 2008) (numbers in parenthesis indicate the strength of the evidence for this recommendation (A = strong … E = consensus))

- For individuals with type I diabetes, insulin therapy should be integrated into an individual's dietary and physical activity pattern. (E)
- Individuals using rapid-acting insulin by injection or an insulin pump should adjust the meal and snack insulin doses based on the carbohydrate content of the meals and snacks. (A)
- For individuals using fixed daily insulin doses, carbohydrate intake on a day-to-day basis should be kept consistent with respect to time and amount. (C)
- For planned exercise, insulin doses can be adjusted. For unplanned exercise, extra carbohydrate may be needed. (E)

The following statements are based on expert consensus (E):

- Carbohydrate and monounsaturated fat together should provide 60–70% of energy intake.
- Recommendations for total fat, saturated fat, cholesterol, vitamins, and minerals are similar for individuals with diabetes and the general population.
- Protein intake can range from 15 to 20% of calories from animal and vegetable protein sources.
  - If the patient has nephropathy, lower protein intake may be desired.
- Carbohydrate intake should be at least 130 g/day.
  - Sucrose may be eaten with meals, but must be counted into the carbohydrate total.
  - Blood glucose is not affected by moderate alcohol usage unless the alcohol is taken with carbohydrate.
  - Alcohol calories should be considered in addition to (not a substitute for) recommended foods.

Consideration of a child's appetite must be given when determining energy requirements and the nutrition prescription. Adequacy of energy intake can be evaluated by following weight gain and growth patterns on the Centers for Disease Control and Prevention (CDC) pediatric growth charts on a regular basis. (Growth charts can be obtained in pdf format at http://www.cdc.gov/growthcharts/.)

- Many children with type I diabetes present at diagnosis with weight loss. Return to normal weight must be restored with insulin initiation, hydration, and adequate energy intake.

As energy requirements change with age, physical activity, and growth rate, an evaluation of height, weight, BMI, and nutrition plan is recommended at least every year. Good metabolic control is essential for normal growth and development. However, withholding food or having the child eat consistently without an appetite for food in an effort to control blood glucose is discouraged. BMI should be monitored and calories restricted if the child becomes overweight.

3.4.3 Carbohydrate Counting

Carbohydrate counting is an outgrowth of the Diabetic Exchange System developed in 1950 to provide uniformity in meal planning. It uses the concept of "exchange" to educate patients on the macronutrient content of commonly consumed foods. Each food portion on a particular list can be exchanged with any other food portion on the same list. The registered dietitian assesses individual requirements and determines an energy prescription. This prescription is translated into macronutrient exchanges to reach goal percentages for each. In this way the patient is assured that their exchange prescription will allow them to reach their individualized requirements for carbohydrate (CHO), fat, and protein in self-selected meals throughout the day. It should be noted that the exchange system is written at the 9–10th grade education level. Patients must be capable of reading at this level and understand the concept of "exchanging" foods. Patient education on the exchange system may require several sessions and good patient compliance.

The concept of carbohydrate (CHO) counting is based on the observation that the CHO content of a meal is the major factor contributing to postprandial glucose variation and, as such, influences premeal insulin requirements. The amount, not the type, of CHO intake is the focus of this meal planning approach. Patients can be referred to the Mayo Clinic educational site for an easy-to-follow diabetic exchange system. (Mayo Clinic Diabetic Exchange system can be used by following the links at this site: http://www.mayoclinic.com/health/diabetes-diet/DA00077/.) In general, food CHO sources are starch, fruit, milk/yogurt, and sweets. Nonstarch vegetables do not need to be counted unless their total at any meal exceeds 15 g CHO. CHO is counted in one of two ways.

- The amount of food containing 15 g CHO counts as one CHO choice.
- Total grams of CHO in a meal or snack can be counted.

3.4.4 Special Nutrition Issues for the Diabetic Patient (American Diabetic Association, 2003)

- Patients requiring *clear or full liquid diets* should receive ~200 g of carbohydrate per day in equally divided amounts, at meal and snack times. Liquids should not be sugar-free. Patients require carbohydrate and calories, and sugar-free liquids do not meet these nutritional needs.

Diabetes medications may need to be adjusted to achieve and maintain metabolic control.

- After *surgery*, food intake should be initiated as quickly as possible. Progression from clear liquid to full liquid to solid foods should be completed as rapidly as tolerated. Adequate carbohydrate and calories should be provided.
- During *catabolic illness*, careful and continuous monitoring of nutritional and glycemic status is critical to ensure that increased nutritional needs are being met and that hyperglycemia is prevented.
  - Caloric needs for most patients are in the range of 25–35 kcal/kg every 24 hour. Care must be taken not to overfeed patients, because excess caloric intake can exacerbate hyperglycemia.
  - For patients with normal hepatic and renal function, protein needs range from 1.0 to 1.5 g/kg body weight, depending on the degree of stress.
- As in a solid diet, the total grams of carbohydrate provided by *enteral or parenteral formulations* will have the greatest impact on blood glucose response. Use of the enteral versus the parenteral route of feeding provides several advantages: a more physiological route, avoidance of central-catheter-related complications, the trophic effect of gastrointestinal cells, and lower costs.
  - For tube feedings, either a standard enteral formula (50% carbohydrate) or a lower-carbohydrate-content formula (33–40% carbohydrate) may be used.
  - Regardless of the type of feeding used, blood glucose monitoring is required to guide adjustments in diabetes medication and maintenance of glycemic control.

3.5 EVIDENCE FOR TARGETED NUTRIENT MODIFICATION IN TYPE I DM

3.5.1 Vitamin A

There is no doubt that even marginal or subclinical vitamin A depletion can have deleterious effects on the immune system (Trasino and Gudas, 2015). The authors also reviewed arguments suggesting that vitamin A status is compromised in type I DM. The few studies that have examined vitamin A status in type I DM patients have found that, compared with nondiabetics, individuals with type I DM have decreased circulating levels of vitamin A bound to retinal binding protein (RBP4) (refer to Essentials II: Heavy Metals, Retinoids and Precursors: Vitamin A (Retinoids)). Animal models have shown that retinyl-ester hydrolase, the enzyme that releases retinol for transfer as RBP4 is reduced in experimental type I DM. Further work with animal models revealed that supplements of all-trans RA more consistently reduced type I DM in NOD mice and delayed loss of β cell mass. These observations suggest that vitamin A metabolism may be dysregulated in type I DM, especially those pathways that convert retinol to RA.

The fundamental question is whether the individual at risk for type I DM suffers from an underlying anomaly in vitamin A absorption, distribution, utilization, or loss that increases risk for subclinical vitamin A deficiency. Depending on the cause of the dysregulation, supplemental vitamin A could be beneficial or have a toxic effect. In fact, BB rats were protected from type I DM by vitamin A deficiency and supplements exacerbated their disease. Given the potentially toxic effects from excess vitamin A, it would appear safer and more prudent to avoid supplemental retinoids and obtain vitamin A and its precursors from green and yellow fruits and vegetables until more work is done in this area.

3.5.2 Vitamin D

Epidemiologic studies have generally shown that regular vitamin D obtained from sunlight, dietary, or supplemental sources beginning in early childhood is protective against autoimmune disorders as reviewed (Takiishi et al., 2014). Other environmental candidates that increase risk for type I DM include exposure to enteroviruses and early introduction of wheat, with their potential to damage enterocytes and increase gut permeability in susceptible individuals. Taken together with reports suggesting that vitamin D in the infant is important for gut maturation, reducing gut permeability to agents/proteins with antigenic potential, it is possible that adequate intestinal vitamin D early in life can reduce gut permeability and minimize antigen translocation and dysregulated stimulation of the innate and adaptive immune response.

A meta-analysis of case control and cohort studies revealed that risk for type I DM was significantly reduced (29%) in infants supplemented with vitamin D (eg, with cod liver oil) versus those not supplemented (Zipitis and Akobeng, 2008). One case-control study reported that cod liver oil taken by the mother during pregnancy was associated with a reduced risk for type I DM in the offspring. A case-control study nested within a 29,072 woman birth cohort in Norway measured 25D content in samples taken from 109 women in late pregnancy who delivered a child who subsequently developed type I DM before 15 years of age. The odds of developing type I DM were more than twofold higher for offspring of women with the lowest 25D compared with the offspring of those with 25D in the upper quartile (Sørensen et al., 2012).

Several studies are ongoing (www.clinicaltrials.gov) to address questions regarding the safety and efficacy of supplemental vitamin D in early life. However, since persuasive evidence on the deleterious effects of vitamin

D deficiency in pregnancy and early life has led many women to take supplemental vitamin D during pregnancy and to supplement their infants as well, it may be difficult to recruit appropriate subjects for a large intervention trial. Recently, a randomized, double-blind placebo-controlled trial evaluated oral D_3 (cholecalciferol: 70 IU/kg per day) supplementation versus placebo in 30 young patients with new-onset type I DM for 12 months (Treiber et al., 2015). T^{regs} were determined by FACS analysis and lymphocyte functional tests were assessed with ex vivo suppression cocultures at months 0, 3, 6, and 12. Suppressive capacity of T^{regs} increased ($p < 0.001$) with cholecalciferol from baseline ($-1.59 \pm 25.6\%$) to 12 months ($37.2 \pm 25.0\%$) and change of suppression capacity from baseline to 12 months was significantly higher with cholecalciferol ($22.2 \pm 47.2\%$) than placebo ($-16.6 \pm 21.1\%$). Despite higher than usual doses of vitamin D (~ 3000 IU/day) there was no change in serum calcium and parathyroid hormone between groups and all were within the normal range. The authors concluded that these data support use of supplemental vitamin D as part of a combined treatment regimen for early-onset type I DM patients.

3.5.3 Pyridoxine (Vitamin B6) and Folate

An additional dietary connection between micronutrient adequacy and optimal immune system function can be seen in the immunomodulatory activities of vitamin B6 and folate described below. The requirement for vitamin B6 (Fig. 3.7) as a cofactor for lymphocyte egress from the thymus has been reported (Selhub et al., 2013). T cells leave the thymus on a S1P gradient regulated by S1P lyase, an enzyme that breaks down S1P and allows the lymphocytes to enter the bloodstream. Vitamin B6 (pyridoxine) is required for S1P lyase activity. Interestingly, a study that tested the safety of the commonly used caramel color food additive, 2-acetyl-4-tetrahydrobutylimidazole (THI) found that THI inhibited the S1P lyase enzyme and caused lymphopenia in mice (Schwab et al., 2005). Supplemental vitamin B6 ameliorated lymphocyte sequestration in the thymus and allowed cells to home to their target sites.

The RDA for vitamin B6 is 1.3 mg/day for adults, with higher amounts required during pregnancy and lactation (Food and Nutrition Board, 1998). While meat, fish, and organ meats are high in vitamin B6, the

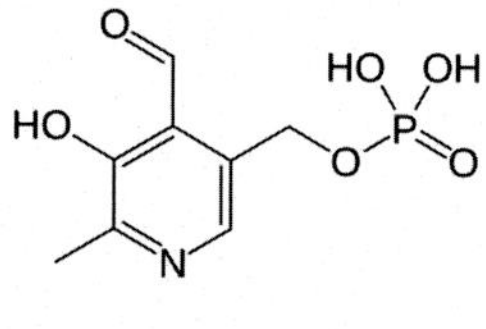

FIGURE 3.7 **Pyridoxal phosphate.** *Source: By Fvasconcellos 23:46, 30 April 2007 (UTC) (own work) (Public domain), via Wikimedia Commons.*

vitamin is labile to heat, thus cooking processes can significantly reduce the vitamin content. Milling and grain processing also cause vitamin B6 loss; foods are not routinely fortified with this vitamin. Vitamin B6 is also found in potatoes and starchy vegetables and some noncitrus fruit. Additional information on this nutrient is available (https://ods.od.nih.gov/factsheets/VitaminB6-HealthProfessional/).

Folate is also essential for lymphocyte proliferation and survival. Information about this nutrient is available and further discussed elsewhere (refer to Chapter 6: Atherosclerosis and Arterial Calcification: Strategies to Repair Endothelial Dysfunction. Nutrient information provided at https://ods.od.nih.gov/factsheets/Folate-HealthProfessional/). Natural T^{reg} cells constitutively express high amounts of folate receptor 4 (FR4). This marker distinguishes them from $CD4^+$ naïve, effector, and memory lymphocyte subtypes. FR4 is functionally essential for T^{reg} cells because these cells were more proliferative in vivo in naïve mice, likely responding to self-antigens and commensal microbes (Yamaguchi et al., 2007). Since folate is essential for DNA and protein synthesis, as well as for epigenetic methylation, the authors proposed that high levels of FR4 allow T^{reg} cells to efficiently incorporate folate for proliferation. It should be noted that serum folate levels provide no information on cellular folate uptake.

3.6 AUTOIMMUNE CONDITIONS THAT COMPLICATE TYPE I DM

Type I DM patients can also suffer from other autoimmune conditions. A recent analysis of data from 28,671 patients <30 years of age with type I DM from 242 specialized centers in Germany and Austria revealed at least one β-cell autoantibody present in 81.6% of patients (Warncke et al., 2010). A total of 19.6% had positive thyroid antibodies with female predominance (62%, $p < 0.0001$). Antibodies to tissue transglutaminase, suggesting celiac disease (CD) were present in 10.7%; CD antibodies were associated with a significantly longer duration of diabetes ($p < 0.0001$). Parietal cell antibodies were found in 283 patients, associated with older age ($p < 0.001$), and adrenal antibodies were present in 94 patients. Thus, at least three different autoimmune phenomena were present in these type I DM patients. Thyroid autoimmunity and antibodies suggestive for celiac disease are the most prevalent additional immune phenomena in type I DM. Parietal/adrenal antibodies are less common.

The prevalence of CD in type I DM ranges from 3% to 16% with a mean of ~8% (Volta et al., 2011). In approximately half the cases, CD does not present with symptoms, but further investigations often reveal suggestive gastrointestinal signs. Both conditions share a common genetic background and an abnormal small intestinal response with variable degrees of enteropathy. If serological screening is positive, the type I DM patient must be treated with a gluten-free diet.

3.6.1 Autoimmune Celiac Disease

3.6.1.1 Pathogenesis and Clinical Presentation

Celiac disease (CD; celiac sprue) is an autoimmune disease triggered by the gliadin fraction of the wheat protein, gluten. Intestinal T cells recognize gliadin peptides presented by the unique HLA-DQ markers of affected patients. Clinical manifestations can include chronic diarrhea, bloating, chronic fatigue, and malabsorption due to reduction in intestinal absorptive capacity. Iron-deficiency anemia is common, especially in premenstrual women whose monthly blood loss exacerbates iron deficiency caused by malabsorption. If the condition is not treated, chronic loss of calcium and other nutrients required for bone integrity can predispose to osteoporosis (refer to Essentials: Nutrients for Bone Structure and Calcification). CD presentation can be atypical. Some patients with markers of genetic susceptibility and autoimmune antibodies for CD are asymptomatic with or without small bowel lesions. Some patients have evidence of increased intestinal permeability with markers of genetic susceptibility and autoimmune antibodies and immune destruction, infiltration of inflammatory cells, and tissue remodeling (Fasano, 2011; Serena et al., 2015) (Fig. 3.8).

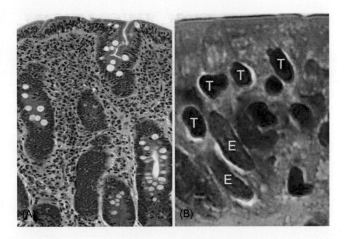

FIGURE 3.8 **Celiac lesions in the small intestine.** Normal small intestinal morphology (left) is compared with a biopsy taken from a CD patient (right). Celiac lesions are most often seen in the duodenum and upper jejunum; these areas are exposed to the highest gliadin concentrations. In addition to villous atrophy, the biopsy shows intraepithelial lymphocytosis, characterized by increased cytotoxic $CD8^+$ T lymphocytes (T); infiltrating cells also include plasma cells, mast cells, and eosinophils (E). Crypt hyperplasia and loss of goblet cells are also evident in the patient biopsy. *Source: Reproduced from Robbins and Cotran Pathologic Basis of Disease, 9th Ed. With permission from Elsevier Saunders.*

The immunopathology of CD is not yet fully understood, as reviewed in Abadie and Jabri (2015). In brief, gluten is normally digested into amino acids and peptides for absorption. In patients with CD, one of these fractions, a 33-amino-acid α-gliadin peptide resists further digestion and translocates to the lamina propria where it is deaminated by tissue transglutaminase. Deaminated gliadin is taken up by antigen-presenting cells (dendrites and macrophages) and presented in the context of HLADQ2 or HLADQ8 self-antigens to T cells. Binding activates the T cell to secrete IFNγ and to help the B cells make antibodies (Fig. 3.9).

While gliadin and other wheat components can lead to enterocyte death and villous atrophy through the immunopathological processes described above, it is also known that gliadin loosens the tight junctions that control paracellular molecular transport (refer to Chapter 2: The Obese Gunshot Patient: Injury and Septic Shock: Innate Host Defense), thereby increasing intestinal permeability. Gliadin has been shown to bind to the chemokine receptor, CXCR3 on enterocytes (Lammers et al., 2008) and elicit release of the tight junctional protein, zonulin 1 (ZO-1) thereby increasing intestinal permeability. ZO-1 is a physiological modulator of intercellular tight junctions and controls paracellular macromolecule trafficking and thus may play

a major role in maintenance of oral tolerance and immune response balance (Fasano, 2011). While gluten, gliadin, and wheat digest can all modify ZO-1 and increase paracellular transport, in vitro studies with Caco2 cells revealed that probiotic microbial strains such as *Lactobacillus rhamnosus* GG (Orlando et al., 2014) and *Bifidobacterium lactis* (Lindfors et al., 2008) inhibited gliadin-induced damage as measured by permeability (transepithelial resistance), actin cytoskeletal rearrangement (extent of membrane ruffling), and expression of the tight junctional protein zonulin 1 (ZO-1).

To complicate matters further, members of the non-gluten α-amylase/trypsin inhibitor (ATI) family contained in wheat and other grains have been shown to strongly induce the innate immune response by binding toll-like receptor 4 (TLR4) (Junker et al., 2012). ATIs are produced in cereal grains to repel pests and parasites. Genes for these and other inhibitors have also been transferred to high-yield, pest-resistant grains by traditional hybridization and by genetic modification techniques. Thus modern grains are likely to contain greater amounts of ATIs capable of eliciting an innate immune response (Ryan, 1990). It is not clear from these studies the extent to which ATIs are inactivated by food preparation processes. However, the

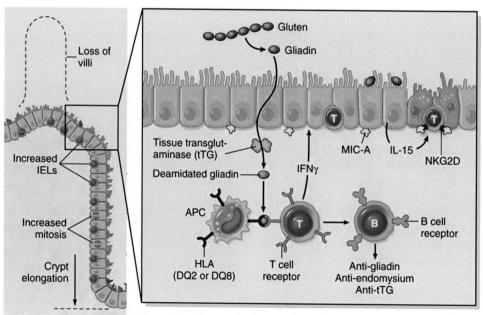

FIGURE 3.9　**Gliadin-induced villous atrophy.** Gliadin induces enterocytes to express the cytokine, IL-15 which induces translocation of MIC-A, a marker related to HLA class 1 molecules, to the surface of injured cells. Intraepithelial lymphocytes (IELs) express coactivating natural killer (NK) cell receptors. One of these receptors (NKG2D) binds MIC-A; binding triggers direct activation and stimulation of intraepithelial lymphocytes and natural killer (NK) cells and leads to enterocyte cytotoxicity and villous atrophy (Hüe et al., 2004). Note that the NKG2D⁺CD8⁺ T cells do not actually recognize gliadin. Meanwhile, the activated T cell stimulates the B cells to produce antigliadin, antiendomysium, and antitransglutaminase antibodies. *Source: Reproduced from Robbins and Cotran Pathologic Basis of Disease, 9th Ed. With permission from Elsevier Saunders.*

ability of cereal digests containing ATI to elicit strong activation of the TLR4 receptor and innate immune system in patients with celiac disease requires further study as discussed elsewhere (refer to Essentials IV: Diet, Microbial Diversity and Gut Integrity: Influence of Diet on Gut Barrier Function: Wheat and Grains).

As indicated above, the etiology of both type I DM and CD involves interaction of environmental and genetic factors. However, the prevalence of CD has increased in the past half century, suggesting that environmental factors may play a more potent role than originally thought. In a recent study, sera collected 50 years ago were compared with sera from a recent cohort of age-matched subjects and analyzed for tissue transglutaminase and endomysial antibodies. Data indicated that the prevalence of undiagnosed CD had increased 4.5-fold over the past 50 years. Further, undiagnosed CD was associated with a nearly fourfold increased risk of death in these human subjects (Rubio-Tapia et al., 2009).

The causes of increased mortality with undiagnosed CD included cardiovascular disease and cancer. A recent report described a gut–vascular barrier (GVB) in mice as well as humans that controls the translocation of antigens into the bloodstream (Spadoni et al., 2015). Construction of this complex barrier is similar to that seen in the blood–brain barrier (BBB). The BBB consists of vascular endothelial cells as well as astroglial cells and pericytes; in the gut, enteroglial cells and pericytes interact with intestinal vascular endothelial cells. The GVB is modified in CD patients with elevated serum transaminases. The authors suggest that the disrupted GVB, even in the symptomless, undiagnosed CD patient could allow antigens to reach peripheral tissues and induce a systemic immune response. Further studies are needed to understand the cause of GVB disruption in CD patients and the long-term consequences of chronic immune stimulation.

3.6.2 The Importance of Screening for Celiac Disease in Type I Diabetic Patients

Most cases of CD with DM are reported to be asymptomatic or silent and are detected by serologic screening. Screening and treating asymptomatic CD is motivated by the demonstration of an increased risk for gastrointestinal malignancy, such as non-Hodgkin lymphoma and cancers of the mouth, pharynx, and esophagus with untreated CD. Undiagnosed CD in children may be the underlying cause of refractory iron-deficiency anemia, short stature, low bone mineral density, and pubertal delay. Underachievement in education and working life has also been observed in adults with silent CD; this may be related causally to the increased prevalence of depressive and disruptive behavioral disorders that have been described in teenagers with untreated CD. In addition, the duration of exposure to gluten in patients with CD may be associated with an increased prevalence of other autoimmune disorders. Therefore, early diagnosis is essential for the prevention of serious complications.

3.6.3 Nutrient Malabsorption in Celiac Disease

Anemia is a presenting feature of about half of newly diagnosed adults in the United States. This presentation characteristically results from proximal disease with compromised iron, folate, and calcium absorption, while absorption of other nutrients in the distal bowel is often adequate. Watery diarrhea and weight loss can also result from significant mucosal involvement. In this case, malabsorption of all nutrients, but especially carbohydrate, fat, fat-soluble vitamins, calcium, magnesium, iron, folate, and zinc, is present. Cobalamin deficiency is more common than thought ($\sim 10\%$ of CD patients) and usually corrects with improved mucosal integrity on a gluten-free diet, especially with improved folate status. (Folate depletion severe enough to cause megaloblastosis of the intestinal mucosa has been associated with malabsorption of food cobalamin.) Iron deficiency in celiac disease can also result from occult blood loss; approximately 50% of celiac patients were found to have guaiac-positive results. The combination of low serum levels of phosphorus and calcium is characteristic of vitamin D deficiency, likely to be a consequence of malabsorption caused by the patient's mucosal disease. Elevated alkaline phosphatase level probably reflects increased turnover of bone from secondary hyperparathyroidism.

3.6.4 Differential Diagnosis of Nutrient Malabsorption

The diagnosis of celiac disease is made by characteristic changes seen on a small intestinal biopsy specimen (Fig. 3.8) and histological improvement when a gluten-free diet is instituted. Features include the absence of villi, crypt hyperplasia, increased intraepithelial lymphocytes, and infiltration of the lamina propria with plasma cells and lymphocytes (Narula et al., 2009). Serologic markers are useful for screening patients and first-degree relatives and in monitoring response to a gluten-free diet. Antigliadin (AGA), anti-endomysial (EMA), and antitissue transglutaminase (anti-tTG) antibodies (Ab) have been used to detect about a 1% prevalence in Caucasian populations, although higher prevalence (2–5%) have been reported in patients with type I diabetes and their first-degree relatives. (AGA immunoglobulin A (IgA) and IgG Ab are sensitive but not specific and should not be used for screening of adults. EMA IgA Ab are highly sensitive and specific for active disease. Anti-tTG detected using an enzyme-linked immunosorbent assay (ELISA) has equal sensitivity to the EMA test. It should be noted that patients with mild disease may have negative antibody studies.)

It can be difficult to differentiate on the basis of suspected malabsorption between inflammatory bowel disease and celiac sprue. Both disorders may have extraintestinal manifestations and can cause malabsorption and iron deficiency. A family history, elevated sedimentation rate, and clubbing are more suggestive of inflammatory bowel disease, whereas vitamin D malabsorption and autoimmune thyroiditis are more typically associated with celiac disease. Iron deficiency may occur in either condition (Fine, 1996). Iron loss is ameliorated in some, but not all of the patients who respond to a gluten-free diet.

3.7 MEDICAL NUTRITION THERAPY FOR CELIAC DISEASE

Treatment consists of life-long gluten-free diet, excluding wheat, rye, and barley and products derived from these grains. Secondary lactase deficiency is often present in patients with active disease, and a lactose-free diet should be followed until intestinal improvement is seen. All individuals with celiac disease should be screened for vitamin and mineral deficiency and have bone densitometry performed. Osteopenia has been reported in 70% of patients with celiac disease. Vitamin and mineral deficiencies should be replenished (see Table 3.1) and women of childbearing age should take folic acid supplements. Since this dietary modification is complicated and requires special knowledge of foods, many patients benefit from nutritional counseling and celiac support groups. (Diet and lifestyle information for the celiac patient can be found at http://www.csaceliacs.org/).

3.7.1 Vital Wheat Gluten in Modern Food Processing

Proline-rich wheat gliadins and other prolamins in rye and barley trigger enterocyte damage. Oat prolamins are rich in glutamine, but not proline, and are rarely toxic. As discussed elsewhere (refer to Essentials IV: Diet, Microbial Diversity and Gut Integrity), and comprehensively reviewed (Kucek et al., 2015), wheat is a major source of antigen. The authors review evidence that grain modification by hybridization and genetic modification has resulted in wheat of higher protein (gluten) content.

TABLE 3.1 Signs, Symptoms and Conditions Suggestive of CD[a]

Signs and Symptoms	Conditions
• Chronic or intermittent diarrhea • Failure to thrive or faltering growth (in children) • Persistent or unexplained gastrointestinal symptoms including nausea and vomiting • Prolonged fatigue ("tired all the time") • Recurrent abdominal pain, cramping, or distension • Sudden or unexpected weight loss • Unexplained iron-deficiency or other unspecified anemia	• Autoimmune thyroid disease • Dermatitis herpetiformis • Irritable bowel syndrome • Type I diabetes • First-degree relatives (parents, siblings, or children) with celiac disease

[a]Adapted from information found at http://www.csaceliacs.org/.

More insidiously, in the last 60 years, the food industry has increased use of gluten separated from wheat (vital wheat gluten) or modifications thereof (isolated wheat proteins). Vital wheat gluten improves the structural integrity of baked goods yet costs less per ton than soy, whey, or casein. In the United States, vital wheat gluten is added to bind multigrain bread; wheat proteins also bind and add protein to processed meat products, reconstituted seafood, and vegetarian meat substitutes. These wheat products are used as thickeners, emulsifiers, and gelling agents; wheat compounds were found in 86% of dry soup powders, 65% of canned soups, 63% of candies, 61% of ice cream, and even in vinegars, marinades, and salad dressings; an extensive list of products tested that contain wheat has been published (Atchison et al., 2010). The authors estimate that nearly 30% of supermarket food products contain wheat. Further, hair and cosmetic products can also contain wheat. Thus, the ubiquity of this potential allergen as an industrial additive can specifically impact celiac disease and nonceliac patients with sensitivity to wheat. It should be noted that in the United States, any product that uses wheat in preparation of "modified food starch" must be labeled as "wheat." Most modified food starch is made from potato, corn, or tapioca (Tables 3.2 and 3.3).

3.7.2 Prognosis

Patients with celiac disease compliant with a gluten-free diet usually experience reduction of symptoms within 2 weeks. The most common reason for treatment failure is continued consumption of gluten. Patient education is essential since wheat products are so widely used in processed foods as discussed above. Other possibilities include missed diagnosis, such as infection, bacterial overgrowth, and pancreatic

TABLE 3.2   Basic Modifications for a Gluten-Free Diet[a]

Gluten-Free	Contains Gluten	Questionable[b]	May Contain Gluten	Contamination
Amaranth	Ale	Brown rice (can be made with barley)	Breading	Oats contain prolamins that do not ordinarily cause symptoms, but they can be contaminated with wheat during harvesting, transport, milling, and processing
Arrowroot	Barley	Caramel color	Broth	
Bean	Beer	Dextrin (can be made from wheat)	Coating mixes	
Buckwheat	Farina	Flour or cereal products	Communion wafers	
Corn	Kamut (pasta wheat)	Hydrolyzed vegetable protein (HVP)	Croutons	
Distilled alcohol	Malt vinegar	Vegetable protein	Imitation bacon	Gluten-free foods can be contaminated if prepared on common surfaces or with utensils used for gluten-containing foods, eg, toasters, flour sifters, deep fat fryers
Distilled vinegar	Rye	Hydrolyzed plant protein (HPP)	Imitation seafood	
Miller	Spelt	Textured vegetable protein (TVP)	Marinades	
Nut flours	Triticale	Malt or malt flavorings (if made from wheat	Pastas	
Potato	Wheat	Modified food starch	Processed meats	
Quinoa	Durum/semolina	Natural/artificial colors	Roux	
Rice		Soy sauce or soy sauce solids (may contain wheat)	Sauces	
Sorghum			Self-basting poultry	
Soy			Soup base	
Tapioca			Stuffing mix	
Tef			Thickeners	

[a]Table abstracted from Nelms et al., 2011. Medical Nutrition Therapy: A Case Study Approach 2nd Ed. pp. 481–484. Publ: Cengage Learning. For more detailed information on foods containing gluten, see http://www.celiac.com/articles/182/1/Unsafe-Gluten-Free-Food-List-Unsafe-Ingredients/Page1.html.
[b]Verify the absence of gluten-containing grains.

insufficiency. Food allergy to milk, wheat, eggs, and other commonly consumed foods can also confound the results. Some patients do not respond despite compliance with the diet; this is termed refractory sprue and is often associated with autoimmune enteritis.

Complications of untreated or poorly controlled celiac disease can include intestinal T-cell lymphoma, gastrointestinal tract carcinomas (small bowel adenocarcinoma, esophageal and oropharyngeal squamous carcinomas), and increased mortality. These risks can be lessened by strict adherence to the diet. Intestinal lymphoma should be suspected in individuals who initially responded to the diet and have recurrent symptoms and abdominal pain.

3.7.3 Glycemic Control With a Gluten-Free Diet

A longitudinal case-controlled study in children with type I DM demonstrated that silent celiac disease resulted in poor weight gain and lower HbA_{1c}

independent of daily insulin dose and regimen (Narula et al., 2009). Intervention with a GFD caused an increase in BMI and glycated hemoglobin (HbA_1C) in case subjects compared with their baseline measures. An earlier report found that the GFD increased weight and height, as well as hemoglobin and serum ferritin levels (Hansen et al., 2006). Other reports found little difference in glycemic control or nutritional status between children with type I DM who had CD or who did not (Taler et al., 2012). These authors questioned the need to subject their asymptomatic patients to life-long GFD.

To resolve these inconsistent results, a randomized multicenter dietary intervention trial, the Celiac Disease and Diabetes-Dietary Intervention and Evaluation Trial (CD-DIET) was initiated in 2012. This trial has potential to provide data on the impact of a gluten-free diet on clinically relevant outcomes including metabolic control (HbA_1C), bone health, glycemic control, and quality of life in patients with type I DM and asymptomatic celiac disease. Children and adults (8–45 years) with type I

TABLE 3.3 Vitamin and Mineral Doses used in Treatment of Malabsorption[a]

Nutrient	Oral Dose	Parenteral Dose
Vitamin A[b]	Water-soluble A, 25,000 U/day[c]	
Vitamin E	Water-soluble E, 400–800 U/day[c]	
Vitamin D[d]	25,000–50,000 U/day	
Vitamin K	5 mg/day	
Calcium[e]	1500–2000 mg elemental calcium/day Calcium citrate, 500 mg calcium/tab[c] Calcium carbonate, 500 mg calcium/tab[c]	
Magnesium	Liquid magnesium gluconate[c] 1–3 tbsp (12–36 mEq magnesium) in 1–2 L of ORS or sports drink sipped throughout day Magnesium chloride hexahydrate[c] 100–600 mg elemental magnesium/day	2 mL of a 50% solution (8 mEq) both buttocks IM
Zinc	Zinc gluconate,[c] 20–50 mg elemental zinc/day[f]	
Iron	150–300 mg elemental iron/day. Polysaccharide iron complex[c] Iron sulfate or gluconate	Iron sucrose[g] Sodium ferric gluconate complex Iron dextran (more complications)
B complex vitamins	1 megadose tablet/day	
Vitamin B12	2 mg/day	1 mg IM or SC/month

[a]Adapted from Table 143-8 in Cecil Textbook of Medicine, 23rd Edition.
[b]Monitor serum vitamin A to avoid toxicity, especially in patients with hypertriglyceridemia.
[c]Form best absorbed or with least side effects.
[d]Monitor serum calcium and 25-OH vitamin D levels to avoid toxicity.
[e]Monitor 24-h urine calcium to assess adequacy of dose.
[f]If intestinal output is high, additional zinc should be given. Monitor copper deficiency for high doses.
[g]Parenteral therapy should be given in a supervised outpatient setting because of the risk of fatal reactions.

DM will be screened for asymptomatic CD. Eligible patients with biopsy-proven CD will be randomly assigned in a 1:1 ratio to treatment with a GFD for 1 year, or to continue with a gluten-containing diet. The primary outcome will evaluate the impact of the GFD on change in glycated hemoglobin. Secondary outcomes will evaluate changes in bone mineral density, blood glucose variability, and health-related quality of life between GFD-treated and the regular diet group over a 1-year period. The study has subsequently been expanded to multiple pediatric and adult centers in Ontario, Canada.

3.8 FUTURE RESEARCH NEEDS

Several lines of research suggest that diet plays a major role in prevention and control of autoimmune diseases such as type I DM and CD. Not only is diet critical in determining and maintaining microbial diversity and metabolism in the gut lumen, but components in the food stream have been shown to modify the integrity of the intestinal barrier, as well as the recently described gut–vascular barrier. Diet-derived microbial products such as gliadin peptides and ATI have been shown to interact with intestinal barrier components and increase translocation of bacteria and antigens into the gut lamina propria where they elicit both innate and adaptive immune responses. Probiotics and other food components have potential to ameliorate this toxic effect. Thus, the gluten-free diet may require modification to improve its efficacy in treating CD. Data supporting an immunomodulary impact of dietary vitamin D have been provided for type I DM, but vitamin D has been less well studied with respect to CD. It must be remembered that untreated (and even asymptomatic) CD can predispose to malabsorption and micronutrient depletion that can compromise host defense. The impact of maternal diet on development of autoimmune diseases in the offspring requires further study. Finally, more work is required to understand the impact of our changing diet composition on the increasing prevalence of autoimmune diseases including type I DM and CD.

References

Abadie, V., Jabri, B., 2015. Chapter 80—immunopathology of celiac disease. In: Mestecky, J., Strober, W., Russell, M.W., Cheroutre, H., Lambrecht, B.N., Kelsall, B.L. (Eds.), Mucosal Immunology, Fourth ed. Academic Press, Boston, MA, pp. 1551–1572.

Akimova, T., Ge, G., et al., 2010. Histone/protein deacetylase inhibitors increase suppressive functions of human FOXP3+ Tregs. Clin. Immunol. 136 (3), 348–363.

Ali, M.K., Bullard, K.M., et al., 2013. Achievement of goals in U.S. diabetes care, 1999–2010. N Engl. J. Med. 368 (17), 1613–1624.

American Diabetes Association, 2003. Translation of the diabetes nutrition recommendations for health care institutions. Diabetes Care. 26 (Suppl. 1), s70–s72.

American Diabetes Association, 2008. Nutrition recommendations and interventions for diabetes: a position statement of the American Diabetes Association. Diabetes Care. 31 (Supplement 1), S61–S78.

American Diabetes Association, 2010. Standards of medical care in diabetes—2010. Diabetes Care. 33 (Supplement 1), S11–S61.

American Diabetes Association, 2016a. 2. Classification and diagnosis of diabetes. Diabetes Care. 39 (Supplement 1), S13–S22.

American Diabetes Association, 2016b. 3. Foundations of care and comprehensive medical evaluation. Diabetes Care. 39 (Supplement 1), S23–S35.

Atchison, J., Head, L., et al., 2010. Wheat as food, wheat as industrial substance; comparative geographies of transformation and mobility. Geoforum. 41 (2), 236–246.

Baeke, F., Takiishi, T., et al., 2010. Vitamin D: modulator of the immune system. Curr. Opin. Pharmacol. 10 (4), 482−496.

Cahill, G.F., 1971. Physiology of insulin in man. Diabetes. 20, 785−799.

Conney, A.H., 2003. Enzyme induction and dietary chemicals as approaches to cancer chemoprevention: the Seventh DeWitt S. Goodman Lecture. Cancer Res. 63 (21), 7005−7031.

Coombes, J.L., Siddiqui, K.R.R., et al., 2007. A functionally specialized population of mucosal CD103+ DCs induces Foxp3+ regulatory T cells via a TGF-β− and retinoic acid−dependent mechanism. J. Exp. Med. 204 (8), 1757−1764.

Dabelea, D., Mayer-Davis, E.J., et al., 2014. Prevalence of type 1 and type 2 diabetes among children and adolescents from 2001 to 2009. JAMA. 311 (17), 1778−1786.

Denison, M.S., Nagy, S.R., 2003. Activation of the aryl hydrocarbon receptor by structurally diverse exogenous and endogenous chemicals. Annu. Rev. Pharmacol. Toxicol. 43 (1), 309−334.

Donath, M.Y., Halban, P.A., 2004. Decreased beta-cell mass in diabetes: significance, mechanisms and therapeutic implications. Diabetologia. 47 (3), 581−589.

Egro, F.M., 2013. Why is type 1 diabetes increasing? J. Mol. Endocrinol. 51 (1), R1−R13.

Fasano, A., 2011. Zonulin and its regulation of intestinal barrier function: the biological door to inflammation, autoimmunity, and cancer. Physiol. Rev. 91 (1), 151−175.

Ferreira, G.B., Vanherwegen, A.-S., et al., 2015. Vitamin D3 induces tolerance in human dendritic cells by activation of intracellular metabolic pathways. Cell Rep. 10 (5), 711−725.

Fine, K.D., 1996. The prevalence of occult gastrointestinal bleeding in celiac sprue. N Engl. J. Med. 334 (18), 1163−1167.

Food and Nutrition Board, 1998. Dietary Reference Intakes: Thiamin, Riboflavin, Niacin, Vitamin B6, Folate, Vitamin B12, Pantothenic Acid, Biotin, and Choline. Institute of Medicine. National Academy Press, Washington, DC.

Franz, M.J., Bantle, J.P., et al., 2002. Evidence-based nutrition principles and recommendations for the treatment and prevention of diabetes and related complications. Diabetes Care. 25 (1), 148−198.

Hansen, D., Brock-Jacobsen, B., et al., 2006. Clinical benefit of a gluten-free diet in type 1 diabetic children with screening-detected celiac disease: a population-based screening study with 2 years' follow-up. Diabetes Care. 29 (11), 2452−2456.

Hla, T., 2005. Dietary factors and immunological consequences. Science. 309 (5741), 1682−1683.

Hoffman, W.H., Passmore, G.G., et al., 2013. Increased systemic Th17 cytokines are associated with diastolic dysfunction in children and adolescents with diabetic ketoacidosis. PLoS ONE. 8 (8), e71905.

Howson, J.M.M., Krause, S., et al., 2012. Genetic association of zinc transporter 8 (ZnT8) autoantibodies in type 1 diabetes cases. Diabetologia. 55 (7), 1978−1984.

Hu, C., Wong, F.S., et al., 2015. Type 1 diabetes and gut microbiota: friend or foe? Pharmacol. Res. 98, 9−15.

Hüe, S., Mention, J.-J., et al., 2004. A direct role for NKG2D/MICA interaction in villous atrophy during celiac disease. Immunity. 21 (3), 367−377.

Jeker, L.T., Bour-Jordan, H., et al., 2012. Breakdown in peripheral tolerance in type 1 diabetes in mice and humans. Cold Spring Harb. Perspect. Med. 2 (3), a007807.

Junker, Y., Zeissig, S., et al., 2012. Wheat amylase trypsin inhibitors drive intestinal inflammation via activation of toll-like receptor 4. J. Exp. Med. 209 (13), 2395−2408.

Kerkvliet, N.I., Wong, C.P., 2009. Activation of aryl hydrocarbon receptor by TCDD prevents diabetes in NOD mice and increases Foxp3+ T cells in pancreatic lymph nodes. Immunotherapy. 1 (4), 539−547.

Kinoshita, M., Takeda, K., 2014. Microbial and dietary factors modulating intestinal regulatory T cell homeostasis. FEBS Lett. 588 (22), 4182−4187.

Knip, M., Siljander, H., 2016. The role of the intestinal microbiota in type 1 diabetes mellitus. Nat. Rev. Endocrinol. Advance online publication.

Kucek, L.K., Veenstra, L.D., et al., 2015. A grounded guide to gluten: how modern genotypes and processing impact wheat sensitivity. Compr. Rev. Food Sci. Food Safety. 14 (3), 285−302.

Lammers, K.M., Lu, R., et al., 2008. Gliadin induces an increase in intestinal permeability and zonulin release by binding to the chemokine receptor CXCR3. Gastroenterology. 135 (1), 194−204, e193.

Li, X., Atkinson, M.A., 2015. The role for gut permeability in the pathogenesis of type 1 diabetes—a solid or leaky concept? Pediatr. Diabetes. 16 (7), 485−492.

Lindfors, K., Blomqvist, T., et al., 2008. Live probiotic Bifidobacterium lactis bacteria inhibit the toxic effects induced by wheat gliadin in epithelial cell culture. Clin. Exp. Immunol. 152 (3), 552−558.

Lukasova, M., Malaval, C., et al., 2011. Nicotinic acid inhibits progression of atherosclerosis in mice through its receptor GPR109A expressed by immune cells. J. Clin. Invest. 121 (3), 1163−1173.

Mathieu, C., 2015. Vitamin D and diabetes: where do we stand? Diabetes Res. Clin. Pract. 108 (2), 201−209.

Mucida, D., Park, Y., et al., 2007. Reciprocal TH17 and regulatory T cell differentiation mediated by retinoic acid. Science. 317 (5835), 256−260.

Narula, P., Porter, L., et al., 2009. Gastrointestinal symptoms in children with type 1 diabetes screened for celiac disease. Pediatrics. 124 (3), e489−495.

Newman, J.C., Verdin, E., 2014. Ketone bodies as signaling metabolites. Trends Endocrinol. Metab. 25 (1), 42−52.

Okey, A.B., 2007. An aryl hydrocarbon receptor odyssey to the shores of toxicology: the Deichmann Lecture, International Congress of Toxicology-XI. Toxicol. Sci. 98 (1), 5−38.

Orlando, A., Linsalata, M., et al., 2014. Lactobacillus GG restoration of the gliadin induced epithelial barrier disruption: the role of cellular polyamines. BMC Microbiol. 14, 19.

Quintana, F.J., Basso, A.S., et al., 2008. Control of Treg and TH17 cell differentiation by the aryl hydrocarbon receptor. Nature. 453 (7191), 65−71.

Rabinovitch, A., Suarez-Pinzon, W., 2007. Roles of cytokines in the pathogenesis and therapy of type 1 diabetes. Cell Biochem. Biophys. 48 (2−3), 159−163.

Round, J.L., Mazmanian, S.K., 2010. Inducible Foxp3+ regulatory T-cell development by a commensal bacterium of the intestinal microbiota. Proc. Natl. Acad. Sci. 107 (27), 12204−12209.

Rubio-Tapia, A., Kyle, R.A., et al., 2009. Increased prevalence and mortality in undiagnosed celiac disease. Gastroenterology. 137 (1), 88−93.

Ryan, C.A., 1990. Protease inhibitors in plants: genes for improving defenses against insects and pathogens. Annu. Rev. Phytopathol. 28 (1), 425−449.

Safe, S., Lee, S.-O., et al., 2013. Role of the aryl hydrocarbon receptor in carcinogenesis and potential as a drug target. Toxicol. Sci. 135 (1), 1−16.

Schwab, S.R., Pereira, J.P., et al., 2005. Lymphocyte sequestration through S1P lyase inhibition and disruption of S1P gradients. Science. 309 (5741), 1735−1739.

Selhub, J., Byun, A., et al., 2013. Dietary vitamin B(6) intake modulates colonic inflammation in the IL10(−/−) model of inflammatory bowel disease. J. Nutr. Biochem. 24 (12), 2138−2143.

Serena, G., Camhi, S., et al., 2015. The role of gluten in celiac disease and type 1 diabetes. Nutrients. 7 (9), 5329.

Singh, N., Gurav, A., et al., 2014. Activation of Gpr109a, receptor for niacin and the commensal metabolite butyrate, suppresses colonic inflammation and carcinogenesis. Immunity. 40 (1), 128–139.

Smith, P.M., Howitt, M.R., et al., 2013. The microbial metabolites, short-chain fatty acids, regulate colonic Treg cell homeostasis. Science. 341 (6145), 569–573.

Sørensen, I.M., Joner, G., et al., 2012. Maternal serum levels of 25-hydroxy-vitamin D during pregnancy and risk of type 1 diabetes in the offspring. Diabetes. 61 (1), 175–178.

Spadoni, I., Zagato, E., et al., 2015. A gut-vascular barrier controls the systemic dissemination of bacteria. Science. 350 (6262), 830–834.

Stoffels, K., Overbergh, L., et al., 2006. Immune regulation of 25-hydroxyvitamin-D3-1α-hydroxylase in human monocytes. J. Bone Miner. Res. 21 (1), 37–47.

Sugimura, T., Wakabayashi, K., et al., 2004. Heterocyclic amines: mutagens/carcinogens produced during cooking of meat and fish. Cancer Sci. 95 (4), 290–299.

Takiishi, T., Ding, L., et al., 2014. Dietary supplementation with high doses of regular vitamin D3 safely reduces diabetes incidence in NOD mice when given early and long term. Diabetes. 63 (6), 2026–2036.

Taler, I., Phillip, M., et al., 2012. Growth and metabolic control in patients with type 1 diabetes and celiac disease: a longitudinal observational case-control study. Pediatr. Diabetes. 13 (8), 597–606.

Tamblyn, J.A., Hewison, M., et al., 2015. Immunological role of vitamin D at the maternal–fetal interface. J. Endocrinol. 224 (3), R107–R121.

Trasino, S.E., Gudas, L.J., 2015. Vitamin A: a missing link in diabetes? Diabetes Manag. (London, England). 5 (5), 359–367.

Treiber, G., Prietl, B., et al., 2015. Cholecalciferol supplementation improves suppressive capacity of regulatory T-cells in young patients with new-onset type 1 diabetes mellitus—a randomized clinical trial. Clin. Immunol. 161 (2), 217–224.

Volta, U., Tovoli, F., et al., 2011. Clinical and immunological features of celiac disease in patients with type 1 diabetes mellitus. Expert Rev. Gastroenterol. Hepatol. 5 (4), 479–487.

Wang, L., de Zoeten, E.F., et al., 2009. Immunomodulatory effects of deacetylase inhibitors: therapeutic targeting of FOXP3+ regulatory T cells. Nat. Rev. Drug Discov. 8 (12), 969–981.

Warncke, K., Fröhlich-Reiterer, E.E., et al., 2010. Polyendocrinopathy in children, adolescents, and young adults with type 1 diabetes. Diabetes Care. 33 (9), 2010–2012.

Yamaguchi, T., Hirota, K., et al., 2007. Control of immune responses by antigen-specific regulatory t cells expressing the folate receptor. Immunity. 27 (1), 145–159.

Zipitis, C.S., Akobeng, A.K., 2008. Vitamin D supplementation in early childhood and risk of type 1 diabetes: a systematic review and meta-analysis. Arch. Dis. Childhood. 93 (6), 512–517.

Further Reading

Albert, et al., 2014. Chapter 23. Pathogens and infection; Chapter 24. The innate and adaptive immune system, Molecular Biology of the Cell, sixth Ed. Garland Science, New York, NY.

Goldman, L., Schafer, A.I., 2012. Chapter 247 (Type I diabetes mellitus), Chapter 143: Approach to the patient with diarrhea and malabsorption (celiac disease). Goldman Cecil Medicine. Elsevier Saunders, Philadelphia, PA.

Kumar, V., et al., 2014. Chapter 24. The endocrine system: the endocrine pancreas. Chapter17. The gastrointestinal tract: malabsorption and diarrhea. celiac disease., Robbins and Cotran Pathologic Basis of Disease, ninth Ed. Elsevier Saunders, Philadelphia, PA.

Nelms, M.N., et al., 2011. Nutrition therapy and pathophysiology. Diseases of the Endocrine System. Wadsworth Cengage Learning, Belmont, CA.

4

Type II Diabetes, Peripheral Neuropathy, and Gout

CHIEF COMPLAINT (ADAPTED FROM TRENCE, D.L., 2002)

DR, a 55-year-old male, visits a new family physician because he has been very tired and generally not feeling well. He has been urinating frequently and also suffers from increasing pain and numbness in his feet and legs. He has been troubled by a swollen and painful knot in the joint of his big toe on frequent occasions.

PAST MEDICAL HISTORY AND MEDICATION

- Five years ago, patient was diagnosed as having diabetes and was placed on metformin twice a day with meals. He had been visiting his regular family physician for twice-yearly checkups and prescriptions until the doctor retired over a year ago. Patient advised that he was laid off 6 months ago and cannot afford to find a new doctor. Patient advised that his prescription ran out over a month ago.
- Mr. R has no history of previous hospitalizations or other major illness. He acknowledged that his weight has gradually increased, especially around the waist since he stopped playing sports in college and entered a sedentary career as an accountant. He advised that he likes a pint or two of Guinness in the pub several evenings a week. Patient advised that he often resorts to antacids for heartburn after pub meals and that he does not take vitamins or supplements, but has been taking capsules of cinnamon because he heard on TV that this spice is good to prevent diabetes.
- Three years ago, patient first noticed intermittent numbness in the toes of his right foot; the numbness slowly spread to both feet and then to the lower legs. The pain was worse at night and awakened him with a burning sensation that involved both legs nearly up to the knees.

- One year ago, he was referred for an electromyographic examination and nerve-conduction studies of the arms and legs. These tests were normal, as were motor and sensory conduction studies performed 4 months ago.
- Patient was prescribed tramadol and capsaicin cream as needed for the pain.
- Since then patient has noted additional symptoms including "pins and needles" and aching in the legs and feet, and swelling in his toe joints.
  - He has no difficulty walking and no change in bladder or bowel function.
  - Patient advised that he has noticed an increase in urination, especially in the last month.

FAMILY MEDICAL HISTORY

Patient's maternal grandmother developed type II diabetes at age 60 and died of kidney failure at age 75. Two sisters also have diabetes and several maternal relatives have been treated for colon cancer. Paternal relatives suffered from cardiovascular disease (CVD); patient's father died at age 85 of a heart attack. His mother remains in good health at age 80.

SOCIAL HISTORY

Patient is an accountant, single, lives alone, and does not follow a special diet. He does not smoke, but reports drinking 2–3 beers every day after work. He does not cook for himself. He eats dinner out at least 2–3 times per week, orders take-out from local restaurants and heats up prepared meals at other times. His job involves considerable desk work, although he does walk to and from the subway (three blocks) daily. He does not engage in regular exercise.

Nutritional Pathophysiology of Obesity and its Comorbidities
DOI: http://dx.doi.org/10.1016/B978-0-12-803013-4.00004-1

PRESENTING SYMPTOMS

General	Fatigue, weakness, numbness and tingling in extremities, pain in joints
Endocrine	Increased thirst (polydipsia) and frequent urination (polyuria)

PHYSICAL EXAMINATION

General	Patient is a well-nourished, well-developed obese male in no acute distress who looks older than his age
Vital signs	Temp. 98.8°F; HR 72 bpm; Resp. 14 bpm; BP 135/85 mm Hg
Anthropometrics	Ht. 5′9″ (175 cm); current weight 200 lb (91 kg); BMI 30 kg/m²
	Waist circumference 42 inches
Fundus exam	Mild diabetic retinopathy
Foot exam	Onychomycosis (toenail fungal infection), mild swelling and erythema in the first metatarsal joint of the right toe
Neurological	Neurologic examination shows normal mental status and cranial-nerve function
	Motor examination shows normal tone, bulk, and strength throughout
	Sensory examination shows marked diminution of the patient's perception of vibration at the toes and ankles; the diminution is less evident at the knees
Chest X-ray	No abnormalities

Laboratory data (fasting values)		Normal values
Serum glucose	285 mg/dL	70–100 mg/dL
HbA1c	9.5%	4–6%
Total cholesterol	240 mg/dL	<200 mg/dL
High-density lipoprotein cholesterol	32 mg/dL	40–50 mg/dL
Triglycerides	275 mg/dL	<150 mg/dL
Uric acid	9.2 mg/dL	3.5–7.0 mg/dL
Serum ferritin	350 ng/mL	24–336 ng/mL
Vitamin B12 level	210 pg/L	>200 pg/L
Urinary glucose	2+	neg.
Urinary protein	1+	neg.
Urinary ketones	2+	neg.
ESR	28 mm/hr	0–22 mm/hr

24-HOUR DIET RECALL

Breakfast (home)	Large plain bagel	1 large (4 oz)
	Orange juice	16 oz glass
	Regular cream cheese	1 oz (2 tbsp.)
	Decaffeinated coffee	2 cups
	Half/half in coffee	2 tbsp.
Snack (office)	Large banana	1 whole
Lunch (office)	Turkey luncheon meat	4 oz
	Swiss cheese	1 slice (1 oz)
	White bread	2 slices
	Potato chips	1 small bag
	Cranberry juice cocktail	16 oz bottle
Snack (office)	Fat-free cookies	6
Dinner (restaurant)	Fried rice (Chinese)	1 cup
	Hunan chicken (fried)	4 oz
	Shrimp and cashews	4 oz
	Beer	2 bottles
	Orange	1 large
Snack (home)	Fat-free ice cream	1 cup

DIET ANALYSIS

Total calories	3460 kcal per day (requirements = 2177)[a]
Protein	155.7 g (18% of total calories)
Fat	115 g (30% of total calories)
Carbohydrate	450 g (52% of total calories)

[a]*Calorie requirements obtained using Harris Benedict equation and activity factor of 1.2 (sedentary).*

RESOURCES

4.1 TYPE II DIABETES AND ENERGY DYSREGULATION

Type II DM has reached epidemic proportions both nationally and internationally. Estimates suggest that almost 8% (24 million persons) of the US population suffer from the disease and that an additional 57 million Americans suffer from prediabetes (defined as fasting blood glucose between 100 and 125 mg/dL) (Patti and Corvera, 2010). Health consequences of diabetic dysregulation include higher risk of death, heart disease, stroke, kidney disease, blindness, amputations, neuropathy, and pregnancy-related complications. Together, diabetes and its comorbidities amount to over $174 billion in health care costs in the United States. More ominously, medical control of diabetes dysregulation is not yet effective, leading to a projected increase in the incidence of diabetes from 2000 to 2050 by over 165%.

β-Cell dysfunction in type II DM produces a panoply of metabolic dysfunctions as reviewed by Saad et al. (2015). Insulin resistance (defects of insulin signaling) increases adipocyte lipolysis with production of potentially toxic nonesterified fatty acids (NEFA). Dysregulated adipokine signaling, abnormal regulation of incretin and gut hormones, hyperglucagonemia, chronic inflammation with increased production of inflammatory mediators, increased hepatic glucose production and kidney glucose resorption, as well as chronic exposure to oxidative stress are seen in type II DM. Dysregulated insulin signaling in the brain results in altered balance of central nervous system pathways controlling food intake and energy expenditure.

Evidence is accumulating that mitochondrial abnormalities may be central to these pathologies (Schilling, 2015). While mitochondria are recognized for their role in generating most cellular adenosine triphosphate (ATP) via the electron transport chain (ETC), they also control for other metabolic activities including (refer to Essentials I: Life in an Aerobic World: Nutrients, Mitochondrial Oxidants, and Antioxidant Protection and Chapter 5: Cardiomyopathy and Congestive Heart Failure: Cardiomyocyte Excitation–Contraction):

- Generation of tricarboxylic acid (TCA) cycle metabolites that function in cytosolic pathways;
- Oxidative catabolism of amino acids; nitrogen excretion via ornithine cycle activity ("urea cycle");
- Generation of reactive oxygen species (ROS), some with important signaling functions;
- Control of cytoplasmic calcium availability;
- Synthesis of cellular Fe/S clusters and protein cofactors for enzyme activity;
- Generation of acetyl CoA for the rate-limiting first steps in steroidogenesis and fatty acid biosynthesis.

This multiplicity of function closely links mitochondrial integrity to endocrine homeostasis. It also explains the variability in pathophysiology, severity, and age of onset of the increasing number of diseases recognized to arise from mitochondrial dysfunction and limitations. Several mechanisms are under investigation as potential causes of insulin resistance and/or diabetes progression:

- Reduced mitochondrial numbers and mitochondrial biogenesis;
- Impaired mitochondrial capacity and/or function including control of oxidative stress;
- Altered insulin signaling due to cellular lipid accumulation, proinflammatory signals, and endoplasmic reticulum stress;
- Reduced incretin-dependent and -independent β-cell insulin secretion.

4.2 DYSREGULATION OF INSULIN SECRETION AND SIGNALING

Abundant data suggest that intrinsic defects or altered regulation of mitochondrial oxidative metabolism may contribute to the development of both obesity and insulin resistance. In general, individuals with high energy requirements and high oxidative capacity remain insulin-sensitive even with high energy intake. In contrast, reduction in ATP utilization with excess substrate available for the ETC results in generation of ROS and insulin resistance with lipid accumulation in inappropriate tissues. Initially, balance can be regained either by increasing oxidative capacity (exercise) or decreasing the fuel load (calorie restriction) with resolution of oxidative stress. If balance is not regained, chronic insulin resistance continues, leading to exacerbation of other symptoms of the metabolic syndrome (hypertension; dyslipidemia) and risk for type II DM. Thus, the pathogenesis of type II diabetes can be viewed as a vicious cycle in which an individual's mitochondrial oxidative capacity, determined by genetic background, ethnic environment, intrauterine exposure, diet, and age is overwhelmed by chronic fuel excess.

High fat intake provides abundant oxidative substrate for nicotinamide adenine dinucleotide (NADH) production that can overload the mitochondrial electron transport chain (ETC), allow escape of high-energy electrons, and lead to increased oxidative stress and ROS generation with further impairment of oxidative capacity (refer to Chapter 5: Cardiomyopathy and Congestive Heart Failure: Mitochondrial Production and Management of Reactive Oxygen Species). ROS signals cellular glucose uptake; high intracellular glucose can stimulate both ROS production and ROS scavenging. Imbalance of these two functions (ROS production and scavenging) leads to a toxic cycle and ultimately, cell death (Liemburg-Apers et al., 2015). With time, mitochondrial damage ensues, not only in muscle and adipose tissue, but also in the pancreas, leading to reduced insulin secretion. Together, these factors contribute to progressively impaired insulin sensitivity and increased risk for DM.

4.2.1 Stimuli for Insulin Secretion and Release

Although insulin production has been reported in other cells such as the adipocyte (Sanjabi et al., 2015), the pancreatic β cell is the major systemic producer of insulin as comprehensively reviewed by Fu et al. (2013). Following transcription, the 110 amino acid, biologically inactive preproinsulin is translocated into the endoplasmic reticulum where the signal peptide is removed. Proinsulin undergoes folding and formation with three disulfide bonds and enters the Golgi where it is cleaved to the 51-amino-acid insulin and the remnant, C-peptide. Recent evidence suggests that overstimulation of proinsulin production can induce unfolded protein stress response and reduce production of mature insulin (Szabat et al., 2016). Both fractions are stored in secretory vesicles (zymogen granules) together with other components such as amyloid polypeptide (islet associated polypeptide (IAPP) or amylin). Insulin is complexed to two Zn ions as a Zn-insulin hexamer. In response to physiologic signals the vesicles are released from the β cell by exocytosis and insulin is secreted with C-peptide in equimolar concentrations. Zn bound to insulin is also released into the blood at the time of β-cell degranulation. Because it is secreted in equimolar amounts with insulin, C-peptide has been used as a surrogate marker for β-cell function.

Glucose oxidation is the primary stimulus for insulin secretion, thus, mutations in mitochondrial DNA that compromise these pathways have been linked to type II DM (Nile et al., 2014). Briefly, diet-derived glucose transits the enterocyte, enters the portal vein and equilibrates across the human β-cell membrane on GLUT 1 and GLUT 2 transporters where it is phosphorylated by the high Km hexokinase IV enzyme, glucokinase. As the rate-limiting first step in glycolysis, glucokinase has been proposed as the pancreatic "glucose sensor" (Wollheim, 2000). Pyruvate, the product of glycolysis, enters the mitochondria where it is converted to acetyl CoA by mitochondrial pyruvate dehydrogenase and the acetate moiety oxidized by the TCA cycle. The high-energy electrons are transported by reducing equivalents (NADH and flavin adenine dinucleotide (FADH2)) to the mitochondrial ETC and used to generate the proton-motive force across the inner mitochondrial membrane. This electrochemical gradient drives both ATP synthesis and ATP transport to the cytoplasm. In the β cell, the increased cytosolic ATP:ADP ratio inhibits the ATP-sensitive K^+ (ATPK) channels, depolarizing the plasma membrane and evoking the opening of voltage-sensitive Ca^{2+} channels. Ca^{2+} influx into the β cell stimulates exocytosis of proformed insulin contained in zymogen granules. This is the rapid first phase of insulin secretion in response to a meal. Continued secretory stimulus is accompanied by a delayed insulin response representing newly synthesized insulin.

In addition to glucose, drugs and macronutrients can stimulate insulin secretion. The ATP-K channel is comprised of two subunits, one of which is a sulfonylurea receptor, targeted by the oral hypoglycemic agents, the sulfonylureas. Other macronutrients enhance glucose-stimulated insulin secretion (GSIS). As recently reviewed (Moore et al., 2015), free fatty acids and amino acids, especially the branched chain amino acids (valine, leucine, isoleucine) also provide substrate for ATP production. Amino acids increase GSIS by contributing substrate to the TCA cycle and also stimulate Ca^{2+} entry into the β cells. Hormones, including melatonin, estrogen, leptin, growth hormone, and glucagon-like peptide (GLP), regulate GSIS by different pathways.

4.2.2 Incretins

Food in the gastrointestinal tract stimulates signals from the gut that promote β-cell insulin secretion. Of these, two incretins have been most well studied. Glucose-dependent insulinotropic polypeptide (GIP) is secreted by enteroendocrine "K cells" in the proximal small bowel. The second incretin is GLP-1, secreted by "L cells" in the distal ileum and colon. These hormones also travel through the portal vein to the pancreas (Baggio and Drucker, 2007). GIP binds to its receptor and increases intracellular cAMP and Ca^{2+} levels in the β cell, stimulating insulin secretion. It also increases insulin action at the target tissue by stimulating lipoprotein lipase (LPL) activity and increasing insulin-stimulated triglyceride (TG) reesterification in the adipocyte. In addition, it modulates fatty acid synthesis and promotes pancreatic β-cell proliferation and survival.

GLP-1 is a posttranslational cleavage product of the proglucagon gene; this gene also produces another peptide, noninsulinotropic GLP-2. GLP-1, like GIP, binds to its receptor on β cells and increases intracellular cAMP and Ca^{2+} levels. GLP-1 also inhibits gastric emptying, decreases food intake, inhibits glucagon secretion, and reduces endogenous glucose production. Pancreatic β cells are protected from apoptosis and are stimulated to proliferate by GLP-1 due to upregulation of the β-cell transcription factor, pancreatic duodenal homeobox-1 protein. This factor binds the insulin gene promoter, stimulating insulin gene transcription. GLP-1 upregulates glucokinase and glucose transporter 2 (GLUT2), enhancing glucose uptake and phosphorylation in the β cell. GIP and GLP-1 are degraded in the circulation by dipeptidyl peptidase (DPPs), especially DPP-4.

Together these hormones produce the "incretin effect" that includes enhanced β-cell insulin secretion, reduced α-cell glucagon secretion, and delayed gastric emptying, thereby promoting satiety (Kim and Egan, 2008). The "incretin effect" is significantly blunted in patients with type II diabetes. While defective secretion of GLP-1 does not appear to be a cause of type II diabetes, treatment with a GLP-1 agonist has been shown to normalize blood glucose, improve β-cell function, and restore first-phase insulin secretion, restoring "glucose competence" to β cells. GLP-1 has a short half-life (1.5–2 minutes) due to rapid degradation by DPP4.

Because of its short half-life, GLP-1 must be administered by continuous subcutaneous infusion. This difficulty has been addressed by development of long-lasting GLP-1 analogs approved for use in the United States. These include exenatide (Byetta) and liraglutide (Victoza), both of which can be administered by subcutaneous injection. Alternatively, DPP4 inhibitors that can be taken orally have been developed. Although continuous stimulation of the pancreas can produce potentially deleterious changes, such as pancreatitis and β-cell transformation in experimental models (Gale, 2013), these complications have not been seen to date in clinical trials (Egan et al., 2014).

4.2.3 Mechanisms of Insulin Signaling

Insulin signaling at the target tissue has been comprehensively reviewed (Muniyappa et al., 2007; Munir et al., 2013). Secreted insulin binds to specific cell surface receptors located on target cells. Insulin binding to its cell surface receptor promotes autophosphorylation of the cytoplasmic insulin receptor tyrosine kinase with subsequent phosphorylation of insulin receptor substrates such as IRS-1. This action activates the lipid kinase, phosphatidylinositol 3-kinase (PI3K), thereby triggering a serine kinase cascade (PKD-1, Akt) that culminates in the translocation of the insulin-dependent glucose transporter (GLUT4) from its cytoplasmic storage vesicle to the plasma membrane and facilitates glucose uptake.

Insulin, by activating PI3K, also modulates vascular tone. It stimulates endothelial nitric oxide synthase (eNOS) to produce the "gasotransmitter," nitric oxide (NO). NO plays many roles in vascular homeostasis as described elsewhere (refer to Chapter 6: Atherosclerosis and Arterial Calcification: Metabolic Targets for Diet Modification). It increases vasodilation and increases blood flow; it also directs capillary recruitment. These actions facilitate glucose utilization in classical target tissues such as the skeletal muscle.

By triggering phosphorylation of IRS-1, insulin also activates a mitogen-activated protein kinase (MAPK) signaling pathway. MAPK activation triggers a cascade that regulates the effects of insulin on mitogenesis, growth, and differentiation. MAPK-dependent insulin signaling pathways are for the most part unrelated to the metabolic actions of insulin. They regulate secretion of the vasoconstrictor, endothelin-1, from the endothelium. In health, by simultaneously stimulating these distinct signaling pathways (PI3K and MAPK), insulin couples metabolic and hemodynamic homeostasis.

4.2.4 AMPK Activators

If ATP availability is optimal, metabolic controls reduce production of substrate for the ETC and enhance ATP utilization for growth and biosynthetic purposes as described elsewhere (refer to Chapter 1: Obesity and Metabolic Syndrome: Nutrient Sensors and Downstream Targets). In fact, insulin resistance has been proposed as a defense against toxic accumulation of glucose and NEFA in metabolizing cells. The master regulator for fuel availability is AMP-activated protein kinase (AMPK) that senses when cellular energy levels are low and generates signals that stimulate ATP production by increasing cellular glucose uptake and fatty acid oxidation and inhibit ATP utilization in processes such as hepatic glucose production and lipogenesis (refer to Chapter 1: Obesity and Metabolic Syndrome: Nutrient Sensors and Downstream Targets). Nutrient excess (glucose, branched chain amino acids; fatty acids) increases ATP availability and inhibits AMPK, reducing insulin signaling and glucose uptake, resulting in insulin resistance (Coughlan et al., 2013). A recent study using human umbilical vein endothelial cells demonstrated that exposure to short-term high-glucose media reduces insulin signaling even in the presence of physiological insulin levels (De Nigris et al., 2015). High extracellular glucose suppressed the PI3K/eNOS arm of insulin signaling while insulin-stimulated MAPK stimulation of endothelin 1 was unchanged.

Abundant evidence describing AMPK dysregulation in obesity, metabolic syndrome, and type II DM has been reviewed (Coughlan et al., 2014). Downstream events in insulin signaling are under intensive investigation to understand the root causes of diabetes and to develop effective therapies (Samuel and Shulman, 2016). Because AMPK regulates insulin sensitivity it is considered a therapeutic target for type II DM as reviewed by Coughlan et al. (2014). Activation of AMPK by caloric restriction and exercise is described below. Several drugs have been developed that activate AMPK by diverse mechanisms; these drugs include the widely used biguanide, metformin (Shaw et al., 2005), GLP-1, and the thiazolidinediones. The mechanisms through which AMPK activators work

FIGURE 4.1 *Galega officinalis* **(French lilac).** Galegine, derived from the medicinal plant, *Galega officinalis* (French lilac) is the toxic parent product of metformin, approved for use in the United States in 1995, and frequently prescribed for Type II DM. This plant was classified as a noxious weed in the United States because it is toxic to grazing animals. Its common name is goat's rue. *Source: Figure by Epibase (own work) (CC BY 3.0 <http://creativecommons.org/licenses/by/3.0>), via Wikimedia Commons.*

and their effective timeframe have been comprehensively reviewed (Grahame Hardie, 2016). These authors also describe the plethora of natural products that have been used as AMPK activators, most of which inhibit the mitochondrial ETC. It is of interest that metformin and its parent compound, phenformin were derived from natural products (Fig. 4.1).

Of the 100 or more natural products shown to activate AMPK (Grahame Hardie, 2016) most are polyphenols, but many have varied structures. Many have been used in traditional Asian medicine and in herbal remedies and the mechanisms through which most act are unknown. Most are produced in plants as toxic defenses against predators and pathogens. Examination of the list of natural products that inhibit the mitochondrial ETC and thus activate AMPK reveals several well-known bioactive components: quercetin, anthocyanin, apigenin, resveratrol, caffeic acid, curcumin, epigallocatechin gallate, eugenol, sulforaphane, theaflavins, and cinnamic acid.

4.2.5 Does Cinnamon Improve Insulin Signaling?

Cinnamon has been recommended to diabetic patients to improve insulin sensitivity and lipid profile. As recently reviewed (Kopp et al., 2014) hydroxychalcone, a bioactive compound in cinnamon, enhanced glucose uptake and glycogen synthesis in 3T3-L1 adipocytes; mice treated with an extract of cinnamon bark also increased adiponectin secretion from adipocytes. Among its other actions, adiponectin activates AMPK. Cinnamaldehyde (CA) increased AMPK activation in 3T3-L1 cells and in high-fat-diet-fed mice

(Huang et al., 2011). Cinnamon may improve insulin sensitivity by other mechanisms. Trans-cinnamic acid (tCA) was identified as a ligand by the G-protein-coupled receptor (GPR109A) expressed in activated macrophages and adipocytes. As described elsewhere (refer to Chapter 3: Type I Diabetes and Celiac Disease: Type I DM as Dysregulated Immune Response: Fig. 3.2) GPR109A activation by niacin and butyrate reduces intestinal inflammation; GPR109A activation by niacin was shown to increase adiponectin secretion from primary adipocyte, but not in the murine cell line, 3T3-L1, or in cultures from wild-type mice (Plaisance et al., 2009). A recent report extended these findings (Kopp et al., 2014). Treatment with tCA stimulated adiponectin secretion and the pAMPK/AMPK ratio in 3T3-L1 cultures; the effect was inhibited by pertussis toxin, suggesting that GPR109A may play a role in tCA stimulated signaling.

Despite clarification of the biological plausibility for cinnamon as an adjuvant therapy for diabetic patients, the effective dose is likely to be far higher than can be consumed in the diet. Furthermore, the safety and efficacy of supplemental cinnamon has not been established and cannot be recommended as noted below. The plethora of bioactive dietary components with similar AMPK activation properties suggests that cinnamon should be consumed, perhaps on morning oatmeal or as a flavor on cappuccino, as part of a diet containing other bioactive components with potential to improve insulin sensitivity.

4.2.6 Insulin Action at the Adipocyte

Insulin stimulation and membrane translocation of GLUT 4 at the adipocyte enhances glucose uptake and glycolytic metabolism to glycerol-3-phosphate. In addition, insulin stimulates lipoprotein lipase (LPL), facilitating uptake of long-chain fatty acids from TG-rich lipoproteins (chylomicrons; very-low-density lipoproteins) by membrane proteins such as fatty acid binding protein, fatty acid translocase (FAT; CD36) and fatty acid transporter protein (FATP). Thus, insulin stimulates TG uptake, esterification and storage in the adipocyte lipid droplet (Fig. 4.2).

The role of aquaglyceroporin in permeation of glycerol to provide a substrate for hepatic gluconeogenesis in the insulin-resistant patient is under intense investigation (Madeira et al., 2015). The authors also provide preliminary evidence to support a model linking dysregulation of aquaglyceroporins on visceral and subcutaneous adipocytes as well as the liver and other tissues in obesity-associated type II DM with ectopic fat deposition. This work is in its very early stages and requires further investigation.

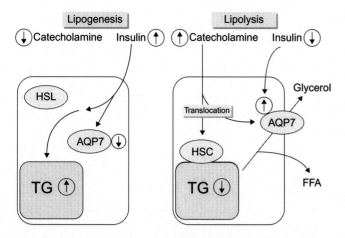

FIGURE 4.2 **Regulation of adipocyte lipogenesis and lipolysis.** Under the influence of insulin, fatty acids are esterified to triglyceride and aquaglyceroporin 7 (AQP7) synthesis is downregulated. Catecholamine-stimulated lipolysis results in hormone-sensitive lipase-mediated triglyceride hydrolysis by adipocyte triglyceride lipase (ATGL) and monoacylglycerol lipase (MAG) to fatty acids and glycerol. AQP7, the main glycerol efflux channel, is upregulated, translocated to the plasma membrane and allows glycerol permeation to the blood stream (Maeda, 2012). *Source: Reproduced from Maeda, N., 2012. Implications of aquaglyceroporins 7 and 9 in glycerol metabolism and metabolic syndrome. Mol. Aspects Med. 33 (5–6), 665–675 with permission from Science Direct.*

4.2.7 Influence of Redox Signaling on Insulin Signaling in Fuel Overload

Redox activities that control metabolic signaling in obesity and nutrition overload have been recently reviewed (Jankovic et al., 2015). Fuel overload from any source (diet or lipolysis) can overwhelm the mitochondrial ETC by generating excess high-energy electrons (NADH and FADH2). As discussed elsewhere (refer to Chapter 6: Atherosclerosis and Arterial Calcification: Metabolic Targets for Diet Modification), an over-reduced ETC can allow electrons to escape, especially from complexes I and III, and directly reduce molecular oxygen to superoxide ($O_2^{\bullet-}$); $O_2^{\bullet-}$ is rapidly dismuted to H_2O_2. While high levels of H_2O_2 are dangerous, low levels engage in redox signaling important for metabolic homeostasis. H_2O_2 can oxidize specific sites (usually cysteine residues) on a variety of enzymes thereby regulating their activities. Catalase (2 $H_2O_2 \rightarrow$ 2 $H_2O + O_2$) and other antioxidant enzymes have been proposed as physiological modulators to maintain H_2O_2 in a range optimal for redox signaling but below the threshold for oxidative damage (Rindler et al., 2013).

H_2O_2 has been shown to exert stimulatory or inhibitory effects on insulin signaling. The authors cite evidence that under conditions of fuel overload and excess H_2O_2 production, the high-fat-fed animal and murine 3T3 cells cultured in high-fat media compensated by upregulating antioxidant defense mechanisms and increased production of uncoupling proteins that bleed off the mitochondrial H^+ gradient. High levels of H_2O_2 reduce insulin sensitivity in insulin-dependent tissues (adipocytes and muscles), reducing uptake of glucose through the insulin-dependent GLUT4 transporter and downregulating insulin-dependent LPL needed for lipid uptake (Wang and Eckel, 2009). Chronic fuel overload can exceed the adaptive mitochondrial capacity, leading to irreversible damage of proteins and signaling pathways, resulting in even more severe insulin resistance. In the adipocyte, insulin resistance prevents glucose and lipid uptake as substrates for TG storage, creating a "dysfunctional" adipocyte unable to form and store fat.

4.3 NUTRIENTS, INSULIN SECRETION, INSULIN SIGNALING, AND TYPE II DM RISK

While some bioactive components activate AMPK and sensitize tissues to insulin action, diet and nutrients also influence insulin secretion and tissue resistance by other mechanisms as reviewed by Moore et al. (2015). Nutrient actions of vitamin D (Wolden-Kirk et al., 2011), leucine, a known insulin secretagog (Yang et al., 2006) and phytoestrogens such as the isoflavone, genestein (Liu et al., 2006a) are being investigated. Impaired carnitine metabolism in obesity and its comorbid conditions have been discussed elsewhere (refer to Chapter 5: Cardiomyopathy and Congestive Heart Failure: Targets for Myocardial Protection: Macronutrient Dysregulation: Carnitine) and reviewed by Flanagan et al. (2010). Additionally, several complementary and alternative remedies to enhance insulin sensitivity have been proposed for which a mechanism has not been identified (Birdee and Yeh, 2010). Selected nutrients are discussed below.

4.3.1 Thiamin (Vitamin B1)

High levels of thiamin (vitamin B1) in the pancreas are maintained by transcriptional regulation of thiamin carriers, THTR-1 and THTR-2. Thiamin deficiency leads to a marked impairment in insulin synthesis and secretion in rats as reviewed by Luong and Nguyen (2012). The precise mechanism through which thiamin acts to modulate insulin secretion in human type II DM has not been clarified. Thiamin is required as coenzyme for the pyruvate dehydrogenase complex that converts cytosolic pyruvate to acetyl CoA in the mitochondrial matrix and for α-ketoglutarate dehydrogenase in the TCA cycle, thus thiamin is essential for providing substrate for ATP production. It is also needed for transketolase (Tk) activity in the hexose monophosphate shunt (HMS), thereby providing ribose substrate for nucleotide biosynthesis and nicotinamide adenine dinucleotide phosphate (NADPH) cofactor for glutathione regeneration (by glutathione reductase). Since the pancreatic β cell is especially vulnerable to oxidative stress and NADPH is a key component of mitochondrial protection, it is possible that Tk activity plays a role in preventing β-cell damage. While these observations suggest that thiamin is required for GSIS, studies that test this possibility at the molecular level are lacking. Mechanistic and clinical studies are needed to identify molecular actions of thiamin with potential to ameliorate the comorbid conditions that result from hyperglycemia (Thornalley, 2005; Pácal et al., 2014).

4.3.2 Metals and Insulin Signaling

Hyperglycemia alters the distribution and urinary excretion of several metals. A 6-day balance study was conducted with 20 healthy male diabetic subjects and 20 controls. Seven minerals were studied: copper (Cu), iron (Fe), zinc (Zn), calcium (Ca), manganese (Mn), selenium (Se), and chromium (Cr). Resident subjects consumed diets constructed according to current diabetes guidelines; all food and beverage items were provided by the investigators. Elemental content was directly measured by inductively coupled plasma—mass spectrometry in duplicate diets. Urine and fecal metals were measured to determine the balance of each element (Cooper et al., 2005). Although dietary elemental intake was not different between subjects and controls, metal excretion was higher in diabetic subjects; urinary Cu excretion closely correlated with that of urinary ferritin. While urinary volume was higher in diabetic patients, statistical modeling excluded urinary volume as a factor in urinary element losses. Serum concentrations of the elements were not different between diabetics and controls despite higher urinary losses, suggesting that hyperglycemia or some aspect of the diabetic condition altered tissue distribution of these elements. This study also reinforces the clinical observations that, in general, use of serum values to assess nutrient status, especially metal status, is not informative.

4.3.3 Copper

As is described elsewhere (refer to Essentials II: Heavy Metals, Retinoids and Precursors: Copper), copper is a toxic metal that is essential for life. Thus, its transport and distribution must be tightly regulated to prevent tissue damage. Copper is an essential cofactor for metalloenzymes, and for the terminal enzyme complex in the mitochondrial respiratory chain, cytochrome c oxidase (COX). If COX activity is compromised by copper depletion, the ETC electrochemical gradient is not generated, ATP cannot be synthesized, and insulin secretion is compromised. Additionally, inhibition of the ETC by copper deficiency can allow electron escape with the generation of ROS as discussed elsewhere (refer to Chapter 5: Cardiomyopathy and Congestive Heart Failure: Production and Management of Reactive Oxygen Species).

The importance of copper in control of blood glucose in rodents was first described over 80 years ago (Keil and Nelson, 1934). Copper deficiency combined with a high fructose (but not starch) intake was associated with an increased frequency of glucose intolerance, electrocardiographic abnormalities, hypertension, and disturbances in lipid metabolism in normal human subjects (Reiser et al., 1985). Fructose (and sucrose) can compromise transcellular calcium absorption by reducing enterocyte ATP content. (Refer to Essentials III: Nutrients for Bone Structure and Calcification: Calcium.) Although not measured in humans, copper transport across the basolateral enterocyte border by Cu-ATPase (ATP7A) could also be compromised by reduced enterocyte ATP. In rats, copper deficiency resulted in increased hepatic and pancreatic lipid deposition and apoptosis of pancreatic acinar cells (Weksler-Zangen et al., 2013). Copper deficiency also increased macrophage infiltration into the pancreas with activation of inflammatory nitric oxide synthase (iNOS) and secretion of IL 1β, creating a β-cell vulnerability and inhibition of GSIS. The authors proposed that copper depletion in rats compromises GSIS in two ways. It limits COX activity thereby reducing ATP needed for insulin secretion, and it enhances inflammation and iNOS expression, resulting in oxidative β-cell damage.

An additional gene—nutrient interaction was recently observed (Weksler-Zangen et al., 2014). Animals genetically predisposed to low pancreatic COX activity responded poorly to high-sucrose, low-copper diets. Animals with normal COX activity were more resistant to the low-copper diet, but GSIS was ultimately compromised in these animals as well. Although the relationship between copper status and glucose tolerance needs further study, it is clear that in this animal model, adequate intracellular copper is essential for optimal β-cell function. Another recent study confirmed that administration of copper sulfate to mice prior to streptozotocin-induced diabetes reduced the incidence of overt diabetes, oxidative stress, lipid peroxidation, and expression of cytokines in mononuclear cells (Schlesinger et al., 2014).

The hyperglycemia that characterizes type II DM appears to compromise copper status. Although the mechanisms are as yet unclear, hyperglycemia can alter copper distribution as reviewed by Cooper et al. (2005)

and Cooper (2012). The two major copper-binding proteins in the serum are ceruloplasmin and albumin. Fragmentary evidence suggests that hyperglycemia impairs the Cu-binding capacity of each; incubation of ceruloplasmin with glucose caused protein fragmentation and time-dependent release of its bound copper. The released copper was available to participate in Fenton-type reactions and produce ROS. In addition, glycated serum albumin has been reported to become pro-oxidant in the presence of Cu. The authors also report that diabetes increased extracellular matrix Cu. These observations are consistent with a mechanistic paradigm described elsewhere (refer to Chapter 5: Cardiomyopathy and Congestive Heart Failure: Targets for Myocardial Protection: Cardiac Remodeling) that outlines the pathways through which oxidative stress results in copper efflux from the cell with the formation of a homocysteine–copper complex. Increased non-ceruloplasmin-bound extracellular Cu and/or homocysteine–Cu complexes have been shown to downregulate cellular Cu uptake through the copper transporter, resulting in intracellular copper depletion. This paradigm could also explain the elevations in serum chelatable Cu (not bound to ceruloplasmin) and increased urinary Cu excretion observed in diabetic patients.

4.3.4 Zinc

The pancreatic β cell requires high intracellular zinc (Zn) concentrations for insulin synthesis, secretion, and for cell viability. Dietary sources and regulation of Zn distribution and transport are described elsewhere (refer to Essentials II: Heavy Metals, Retinoids and Precursors: Zinc) and were recently reviewed (Ranasinghe et al., 2015). As indicated above, mature insulin is complexed with Zn for storage in zymogen granules. Mutations in the zinc transporter, ZnT8, localized with zinc on zymogen granules, reduce insulin secretion. Since Zn content in the β cell is in excess of this storage function, it is likely that intracellular Zn is required for other functions including its known role in DNA replication and protein biosynthesis. It is of interest that while basal insulin secretion was not altered in isolated mouse islets exposed to high levels of extracellular Zn^{2+} alone (control), GSIS was significantly reduced. Removal of extracellular Zn by chelation significantly increased GSIS. This observation is consistent with an autocrine regulation of insulin release; thus zinc released from the β cell with insulin controls subsequent insulin secretion (Slepchenko et al., 2015).

Zinc also enhances insulin signaling at the peripheral tissues. The role of Zn in insulin-stimulated glucose uptake in adipocytes has been reviewed (Moore

et al., 2015). Zinc enhances insulin action by stimulating the PI3K pathway that promotes the GLUT4 transporter translocation from the cytoplasm to the plasma membrane. Examination of data on 82,297 women in the Nurses Health Study revealed that subjects in the highest quintile of dietary zinc had the lowest relative risk of developing type II DM (Sun et al., 2009). Furthermore, improved fasting blood glucose and lowered insulin levels were seen in a mouse model as well as in type II DM patients supplemented with 30 mg elemental Zn/day (Al-Maroof and Al-Sharbatti, 2006).

4.3.5 Iron

As described elsewhere (refer to Essentials II: Heavy Metals, Retinoids and Precursors: Iron), iron is essential for oxygen transport and other biological functions, but can induce oxidative damage if not tightly controlled. The French pharmacist, Apollinaire Bouchardat (1809–86) first characterized the deleterious effects of iron overload on pancreatic function as "bronze diabetes." The unusually high requirement for insulin in these cases was characterized as "insulin resistance" (Root, 1929). Since that time, the link between iron overload, metabolic syndrome, and type II DM has been intensively studied as recently reviewed (Simcox and McClain, 2013; Fernández-Real and Manco, 2014).

In brief, intracellular iron accumulation in obese and diabetic subjects has been associated with insulin resistance. Studies using an iron-chelating agent, deferoxamine, both in HepG2 cells and a rat model demonstrated that iron reduction increased insulin signaling through the PI3K/Akt signaling pathways, downregulating hepatic gluconeogenesis and upregulating glycogen synthesis. Iron depletion in the rat model increased hypoxia inducible factor (HIF1α), improved glucose tolerance, upregulated the insulin receptor, and improved glucose signaling (Dongiovanni et al., 2008). The authors attributed increased hepatic glucose output and decreased cellular glucose uptake, exacerbating hyperglycemia to the increased iron load. Studies using the iron chelator, deferoxamine, also demonstrated stabilized HIF-1α with induction of the constitutive glucose transporter GLUT 1 and the insulin receptor in iron saturated HepG2 cells and in liver from rats fed a high-iron diet. Iron also reduced glucose uptake by isolated rat adipocytes (Green et al., 2006) (Fig. 4.3).

Circulating ferritin, the iron storage protein, is often elevated in obesity and in patients with type II DM, reflecting a state termed the dysmetabolic-hepatic iron-overload syndrome (DIOS). The mechanism appears to involve an adaptive response to chronic inflammation. Intracellular iron accumulation is mediated by

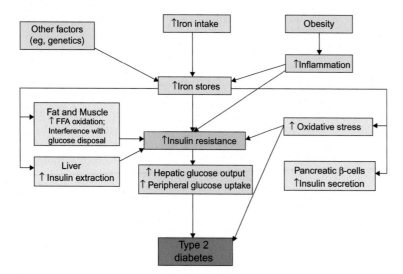

FIGURE 4.3 **Iron overload and type II DM.** Several mechanisms can increase iron stores including increased dietary intake, obesity, and chronic inflammation. Iron, especially unbound iron, is a known pro-oxidant and not only increases insulin resistance, but reduces insulin secretion by the pancreatic β cells. Increased iron stores interfere with glucose disposal in fat and muscle and reduce insulin extraction by the liver. These combined effects increase the risk for type II DM (Rajpathak et al., 2009). *Source: Reproduced from Rajpathak, S.N., J.P. Crandall, et al., 2009. The role of iron in type 2 diabetes in humans. Biochimica et Biophysica Acta (BBA)—General Subjects 1790 (7), 671–681 with permission from Science Direct.*

hepcidin secreted by the liver, but also other cells including pancreatic islet cells, adipocytes, and macrophages, as described elsewhere (refer to Essentials II: Heavy Metals, Retinoids and Precursors: Iron) (Datz et al., 2013). Since ferritin is posttranscriptionally regulated by intracellular iron content, intracellular ferritin accumulates with the iron load; a fraction of the ferritin leaves the cell and is assessed as serum ferritin.

In a cell culture model using human liver SK-HEP-1 cells, intracellular iron accumulation was shown to damage the mitochondria and induce oxidative stress, resulting in hepatic insulin resistance and gluconeogenesis. Mitochondrial dysfunction reduced iron utilization by ROS-damaged iron centers in the ETC, both reducing ATP production and further increasing cellular iron content (Lee et al., 2015). Iron overload in a mouse model reduced LPL activity (Kim et al., 2013), contributing to the dyslipidemia observed with metabolic syndrome and diabetes. Iron-rich adipose tissue displayed insulin resistance associated with reduced adipose tissue mass, cell volume, and importantly, reduced lipid storage capacity.

Pancreatic β cells, like all other cells, take up iron via transferrin receptors expressed on the plasma membrane. The pancreatic islets avidly take up iron, increasing their cell mass and increasing insulin secretion. Rat islet β cells exposed to high glucose in vitro stored 4–8-fold more ferritin than low-glucose controls. Under the influence of hepcidin, β cells cannot release iron. It has been speculated that since β cells are highly vulnerable to iron-induced oxidative stress, the upregulation of ferritin biosynthesis provides a measure of antioxidant protection (Fernández-Real et al., 2015).

At the same time, in spite of DIOS, many patients, especially obese patients, suffer from microcytic anemia. Because hepcidin is elevated and ferroportin is degraded, enterocytes cannot transfer iron to transferrin and iron absorption is compromised. At the same time, macrophages in the bone marrow cannot release iron to developing erythroblasts, contributing to microcytic anemia in the face of tissue iron overload common in obese patients. Since obese patients are relatively iron-overloaded, this microcytic anemia is not due to iron deficiency. Unless iron deficiency is specifically demonstrated, obese and diabetic patients should not be treated with iron supplements.

4.3.6 Vitamin A (Retinoids)

Vitamin A (VA) is known to be essential for epithelial cell maintenance, immune competence, reproduction, and embryonic development (Cunningham and Duester, 2015) as described elsewhere (refer to Essentials II: Heavy Metals, Retinoids and Precursors: Retinoids). The requirement of VA in pancreatic function has been reviewed (Brun et al., 2015). For example, genes that encode binding proteins for retinol and retinoic acid as well as retinoic acid transcription factors, retinoic acid receptor and retinoid X receptor (RAR; RXR) have been localized to pancreatic β cells. Furthermore, isolated pancreatic islets from rats fed a VA-deficient diet secreted less insulin in response to a glucose challenge. Retinoic acid was shown to promote generation of pancreatic progenitor cells and facilitate their differentiation into β cells (Öström et al., 2008). The authors also present data to establish that all-trans-retinoic acid/RAR signaling is critical for maintaining normal pancreatic β-cell activities.

The requirement of VA for maintenance and function of the pancreatic islets was recently demonstrated in mice

(Trasino et al., 2015a). Adult mice depleted in dietary VA (all-trans-retinol) exhibited reduced pancreatic VA levels as well as impaired remodeling of the endocrine pancreas seen as decreased β-cell mass, increased α-cell mass, and a shift to smaller islets. GSIS was decreased in these animals, resulting in hyperglycemia and hyperglucagonemia. VA deficiency affected all organs, but pancreatic β cells appeared exquisitely sensitive to apoptosis. The authors also noted major reductions in the VA cellular binding proteins and in the retinoic acid metabolizing enzyme, CYP26A1, especially in larger islets; reintroduction of VA reversed these changes. Of interest was the high circulating level of retinol bound to retinol binding protein (RBP4) in depleted mice.

While these effects were elicited by dietary VA depletion, a second study by these authors suggested that obesity may potentiate tissue VA depletion, especially in critical organs such as liver, lungs, pancreas, kidney, and adipose tissue. The impact of obesity on tissue VA content was seen in genetically obese mice and in mice made obese by a high-fat diet. VA organ depletion did not coincide with the initiation of the high-fat diet regimen; tissue VA only became apparent after the mice had become obese. This observation suggested that tissue VA depletion occurred because of a metabolic dysregulation associated with increased fat mass, and was not directly induced by the consumption of a high-fat diet. In this trial, despite tissue VA depletion, serum VA (RBP4) was high, leading the authors to characterize tissue VA deficiency in obesity as "silent" (Trasino et al., 2015b).

As discussed elsewhere (refer to Essentials II: Heavy Metals, Retinoids and Precursors: Iron), and recently reviewed (Datz et al., 2013), obesity has been associated with intracellular iron retention. Chronic inflammation, cytokine and leptin signaling through the JAK2/STAT3 pathway (Chung et al., 2007) all upregulate hepcidin transcription and secretion and reduce iron efflux from the cells. Upregulation of ferritin biosynthesis in the iron-overloaded cell is reflected by elevated serum ferritin; explaining the elevated serum ferritin reported in obese and type II DM patients. A positive association between high iron stores and adipocyte-derived circulating RBP4 has been reported (Tamori et al., 2006). Taken together with the studies listed above (Trasino et al., 2015b) and described elsewhere (refer to Essentials II: Heavy Metals, Retinoids and Precursors: Retinoids: Retinoids and Iron Interactions), an intriguing mechanistic link between obesity, iron retention, high serum ferritin, and high serum RBP4 in conjunction with inadequate tissue VA content and dysregulation of macronutrient oxidation has been proposed. The possibility that "silent" VA deficiency in pancreatic β cells also impairs insulin secretion in humans requires further study.

4.3.7 Food-Derived Polyphenols

Diets rich in polyphenols have been reported to increase insulin secretion and reduce insulin resistance in human subjects (Bozzetto et al., 2015), but their mechanism of action is not well clarified. While the conjugated ring structure suggests an antioxidant activity, there is little evidence that this is their function in vivo. In fact, some polyphenols actually exert a pro-oxidant effect. Targeted diet modification will require careful analysis of the molecular actions of dietary components as has been presented below (Munir et al., 2013).

The authors also point out several problems in linking diet-derived nonnutritive components to the pathology of a specific disease. The dose provided by the usual intake is far less than the dose at which metabolic changes are observed in experimental models. Furthermore, polyphenols and other bioactive food components are bound into a food matrix which can enhance or inhibit the bioavailability and/or action of the component. Additionally, these components are exposed to microbial metabolism in the intestine and to metabolic modification by the host organs such as the liver. In a comprehensive review of the actions of the polyphenols in green tea, chocolate, and citrus, these authors identify specific actions of these bioactive dietary components on insulin signaling. As an example, the actions of the polyphenol in green tea, epigallocatichin gallate, are presented here.

4.3.8 Green Tea and Epigallocatichin Gallate

The action of epigallocatichin gallate (EGCG) has been reviewed (Munir et al., 2013). Briefly, in the rat, EGCG inhibits hepatic gluconeogenesis through activation of AMPK and stimulates glucose uptake in skeletal muscle by a PI3K-dependent mechanism that upregulates GLUT 4 transport. EGCG also stimulates eNOS using H_2O_2 redox signaling. Both actions mimic some aspects of insulin action. NO production lowers peripheral vascular resistance by its vasodilatory action and increases blood flow. It also recruits nutritive capillaries to the skeletal muscle. EGCG decreases MAPK-dependent endothelin 1 expression, possibly by regulating FOX01. Together, NO production and inhibition of endothelin 1 reduce vasoconstrictor tone (Fig. 4.4).

Using the spontaneously hypertensive rat model, the authors demonstrated that 3 weeks of EGCG administration reduced insulin resistance, improved endothelial dysfunction, opposed hypertension, protected against ischemia/reperfusion injury, and raised plasma adiponectin levels (Munir et al., 2013). They cite evidence that EGCG lowered lipid and glucose levels in rat models and in diabetic (db/db) mice; EGCG increased the size

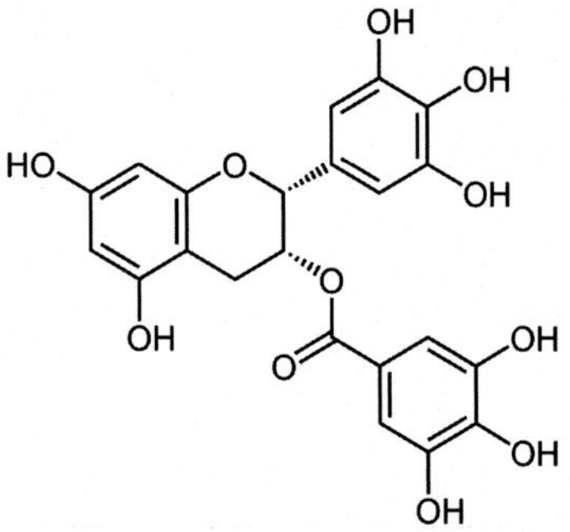

FIGURE 4.4 **Epigallocatichin gallate.** This molecule is found in high concentrations in the dried leaves of green tea (~7000 mg/100 g) with smaller amounts in white and black tea. Fresh tea leaves are used to make green tea. To make black tea, leaves are oxidized under controlled heat and temperature, converting the catechins to theaflavins and other metabolites. *Source: Reproduced from https://commons.wikimedia.org/wiki/File:Epigallocatechin_gallate_structure.svg (figure in public domain).*

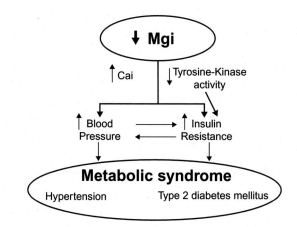

FIGURE 4.5 **Hypothesis: Mg and metabolic syndrome.** Reduced intracellular Mg (Mgi) alters electrolyte channels, increasing intracellular calcium and reducing phosphorylation activities including the cytoplasmic tyrosine kinase of the insulin receptor. These changes raise blood pressure and increase insulin resistance, leading to metabolic syndrome and comorbid conditions including type II DM (Barbagallo and Dominguez, 2007). *Source: Reproduced from Barbagallo, M., L.J. Dominguez, 2007. Magnesium metabolism in type 2 diabetes mellitus, metabolic syndrome and insulin resistance. Arch. Biochem. Biophys. 458 (1), 40–47 with permission from Science Direct.*

and number of pancreatic islets, improved β-cell morphology and led to glucose tolerance comparable to that observed with rosiglitazone. EGCG has also been effective in humans. Meta-analyses of clinical trials concluded that green tea consumption lowered both total and LDL cholesterol levels and improved endothelial function assessed by flow-mediated dilation of the brachial artery (Zheng et al., 2011). Green tea extract also lowered blood pressure, improved insulin sensitivity, and lowered blood cholesterol in obese, hypertensive patients (Suliburska et al., 2012).

4.3.9 Magnesium

Intracellular magnesium (Mg) is a cofactor for several enzymes in carbohydrate metabolism as described elsewhere (refer to Essentials III: Nutrients for Bone Structure and Calcification: Magnesium) and reviewed by Barbagallo and Dominguez (2007) and Mooren (2015). Briefly, as an integral part of Mg-ATP, Mg is required for glycolysis and regulates all enzymes involved in phosphorylation, including phosphorylation of tyrosine kinase on the insulin receptor. It is also required for ATP and phosphate transfer enzymes that regulate ion transport such as Ca-ATPases in the plasma membrane, endoplasmic and sarcoplasmic reticulum. Thus, reduced intracellular Mg can reduce insulin signaling via tyrosine kinase, impair insulin action, and worsen insulin resistance. In fact, Mg deficit has been proposed as a common underlying mechanism of "insulin resistance" seen in metabolic syndrome, obesity, and other metabolic conditions.

Most Mg is found in the bone and soft tissue; only ~1% is found in the serum and only a fraction of that value is ionized (Mg^{2+}). Since Mg^{2+} is difficult to measure, total serum Mg is often used to assess Mg status. However, total serum Mg has been shown to be an inaccurate measure of intracellular Mg (Mgi). Careful measurement of serum ionized Mg^{2+} using a Mg-specific selective ion electrode apparatus and intracellular Mg (Mgi) analyzed by ^{31}P-NMR spectroscopy revealed that Mg^{2+} and Mgi concentrations, but not total serum Mg levels, were significantly reduced in type II DM patients compared with nondiabetic control subjects. A close relationship was observed between serum ionized Mg^{2+} and Mgi (Resnick et al., 1993). The authors suggested that type II DM is characterized by intracellular and serum ionized Mg deficiency and that this deficit may exacerbate glucose intolerance and contribute to cardiovascular morbidity associated with the diabetic state. Since Mg status is assessed more frequently using total Mg than serum ionized Mg^{2+}, depletion of this critical element may be missed in nutrition assessment of diabetic patients (Fig. 4.5).

Mg depletion has been reported in type II DM as comprehensively reviewed (Mooren, 2015). Dietary Mg depletion has been reported (Ford and Mokdad, 2003); a meta-analysis of seven prospective cohort studies on 286,688 patients and 10,912 cases of type II DM over 6–17 years of follow-up found that an increase in dietary Mg of 100 mg/day significantly decreased the risk of diabetes by 15% (Larsson and Wolk, 2007). Further meta-analyses identified a significant and dose related inverse association (relative risk 0.78) between Mg in food and risk for type II DM (Schulze et al., 2007; Dong et al., 2011). It should be remembered that conditioned Mg depletion occurs secondary to gastrointestinal and renal losses, especially with use of diuretics such as furosemide and hydrochlorthiazide and other drugs that may be taken by patients with metabolic syndrome and associated conditions. The mechanisms for conditioned Mg loss have been reviewed (de Baaij et al., 2015) and are discussed elsewhere (refer to Essentials III: Nutrients for Bone Structure and Mineral Metabolism: Magnesium).

4.3.10 Physical Activity

Exercise has been shown to partially restore mitochondrial dysfunction in type II DM by improving insulin-mediated glucose disposal. Exercise and caloric restriction increase the AMP:ATP ratio, thereby activating AMPK as reviewed by Coughlan et al. (2014). The authors cite studies in rodents and humans that have identified AMPK as the critical enzyme through which insulin sensitivity is enhanced. It is also possible that exercise-induced restoration of metabolic flexibility in type II DM is at least partly accounted for by increased mitochondrial biogenesis and possibly by intrinsic mitochondrial adaptations such as increasing ETC activity and preventing excessive ROS generation as discussed elsewhere (refer to Chapter 5: Cardiomyopathy and Congestive Heart Failure: Cardiomyocyte Excitation—Contraction). In a recent human study, the insulin-sensitizing effect of exercise training occurred in the absence of major changes in body mass and was not restricted to muscle, but also improved hepatic and adipose tissue insulin sensitivity (Meex et al., 2010).

4.3.11 Protein Intake and Resistance Exercise

Increasing the protein content of the diet from 15% to 30% of calories has been shown to lower postprandial blood glucose in persons with type II diabetes and to improve overall glucose control. Additional trials were conducted further reducing carbohydrates from 40% to 20% of the diet (50% fat with ~10% of calories from saturated fat). Plasma and urinary β-hydroxybutyrate were similar on both diets, but the mean 24-hour integrated serum glucose, glycohemoglobin, and serum insulin decreased while plasma glucagon increased. Serum cholesterol was unchanged (Gannon and Nuttall, 2004). Thus, substituting carbohydrates with mono- and polyunsaturated fat for 5 weeks dramatically reduced the circulating glucose concentration in patients with untreated type II DM. A calorie-controlled high-protein diet with resistance exercise (~45 minutes on three consecutive days/week) achieved greater weight loss and more favorable changes in body composition with

similar improvements in glycemic control and CVD risk markers in overweight/obese patients with type II diabetes (Wycherley et al., 2010). These studies are consistent with results from studies that demonstrate increased kidney gluconeogenesis without an increase in global endogenous glucose production in rats fed a high-protein diet (Pillot et al., 2009). The authors proposed that protein feeding improved insulin suppression of hepatic gluconeogenesis.

It should be noted that concern has been raised over increased protein intake in patients with diabetes because of their increased risk for chronic kidney failure. Early studies suggested that protein restriction reduced progression to end-stage disease as discussed elsewhere (refer to Chapter 7: Diabetic Nephropathy and Chronic Kidney Disease: Targeted Macronutrient Modifications to Slow CKD Progression). However, current preliminary trials have not demonstrated that a modest increase in protein intake increases risk for CKD progression in patients with diabetes (Parry-Strong et al., 2013). Long-term human trials are needed to test for possible renal complications in type II DM patients who consume a high-protein, low-carbohydrate diet.

4.4 RECOMMENDED NUTRITION THERAPY TO MANAGE DIABETES

On the basis of experimental and epidemiological studies, general recommendations have been proposed to prevent and manage diabetes (O'Keefe et al., 2008). These authors suggest a diet high in minimally processed, high-fiber, plant-based foods such as vegetables and fruits, whole grains, legumes, and nuts predicted to markedly blunt the postmeal increase in glucose, TGs, and inflammation. Additionally, lean protein, vinegar, fish oil, tea, cinnamon, calorie restriction, weight loss, exercise, and low-dose to moderate-dose alcohol have been proposed to modulate postprandial dysmetabolism. Diet patterns such as the traditional Mediterranean or Okinawan diets that incorporate these foods are recommended.

The American Diabetes Association (ADA) position statement of 2013 for patients living with diabetes describes the most recent guidelines (Evert et al., 2013). In general, the ADA recommends a healthful eating pattern, regular physical activity, and appropriate pharmacotherapy for effective diabetes management. The ADA recognizes the integral role of nutrition therapy in overall diabetes management and the necessity that the person with diabetes be actively engaged in self-management, education, and treatment planning with his or her health care provider.

A grading system, developed by the ADA and modeled after existing methods, was utilized to clarify and codify the evidence that forms the basis for the recommendations. The level of evidence that supports each recommendation is listed after the recommendation using the letters A, B, C, or E. A table linking recommendations to evidence can be reviewed at http://professional.diabetes.org/nutrition.

4.4.1 General Recommendations

- Individuals with prediabetes or diabetes should receive individualized medical nutrition therapy (MNT), preferably provided by a registered dietitian familiar with the components of diabetes MNT.
- In overweight and obese insulin-resistant individuals, modest weight loss is recommended (~7% of BW).
  - For weight loss, either low-carbohydrate or low-fat calorie-restricted diets may be effective in the short term (up to 1 year).
  - For patients on low-carbohydrate diets, monitor lipid profiles, renal function, and protein intake (in those with nephropathy), and adjust hypoglycemic therapy as needed.
  - Physical activity (150 minutes/week) and behavior modification are important components of weight-loss programs and are most helpful in maintenance of weight loss.

4.4.2 Nutrient Recommendations

4.4.2.1 Carbohydrate Recommendations

- Evidence is inconclusive for an ideal amount of carbohydrate intake for people with diabetes. Therefore, collaborative goals should be developed with the individual with diabetes. (C)
- The amounts of carbohydrates and available insulin may be the most important factors influencing glycemic response after eating and should be considered when developing the eating plan. (A)
- Monitoring carbohydrate intake, whether by carbohydrate counting or experience-based estimation, remains a key strategy in achieving glycemic control. (B)
- For good health, carbohydrate intake from vegetables, fruits, whole grains, legumes, and dairy products should be advised over intake from other carbohydrate sources, especially those that contain added fats, sugars, or sodium. (B)
- Substituting low-glycemic load foods for higher-glycemic load foods may modestly improve glycemic control. (C)
- People with diabetes should consume at least the amount of fiber and whole grains recommended for the general public (14 g fiber/1000 kcal) and foods containing whole grains (one-half of grain intake). (C)

4.4.3 Protein

- For people with diabetes and no evidence of diabetic kidney disease, evidence is inconclusive to recommend an ideal amount of protein intake for optimizing glycemic control or improving one or more CVD risk measures; therefore, goals should be individualized. (C)
- For people with diabetes and diabetic kidney disease (either micro- or macroalbuminuria), reducing the amount of dietary protein below the usual intake is not recommended because it does not alter glycemic measures, cardiovascular risk measures, or the course of glomerular filtration rate decline. (A)
- In individuals with type II diabetes, ingested protein appears to increase insulin response without increasing plasma glucose concentrations. Therefore, carbohydrate sources high in protein should not be used to treat or prevent hypoglycemia. (B)

4.4.4 Fat

- Evidence is inconclusive for an ideal amount of total fat intake for people with diabetes; therefore goals should be individualized (C); fat quality appears to be far more important than quantity. (B)
- In people with type II diabetes, a Mediterranean-style, monounsaturated fatty acid-rich eating pattern may benefit glycemic control and CVD risk factors and can, therefore, be recommended as an effective alternative to a lower-fat, higher-carbohydrate eating pattern. (B)
- Evidence does not support recommending omega-3 (eicosapentaenoic acid (EPA) and docosahexaenoic acid (DHA)) supplements for people with diabetes for the prevention or treatment of cardiovascular events. (A)
- As recommended for the general public, an increase in foods containing long-chain omega-3 fatty acids (EPA and DHA) (from fatty fish) and omega-3 linolenic acid (ALA) is recommended for individuals with diabetes because of their beneficial effects on lipoproteins, prevention of heart disease, and associations with positive health outcomes in observational studies. (B)
- The recommendation for the general public to eat fish (particularly fatty fish) at least two times (two servings) per week is also appropriate for people with diabetes. (B)
- The amount of dietary saturated fat, cholesterol, and *trans* fat recommended for people with diabetes is the same as that recommended for the general population. (C)

- Individuals with diabetes and dyslipidemia may be able to modestly reduce total and LDL cholesterol by consuming 1.6–3 g/day of plant stanols or sterols typically found in enriched foods. (C)

4.4.5 Micronutrients and Herbal Supplements

- There is no clear evidence of benefit from vitamin or mineral supplementation in people with diabetes who do not have underlying deficiencies. (C)
- Routine supplementation with antioxidants, such as vitamins E and C and carotene, is not advised because of lack of evidence of efficacy and concern related to long-term safety. (A)
- There is insufficient evidence to support the routine use of micronutrient supplements such as chromium, Mg, and vitamin D to improve glycemic control in people with diabetes. (C)
- There is insufficient evidence to support the use of cinnamon or other herbs/supplements for the treatment of diabetes. (C)
- It is recommended that individualized meal planning include optimization of food choices to meet recommended dietary allowance (RDA)/dietary reference intake (DRI) for all micronutrients. (E)

4.4.6 Other Recommendations

- Sugar alcohols and nonnutritive sweeteners are safe when consumed within the acceptable daily intake levels established by the Food and Drug Administration.
- Daily intake of alcohol should be limited to a moderate amount (one drink per day or less for adult women and two drinks per day or less for adult men).

4.5 DIFFERENTIAL DIAGNOSIS OF PERIPHERAL NEUROPATHY IN DIABETES

4.5.1 Hyperglycemia and Other Risk Factors

Peripheral neuropathy (PN) is a common and intractable complication of diabetes. In addition to autonomic nerves, PN can involve both somatosensory and motor nerves. Neuropathy can compromise cardiovascular autonomic nerves, shortening the patient's life expectancy; compromised nerve function in the lower limbs increases risk and subsequent costs of amputation (Yagihashi et al., 2010). The diabetic patient undergoes significant alteration of the nerve fibers of the skin epidermis. Studies using punched skin biopsy revealed that affected nerve fibers in the diabetic patient are distorted and twisted, with focal swelling or beading. Changes are found even in subjects with impaired glucose tolerance and prediabetes, and become marked in patients with established diabetes. In more severe diabetic cases, the nerve fibers actually disappeared. In patients with small-fiber neuropathy in whom DM or impaired glucose tolerance is detected, the duration of neuropathic symptoms before the diagnosis ranges from 3 to 240 months (median, 54 months). These data suggest that an individual can develop neuropathy even in the absence of long-standing diabetes.

The pathology of PN is unknown and thought to involve multiple cascades of injury, including hypoxia, proinflammatory processes, oxidative stress (Vincent et al., 2004), in addition to direct injury to endoneureal microvessels and neural tissues. Damage is exacerbated by risk factors including hypertension, hyperlipidemia, smoking, and insulin resistance. For these reasons, rigorous glycemic control has been proposed to reduce the incidence of neuropathy in patients with known and suspected DM. The Diabetes Control and Complications Trial Research Group (1993) confirmed the efficacy of tight glucose control for PN in some, but not all studies. Even if a patient's neuropathy is unrelated to newly diagnosed diabetes, improved glycemic control may prevent or slow nerve damage resulting from a superimposed PN. Unfortunately, on a global level, there are no effective treatments.

4.6 NUTRIENT-ASSOCIATED NEUROPATHY

Several nutrients are known to be essential in maintaining the structure and function of the nervous system. Neuropathy has been reported as a result of abnormal availability and metabolism of vitamins B1, B6, and B12, as well as copper and other nutrients (Gorson and Ropper, 2006). Type II DM is complicated by obesity and bariatric surgery potentiates nutrient malabsorption and is associated with several types of neuropathy (Goodman, 2015). The PN of type II DM is associated with esophageal erosion and gastroesophageal reflux disease (GERD); nutrient depletion has been reported with use of drugs for GERD control (Zdilla, 2015). Several investigators have noted that the paresthesias, impaired vibration sense, and impaired proprioception symptoms of diabetic PN overlap with those due to specific nutrient depletion. This observation suggests that both factors could be superimposed on injury due to hyperglycemia-associated injury. Since neuropathy secondary to nutrient deficiency is often at least partially reversible, these possibilities should be investigated in the patient with diabetic PN.

4.6.1 Vitamin B12 (Cobalamin) Deficiency

Vitamin B12 is produced by microbial metabolism and is found primarily in animal products. The RDA for vitamin B12 is 2.4 µg daily for all individuals aged 14 and older. Because vitamin B12 absorption declines with age, it has been recommended that individuals over 50 years of age should meet the RDA by eating foods reinforced by vitamin B12 or by taking vitamin B12 supplements (Food and Nutrition Board, 1998). Vitamin B12 can also be found in some plant foods, especially seaweed, possibly as a result of synergistic microbial growth. Additional information is available (http://ods. od.nih.gov/factsheets/VitaminB12-HealthProfessional).

Vitamin B12 deficiency is classically associated both with megaloblastic anemia and with subacute combined degeneration involving demyelination of large spinal nerves. It is increasingly recognized that B12 deficiency can present as an exclusive PN, typically manifesting as axonal neuropathy based on electrophysiology and pathology (Saperstein et al., 2003). Megaloblastic anemia can result from deficiency of either folate or vitamin B12. Effective treatment requires diagnostic assessment to identify the precise cause of the condition. It must be noted that treatment of megaloblastic anemia due to vitamin B12 deficiency with only folate supplements can cure the anemia, but is likely to exacerbate PN due to ongoing vitamin B12 deficiency.

Unfortunately, cobalamin deficiency is difficult to diagnose, since there is no gold standard for its confirmation (Carmel, 2002; Carmel et al., 2003). Although low serum cobalamin levels (<200 pg/mL; <148 pmol/L) identify >97% of patients with clinical deficiency, patients with subclinical deficiency often have normal serum B12 levels (250–350 pg/mL; 185–258 pmol/L). Vitamin B12 is carried in the serum on two distinct carriers, transcobalamin and haptocorrin; only 20–30% of serum vitamin B12 is bound to transcobalamin, the protein that facilitates vitamin B12 uptake by the cell. Serum cobalamin levels are also known to fluctuate widely in a single patient; thus multiple repeat tests may be required to confirm the diagnosis. Functional tests for vitamin B12 deficiency, such as serum homocysteine and methylmalonic acid, are not specific for cobalamin and can falsely skew the diagnosis. Finally, patients with PN (also with polyneuropathy or cognitive decline) due to vitamin B12 deficiency may not present with classic megaloblastic anemia symptoms. This could occur if the patient was also being treated with folic acid.

Vitamin B12 deficiency is seldom due to reduced nutrient intake. Table 4.1 gives examples of clinical conditions that increase risk for vitamin B12 malabsorption. For this reason, it is important that a careful clinical history and evaluation be conducted to identify

TABLE 4.1   Examples of Clinical Risks for Vitamin B_{12} Absorption and Malabsorption

Absorption/ utilization site	Cause of depletion	Presumptive causes
Stomach	Low gastric secretion	Age-associated gastric atrophy; long-term ingestion of antacids (H2-receptor antagonists/proton pump inhibitors)
		Primary hypothyroidism
		Chronic *Helicobacter pylori* infection
	Lack of intrinsic factor	Pernicious anemia; aging (reduced parietal cell mass); atrophic gastritis
Small bowel	Low pancreatic secretion	Cystic fibrosis; pancreatitis; acid chime; deficiency of folate (megaloblastic gut)
	Microbial overgrowth	AIDS, bariatric surgery; blind loop
Terminal ileum	Uptake failure	Ileal resection; metformin (biguanides)
Cellular utilization	Inhibition of methionine synthase	Nitrous oxide exposure; unmetabolized folic acid (from fortification and supplements) (Morris et al., 2003)

risks for malabsorption. Family history is also important to identify the possibility of autoimmune or genetic tendencies that can compromise B12 absorption and utilization (Andres et al., 2004). Finally, once absorbed, cobalamin utilization can be compromised by a variety of factors that inhibit function of the key cobalamin dependent enzyme *methionine synthase*. This enzyme transfers a methyl group from serum 5-methyl folate to homocysteine to form methionine, substrate for S-adenosylmethionine, the universal methyl donor (refer to Chapter 6: Atherosclerosis and Arterial Calcification: Fig. 6.12). High-dose folic acid supplements can correct megaloblastic anemia and mask vitamin B12 depletion, but do not transfer the methyl group to homocysteine, thus myelin methylation is impaired. Under these conditions, neurological defects continue and can be exacerbated (Morris et al., 2010). Failure to methylate myelin has been implicated in demyelination seen with subacute combined systems disease and irreversible nerve damage.

The appropriate treatment of vitamin B12 deficiency requires investigation into the cause of the deficiency. Patients who malabsorb the vitamin due to pernicious anemia (the autoimmune destruction of parietal cells) lack intrinsic factor required to transport vitamin B12 through the intestine. Similarly, gastrointestinal surgery can compromise ileal receptor activity; both must be treated with parenteral vitamin. Injectable vitamin B12 is available in 1000 µg doses and can be safely given every 1−3 months as needed. Elderly patients without gastrointestinal pathology can become deficient secondary to age-associated gastric atrophy; in this case, food protein-bound B12 will not be released and bound to intrinsic factor. These patients can be safely treated orally with 100−1000 µg crystalline B12/day (Herbert, 1997; Stabler, 2013) and monitored for vitamin B12 status.

Bariatric surgery can result in absent or reduced gastric acid and/or lack of intrinsic factor and compromise vitamin B12 status. If the patient was vitamin-replete prior to surgery, hepatic stores can maintain the patient for several months postsurgery as reviewed by Goodman (2015). Patients typically present with parathesias, decreased muscle stretch reflexes, weakness associated with spasticity and loss of position and vibratory sense. Vision loss, dementia, psychosis, and altered mood have also been reported. Autonomic symptoms are rare. All patients who undergo bariatric surgery should take vitamin B12 supplements pre- and postsurgery as indicated by serum vitamin B12 levels.

Drug−nutrient relationships have been reported. Nitrous oxide, used as an anesthetic agent since 1844 and currently used for dental anesthesia oxidizes Co(I) to Co(II) and inactivates vitamin B12 required for methionine synthase activity (Ferner et al., 2014). Its long-term use, especially for recreational purposes, causes megaloblastic anemia and also subacute combined degeneration of the spinal cord with demyelination.

Of considerable interest for the diabetic is the literature suggesting increased risk for vitamin B12 deficiency and worsened diabetic neuropathy in patients on the biguanide, metformin (Liu et al., 2006b). Progressive risk with extended drug treatment has been documented (de Jager et al., 2010), but the mechanism for risk is less clear. Liu et al. suggest that metformin binds calcium needed for ileal receptor transport, while others have speculated that H2 blockers and other "over-the-counter" remedies taken to relieve gastrointestinal side effects of metformin may cause B12 malabsorption.

4.6.2 Vitamin B12 Deficiency in Diabetics

The risk for vitamin B12 deficiency in type II DM patients was recently reviewed (Zdilla, 2015) The author cited 2011 Data from the Centers for Disease Control and Prevention that estimated >85% of patients with type II DM were overweight or obese and that 60−70% had some degree of nerve damage that put them at risk for erosive esophagitis with

~40% reporting symptomatic GERD. At the same time, 70% of diabetics were taking oral antidiabetic medication, primarily metformin. These statistics suggest that a large number of patients simultaneous take both antireflux medication and metformin, each with potential to reduce vitamin B12 availability.

Metformin can impair vitamin B12 status in at least three ways: it can decrease bile secretion, promoting intestinal bacterial overgrowth and lead to malabsorption of an intrinsic factor−vitamin B12−bacteria complex. It can also decrease intrinsic factor secretion and antagonize calcium-dependent B12 absorption via receptors in the ilium (Ting et al., 2006; Wile and Toth, 2010). The action of acid-suppressor medications varies with the drug and may involve decreased gastric acid, pepsin, and intrinsic factor secretion (Zdilla, 2015).

Vitamin B12 deficiency was detected in 22% of outpatient diabetic patients (Pflipsen et al., 2009). Since many type II DM patients are older than 60 years, the age-associated decrease in vitamin B12 (Herbert, 1997) absorption must also be considered (Lindenbaum et al., 1994). Because the neuropathy of vitamin B12 deficiency is associated with demyelination and subacute combined systems disease and reversal is uncommon, early diagnosis and treatment are critical to arrest the process.

4.6.3 Thiamin Deficiency

4.6.3.1 Vitamin B1 (Thiamin)

In its coenzyme form (thiamin pyrophosphate (TDP)), thiamin plays a key role in energy production from carbohydrate via reactions that include the decarboxylation of pyruvic acid and α-ketoglutamic acid and the utilization of pentose in the hexose monophosphate shunt (HMS; pentose phosphate pathway, PPP). By bridging the glycolytic and the HMS pathways critical for creating chemical reducing power in cells, the vitamin is proposed to play a role in several pathways including energy production and utilization as well as host defense. Dietary thiamin deficiency may be a problem in some populations including the elderly and chronically malnourished individuals subsisting at the poverty level.

Thiamin is localized to the seed coat of grains, seeds, and legumes. Pork is a major source but heating destroys some of the thiamin. Since thiamin is water-soluble, pork drippings should be made into a pan sauce and any cooking water should be retained. Green vegetables are good sources, but baking soda, often added to stabilize the green chlorophyll, reduces thiamin availability. Dairy products and fruits are low in thiamin. Milling and processes can reduce thiamin in grains; thus in the United States, flour and other staple foods are supplemented with thiamin to reduce the risk of thiamin deficiency.

Dietary thiamin is absorbed by active transport at physiological doses and by paracellular transport at pharmacological doses. Most dietary thiamin is phosphorylated and must be hydrolyzed by intestinal phosphatases prior to absorption. Small amounts of absorbed thiamin are stored in the liver, but stores must be continually replaced by dietary intake. While gut microbiota synthesize thiamin, the degree to which this is absorbed is not yet clear.

Thiamin deficiency occurs because of poor thiamin intake and/or bioavailability, altered metabolic handling, and increased renal excretion as reviewed by Pácal et al. (2014). Additionally, some cultures consume a diet containing foods (fermented fish and betal nuts) that contain antithiamin factors (Vimokesant et al., 1982). Diets high in absorbable, simple carbohydrates increase the utilization of thiamin. Chronic alcohol intake reduces thiamin absorption (Subramanya et al., 2010); patients with malabsorptive conditions and renal failure also malabsorb/hyperexcrete thiamin (Bukhari et al., 2011).

In general, the estimated thiamin requirement for adults is 1−1.2 mg/day (Food and Nutrition Board, 1998). Serum thiamin levels are minimally useful for assessing thiamin status. Urinary thiamin excretion reflects dietary intake, but not tissue stores. Thiamin status can be indirectly measured by assaying the function of the thiamin-dependent enzyme, transketolase (Tk). This enzyme depends on thiamin diphosphate (TDP), the primary active thiamin cofactor. Erythrocyte hemolysates are cultured in the absence and presence of added TDP and the percent increase in the metabolic product is quantitated. The result reflects the extent of unsaturation of Tk with TDP as shown in Table 4.2. Minimal toxicity has been reported for supplemental thiamin given to normalize indices of thiamin status.

TABLE 4.2 Biochemical and Functional Indicators of Thiamin Deficiency

Indicator	Subclinical deficiency	Severe deficiency
Erythrocyte transketolase activity	1.20−1.25	>1.25
Erythrocyte thiamin (nmol/L)	70−90	<70
Thiamin pyrophosphate effect (%)	15−24	≥25
Urinary thiamin (nmol (μg)/g creatinine)	90−220 (27−66) <27 (nmol (μg)/day)	133−333 (40−100) <40 (nmol (μg)/day)

Individuals at risk for thiamin deficiency include chronic alcohol abusers, older adults, patients with diabetes, and those with malabsorption including patients who have undergone bariatric surgery.

4.6.4 Thiamin Depletion in Diabetics

Thiamin depletion has been implicated in the pathogenesis of diabetes by decreasing insulin biosynthesis and secretion and by inducing oxidative stress and β-cell damage as reviewed by Pácal et al. (2014) and described above. Pancreatic cells import thiamin via pH-dependent, carrier-mediated transporters (THTR-1 and THTR-2) that are adaptively regulated by extracellular thiamin level via transcriptionally mediated mechanism(s).

Several mechanisms have been suggested to explain this relationship. Thiamin is required for Tk, a critical enzyme in the pentose phosphate shunt. Supplemental thiamin inhibited three of the major biochemical pathways activated in hyperglycemia-induced metabolic dysfunction: the hexosamine pathway, the advanced glycation end product formation pathway, and the diacylglycerol−protein kinase C pathway (Hammes et al., 2003) (refer to Chapter 7: Diabetic Nephropathy and Chronic Kidney Disease: Fig. 7.8). Furthermore, thiamin reuptake by renal tubules is impaired with hyperglycemia as reviewed by Larkin et al. (2012). Human primary proximal tubule epithelial cells incubated with high glucose concentrations expressed less THTR-1 and THTR-2 mRNA and protein as well as transcription factor specificity protein-1, resulting in a 37% decrease in apical to basolateral thiamin transport across cell monolayers.

Clinical trials that provided glucose intolerant patients with thiamin supplements (∼300 mg/day) have demonstrated reduced blood glucose levels. A 6-week randomized, double-blind crossover trial of subjects with glucose intolerance revealed a significant reduction in 2-hour postprandial glucose levels with no change in fasting plasma glucose, insulin, or HOMA-IR values (Alaei et al., 2013). No adverse effects from supplementation far in excess of the RDA (1.2 mg/day) were reported. A comprehensive review of experimental and clinical research in this area has been published (Pácal et al., 2014). The authors conclude that translational research is required to clarify the risk factors leading to thiamin deficiency in diabetic patients as well as the safe and effective dose of supplemental thiamin in these patients. Subclinical thiamin deficiency has been associated with various combinations of PN manifest primarily as symmetrical axonal lower extremity predominant with sensorimotor neuropathy as reviewed by Goodman (2015), thus neurological effects of thiamin depletion could be superimposed on diabetic neuropathy in a subpopulation of diabetics.

4.7 NUTRIENT-ASSOCIATED GOUT

Gout typically presents as a rapid-onset, severe, self-limiting joint pain reaching its maximum over 6–12 hours, with swelling and erythema, usually involving the first metatarsophalangeal (MTP) joint at some point (Rider and Jordan, 2010). The first MTP joint is affected in 90% of cases and is the first joint affected in 70% of cases. Diagnostic predictors include male sex, previous patient-reported arthritis attack, onset within 1 day, joint redness, first MTP joint involvement, hypertension or one or more CVDs, and serum uric acid level exceeding 5.88 mg/dL (>350 mM/L) as described (Janssens et al., 2010). The presence of tophi supports the diagnosis and evidence of monosodium urate crystals from a joint aspirate (during or between attacks) would be the gold standard.

Humans and other primates have high normal uric acid due to a missense mutation in the uricase gene. Susceptibility is known to be highly heritable and related to polymorphisms in genes in the purine salvage pathway. Normal uric acid levels in the human are maintained by a balance between production and renal excretion. Whatever the cause, hyperuricemia is strongly associated with the development of gout, and gout is strongly associated with metabolic syndrome. A recent analysis of NHANES III data revealed that ~60% of the individuals with gout had metabolic syndrome (Choi et al., 2007). The close relationship between hyperuricemia and excess insulin could be mediated through a number of overlapping mechanisms including:

- Higher insulin levels are known to reduce renal excretion of urate.
- Insulin may enhance renal urate reabsorption via stimulation of urate-anion exchanger URAT1 and/or the Na$^+$-dependent anion cotransporter in brush-border membranes of the renal proximal tubule
- Because serum levels of leptin and urate tend to increase together, some investigators have suggested that leptin may affect renal urate excretion/reabsorption.
- In insulin resistance syndrome, impaired oxidative phosphorylation may increase systemic adenosine concentrations by increasing the intracellular levels of coenzyme A esters of long-chain fatty acids.
  - Increased adenosine, in turn, can result in renal retention of sodium, urate, and water.
  - Some researchers have speculated that chronically increased extracellular adenosine concentrations may also contribute to hyperuricemia by increasing urate production.

4.7.1 Fructose, Urate, and Type II Diabetes: Are They Causally Related?

Growing evidence suggests that overconsumption of dietary fructose can lead to hyperuricemia, hypertriglyceridemia, and metabolic syndrome, thereby increasing risk for clinical sequelae associated with obesity and type II diabetes. The intracellular utilization of fructose is different from that of glucose, primarily due to its different transporters and enzymes involved in its metabolism (Johnson et al., 2009, 2013). Metabolic processing of dietary fructose through fructokinase requires ATP, and produces a relative cellular hypoxia. The AMP formed in this reaction is converted to xanthine and finally, by xanthine oxidoreductase, to uric acid. A transient rise in serum uric acid is seen after acute ingestion of fructose; chronic fructose ingestion (several weeks) can even raise fasting serum uric acid levels (Fig. 4.6).

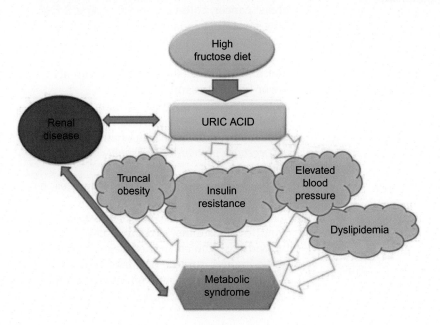

FIGURE 4.6 Dietary fructose is rapidly phosphorylated in the cell at the expense of ATP, leading to ATP depletion and uric acid biosynthesis. At the same time, fructose-1-phosphate is converted to glycerol-3-phosphate and acyl coenzyme A, raising triglycerides (dyslipidemia). Uric acid is also known to inhibit nitric oxide bioavailability (Nakagawa et al., 2006). Uric acid has been incriminated in gout (not shown) and other chronic diseases associated with metabolic syndrome (Kanbay et al., 2016). *Source: Reproduced from Kanbay, M., T. Jensen, et al., 2016. Uric acid in metabolic syndrome: from an innocent bystander to a central player. Eur. J. Intern. Med. 29, 3–8 with permission from Science Direct.*

In summary, dietary fructose is metabolized by pathways that lead to the metabolic syndrome constellation as well as the painful manifestations of gout. Active research in this area is ongoing (Kanbay et al., 2016) to translate metabolic alterations observed with fructose ingestion into recommendations for prevention and control of gout and the metabolic syndrome and to prevent development of type II DM.

4.7.2 Nutritional Therapy for the Patient With Gout

The traditional dietary recommendation for hyperuricemia and gout is centered on adopting low-purine diets, avoiding meat, seafood, and purine-rich vegetables (Rider and Jordan, 2010). Patient compliance is usually poor, since such diets are broadly unappealing. Current literature supports a calorie-restricted diet with low carbohydrate (40% of energy), high vegetable protein (30% of energy) and moderate fat (30% of energy of which <7–10% is saturated).

Additional dietary recommendations have been provided (Choi, 2010) (below) to guide patient consultation.

- *Exercise daily and reduce weight*
  - Increased adiposity is associated with higher uric acid levels and an increased future risk of gout.
    - Weight loss is associated with lower uric acid levels and a decreased risk of gout.
  - Obesity is associated with many important health outcomes, including coronary heart disease (CHD), hypertension, type II diabetes, kidney stones, and gallstones.
- *Limit red meat intake*
  - Red meat intake is associated with higher uric acid levels and increased future risk of gout.
  - The mechanism behind this increased risk may be multifactorial.
    - The urate-raising effect of artificial short-term loading of purified purine has been well demonstrated by metabolic experiments in animals and humans.
    - Red meat is the main source of saturated fats, which are positively associated with insulin resistance, which reduces renal excretion of urate.
    - Saturated fats also increase LDL cholesterol levels more than high-density lipoprotein cholesterol, creating a negative net effect on cardiovascular risk.
    - Higher levels of saturated fats or red meat consumption have been linked to major disorders such as coronary artery disease, type II DM, and certain types of cancer.

- *Tailor seafood intake to the individual and consider omega-3 fatty acid supplements with moderate seafood intake.*
  - Seafood intake has been linked to higher serum uric acid levels and increased future risk of gout, likely due to its high purine content.
  - Increased intake of oily fish, other fish, and shellfish was associated with an increased risk of gout.
    - Given the apparent cardiovascular benefits from fish products, particularly oily fish that are rich in omega-3 fatty acids, it would be difficult to justify a recommendation to avoid all fish intake considering only the risk of gouty flares.
    - Oily fish may be allowed while implementing other lifestyle measures, particularly among gouty patients with cardiovascular comorbidities.
  - Plant-derived omega-3 fatty acids or supplements of EPA and DHA could be considered in the place of fish consumption among patients with gout or hyperuricemia.
    - Diets enriched in both ALA and EPA significantly suppress urate crystal-induced inflammation in a rat model, raising an intriguing potential protective role of these fatty acids against gout flares.
- *Drink skim milk or consume other low-fat dairy products up to two servings daily*
  - Low-fat dairy consumption has been inversely associated with serum uric acid levels and also with a decreased future risk of gout.
    - Low-fat dairy foods have also been linked to a lower incidence of CHD, premenopausal breast cancer, colon cancer, and type II DM.
    - Low-fat dairy foods have been one of the main components of the dietary approaches to stop hypertension (DASH) diet that has been shown to substantially lower blood pressure.
    Note that a high-dairy-fat version of the DASH diet has also been shown to improve metabolic syndrome (Chiu et al., 2016).
- *Consume vegetable protein, nuts, legumes, and purine-rich vegetables*
  - Vegetable protein, although high in purines, has been shown to lower serum uric acid.
  - Nuts and legumes are excellent sources of protein, fiber, vitamins, and minerals.
    - Individuals who consumed vegetable protein in the highest quintile of intake actually had a 27% lower risk of gout compared with the lowest quintile.

- Nut consumption is associated with several important health benefits including
  - lower incidence of CHD, sudden cardiac deaths, gallstones, and type II diabetes.
- Legumes or dietary patterns with increased legume consumption have been linked to
  - lower incidence of CHD, stroke, certain types of cancer, and type II diabetes.
- The recent healthy-eating pyramid recommends consumption of nuts and legumes 1–3 times daily.
  - This recommendation appears readily applicable among patients with gout/hyperuricemia.
- *Reduce alcoholic beverages to a moderate level*
  - That is, one to two drinks per day for men, and no more than one drink per day for women.
  - Beer and liquor, have been associated with higher uric acid levels and increased risk of gout.
    - Overall health benefits of sensible moderate drinking (1–2 drinks/day for men and <1 drink/day for women) likely outweigh the risks, as more than 60 prospective studies have consistently indicated that moderate alcoholic consumption is associated with a 25–40% reduced risk for CHD.
  - Starting drinking is not generally recommended, since similar benefits can be achieved with exercise or healthier eating.
- *Limit sugar-sweetened soft-drinks and beverages*
  - Fructose contained in these beverages increases serum uric acid levels and increases risk of gout.
  - Fructose intake has been linked to increased insulin resistance, a positive energy balance, weight gain, obesity, type II diabetes, an increased risk of certain cancers, and symptomatic gallstone disease.
  - Sweet fruits (ie, apples and oranges) have also been linked to hyperuricemia and the risk of gout.
    - However, given the other health benefits of these food items, it appears difficult to justify restricting whole fruit, even among gout patients.
- *Allow coffee drinking if patient is already drinking coffee*
  - Both regular and decaffeinated coffee drinking have been associated with lower uric acid levels and a decreased risk of gout.
    - Coffee drinking has been linked to a lower risk of type II diabetes, also to lower risks for kidney stones, symptomatic gallstone disease and Parkinson's disease.
  - Caffeine tends to promote calcium excretion in urine.
    - Note that excess coffee intake, about four or more cups per day, may increase the risk of fractures among women.
  - Caffeine, is a xanthine (ie, 1,3,7-trimethyl-xanthine), and likely exerts a protective effect

against gout similar to the action of allopurinol by inhibiting xanthine oxidase.
- This means that intermittent use of coffee or acute introduction of a large amount of coffee may trigger gout attacks as allopurinol introduction does.
- If a patient with gout chooses to try coffee intake to help reduce uric acid levels and the risk of gout, its initiation may need to be similar to that of allopurinol.
- *Consider taking vitamin C supplements*
  - Ascorbate has been found to increase urate excretion, thereby reducing serum uric acid levels in clinical trials and has recently been linked to a reduced future risk of gout.
    - Whereas these data suggest that total vitamin C intake of >500 mg/day is associated with a reduced risk, the potential benefit of lower intake remains unclear.
    - Potential cardiovascular benefit of vitamin C may also be relevant among gout patients, because of their increased risk of cardiovascular morbidity and mortality.
  - Given the general safety profile associated with vitamin C intake, particularly within the generally consumed ranges (eg, tolerable upper intake level of vitamin C <2000 mg in adults according to the Food and Nutrition Board, Institute of Medicine), vitamin C may provide a useful option in the prevention of gout.

4.8 FUTURE RESEARCH NEEDS

The influence of a highly processed, energy-dense, obesogenic food supply and a reduction in energy expenditure in a society increasingly dependent on technology rather than physical activity is widely thought to promote the increased risk for obesity and type II DM in the United States and globally. The food supply is also nutrient-poor, in part due to removal of nutrients with processing and to the addition of sugars and fats to increase consumer acceptance. Because assessment methods and biomarkers are inadequate to identify dysregulation of nutrient distribution in the patient at risk for diabetes, these deficiencies are often overlooked in therapeutic strategies. At the same time, focused mechanistic studies are clarifying the roles of specific nutrients in maintaining metabolic integrity as well as the consequences to the patient when these nutrients are inadequately supplied or dysregulated. Translational studies are urgently needed to bring research insights to the bedside and to derive targeted diet strategies to remediate inadequate nutrient status in the diabetic patient.

References

Alaei, S.F., Soares, M.J., et al., 2013. High-dose thiamine supplementation improves glucose tolerance in hyperglycemic individuals: a randomized, double-blind cross-over trial. Eur. J. Nutr. 52 (7), 1821–1824.

Al-Maroof, R.A., Al-Sharbatti, S.S., 2006. Serum zinc levels in diabetic patients and effect of zinc supplementation on glycemic control of type 2 diabetics. Saudi. Med. J. 3, 344–350.

Andres, E., Loukili, N.H., et al., 2004. Vitamin B12 (cobalamin) deficiency in elderly patients. CMAJ. 171 (3), 251–259.

Baggio, L.L., Drucker, D.J., 2007. Biology of incretins: GLP-1 and GIP. Gastroenterology. 132 (6), 2131–2157.

Barbagallo, M., Dominguez, L.J., 2007. Magnesium metabolism in type 2 diabetes mellitus, metabolic syndrome and insulin resistance. Arch. Biochem. Biophys. 458 (1), 40–47.

Birdee, G.S., Yeh, G., 2010. Complementary and alternative medicine therapies for diabetes: a clinical review. Clin. Diabetes. 28 (4), 147–155.

Bozzetto, L., Annuzzi, G., et al., 2015. Polyphenol-rich diets improve glucose metabolism in people at high cardiometabolic risk: a controlled randomised intervention trial. Diabetologia. 58 (7), 1551–1560.

Brun, P.-J., Grijalva, A., et al., 2015. Retinoic acid receptor signaling is required to maintain glucose-stimulated insulin secretion and β-cell mass. FASEB J. 29 (2), 671–683.

Bukhari, F.J., Moradi, H., et al., 2011. Effect of chronic kidney disease on the expression of thiamin and folic acid transporters. Nephrol. Dial. Transplant. 26 (7), 2137–2144.

Carmel, R., 2002. Measuring and interpreting holo-transcobalamin (holo-transcobalamin II). Clin. Chem. 48 (3), 407–409.

Carmel, R., Green, R., et al., 2003. Update on cobalamin, folate, and homocysteine. Hematology. 2003 (1), 62–81.

Chiu, S., Bergeron, N., et al., 2016. Comparison of the DASH (Dietary Approaches to Stop Hypertension) diet and a higher-fat DASH diet on blood pressure and lipids and lipoproteins: a randomized controlled trial. Am. J. Clin. Nutr. 103 (2), 341–347.

Choi, H.K., 2010. A prescription for lifestyle change in patients with hyperuricemia and gout. Curr. Opin. Rheumatol. 22 (2), 165–172.

Choi, H.K., Ford, E.S., et al., 2007. Prevalence of the metabolic syndrome in patients with gout: the Third National Health and Nutrition Examination Survey. Arthritis Care Res. 57 (1), 109–115.

Chung, B., Matak, P., et al., 2007. Leptin increases the expression of the iron regulatory hormone hepcidin in HuH7 human hepatoma cells. J. Nutr. 137 (11), 2366–2370.

Cooper, G.J.S., 2012. Selective divalent copper chelation for the treatment of diabetes mellitus. Curr. Med. Chem. 19 (17), 2828–2860.

Cooper, G.J.S., Chan, Y.-K., et al., 2005. Demonstration of a hyperglycemia-driven pathogenic abnormality of copper homeostasis in diabetes and its reversibility by selective chelation: quantitative comparisons between the biology of copper and eight other nutritionally essential elements in normal and diabetic individuals. Diabetes. 54 (5), 1468–1476.

Coughlan, K.A., Valentine, R.J., et al., 2013. Nutrient excess in AMPK downregulation and insulin resistance. J. Endocrinol. Diabetes Obes. 1 (1), 1008.

Coughlan, K.A., Valentine, R.J., et al., 2014. AMPK activation: a therapeutic target for type 2 diabetes? Diabetes, Metab. Syndr. Obes. 7, 241–253.

Cunningham, T.J., Duester, G., 2015. Mechanisms of retinoic acid signalling and its roles in organ and limb development. Nat. Rev. Mol. Cell. Biol. 16 (2), 110–123.

Datz, C., Felder, T.K., et al., 2013. Iron homeostasis in the metabolic syndrome. Eur. J. Clin. Invest. 43 (2), 215–224.

de Baaij, J.H.F., Hoenderop, J.G.J., et al., 2015. Magnesium in man: implications for health and disease. Physiol. Rev. 95 (1), 1–46.

de Jager, J., Kooy, A., et al., 2010. Long term treatment with metformin in patients with type 2 diabetes and risk of vitamin B-12 deficiency: randomised placebo controlled trial. BMJ. 340, c2181.

De Nigris, V., Pujadas, G., et al., 2015. Short-term high glucose exposure impairs insulin signaling in endothelial cells. Cardiovasc. Diabetol. 14 (1), 1–7.

The Diabetes Control and Complications Trial Research Group, 1993. The effect of intensive treatment of diabetes on the development and progression of long-term complications in insulin-dependent diabetes mellitus. N Eng. J. Med. 329 (14), 977–986.

Dong, J.-Y., Xun, P., et al., 2011. Magnesium intake and risk of type 2 diabetes: meta-analysis of prospective cohort studies. Diabetes Care. 34 (9), 2116–2122.

Dongiovanni, P., Valenti, L., et al., 2008. Iron depletion by deferoxamine up-regulates glucose uptake and insulin signaling in hepatoma cells and in rat liver. Am. J. Pathol. 172 (3), 738–747.

Egan, A.G., Blind, E., et al., 2014. Pancreatic safety of incretin-based drugs—FDA and EMA assessment. N. Eng. J. Med. 370 (9), 794–797.

Evert, A.B., Boucher, J.L., et al., 2013. Nutrition therapy recommendations for the management of adults with diabetes. Diabetes Care. 36 (11), 3821–3842.

Fernández-Real, J.M., Manco, M., 2014. Effects of iron overload on chronic metabolic diseases. Lancet Diabetes Endocrinol. 2 (6), 513–526.

Fernández-Real, J.M., McClain, D., et al., 2015. Mechanisms linking glucose homeostasis and iron metabolism toward the onset and progression of type 2 diabetes. Diabetes Care. 38 (11), 2169–2176.

Ferner, R.E., Mackenzie, A.A., et al., 2014. The adverse effects of nitrous oxide. Adverse Drug React. Bull. 285, 1099–1102.

Flanagan, J., Simmons, P., et al., 2010. Role of carnitine in disease. Nutr. Metab. (Lond.) 7 (1), 30.

Food and Nutrition Board, 1998. Dietary Reference Intakes: Thiamin, Riboflavin, Niacin, Vitamin B6, Folate, Vitamin B12, Pantothenic Acid, Biotin, and Choline. I. o. Medicine. National Academy Press, Washington, DC.

Ford, E.S., Mokdad, A.H., 2003. Dietary magnesium intake in a national sample of U.S. adults. J. Nutr. 133 (9), 2879–2882.

Fu, Z., Gilbert, E.R., et al., 2013. Regulation of insulin synthesis and secretion and pancreatic beta-cell dysfunction in diabetes. Curr. Diabetes Rev. 9 (1), 25–53.

Gale, E., 2013. Incretin therapy: should adverse consequences have been anticipated? BMJ. 346, f3617.

Gannon, M.C., Nuttall, F.Q., 2004. Effect of a high-protein, low-carbohydrate diet on blood glucose control in people with type 2 diabetes. Diabetes. 53 (9), 2375–2382.

Goodman, J.C., 2015. Neurological complications of bariatric surgery. Curr. Neurol. Neurosci. Rep. 15 (12), 79.

Gorson, K.C., Ropper, A.H., 2006. Additional causes for distal sensory polyneuropathy in diabetic patients. J. Neurol. Neurosurg. Psychiatry. 77 (3), 354–358.

Grahame Hardie, D., 2016. Regulation of AMP-activated protein kinase by natural and synthetic activators. Acta Pharm. Sinica B. 6 (1), 1–19.

Green, A., Basile, R., et al., 2006. Transferrin and iron induce insulin resistance of glucose transport in adipocytes. Metabolism. 55 (8), 1042–1045.

Hammes, H.-P., Du, X., et al., 2003. Benfotiamine blocks three major pathways of hyperglycemic damage and prevents experimental diabetic retinopathy. Nat. Med. 9 (3), 294–299.

Herbert, V., 1997. Vitamin B-12 and folic acid supplementation. Am. J. Clin. Nutr. 66, 1479–1489.

Huang, B., Yuan, H.D., et al., 2011. Cinnamaldehyde prevents adipocyte differentiation and adipogenesis via regulation of peroxisome proliferator-activated receptor-γ (PPARγ) and

AMP-activated protein kinase (AMPK) pathways. J. Agric. Food Chem. 59 (8), 3666–3673.

Jankovic, A., Korac, A., et al., 2015. Redox implications in adipose tissue (dys)function—a new look at old acquaintances. Redox Biol. 6, 19–32.

Janssens, H.J.E.M., Fransen, J., et al., 2010. A diagnostic rule for acute gouty arthritis in primary care without joint fluid analysis. Arch. Intern. Med. 170 (13), 1120–1126.

Johnson, R.J., Perez-Pozo, S.E., et al., 2009. Hypothesis: could excessive fructose intake and uric acid cause type 2 diabetes? Endocr. Rev. 30 (1), 96–116.

Johnson, R.J., Nakagawa, T., et al., 2013. Sugar, uric acid, and the etiology of diabetes and obesity. Diabetes. 62 (10), 3307–3315.

Kanbay, M., Jensen, T., et al., 2016. Uric acid in metabolic syndrome: from an innocent bystander to a central player. Eur. J. Intern. Med. 29, 3–8.

Keil, H.L., Nelson, V.E., 1934. The role of copper in carbohydrate metabolism. J. Biol. Chem. 106 (1), 343–349.

Kim, W., Egan, J.M., 2008. The role of incretins in glucose homeostasis and diabetes treatment. Pharmacol. Rev. 60 (4), 470–512.

Kim, J., Jia, X., et al., 2013. Iron loading impairs lipoprotein lipase activity and promotes hypertriglyceridemia. FASEB J. 27 (4), 1657–1663.

Kopp, C., Singh, S., et al., 2014. Trans-cinnamic acid increases adiponectin and the phosphorylation of AMP-activated protein kinase through G-protein-coupled receptor signaling in 3T3-L1 adipocytes. Int. J. Mol. Sci. 15 (2), 2906.

Larkin, J.R., Zhang, F., et al., 2012. Glucose-induced down regulation of thiamine transporters in the kidney proximal tubular epithelium produces thiamine insufficiency in diabetes. PLoS ONE. 7 (12), e53175.

Larsson, S.C., Wolk, A., 2007. Magnesium intake and risk of type 2 diabetes: a meta-analysis. J. Intern. Med. 262 (2), 208–214.

Lee, H.J., Choi, J.S., et al., 2015. Effect of excess iron on oxidative stress and gluconeogenesis through hepcidin during mitochondrial dysfunction. J. Nutr. Biochem. 26 (12), 1414–1423.

Liemburg-Apers, D., Willems, P.G.M., et al., 2015. Interactions between mitochondrial reactive oxygen species and cellular glucose metabolism. Arch. Toxicol. 89 (8), 1209–1226.

Lindenbaum, J., Rosenberg, I.H., et al., 1994. Prevalence of cobalamin deficiency in the Framingham elderly population. Am. J. Clin. Nutr. 60, 2–11.

Liu, D., Zhen, W., et al., 2006a. Genistein acutely stimulates insulin secretion in pancreatic β-cells through a cAMP-dependent protein kinase pathway. Diabetes. 55 (4), 1043–1050.

Liu, K.W., Dai, L.K., et al., 2006b. Metformin-related vitamin B12 deficiency. Age Ageing. 35 (2), 200–201.

Luong, Kvq, Nguyen, L.T.H., 2012. The impact of thiamine treatment in the diabetes mellitus. J. Clin. Med. Res. 4 (3), 153–160.

Madeira, A., Moura, T.F., et al., 2015. Aquaglyceroporins: implications in adipose biology and obesity. Cell. Mol. Life Sci. 72 (4), 759–771.

Maeda, N., 2012. Implications of aquaglyceroporins 7 and 9 in glycerol metabolism and metabolic syndrome. Mol. Aspects Med. 33 (5–6), 665–675.

Meex, R.C.R., Schrauwen-Hinderling, V.B., et al., 2010. Restoration of muscle mitochondrial function and metabolic flexibility in type 2 diabetes by exercise training is paralleled by increased myocellular fat storage and improved insulin sensitivity. Diabetes. 59 (3), 572–579.

Moore, W., Bowser, S., et al., 2015. Beta cell function and the nutritional state: dietary factors that influence insulin secretion. Curr. Diab. Rep. 15 (10), 1–9.

Mooren, F.C., 2015. Magnesium and disturbances in carbohydrate metabolism. Diabetes, Obes. Metab. 17 (9), 813–823.

Morris, M., Evans, D.A., et al., 2003. Consumption of fish and n-3 fatty acids and risk of incident Alzheimer disease. Arch. Neurol. 60 (7), 940–946.

Morris, M.S., Jacques, P.F., et al., 2010. Circulating unmetabolized folic acid and 5-methyltetrahydrofolate in relation to anemia, macrocytosis, and cognitive test performance in American seniors. Am. J. Clin. Nutr. 91 (6), 1733–1744.

Munir, K.M., Chandrasekaran, S., et al., 2013. Mechanisms for food polyphenols to ameliorate insulin resistance and endothelial dysfunction: therapeutic implications for diabetes and its cardiovascular complications. Am. J. Physiol. Endocrinol. Metab. 305 (6), E679–E686.

Muniyappa, R., Montagnani, M., et al., 2007. Cardiovascular actions of insulin. Endocr. Rev. 28 (5), 463–491.

Nakagawa, T., Hu, H., et al., 2006. A causal role for uric acid in fructose-induced metabolic syndrome. Am. J. Physiol. Renal. Physiol. 290 (3), F625–631.

Nile, D.L., Brown, A.E., et al., 2014. Age-related mitochondrial DNA depletion and the impact on pancreatic beta cell function. PLoS ONE. 9 (12), e115433.

O'Keefe, J.H., Gheewala, N.M., et al., 2008. Dietary strategies for improving post-prandial glucose, lipids, inflammation, and cardiovascular health. J. Am. Coll. Cardiol. 51 (3), 249–255.

Öström, M., Loffler, K.A., et al., 2008. Retinoic acid promotes the generation of pancreatic endocrine progenitor cells and their further differentiation into β-cells. PLoS ONE. 3 (7), e2841.

Pácal, L., Kuricová, K., et al., 2014. Evidence for altered thiamine metabolism in diabetes: is there a potential to oppose gluco- and lipotoxicity by rational supplementation? World J Diabetes. 5 (3), 288–295.

Parry-Strong, A., Leikis, M., et al., 2013. High protein diets and renal disease—is there a relationship in people with type 2 diabetes? Br. J. Diabetes Vascul. Dis. 13 (5–6), 238–243.

Patti, M.-E., Corvera, S., 2010. The role of mitochondria in the pathogenesis of type 2 diabetes. Endocr. Rev. 31 (3), 364–395.

Pflipsen, M.C., Oh, R.C., et al., 2009. The prevalence of vitamin B12 deficiency in patients with type 2 diabetes: a cross-sectional study. J. Am. Board Fam. Med. 22 (5), 528–534.

Pillot, B., Soty, M., et al., 2009. Protein feeding promotes redistribution of endogenous glucose production to the kidney and potentiates its suppression by insulin. Endocrinology. 150 (2), 616–624.

Plaisance, E.P., Lukasova, M., et al., 2009. Niacin stimulates adiponectin secretion through the GPR109A receptor. Am. J. Physiol. Endocrinol. Metab. 296 (3), E549–E558.

Rajpathak, S.N., Crandall, J.P., et al., 2009. The role of iron in type 2 diabetes in humans. Biochimica et Biophysica Acta (BBA)—Gen. Sub. 1790 (7), 671–681.

Ranasinghe, P., Pigera, S., et al., 2015. Zinc and diabetes mellitus: understanding molecular mechanisms and clinical implications. DARU. 23 (1), 44.

Reiser, S., Smith, J.C., et al., 1985. Indices of copper status in humans consuming a typical American diet containing either fructose or starch. Am. J. Clin. Nutr. 42 (2), 242–251.

Resnick, L.M., Altura, B.T., et al., 1993. Intracellular and extracellular magnesium depletion in Type 2 (non-insulin-dependent) diabetes mellitus. Diabetologia. 36 (8), 767–770.

Rider, T.G., Jordan, K.M., 2010. The modern management of gout. Rheumatology. 49 (1), 5–14.

Rindler, P.M., Crewe, C.L., et al., 2013. Redox regulation of insulin sensitivity due to enhanced fatty acid utilization in the mitochondria. Am. J. Physiol. Heart Circ. Physiol. 305 (5), H634–H643.

Root, H.F., 1929. Insulin resistance and bronze diabetes. N. Eng. J. Med. 201 (5), 201–206.

Saad, M., Abdelkhalek, T., et al., 2015. Insights into the molecular mechanisms of diabetes-induced endothelial dysfunction: focus on oxidative stress and endothelial progenitor cells. Endocrine. 50 (3), 537–567.

Samuel, V.T., Shulman, G.I., 2016. The pathogenesis of insulin resistance: integrating signaling pathways and substrate flux. J. Clin. Invest. 126 (1), 12–22.

Sanjabi, B., Dashty, M., et al., 2015. Lipid droplets hypertrophy: a crucial determining factor in insulin regulation by adipocytes. Sci. Rep. 5, 8816.

Saperstein, D.S., Wolfe, G.I., et al., 2003. Challenges in the identification of cobalamin-deficiency polyneuropathy. Arch. Neurol. 60 (9), 1296–1301.

Schilling, J.D., 2015. The mitochondria in diabetic heart failure: from pathogenesis to therapeutic promise. Antioxid. Redox Signal. 22 (17), 1515–1526.

Schlesinger, S., Siegert, S., et al., 2014. Postdiagnosis body mass index and risk of mortality in colorectal cancer survivors: a prospective study and meta-analysis. Cancer Causes Control. 25 (10), 1407–1418.

Schulze, M.B., Schulz, M., et al., 2007. Fiber and magnesium intake and incidence of type 2 diabetes: A prospective study and meta-analysis. Arch. Intern. Med. 167 (9), 956–965.

Shaw, R.J., Lamia, K.A., et al., 2005. The kinase LKB1 mediates glucose homeostasis in liver and therapeutic effects of metformin. Science (New York, NY). 310 (5754), 1642–1646.

Simcox, J.A., McClain, D.A., 2013. Iron and diabetes risk. Cell Metab. 17 (3), 329–341.

Slepchenko, K., Daniels, N., et al., 2015. Autocrine effect of Zn^{2+} on the glucose-stimulated insulin secretion. Endocrine. 50 (1), 110–122.

Stabler, S.P., 2013. Vitamin B12 deficiency. N. Eng. J. Med. 368 (2), 149–160.

Subramanya, S.B., Subramanian, V.S., et al., 2010. Chronic alcohol consumption and intestinal thiamin absorption: effects on physiological and molecular parameters of the uptake process. Am. J. Physiol. Gastrointest. Liver Physiol. 299 (1), G23–G31.

Suliburska, J., Bogdanski, P., et al., 2012. Effects of green tea supplementation on elements, total antioxidants, lipids, and glucose values in the serum of obese patients. Biol. Trace Elem. Res. 149 (3), 315–322.

Sun, Q., van Dam, R.M., et al., 2009. Prospective study of zinc intake and risk of type 2 diabetes in women. Diabetes Care. 32 (4), 629–634.

Szabat, M., Page, M.M., et al., 2016. Reduced insulin production relieves endoplasmic reticulum stress and induces β cell proliferation. Cell Metab. 23 (1), 179–193.

Tamori, Y., Sakaue, H., et al., 2006. RBP4, an unexpected adipokine. Nat. Med. 12 (1), 30–31.

Thornalley, P.J., 2005. The potential role of thiamine (Vitamin B1) in diabetic complications. Curr. Diabetes Rev. 1 (3), 287–298.

Ting, R.Z.-W., Szeto, C.C., et al., 2006. Risk factors of Vitamin B12 deficiency in patients receiving metformin. Arch. Intern. Med. 166 (18), 1975–1979.

Trasino, S.E., Benoit, Y.D., et al., 2015a. Vitamin A deficiency causes hyperglycemia and loss of pancreatic β-cell mass. J. Biol. Chem. 290 (3), 1456–1473.

Trasino, S.E., Tang, X.-H., et al., 2015b. Obesity leads to tissue, but not serum Vitamin A deficiency. Sci. Rep. 5, 15893.

Vimokesant, S., Kunjara, S., et al., 1982. Beriberi caused by antithiamin factors in food and its prevention. Ann. N.Y. Acad. Sci. 378 (1), 123–136.

Vincent, A.M., Russell, J.W., et al., 2004. Oxidative stress in the pathogenesis of diabetic neuropathy. Endocr. Rev. 25 (4), 612–628.

Wang, H., Eckel, R.H., 2009. Lipoprotein lipase: from gene to obesity. Am. J. Physiol. Endocrinol. Metab. 297 (2), E271–E288.

Weksler-Zangen, S., Jörns, A., et al., 2013. Dietary copper supplementation restores β-cell function of Cohen diabetic rats: a link between mitochondrial function and glucose-stimulated insulin secretion. Am. J. Physiol. Endocrinol. Metab. 304 (10), E1023–E1034.

Weksler-Zangen, S., Aharon-Hananel, G., et al., 2014. IL-1β hampers glucose-stimulated insulin secretion in Cohen diabetic rat islets through mitochondrial cytochrome c oxidase inhibition by nitric oxide. Am. J. Physiol. Endocrinol. Metab. 306 (6), E648–E657.

Wile, D.J., Toth, C., 2010. Association of metformin, elevated homocysteine, and methylmalonic acid levels and clinically worsened diabetic peripheral neuropathy. Diabetes Care. 33 (1), 156–161.

Wolden-Kirk, H., Overbergh, L., et al., 2011. Vitamin D and diabetes: its importance for beta cell and immune function. Mol. Cell. Endocrinol. 347 (1–2), 106–120.

Wollheim, C.B., 2000. Beta-cell mitochondria in the regulation of insulin secretion: a new culprit in Type II diabetes. Diabetologia. 43 (3), 265–277.

Wycherley, T.P., Noakes, M., et al., 2010. A high-protein diet with resistance exercise training improves weight loss and body composition in overweight and obese patients with type 2 diabetes. Diabetes Care. 33 (5), 969–976.

Yagihashi, S., Mizukami, H., et al., 2010. Mechanism of diabetic neuropathy: where are we now and where to go? J. Diabetes Invest. 2 (1), 18–32.

Yang, J., Wong, R.K., et al., 2006. Leucine regulation of glucokinase and ATP synthase sensitizes glucose-induced insulin secretion in pancreatic β-cells. Diabetes. 55 (1), 193–201.

Zdilla, M.J., 2015. Metformin with either histamine H2-receptor antagonists or proton pump inhibitors: a polypharmacy recipe for neuropathy via Vitamin B12 depletion. Clin. Diabetes. 33 (2), 90–95.

Zheng, X.-X., Xu, Y.-L., et al., 2011. Green tea intake lowers fasting serum total and LDL cholesterol in adults: a meta-analysis of 14 randomized controlled trials. Am. J. Clin. Nutr. 94 (2), 601–610.

Further Reading

Goldman, L., Schafer, A.I., 2012. Goldman cecil medicine. Elsevier Saunders, Philadelphia, PA, Chapter 446: Peripheral Neuropathies; Chapter 294: Crystal Deposition Diseases; Chapter 248: Type 2 Diabetes Mellitus.

Kumar, V., et al., 2014. Robbins and cotran pathologic basis of disease, ninth ed. Elsevier Saunders, Philadelphia, PA, Chapter 24. The Endocrine System: Type II Diabetes; Chapter 27. Peripheral Nerves and Skeletal Muscle: Peripheral Neuropathy; Chapter 26. Bones, Joints and Soft Tissue Tumors: Crystal-Induced Arthritis.

Trence, D.L., 2002. Case study: peripheral neuropathy in diabetes: is it diabetic neuropathy? Clin. Diabetes. 20 (2), 103–104. Available from: http://dx.doi.org/10.2337/diaclin.20.2.103.

5

Cardiopathy and Congestive Heart Failure

CHIEF COMPLAINT (ADAPTED FROM GORE, J.M. AND FALLON, J.T., 1994)

Mr. W., a 45-year-old white male presented complaining of increasing fatigue and a vague feeling of being "unwell." He complained of "heart palpitations" and was concerned that control of his recently diagnosed diabetes might not be optimal.

PAST MEDICAL HISTORY AND MEDICATIONS

Mr. W. had been well until 4 months earlier, when he had a severe upper respiratory tract infection. The viral infection resolved with no treatment except a week of bed rest and fluids at home. Two months before the current admission he was admitted to another hospital with polydipsia, polyuria, polyphagia, and a weight loss of 9 kg. Pertinent tests included high blood glucose concentration (437 mg/dL), negative chest X-ray, and ultrasonic abdominal examination. Laboratory values included serum cholesterol 222 mg/dL; serum triglycerides 297 mg/dL; and normal serum protein at 6.8 g/dL (albumin, 4.0 g; globulin, 2.8 g/dL). Electrocardiogram showed normal rhythm with abnormal T waves in leads 1, 2, aVL, and aVF and inverted T waves in leads V_3 through V_6; three subsequent electrocardiograms showed no change. The thyroxine concentration was 3.6 μg/dL, and the thyrotropin concentration was 2.5 mU/L. The patient was treated with insulin and discharged under close diabetic control on the seventh hospital day.

Several weeks prior to this admission, Mr. W. reported the onset of fatigue and exertional dyspnea. In the 2 weeks before the current admission, he noted an increase in weight (6.4 kg), accompanied by a sensation of increasing abdominal girth. Two days before

this admission palpitations developed, the patient was readmitted to the same hospital, where atrial fibrillation was detected. Cardiac ultrasonographic examination showed dilatation of the left ventricle with global hypokinesis; the estimated left ventricular ejection fraction was calculated to be 10–15%; there was + to ++ mitral regurgitation with +++ tricuspid regurgitation. Radiographs of the chest showed slight cardiomegaly.

Three years earlier he had been evaluated elsewhere because of epistaxis and was diagnosed with anemia, leukopenia, and thrombocytopenia. Studies included a bone marrow examination, but no definitive diagnosis was established. The medical record does not indicate a history of chest pain, fever, chills, cough, or syncope.

LIFESTYLE AND FAMILY HISTORY

Mr. W. is a computer programmer who has lived alone following his divorce 5 years ago. He has always been active; he is a member of several sports teams and works out at the gym three times a week. Since his diagnosis of diabetes, he has greatly reduced his activity, largely because he felt vaguely "unwell." Mr. W. does not smoke or use recreational drugs. He does have a few drinks (usually whiskey) 3–4 nights a week after work. Mr. W takes a high-potency vitamin/mineral supplement daily to boost his energy levels.

Mr. W's parents were killed in a car accident 15 years ago. To his knowledge, both were in good health. His father's brother, a bachelor who lived alone, died at age 60 from cirrhosis. The cause of the liver failure was unclear. Mr. W. does not remember his uncle as an alcoholic and noted that he kept himself fit and was a great believer in health foods and exercise. His paternal grandmother had diabetes mellitus; there was no family history of cardiac or hematologic disease.

Nutritional Pathophysiology of Obesity and its Comorbidities
DOI: http://dx.doi.org/10.1016/B978-0-12-803013-4.00005-3

NUTRITION HISTORY

General

- Because he lives alone, Mr. W. eats dinner out 5–6 times a week. A typical meal includes steak and potatoes, followed by dessert and brandy. He seldom eats breakfast at home and stops at the local diner for coffee and a sweet roll. Lunch, usually a roast beef or pastrami sandwich from the local deli, with a pickle, chips and soda, is eaten at his desk. Several times a week he substitutes a liquid meal (Ensure Plus) for lunch. On weekends, Mr. W. has a roll and coffee for breakfast, and spends the rest of the day with friends. These outings usually involve dinner out.

Diet Analysis

- The dietitian conducted a 24-hour recall and assessment and obtained the following information:

• Total calories	2806 kcal/day	
• Protein	97.8 g/day	(10% of calories)
• Fat	74 g/day	(35% of calories)
• Carbohydrate	262 g/day	(55% of calories)
Basal caloric requirements based on Harris Benedict equation[a]	1474 kcal/day	
Estimated total energy requirements (activity factor = 1.25)	1843 kcal/day	
Estimated protein requirements (kg × 0.8 g/kg)[b]	57–61 g/day	

[a]Please see Cornell site for calculation: http://www-users.med.cornell.edu/~spon/picu/calc/beecalc.htm.
[b]Based on dietary reference intake (DRI) for men 35–50 years (or 0.8 gm/kg).

INTAKE EXAMINATION

General Appearance

Mr. A appears normal weight, with fat distribution to the mid-section and relatively thin arms and legs.

• Height	5′9″ (175 cm)	(Height at exam 3 years ago 5′10″ (179.6 cm))
• Weight	168 lb (76.2 kg)	(±5 lbs from most recent exam; +12 lbs since exam 2 years ago)
• BMI	24.9 kg/m^2	

Vitals

- Temperature: 36.4°C

- Blood pressure: 100/60
- Heart rate: 120 bpm
- Respiration: 20 bpm

Physical Examination

Laboratory Data	Patient's Values	Normal Values
• Albumin	3.8 g/dL	3.5–5.8 g/dL
• Hemoglobin	11.5 g/dL	11.8–15.5 g/dL
• Hematocrit	44%	36–46%
• Transferrin saturation	55%	20–50%
• Serum ferritin	305 ng/mL	18–270 ng/mL
• Serum 25-OH D	26 ng/mL	>25 ng/mL (62.5 nmol/L)[a]
• Serum PTH	80 pg/mL	10–65 pg/mL
• Triglycerides	203 mg/dL	<150 mg/dL
• Total cholesterol	215 mg/dL	<200 mg/dL
• HDL-C	32 mg/dL	>40–50 mg/dL

[a]Cecil gives the following normal values for plasma 25-OH D$_3$: 15–80 ng/mL (37.4–200 nmol/L).

CARDIAC CATHETERIZATION DATA

Right-sided dominance with normal vessels. Left ventricular study showed severe global hypokinesis without mitral regurgitation. The left ventricular ejection fraction was 17%; left ventricular end-diastolic volume was 310 mL.

Summary of clinical examination: 25-year-old man with type II diabetes and recent onset of congestive heart failure with atrial fibrillation. Although the symptoms of heart failure were fairly recent in onset, his history suggests that cardiac involvement may have been part of an ongoing systemic process.

RESOURCES

The flesh is consumed and becomes water ... the abdomen fills with water; the feet and legs swell; the shoulders, clavicles, chest and thigh melt away *Hippocratic Corpus, Affections XXII (6th century BC)*.

5.1 THE COMPROMISED HEART

It is likely that in this work, Hippocrates was describing cardiac cachexia, but since his work does not directly reference the heart, it could also have been cachexia from other end-stage diseases such as kidney disease or cancer (Katz and Katz, 1962). In heart failure, abnormalities in cardiac structure and function impair the ability of the heart to effectively fill and eject blood, thereby reducing its capacity to meet the needs of the organism. Ischemic injury reduces adenosine triphosphate (ATP) production required for muscle function and repair, while failure to perfuse the kidneys and excrete a sodium load increases fluid volume, accounting for the edema and muscle wasting clinically observed in cachexia. While clarification of the etiology and pathogenesis of cardiac abnormalities has been the subject of extensive research, the multiple ways in which altered nutrient availability can compromise heart structure and function have been less well studied.

The heart distributes approximately 6000 L of blood through the body each day to maintain optimal tissue perfusion. The heart muscle (myocardium) maintains coordinated contraction (during systole) and relaxation (during diastole). (The cardiac cycle consists of alternating *systole*, originating from New Latin and Ancient Greek words meaning "to send," and *diastole*, from the Greek word meaning dilation.) Congestive heart failure (CHF) describes a progressive reduction in heart function that affects nearly 2% of the US population and is a major contributing factor in morbidity and mortality worldwide (Roger et al., 2012). The 5-year survival may be as low as 50%, worse than many forms of cancer. Death usually results from arrhythmia or progressive pump dysfunction (Firth and Yancy, 1990).

Primary cardiomyopathies can result from genetic lesions or can be acquired, while secondary cardiomyopathies result from complications of systemic disease or multiorgan system disease. An expert panel of the American Heart Association has characterized cardiomyopathies as *"a heterogeneous group of diseases of the myocardium associated with mechanical and/or electrical dysfunction that usually (but not invariably) exhibit inappropriate ventricular hypertrophy or dilation and are due to a variety of causes that frequently are genetic. Cardiomyopathies either are confined to the heart or are part of generalized systemic disorders, often leading to cardiovascular death or progressive heart failure-related disability."* (Maron et al., 2006).

Anatomic abnormalities in the heart can be classed into three patterns: dilated cardiomyopathy, hypertrophic cardiomyopathy, and restrictive cardiomyopathy. Dilated cardiomyopathy is the most common (90% of cases) and is characterized by progressive myocardial damage with dilation of the contractile system and usually associated with hypertrophy and remodeling. Myocardial damage can result from both infectious and noninfectious agents (see Table 12.12 in Robbins and Cotran. Pathologic Basis of Disease, 9th Edition for a comprehensive list of etiologic conditions

associated with cardiomyopathy) that trigger heart inflammation or myocarditis, an inflammatory cellular infiltrate with or without associated myocyte necrosis. If the inflammation does not resolve and becomes chronic, damage can progress to long-term myocarditis and dilated cardiomyopathy with chronic heart failure (Palomer et al., 2013).

Infectious agents can include bacteria, fungi, and parasites (Chagas' disease). Viruses (coxsackievirus B, parvovirus B19) are major causes of dilated cardiomyopathy; infection with *Borrelia burgdorferi* (Lyme disease) and human immunodeficiency virus (HIV) can also cause myocarditis (Cooper, 2009). In developed countries, ischemia following cardiac surgery or myocardial infarction (MI) is a major stimulus for CHF. Long-standing hypertension, hyperthyroid disease, anemia, untreated chronic valvular disease, hemochromatosis, and thromboembolic disorders can increase the cardiac workload and damage heart muscle. Genetic mutations that disrupt the cytoskeleton linking the contractile apparatus (sarcolemma and sarcomere) to its energy source can lead to failure of robust pump activity and cardiomyopathy (Towbin and Bowles, 2002). Genetic mutations have been found in 30–50% of the cases. Toxins such as alcohol or cardiotoxic drugs such as with doxorubicin (Adriamycin) can precipitate an immune hypersensitivity and damage the cardiac muscle.

Cardiomyopathy is also seen as a comorbidity of obesity. Hormonal dysregulation and altered blood flow that increase workload of the heart appear to be precipitating factors (Alpert et al., 2014). If obesity is complicated by diabetes, hyperglycemia can increase myocardial damage and diabetic cardiomyopathy can ensue as described below (Huynh et al., 2014). In many cases, the precipitating cause is not identified (Gore and Fallon, 1994). Several nutrient deficiencies such as thiamin (Sica, 2007) are well known to target the heart; severe thiamin deficiency causes the deficiency disease, beri beri (Blankenhorn et al., 1946). Other nutrient deficiencies including magnesium, zinc, copper, carnitine, and coenzyme Q10; depletion of amino acids such as taurine also damage the myocardium. Cardiomyopathy secondary to poor diet is often complicated by other deficiency symptoms such as pellagra (niacin deficiency). Damage is potentiated by aging and by concomitant use of recreational drugs, excess iron supplements, and alcohol (Soukoulis et al., 2009). It should be noted that alcohol use reduces the bioavailability and utilization of many nutrients including thiamin, niacin, and vitamin A (see below). The impact of micronutrient status on cardiomyopathy is infrequently assessed, in part because micronutrients are differentially distributed among body compartments and because biomarkers that target the appropriate nutrient form and compartment have not yet been established.

Processes required for optimal myocardial function are reviewed below to facilitate the discussion of mechanisms through which bioactive components in the diet can modify the initiation and progression of cardiomyopathy.

5.2 CARDIOMYOCYTE EXCITATION–CONTRACTION

5.2.1 Mitochondria and ATP Production

The myocardium requires enormous amounts of ATP for excitation–contraction (EC). Almost 2% of cellular ATP is consumed in a single heart beat; the total ATP production in the heart is turned over within 1 minute (Nickel et al., 2013). ATP is produced in the mitochondria by step-wise oxidation of fuels, primarily fat and carbohydrate. As illustrated in Fig. 5.1, glucose-derived pyruvate is converted to acetyl CoA by mitochondrial pyruvate dehydrogenase and long-chain fatty acids (FAs) are transported into the mitochondria and converted to acetyl CoA in the β-oxidation pathway. Following conversion to acetyl CoA, fuels are oxidized through the citric acid cycle (Krebs; tricarboxylic acid cycle); this pathway produces high-energy electrons and releases

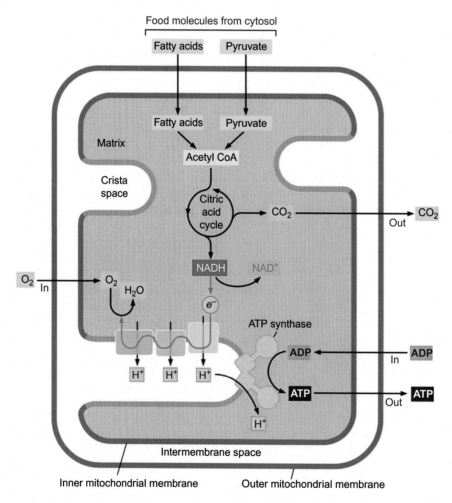

FIGURE 5.1 **Nutrient requirements for ATP generation.** Glucose-derived pyruvate is converted to acetate by thiamin-dependent pyruvate dehydrogenase and bound to coenzyme A (CoA) comprised largely of pantothenate. Retinol (vitamin A) is required for carnitine-dependent transporters that facilitate entry of long-chain fatty acids into the mitochondrial matrix where they are cleaved to acetate by β oxidation. High-energy electrons produced by oxidation of acetyl CoA in the citric acid cycle are transported by NADH (niacin) and FADH2 (riboflavin) to the ETC. Biotin is a cofactor for enzymes that remove energy-poor CO_2 from the mitochondrion. Iron and copper cofactors in the ETC facilitate electron transport and generation of the electrochemical gradient. Ubiquinone (coenzyme Q) is an endogenously produced component of the ETC, but is often low in CHF patients. Cytochrome oxidase, the terminal ETC complex holds binds poor electrons in a heme–copper queue to await binding to molecular O_2. Molecular O_2 is secured at a bimetallic site formed by another heme and a closely annexed copper atom until it has been reduced to water by four protons pumped from the matrix; As it is reduced, O_2 passes sequentially through superoxide ($O_2^{\bullet-}$), hydrogen peroxide (H_2O_2), and hydroxyl radical ($OH^{\bullet}$) intermediates. Only then is the oxygen released in the form of two molecules of water. *Source: Reproduced with permission from Molecular Biology of the Cell, 6th Edition, Garland Science.*

energy-depleted CO_2. Electrons are transferred to transport vehicles, nicotinamide adenine dinucleotide (NADH) and flavin adenine dinucleotide (FADH2) and uploaded onto the electron transport chain (ETC) located on the inner mitochondrial membrane. ETC complexes use energy extracted from the high-energy electrons to translocate protons (H^+) across the inner mitochondrial membrane, generating an electrochemical gradient comprised of a pH gradient (matrix pH > intermembranous space/cytosol) and a voltage gradient (matrix-negative, crista-positive). The resultant proton motive force ($\Delta\mu_H$) across the inner mitochondrial membrane drives ATP synthesis and transports selected substrate molecules (ADP) into the mitochondrial matrix (Mitchell, 1961).

5.2.2 Cardiomyocyte EC

Cardiomyocyte EC is initiated with distribution of an action potential wave through gap junctions in the cardiomyofibers (Fig. 5.2). This signal stimulates the cardiomyocyte to develop tension by sarcomere shortening. The sarcomere unit consists of a precisely ordered array of thick and thin protein filaments (Fig. 5.3). The thin filaments contain actin helices and associated proteins attached at their plus end to a Z-disk. Unattached and capped actin filaments extend toward the center of the sarcomere, where they interact with the motor protein, myosin. Myosin head groups bind actin and use the energy from ATP hydrolysis to "walk" toward the plus end of actin, thereby

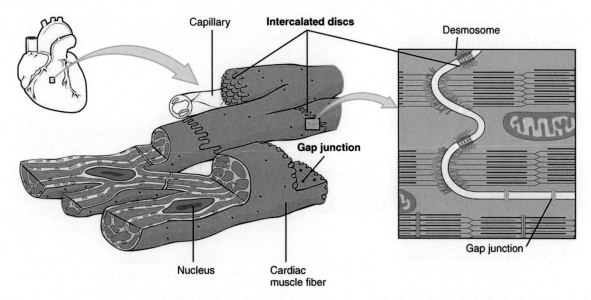

FIGURE 5.2 Cardiomyocyte structure. Cardiac muscle fibers consist of myofilaments arranged in sarcomeres. In contrast to skeletal muscle, cardiomyofibers are highly branched, linked at intercalated disks by desmosomes and punctuated with gap junctions that allow propagation of the action potential. T-tubules transmit the nerve impulse to voltage-gated Ca^{2+} channels at the sarcomere Z-disk. Ca^{2+} influx triggers additional Ca^{2+} release from the adjacent sarcoplasmic reticulum, amplifying the calcium signal required for effective contraction. *Source: Figure By OpenStax College (CC BY 3.0 (http://creativecommons.org/licenses/by/3.0)), via Wikimedia Commons.*

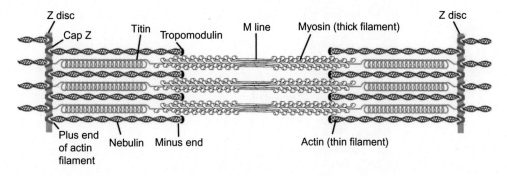

FIGURE 5.3 Sarcomere contraction. In response to Ca^{2+} influx, myosin-binding sites on actin filaments are exposed, allowing myosin head groups to bind with hydrolysis of ATP. Myosin filaments "walk" on actin filaments toward the plus end attached to the Z disk. As the myosin filament walks past the actin filament the sarcomere shortens and the muscle contracts. Accessory proteins maintain structural integrity of the sarcomere. *Source: Reproduced with permission from Molecular Biology of the Cell, Garland Science.*

shortening the sarcomere and contracting the muscle. The precise ordering of the sarcomere is essential since minor missense mutations in the genes for the cardio-specific isoforms of actin and myosin have been implicated in serious cardiomyopathies (Tham et al., 2015). The primary signal for actin—myosin binding is the free calcium (Ca^{2+}) transient (Clusin, 2008) that provides a sharp calcium spike followed by immediate withdrawal of the ion from the cytoplasm to allow the sarcomeres to return to their relaxed position (Fig. 5.4).

5.2.3 Regulation of Cardiomyocyte ATP Production

ATP production is closely tied to the cardiomyocyte workload and mediated by two major signals, the matrix concentrations of ADP and free calcium (Ca^{2+}) in the myocyte. Ca^{2+} is stored in the sarcoplasmic reticulum (SR), a modified endoplasmic reticulum that forms a web-like sheath surrounding each myofibril. The muscle fiber, containing multiple myofibrils, is covered

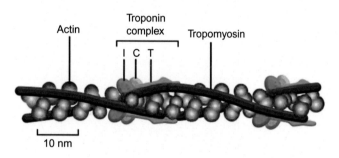

FIGURE 5.4　**Actin and its accessory proteins.** Incoming Ca^{2+} binds troponin, resulting in a structural change that releases tropomyosin and exposes myosin-binding sites on actin. *Source: Reproduced with permission from Molecular Biology of the Cell, Garland Press.*

with a plasma membrane (sarcolemma) that extends in folds called T (transverse)-tubules to the interior of each myofibril bundle to reach the Z-disk of each sarcomere.

Cardiac SR contains much less calcium than skeletal muscle SR, thus cardiac muscle receives most of its calcium from extracellular sources through the voltage-gated L-type calcium channels (LTCC) in the T-tubule sarcolemma.

The cyclic change in myocyte [Ca^{2+}] is regulated by a mechanism called Ca^{2+}-induced Ca^{2+} release (CICR). With every beat, Ca^{2+} enters cardiac myocytes through LTCC in the sarcolemma. This small Ca^{2+} increase in the myocyte triggers a larger release of Ca^{2+} through ryanodine receptors (RyR) in the SR membrane. The CICR floods the sarcomere with Ca^{2+}, which binds to the actin-associated troponin complex. In the resting state, troponin binds tropomyosin, an elongated protein that lies in the groove of the actin helix (Fig. 5.4). When bound to four Ca^{2+} molecules, troponin C triggers tropomyosin release from actin, exposing myosin-binding sites. Myosin "walks" along the actin filament and contracts the sarcomere.

Ca^{2+} is rapidly pumped back into the SR by a Ca^{2+}-ATPase (SERCA2) or out of the cell by sarcolemma Ca^{2+}-ATPase or the Na^{+}—Ca^{2+} exchanger (Tham et al., 2015). Mitochondrial ATP production is tightly linked to EC. With each Ca^{2+} transient, the cardiomyocyte mitochondria also take up the ion; mitochondrial Ca^{2+} uptake stimulates substrate oxidation (Krebs; citric acid cycle) and ATP production as described in Fig. 5.6. To assure that the [Ca^{2+}] is sufficiently high at the mitochondrial permeability transition pore (mPTP; also called MCU), mitochondria are closely annexed to the SR by membrane proteins, mitofusin 1 and 2 (Mfn); Mfn 2 proteins have specific affinity for the endoplasmic/sarcoplasmic reticulum (de Brito and Scorrano, 2008) (Fig. 5.5).

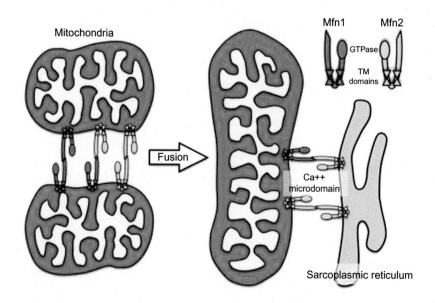

FIGURE 5.5　**Mitofusin (Mfn) tethering.** Mfn are membrane proteins that facilitate mitochondria fusion and fission. Mfn 2 has an affinity for the sarcoplasmic reticulum, allowing the mitochondria direct access to the Ca^{2+} microdomains. *Source: Reproduced from Dorn, G.W., Song, M., et al., 2015. Functional implications of mitofusin 2-mediated mitochondrial-SR tethering. J. Mol. Cell. Cardiol. 78, 123—128., with permission from Science Direct.*

Because the mitochondria are in close proximity to the SR, the [Ca^{2+}] released from the RyR into a "microdomain" (pink area in Fig. 5.6) immediately adjacent to the mitochondrial MCU is many fold higher than elsewhere in the cytosol. High microdomain [Ca^{2+}] triggers opening of the MCU. This protein pore also permits diffusion of small molecules (<1500 Da) such as sodium, into the mitochondrial matrix. Cyclic Ca^{2+} transients into the mitochondria enhance action of the citric acid cycle, activating dehydrogenases that increase NADH and FADH2 substrate for the ETC, raising the electrochemical gradient ($\Delta\mu_H$), generating

force to increase ATP synthesis to levels appropriate to the needs of the cardiomyocyte actin–myosin interactions (Dorn et al., 2015).

5.3 MYOCARDIAL INJURY AND CONSEQUENCES

When the heart is injured, damaged myocytes trigger signaling cascades that activate immune cells resident in the myocardium. The noninjured heart is home to resident immune cells, primarily macrophages but also

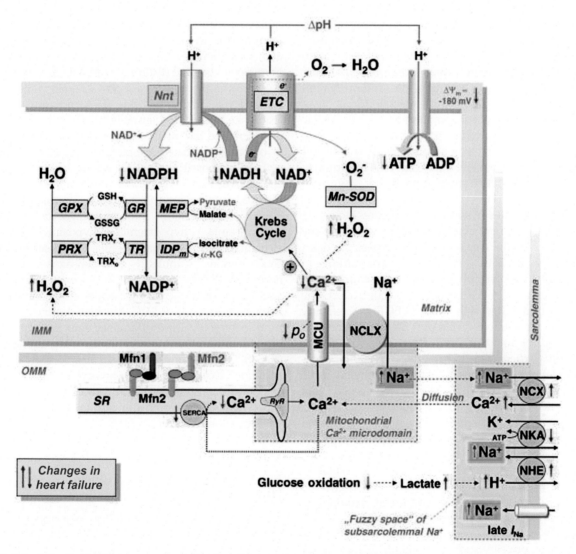

FIGURE 5.6 **Mitochondrial ion transport.** Ca^{2+} released by the sarcoplasmic reticulum (SR) is taken up by the MCU and activates the Krebs (citric acid) cycle dehydrogenases (pyruvate dehydrogenase, isocitrate dehydrogenase, α-ketoglutarate (α-KG) dehydrogenase). Increased oxidation of pyruvate and fatty acyl CoA oxidation produces more energy carried on NADH (and FADH2) to the ETC for ATP production. Ca^{2+} also increases the malate shuttle, responsible for maintaining the cytoplasmic NAD + pool. Calcium is rapidly pumped back into the SR by SERCA (green circle). In heart failure, ion flux is compromised (red arrows): mitochondrial [Ca^{2+}] is lower and [Na$^+$] higher, secondary to increased electrolyte flux (NCX, NKA, NHE). Reduced NADH and ATP synthesis also reduces NADPH, required for regeneration of the endogenous antioxidant, reduced glutathione (GSH) resulting in increased oxidative damage (H$_2$O$_2$). *Source: Reproduced from Bay, J., Kohlhaas, M., et al., 2013. Intracellular Na+ and cardiac metabolism. J. Mol. Cell. Cardiol. 61, 20–27 with permission from Science Direct.*

dendritic cells, located near endothelial cells or within interstitial spaces. Lymphocytes and mast cells are present in small numbers in the unstressed heart, but neutrophils are limited to the vascular compartment (Epelman et al., 2015). Resident immune cells recognize pathogen- and damage-associated molecular patterns (PAMPs; DAMPs) by pattern recognition receptors (PRRs) on their surface. PPRs can also recognize other metabolites released into the extracellular space by injured myocardial cells. Upon recognition of damage, resident immune cells secrete chemotactic signals that recruit neutrophils and other immune cells to the injured area. Activation of proinflammatory cascades such as NF$\kappa\beta$ has also been reported (Palomer et al., 2013) (refer to Chapter 2: The Obese Gunshot Patient, Injury and Septic Shock: Innate Host Defense). Inflammatory signals trigger inflammasome generation and upregulate toll-like and other receptors, expanding myocyte damage and resulting in inflammation-induced cardiomyopathy with mitochondrial dysfunction and cardiomyocyte apoptosis as well as cardiac remodeling and hypertrophy (Palomer et al., 2013).

Chronic ischemic injury is accompanied by reduced coronary circulation. Reduced oxygen delivery triggers release of the master transcription factor, hypoxia-inducible factor-1 (HIF-1) that translocates to the nucleus and regulates a spectrum of genes including those involved in angiogenesis and arteriogenesis (Rey and Semenza, 2010). Angiogenesis increases myocardial capillary density and allows adaptation to deficiency in oxygen supply. Endothelial cell precursor transformation and migration are controlled by vascular endothelial growth factor (VEGF), specifically alteration of the relative levels of VEGFR1 (regression of hypertrophy) and VEGFR2 (induction of hypertrophy) and participation by other factors (angiopointin-1, platelet derived growth factor, endothelial nitric oxide synthase (eNOS)). It should be noted that chronic ischemia is often accompanied by decreased, not increased, capillary density suggesting impaired HIF-1 signaling. Continued ischemia leads to further injury, remodeling, and contractile dysfunction.

Myocyte death is thought to be due to excessive β-adrenergic receptor signaling and Ca^{2+} dysregulation (Nakayama et al., 2007). Catecholamine binding to β-adrenergic receptors activates adenyl cyclase and cAMP-dependent PKA that phosphorylates and alters the function of cardiac proteins such as perilipin 5, specific for myocardial lipid droplets (LDs) (discussed below) and phospholamban, a small transmembrane protein located in the SR. Phospholamban regulates myocyte $[Ca^{2+}]$, thus preventing Ca^{2+} overload. In the unphosphorylated state, phospholamban inhibits the SR Ca^{2+}-ATPase (SERCA2a) that facilitates Ca^{2+} re-uptake. When phospholamban is phosphorylated, SERCA2a inhibition is reversed

(MacLennan and Kranias, 2003). Ca^{2+} overload in the myocyte following ischemic heart disease results in hypercontracture, massive tissue destruction, and enzyme release. Because it upregulates mitochondrial Ca^{2+} permeability pores, Ca^{2+} overload results in osmotic injury from Na^+ uptake and marked reduction in ATP production (Zimmerman et al., 1967; Piper, 2000).

A major cause of myocyte Ca^{2+} overload is ischemia-induced intracellular acidosis that activates the Na^+-H^+ exchanger and causes Na^+ overload. This triggers further influx of Ca^{2+} by activating the Na^+-Ca^{2+} exchanger in the plasma membrane. Ca^{2+} overload in the injured myocyte can also result from upregulation of the sarcolemmal L-type Ca^{2+} channels (LTCC). Other causes of Ca^{2+} overload can include failure to export calcium due to cell swelling, impaired vascular relaxation, formation of white cell plugs, and to other aspects of myocyte damage resulting from oxidative injury.

Studies in the re-perfused postischemic rat heart demonstrated that free fatty acids (FFA) accumulated at the lesion site and caused sustained Na^+ and Ca^{2+} overload that increased the membrane permeability pores, resulting in caspase-3-mediated apoptosis (Fang et al., 2008). The authors proposed that FFAs act as naturally occurring endogenous ionophores and contribute to cell death. Elevated catecholamine activity also promotes translocation of plasma cations into cardiomyocytes, reflected as hypocalcemia, hypokalemia, hypozincemia, hyposelenemia, and hypomagnesemia. Low serum calcium concentration stimulates compensatory parathyroid hormone (PTH) secretion, increasing serum ionized calcium, and thus increasing calcium influx into the injured myocyte through the LTCC. Myocyte death can also result from increased release of angiotensinogen from the liver, triggering the renin–angiotensin–aldosterone system (RAAS). RAAS activation stimulates sodium and water retention. Dead and dying myocytes leak intracellular troponins and stimulate repair processes that replace lost myocytes with fibrous scarring to maintain structural integrity (Borkowski et al., 2011).

5.3.1 Obstructive Sleep Apnea (Intermittent Hypoxia)

Obstructive sleep apnea (OSA) is a common finding in obese patients, especially those with risk factors for metabolic syndrome (Punjabi, 2008). OSA is characterized by recurrent episodes of upper airway collapse leading to repetitive apneas and hyperpneas, producing a pattern of nocturnal intermittent hypoxia (IH). While the mechanistic connection between IH and cardiomyopathy is still unclear, the role of endoplasmic reticular (ER) stress has been recognized for many years (Okada et al., 2004). The ER is the site of protein biosynthesis and maturation with chaperone-dependent protein folding. The extensive cellular reticular network of the ER senses cell homeostasis, including ATP production, nutrient homeostasis, electrolyte balance and other critical homeostatic factors, and elaborates immediate coping responses. Disruption of this network and failure to reconcile ER stress leads to activation of cell death pathways as reviewed by Groenendyk et al. (2013). The ER protein processing can be disturbed by several types of stresses, allowing unfolded proteins to accumulate and aggregate within the ER. To reconcile this stress, the cell upregulates a complex adaptive response (the unfolded protein response) that restores ER homeostasis. Prolonged ER stress leads to cell death by autophagy and apoptosis.

A recent study using a transgenic mouse model demonstrated that chronic IH resulted in increased ER stress and associated cardiomyopathy. The authors suggested that ER stress is the mechanistic link between OSA and cardiomyocyte injury and death (Zhou et al., 2014).

5.3.2 Cardiac Remodeling: Cardiomyocytes and Stromal Cells

The initial response of the heart to chronic stress, myocardial injury, or hypertension is hypertrophic adaptation and an increase in heart mass as shown in Fig. 5.7. Myocytes increase in size with addition of sarcomeres and mitochondria to adapt to the increased workload. Pressure overload (eg, due to hypertension) results in new sarcomeres assembled in parallel, expanding the cross-sectional area of the ventricles. Volume overload, on the other hand, leads to sarcomeres assembled in series, resulting in dilation of the heart. Cardiac hypertrophy, in most instances, signals

a poor prognosis. Evidence suggests that rather than a compensatory response to increased mechanical load, hypertrophy is most often maladaptive and progresses to heart failure (Frey and Olson, 2003). As the hypertrophic heart fails, changes in gene expression cause loss of contractile proteins, downregulation of Ca^{2+} handling proteins, accumulation of extracellular matrix (ECM) elements and changes in growth factors and cytokines including VEGF and tumor necrosis factor (TNF) (Elsherif et al., 2004a). If the stress is prolonged, the contractile elements elongate and the heart undergoes decompensation and dilation with reduced contractile function. While the signaling pathway that triggers cardiomyocyte hypertrophy is not entirely known, it is likely that an integrated network of signals involving mitogen-activated protein kinases (MAPKs) and other stress-activated protein kinases is involved (Molkentin, 2004). Signals for cardiac remodeling can come from damaged and necrotic myocytes that release damage associated molecular pattern fragments (DAMPs), recognized by pattern receptors on immune cells, triggering inflammation and repair.

Cardomyocytes comprise only about 30% of total myocardial cell number but account for 70–80% of the heart's mass. Chronic injury results in focal myocyte death and microscopic scarring. Proteins released from dead and dying myocytes elevate serum troponin levels. A structural analysis of ventricular hypertrophy in human hearts from renal patients with end-stage disease revealed myocyte hypertrophy with increases in both myocyte diameter and length. The total number of myocytes was decreased by ~30%. Thus, some fraction of cardiac hypertrophy can be attributed to myocyte enlargement. In the past, the view was held that cardiomyocytes divided rarely, if at all. However, recent work with adult stem cells has demonstrated cardiomyocyte

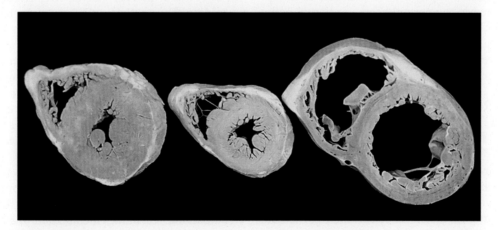

FIGURE 5.7 **Cardiac remodeling.** Hypertrophied ventricle (left) results from chronic pressure overload. Dilated cardiomyopathy (right) shows lengthened fibers that initially increase cardiac output. Ventricular dilation increases wall tension but also increases ATP requirements of the compromised myocardium. Decompensation occurs when the failing heart can no longer pump enough blood to meet the needs of the body. *Source: Reproduced from Robbins and Cotran Pathologic Basis of Disease, 9th Ed. (Figure from Cardiovascular Disease Case 5.) With permission from Elsevier Saunders.*

plasticity and suggests that some degree of myocyte proliferation is possible. This work has prompted tissue engineering investigations using adult stem cells to rescue and replace injured myocytes (Laflamme and Murry, 2011; Leri et al., 2015).

In addition to myocytes, the heart contains a stromal fraction that includes true resident mesenchymal fibroblasts, hematopoietic-derived fibroblast stem cells and those that have arisen from epithelial—mesenchymal transition. True fibroblasts are the best-known cardiac stromal cell type and are thought to be principally responsible for the formation and renewal of ECM (Moore-Morris et al., 2014). Cardiac fibroblasts can also regulate cardiomyocyte proliferation and growth during development. Stromal cell responses to inflammatory signals following injury or pressure loads are likely to trigger the ECM remodeling observed in the failing heart. A current hypothesis holds that reactive scarring of the heart is due to action of resident fibroblasts, whereas hematopoietic-derived myofibroblasts are responsible for myocardial remodeling in heart failure, characterized by excess collagen, muscle fiber entrapment, muscle atrophy, arrhythmias, and electrophysiological changes and increased ventricular stiffness (Crawford et al., 2012; Bani and Nistri, 2014).

In the normal heart, autophagy (refer to Chapter 1: Obesity and Metabolic Syndrome: Nutrient Sensors and Downstream Targets) is maintained at a low basal level to maintain housekeeping functions such as clearance of effete intracellular proteins and organelles and to provide essential nutrient substrate to the heart during periods of starvation. In the stressed heart, autophagy is activated to provide energy substrates; in fact, when autophagy is stimulated by ischemia, apoptosis and other ischemic-related injuries were reduced. If autophagy is inactivated in the normal heart, cardiomyopathy will result. Under ischemic conditions, inhibition of autophagy accelerates ventricular dysfunction and worsens heart failure (Yan et al., 2005).

5.4 TARGETS FOR MYOCARDIAL PROTECTION: MACRONUTRIENT DYSREGULATION

Diabetic cardiomyopathy is characterized by impaired diastolic relaxation time and reduced cardiac contractility. Several metabolic dysregulations characteristic of obesity and the metabolic syndrome can mediate these effects, including systemic insulin resistance and hyperinsulinemia, aberrant activation of the RAAS, and increased oxidative stress (Fig. 5.8). Maladaptive calcium homeostasis can overload the cardiomyocyte, reducing relaxation in diastole, while ER stress can

increase cardiomyocyte death, with subsequent remodeling (Mandavia et al., 2013). The current literature suggests that several of these deleterious changes can be prevented and controlled by targeted dietary modification as described below.

5.4.1 Insulin Resistance

Insulin deficiency and/or insulin resistance characteristic of diabetes results in metabolic stress, including hyperglycemia, hyperlipidemia, and ketonemia (refer to Chapter 1: Obesity and Metabolic Syndrome and Chapter 4: Type II Diabetes, Peripheral Neuropathy and Gout). As reviewed by Hoffman et al. (2013), each of these stressors increases the risk for cardiomyocyte damage. Reduced insulin signaling alters the ratio of macronutrients oxidized in the mitochondria. Fatty acid oxidation is increased and glucose oxidation is decreased. The final common pathway of these obesity-associated metabolic dysregulations appears to be increased oxidative stress. Oxidation products such as oxidized lipoproteins interact with receptors for the innate immune response (refer to Chapter 2: The Obese Gunshot Patient, Injury and Septic Shock) causing a low-grade systemic inflammation, and further metabolic dysregulation and injury.

5.4.2 Macronutrient Substrate Dysregulation in Obesity and Diabetes

While the heart is promiscuous and can utilize a spectrum of macronutrients including ketones, lactate, and amino acids to maintain its continued ATP requirements, fatty acid oxidation normally provides substrate for 50—70% of its ATP needs, with pyruvate oxidation providing the remainder. Differentials that control substrate utilization are primarily the relative FA and pyruvate availability and subsequent conversion into acetyl CoA for entry into the citric acid cycle (Lopaschuk et al., 2010). Acetyl CoA produced by the cardiomyocyte mitochondria can have two alternate fates. If not oxidized by citric acid cycle enzymes, acetyl CoA is exported as acylcarnitine, converted to neutral triglycerides in the cytoplasm and stored in intramyocardial LDs for later use.

Control of nutrient substrate for mitochondrial oxidation may be centrally mediated. Preliminary evidence has been reviewed (Stark et al., 2015) that shows that hypothalamic carnitine metabolism may control substrate utilization. Carnitine facilitates FA transport into the matrix and also controls extrusion of matrix acetyl-CoA for cytoplasmic FA biosynthesis. Since elevated acetyl-CoA levels in the mitochondria allosterically inhibit pyruvate dehydrogenase, glucose oxidation is

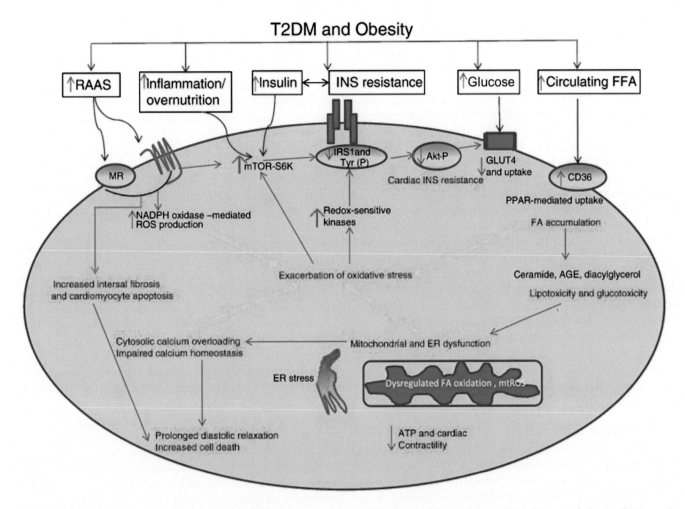

FIGURE 5.8 Obesity and cardiomyopathy. Metabolic dysregulations characteristic of obesity complicated by metabolic syndrome and type II DM have potential to damage the cardiomyocyte. Of note, oxidative stress is central to these deleterious changes (Mandavia et al., 2013). *T2DM,* diabetes mellitus type II; *RAAS,* renin−angiotensin II−aldosterone system; *INS,* insulin; *FFA,* free FAs; *mTOR,* mammalian target of rapamycin; *S6K,* S6 kinase 1 protein; *IRS1,* insulin receptor substrate 1 protein; *GLUT4,* glucose transporter 4; *FA,* fatty acids; *AGE,* advanced glycation end-products; *PPAR,* peroxisome proliferator activator receptor; *CD36,* cluster differentiation protein 36; *ER,* endoplasmic reticulum; *MR,* mineralocorticoid receptor; *ROS,* reactive oxygen species; *mtROS,* mitochondrial ROS; *NADPH oxidase,* nicotinamide adenine dinucleotide phosphate-oxidase; *ATP,* adenosine triphosphate. *Source: Reproduced from Mandavia, C.H., Aroor, A.R., et al. 2013. Molecular and metabolic mechanisms of cardiac dysfunction in diabetes. Life Sci. 92(11): 601−608 with permission from Science Direct.*

reduced. The authors propose that hypothalamic carnitine metabolism senses nutrient availability and integrates hormonal response, thus regulating energy homeostasis. They further propose that the metabolic actions of ghrelin, leptin, and insulin may depend on carnitine-dependent nutrient sensing.

5.4.2.1 Carnitine

The role of carnitine in disease has been reviewed (Flanagan et al., 2010). Approximately 75% of carnitine requirements are obtained from the omnivore diet, especially from animal products; the diet of strict vegetarians contains less carnitine, but consumers absorb it more avidly. Carnitine is also formed by endogenous pathways (Vaz and Wanders, 2002). The liver, kidney,

and brain have the enzymes to form carnitine de novo from lysine and methionine using ascorbate as a cofactor for two α-ketoglutarate-requiring dioxygenase reactions, ε-*N*-trimethyllysine hydroxylase and γ-butyrobetaine hydroxylase (Rebouche, 1991; Otsuka et al., 1999). A variety of primary and secondary carnitine deficiencies due to genetic mutations in carnitine biosynthesis and distribution have been described. Carnitine deficiency can also be acquired in the setting of obesity and its comorbid conditions, including renal disease.

Human skeletal and cardiac muscle contain ∼95% of total body carnitine; this carnitine must be obtained from the plasma, since these organs do not contain carnitine biosynthetic enzymes. Both aging and chronic high-fat feeding decreased expression of mitochondrial carnitine transporters; plasmalemmal carnitine transporter (OCTN2) was decreased in age, but not with high-fat feeding (Noland et al., 2009). It should be noted that these studies were done in rodents competent to produce ascorbate, required for carnitine biosynthesis. Nonetheless, even with adequate ascorbate, human carnitine status was compromised by metabolic stress and associated with increased skeletal muscle accumulation of acylcarnitine esters and diminished hepatic expression of carnitine biosynthetic genes. Marked mitochondrial abnormalities were seen with carnitine insufficiency, including incomplete β-oxidation, and impaired substrate switching from FA to pyruvate.

Results of carnitine supplementation in patients with cardiovascular disease have been reviewed (Flanagan et al., 2010). In general, results have been positive in reducing myocardial injury, preventing ventricular remodeling and normalizing mitochondrial substrate. Additional work is necessary to accurately assess human carnitine status and identify mechanisms that predispose to carnitine decline in aging, obesity, and chronic disease.

5.4.3 The "Fatty Heart" in Obesity and Cardiomyopathy

The concept of the "fatty heart" associated with obesity has been promulgated for centuries (Corvisart des Marets, 1812). Recent work suggests that an obesity-induced metabolic shift overburdens the cardiomyocyte mitochondria with lipid substrate (nonesterified fatty acid (NEFA)). Acetyl CoA not used for ATP synthesis is potential substrate for formation of toxic lipid metabolites including ceramide as recently reviewed by Schilling (2015). Normally, NEFAs are stored in LDs that provide readily available substrate to adjacent mitochondria. The insulin-resistant heart contains many more LDs; excess lipid accumulation is termed "cardiac steatosis". It has been suggested that LD proliferation in the insulin-resistant heart is a defense against lipotoxicity induced by NEFA. Experimental studies have shown that in response to elevated lipid levels, myocytes from a transgenic murine model of metabolic syndrome exhibited upregulated mitochondrial biogenesis together with nuclear receptor transcription factor, PPARγ and its coactivators, including PPARγ-regulated genes involved in FA uptake (CD36), as well as genes for β oxidation and triglyceride biosynthesis (Duncan et al., 2007).

Regulation of LD function is an important strategy used by the heart to minimize damage from excess FA availability as reviewed by Pollak et al. (2015). In a mechanism similar to that of perilipin 1 in adipocytes, perilipin 5 in cardiomyocyte LDs acts as a lipolytic barrier. Fasting induces a rise in cAMP levels and activates PKA, which phosphorylates perilipin 5, releasing CGI-58 (coactivator for adipose triglyceride lipase) allowing TG hydrolysis and FA release. Experimental loss of cardiomyocyte perilipin 5 resulted in uncontrolled lipolysis, myocyte lipotoxicity, and cardiac dysfunction (Kuramoto et al., 2012). In contrast, overexpression of perilipin 5 in mice increased cardiac steatosis but only modestly damaged heart function (Wang et al., 2013). Thus, it appears that LDs actually reduce the oxidative burden to the heart even in the presence of abundant fat accumulation.

Leptin, an adipokine produced by adipose tissue and elevated with increased fat mass, is also expressed in the heart (Green et al., 1995). High leptin levels have been reported in patients with cardiomyopathy and hypertrophy, suggesting that leptin exerts deleterious effects on the heart (Bobbert et al., 2012). However, other investigators have proposed that these effects may be indirect, likely due to leptin's central actions on blood pressure and heart rate. It should be noted that several studies have shown a beneficial action of leptin on the heart, protecting the heart and other organs from excessive lipid accumulation and lipotoxicity in obesity (Unger, 2005). Mechanisms proposed for beneficial actions of leptin include downregulation of enzymes that regulate mitochondrial FA uptake, suggesting that high cardiac leptin levels may compensate for increased FA availability in the obese patient. At this time, data on the influence of leptin on myocardial function in the obese patient are inconclusive and require further investigation (Sweeney, 2010; Hall et al., 2015).

5.5 TARGETS FOR MYOCARDIAL PROTECTION: OXIDATIVE DAMAGE

In the healthy mitochondria the redox state of the electron transport chain (ETC) is neither excessively oxidized nor reduced. As molecular oxygen binds energy-poor electrons it is successively converted to a superoxide radical $O_2^{\bullet-}$, hydrogen peroxide (H_2O_2), hydroxyl radical ($OH^{\bullet}$), and finally to water. Oxygen and its reactive derivatives are effectively constrained by iron and copper ions in the terminal ETC complex, cytochrome oxidase. The mitochondria also contain overlapping antioxidant systems that keep the ROS in optimal balance. Two dysregulations in ETC function can lead to increased production of superoxide ($O_2^{\bullet-}$). The first occurs when high-energy electrons donated by NADH and FADH2 are not efficiently passaged through the ETC. This can be due to abnormalities in substrate oxidation (eg, if coenzyme Q10 is limiting) or if demand for ATP is reduced (eg, in the sedentary individual). Increased fat oxidation with obesity, diabetes, or alcoholism can also increase high-energy electron (NADH and FADH2) presentation to the ETC (Jankovic et al., 2015). In consequence, the ETC is highly reduced and unpaired electrons can "escape" from the ETC and directly reduce molecular O_2, forming $O_2^{\bullet-}$. The second condition conducive for ROS production occurs when mitochondria have a high $\Delta\mu_H$ and a reduced coenzyme Q 10 pool, but ATP is not produced (eg, with copper deficiency). (Coenzyme Q10 shuttles electrons from complex II to complex III in the ETC.) Under these conditions, H_2O_2 is the predominant radical generated (Murphy, 2009).

Because ROS damages mitochondrial proteins, DNA, and other structures, the heart must be protected by antioxidant and radical quenching systems. Any $O_2^{\bullet-}$ produced is efficiently dismutated to H_2O_2 by mitochondrial manganese—superoxide dismutase (Mn-SOD), and by zinc—copper SOD (Zn-Cu SOD) in the intermembranous space and cytoplasm. (Mitochondria contain Mn-SOD as do their primitive bacterial ancestors; eukaryote cells are protected by Zn-Cu SOD.) H_2O_2 is reduced to water by catalase and by the major endogenous antioxidant, reduced glutathione (GSH). To maintain glutathione in its reduced state (GSH), some of the high-energy electrons carried on NADH are transferred to $NADP^+$ by nicotinamide nucleotide transhydrogenase (NNT) (Bay et al., 2013). GSH donates reducing equivalents to H_2O_2 by selenium-dependent glutathione peroxidase (GPX) and thioredoxin (TRX); oxidized forms are in turn reduced by NADPH (Nickel et al., 2014). It should be noted that the riboflavin cofactor, flavin adenine dinucleotide (FAD), is required by glutathione reductase to transport H^+ from NADPH to oxidized glutathione (GSSG), thereby reducing it to its active, sulfhydryl form (Ashoori and Saedisomeolia, 2014).

5.5.1 Antioxidant Protection

Normally, NNT uses proton motive force ($\Delta\mu_H$) across the inner mitochondrial membrane to couple electron transfer from NADH to NADPH, enhancing GSH-dependent antioxidant protection. Ischemia, pressure overload, and other pathological conditions can drive this reaction in reverse. Conditions that inhibit NADH production and/or reduce $\Delta\mu_H$ reverse the NNT pathway. In this case, electrons are removed from NADPH to support NADH-generated oxidative phosphorylation, thereby reducing intracellular antioxidant protection and increasing oxidative damage (Murphy, 2015; Nickel et al., 2015). ROS production that overwhelms detoxification systems compromises not only ATP production, but other mitochondrial functions including signaling and cell-cycle control, and contributes to cell death (McBride et al., 2006). Mitochondrial damage also permeablizes the outer mitochondrial membrane, allowing release of small proteins such as cytochrome c and triggering cardiomyocyte apoptosis (Murphy, 2009).

Although diabetes and associated conditions that contribute to heart failure are characterized by oxidative stress, clinical trials, usually with single-unit antioxidants (eg, vitamin E) have been inconclusive. As is discussed elsewhere (refer to Essentials I: Life in an Aerobic World), the systems that control redox status of the individual are complex and some "antioxidants" are known to exert pro-oxidant effects under certain conditions. Furthermore, the importance of redox signaling in cardiovascular health is becoming more widely recognized. A recent review explores the importance in the heart of reactive oxygen species that, on the one hand can exert tissue damage, and on the other, are required for cardiomyocyte signaling (Hafstad et al., 2013).

If myocardial oxidative stress is not ameliorated by single-unit antioxidants, perhaps enhanced myocyte defense would be more effectively achieved by up-regulating endogenous antioxidants and enzymatic cascades. For example, coenzyme Q10 is an endogenous, lipid-soluble antioxidant molecule produced by all tissues as described elsewhere (refer to Essentials I: Life in an Aerobic World: Single Unit Antioxidant Nutrients: Coenzyme Q10). Myocardial concentration of this molecule is lower in pathological conditions such as diabetes and heart disease (Folkers et al., 1985), and further lowered by drugs that inhibit its biosynthesis (Folkers et al., 1990; Banach et al., 2015). The efficacy of coenzyme Q10 supplements under these conditions requires further investigation.

Enzymatic antioxidant defense is mediated by nuclear factor (erythroid-derived 2)-related factor 2 (Nrf2), a transcription factor that is sequestered in the cytoplasm. As discussed elsewhere (refer to Essentials I: Life in an Aerobic World: Nutrients and Transcriptional Control of Antioxidant Protection), this pathway is triggered by factors associated with oxidative stress that trigger its release and facilitate Nrf2 translocation to the nucleus where it regulates the expression of a spectrum of over 200 cytoprotective genes. The food supply contains bioactive components also known to activate the Nrf2 signaling pathway. Paramount among these agents are members of the vast flavonoid family in fruits and vegetables, including isoflavones in soy beans and phenylpropenoids in green tea, coffee, and chocolate (Boettler et al., 2011). Flavonoids are metabolized to electrophilic quinones that also activate Nrf2 signaling (Stefanson and Bakovic, 2014).

Of interest is the recent investigation of sulforaphane (SFN) found naturally in cabbage, broccoli, and kale, on cardiomyocyte oxidative stress (Singh et al., 2015). Rat cardiomyoblasts (cell line H9c2; in vitro) and wild-type mice (in vivo) were pretreated with SFN with and without exposure to the cardiotoxic chemotherapeutic agent, doxorubicin (DOX). SFN treatment restored cardiac function in DOX-induced cells and reduced murine cardiomyopathy and mortality. Nrf2 activity was increased sixfold in SFN-treated cells versus controls, and fourfold in SFN + DOX-treated cells. Active Nrf2 was 50% higher in SFN-treated murine hearts and remained significantly higher even with DOX treatment. Gene transcripts of major antioxidant enzymes (including superoxide dismutase; catalase; glutathione peroxidase) were elevated with SFN treatment and remained high even with DOX treatment. Additionally, cardiac accumulation of 4-hydroxynonenal (4-HNE) protein adducts was reduced in SFN + DOX hearts, implying that lipid peroxidation induced by DOX treatment was reduced by SFN pretreatment. It is, of course, important to remember that dietary bioactive components undergo significant modification with ingestion. Thus SFN in cabbage, upon exposure to food processing (cooked or raw), the intestinal microbiota and detoxification systems in the enterocyte and liver may be structurally altered and have different actions than pure SFN extracted from the foods (refer to Essentials IV: Diet, Microbial Diversity and Gut Integrity).

5.5.2 Metals and Metallothionein

Metallothionein (MT) comprises a family of low-molecular-weight (500–14,000 Da) proteins localized to the Golgi membrane. By virtue of its high cysteine content (about one-third of amino acids), MT binds heavy metals such as zinc (Zn), cadmium (Cd), copper (Cu), mercury (Hg), and bismuth (Bi). MT is induced by Zn and heavy metals, but also by hormones (dexamethasone; glucagon; epinephrine; norepinephrine), cytokines (IL-1; IL-6; TNFα; INFγ; angiotensin II) and by other physical and chemical stressors. MT is upregulated by metal-binding transcription factor, MFT-1, to increase cellular zinc-binding capacity in response to increased zinc exposure as discussed elsewhere (refer to Essentials II: Heavy Metals, Retinoids and Precursors).

MT is a major contributor to cytoprotection. By binding heavy metals such as cadmium, zinc, and copper, it prevents metal-induced oxidative damage. These potentially toxic ions are safely bound to intracellular MT and donated as needed by the cell. Oxidative stress releases MT-bound metals, providing cofactors for antioxidant metalloproteins such as Zn-Cu superoxide dismutase and ceruloplasmin. Despite its important actions in preventing oxidative damage, regulation of MT biosynthesis and content in diseased tissues is not well characterized; MT appears to be reduced in aging and diseases such as diabetic cardiomyopathy, but also responds to diet (Krizkova et al., 2016). The possibility that MT is one of the genes upregulated by Nrf2 has

recently been confirmed in mice in the context of diabetic nephropathy (Wu et al., 2015).

MT was shown to ameliorate ER stress associated with chronic intermittent hypoxia (IH) in mice; this model was developed to simulate OSA in humans, known to increase risk for cardiomyocyte death and associated IH-induced cardiac remodeling and dysfunction (Zhou et al., 2014). Using the murine transgenic model, the authors demonstrated that MT upregulation protected cardiomyocytes from IH-induced ER stress and subsequent cell death; this protective effect was not seen in MT-knockout mice. The authors noted that in the absence of MT, exposure to IH caused significantly early and severely decreased Akt phosphorylation. This observation is consistent with earlier studies reporting that activation of Akt rescues ER stress-impaired murine cardiac contractile function via glycogen synthase kinase-3β suppression of mitochondrial permeation pore opening (Zhang et al., 2011).

Pretreatment with zinc (or low-dose cadmium) to induce tissue MT abundance reduced symptoms of diabetic cardiomyopathy in streptozotocin (STZ)-treated mice (Wang et al., 2006). A recent trial, also in STZ-treated mice, tested an analog of the diarylheptanoid dietary curcuminoid found in plants of the ginger family (turmeric; ginger) for its ability to prevent diabetes-induced cardiac oxidative damage, fibrosis, ER stress, and apoptosis. The protective effects were mediated by inhibition of c-Jun N-terminal kinase (JNK), a member of the MAPK family. The authors proposed that JNK inhibition may lead to upregulation of MT expression in cardiomyocytes, thus preventing diabetes-induced damage (Wang et al., 2014).

5.5.3 Iron Overload

Although iron is essential to all life, including bacteria, it is potentially toxic and must be tightly controlled as discussed elsewhere (refer to Essentials II: Heavy Metals, Retinoids and Precursors: Iron). Iron overload is widespread globally but often overlooked in the diagnosis of cardiomyopathy. The condition occurs due to a primary genetic alteration in iron control (hereditary hemochromatosis, β thalassemia), to acquired dysregulation of iron distribution (all anemias not caused by iron depletion) and to inadvertent or deliberate intake of iron in excess of needs. Under conditions of iron overload, the iron in circulation exceeds the serum transferrin iron-binding capacity, resulting in nontransferrin-bound iron (NTBI) in circulation. NTBI bypasses the normal feedback regulation that controls iron uptake into the cell, expanding the labile intracellular iron pool. In the presence an oxidative environment and iron overload, Fe^{2+} is available to participate in the Fenton reaction; iron is oxidized to Fe^{3+} with the release of the highly reactive hydroxyl radical ($OH^{\bullet}$) with potential to exceed cellular antioxidant systems, increase peroxidation and damage lipids, proteins, and cell structures.

The process through which iron overload can alter diastolic function and produce dilated cardiomyopathy has been described (Murphy and Oudit, 2010) and is illustrated in Fig. 5.9. More recent studies have indicated that other transport systems may participate in iron uptake and exert even more damage (Wongjaikam et al., 2015). In view of the widespread prevalence of iron overload and its devastating complications to multiple organ systems including liver and pancreas, newer therapies, in addition to phlebotomy and chelation therapy, are urgently being sought.

It is therefore of interest that the Sirtuin/Forkhead box O1 (Sirt/FOXO1) axis has been identified as a key target pathway to protect cardiomyocytes from oxidative stress induced by iron overload (Matsushima and Sadoshima, 2015) and activation of this pathway rescued human cardiomyocytes and cardiofibroblasts. Other authors have demonstrated that resveratrol, a natural polyphenol contained in the skins of red grapes and in red wine, upregulated Sirt1 and protected cells in a murine model of hemochromatosis (Das et al., 2015).

5.6 TARGETS FOR MYOCARDIAL PROTECTION: CALCIUM AND HYPERCONTRACTURE

Transport of ions, molecules, and metabolites across biomembranes is accomplished by specialized transporters and exchangers. The transient rise in intracardiomyocyte calcium is generally agreed to be the signal that triggers cardiomyocyte contraction. The initial Ca^{2+} influx through the L-type Ca^{2+} channels (LTCC) in the sarcolemma triggers a propagating wave that stimulates the larger release of Ca^{2+} from the ryanodine receptors on the SR as reviewed by Eisner et al. (2013). Under pathological conditions, this initial Ca^{2+} release can potentially depolarize the cell membrane and elicit cardiac arrhythmias (Tham et al., 2015). Cardiomyocyte relaxation requires that cytoplasmic Ca^{2+} must be removed to the extracellular fluid by the Na^{+}-Ca^{2+} exchanger (NCX) or by reuptake into the SR by the SR Ca ATPase (SERCA). Any change in myocyte sodium or calcium homeostasis is associated with heart failure, conduction abnormalities, and contractile dysfunction (Wagner et al., 2015).

The highly impermeable inner mitochondrial membrane structure containing cardiolipin, a double phospholipid found in bacteria, is breached by the mitochondrial permeability transition pore (mPTP; also

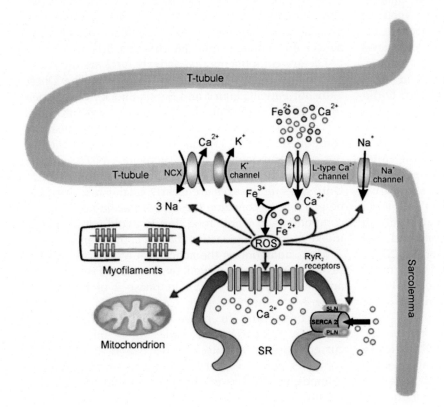

FIGURE 5.9 **Cardiomyocyte damage in iron overload.** Cardiac L-type Ca^{2+} channels (LTCC) appear to be the major entry point for NTBI, although others are under investigation. Fe^{2+} entry slows Ca^{2+} transport, thereby decreasing contractile function and potentially impairing diastolic function as observed in early iron overload. At higher levels, iron—calcium competition can also contribute to systolic dysfunction. Increased ROS impairs cardiac EC coupling, leading to reduced systolic and elevated diastolic Ca^{2+} concentrations in the myocyte. Additionally, nerve conduction is impaired, likely from chronic iron deposition in the node tissue and elevated interstitial fibrosis. Other electrical abnormalities and abnormal nerve conduction reported with iron overload have been linked to arrhymias. Iron also causes increased cardiomyocyte apoptosis, altered cellular metabolism, and stimulation of cardiac fibroblasts, exacerbating myocardial fibrosis. *Source: Reproduced from Murphy, C.J., Oudit, G.Y., 2010. Iron-overload cardiomyopathy: pathophysiology, diagnosis, and treatment. J. Card. Fail. 16 (11), 888–900 with permission from Science Direct.*

called MCU). The MCU is closely annexed to the SR and allows nonselective traffic not only of charged particles (eg, Ca^{2+}; Na^+) but of water and any substance smaller than 1.5 kDa (Hunter and Haworth, 1979). Regulation of the mPTP opening is under intensive investigation (Bernardi and Di Lisa, 2015). Abrupt mPTP induction causes increased mitochondrial permeability (called the permeability transition; PT) accompanied by a sudden loss of mitochondrial membrane potential, reduction in ATP synthesis, loss of ionic homeostasis, matrix swelling, and outer membrane rupture.

Reduction of plasma $[Ca^{2+}]$ resulting from a diet high in sodium and low in calcium in the presence of hypovitaminosis D can stimulate the PTH response. As described elsewhere (refer to Essentials III: Nutrients for Bone Structure and Calcification: Calcium), chronic high intake of fructose or its disaccharide, sucrose, is metabolized in the enterocyte by ATP-dependent ketohexokinase, thereby impairing ATP-dependent transcellular calcium absorption, also resulting in a lowered plasma $[Ca^{2+}]$ and a compensatory PTH response. PTH increases active vitamin D (1,25D) formation in the kidney, thereby stimulating vitamin-D-dependent dietary calcium absorption and osteoclast bone resorption. An overshoot of serum $[Ca^{2+}]$ by these adaptive responses can increase intracardiomyocyte $[Ca^{2+}]$ influx through L-type Ca^{2+} channels on the sarcolemma. Theoretically, high calcium intake, especially from supplemental calcium could also raise serum-ionized calcium and increase the risk for heart disease by a number of mechanisms including increased LTCC uptake. Systematic reviews have identified increased risk with calcium supplements in some, but not all, subject groups (Reid, 2013; Xiao et al., 2013); concomitant intake of vitamin D did not modify the results.

Since excess cardiomyocyte $[Ca^{2+}]$ can induce sarcomere hypercontracture, excess Ca^{2+} that enters through the LTCC must be sequestered in the mitochondria since the SR has limited calcium storage capacity. Excess mitochondrial $[Ca^{2+}]$ elicits oxidative stress and stimulates opening of the mPTP, allowing uptake of sodium and other solutes that result in osmotic swelling, structural degeneration, and cell death, as illustrated in Fig. 5.10 (Yusuf et al., 2012). mPTP induction has been implicated in mitochondrial disruption and apoptosis, not only in cardiomyopathy, but in neurodegeneration and other disease states.

The literature contains preliminary work on the impact of pharmacologic preparations as well as dietary components such as sulforophane (from members of the cabbage family) and resveratrol (from red wine) on maintenance of mitochondrial integrity, mPTP induction,

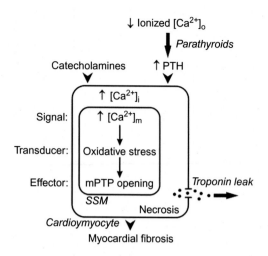

FIGURE 5.10 **Electrolyte alteration and mitochondrial necrosis.** Elevated catecholamine stimulation as well as compensatory serum calcium elevation due to PTH stimulation increases calcium influx through the LTCC. While the mitochondria sequesters excess calcium to protect the EC apparatus, oxidative stress signals opening of the mPTP with subsequent sodium entry, structural alterations, and mitochondrial necrosis. *Source: Reproduced from Yusuf, J., Khan, M.U., et al., 2012. Disturbances in calcium metabolism and cardiomyocyte necrosis: the role of calcitropic hormones. Prog. Cardiovasc. Dis. 55 (1), 77–86 with permission from Science Direct.*

redox signaling, and mitochondria-induced cell death (Negrette-Guzmán et al., 2013; Posadino et al., 2015). This fragmentary work suggests that the fate of injured mitochondria is determined, in part, by Nrf2 activation and subsequent expression of genes that can stimulate mitochondrial biogenesis and/or facilitate mitophagy of irreversibly damaged mitochondria (Fig. 5.11).

Dietary components that activate Nrf2 are discussed below and elsewhere (refer to Chapter 2: The Obese Gunshot Patient, Injury and Septic Shock). Given their involvement in the fate of mitochondria, it is important to consider that the dose of an extracted bioactive component may be critical in tissue response. At low doses, components that activate Nrf2 appear protective, but at high doses could exert toxic effects. Until the safe dosages are clearly established, it would be prudent to consume these pharmacologically active components in whole food, but to avoid supplemental intake of isolated bioactive dietary components.

Mg^{2+} antagonizes the action of Ca^{2+} in the heart at several points as reviewed by Iseri and French (1984) and de Baaij et al. (2015). In phase 2 of the cardiac cycle, Mg^{2+} inhibits L-type Ca^{2+} channels, preventing hypercontracture. In phase 3, delayed rectifier K^+ channels repolarize the cell. Mg^{2+} inhibits these currents and modulates phases 3 and 4. Mg^{2+} also competes with Ca^{2+} for binding sites on troponin and calmodulin, thereby inhibiting the ability of unbound calcium to stimulate excessive contraction. Mg^{2+}

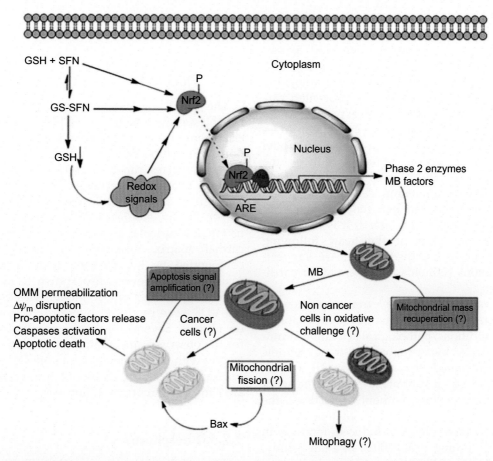

FIGURE 5.11 **Sulforophane and mitochondrial death.** Sulforophane induces Nrf2 activation (refer to Essentials 1: Life in an Aerobic World: Nutrient Transcriptional Control of Antioxidant Protection) and increases expression of phase II detoxification enzymes. It also activates mitochondrial biogenesis (MB). Normal cells segregate mitochondria damaged by oxidants (yellow mitochondria) and degrade them by mitophagy. Intact mitochondria (orange mitochondria) repopulate the cell by MB. The damaged mitochondria and dysfunctional mitochondria in cancer cells induce Bax oligomerization and outer-membrane permeabilization, disrupting the membrane potential and stimulating apoptosis. Question marks indicate areas in need of confirmation. *Source: Reproduced from Negrette-Guzmán, M., Huerta-Yepez, S., et al., 2013. Modulation of mitochondrial functions by the indirect antioxidant sulforaphane: a seemingly contradictory dual role and an integrative hypothesis. Free Radic. Biol. Med. 65, 1078–1089 with permission from Science Direct.*

facilitates rapid Ca^{2+} return to the SR by stimulating Ca^{2+}-ATPase (SERCA2) activity and efflux from the myocyte by stimulating sarcolemma Ca^{2+}ATPase and the Na^+-Ca^{2+} exchanger (NCX). Mg^{2+} reduces Ca^{2+} impact on mitochondrial activity and modulates myocardial contractility through its interaction with calcium, potassium, and sodium at the SR membrane. The multiple actions of Mg^{2+} in regulating electrolytes and the cardiac cycle dictate that Mg concentration in the cardiomyocyte must be tightly regulated (Tashiro et al., 2013). Since Mg depletion exacerbates, and repletion ameliorates muscle spasms in the heart and vasculature, Mg^{2+} has been called "nature's physiologic calcium channel blocker."

Because intracardiomyocyte zinc concentration is also important in calcium signaling, Zn transport into the cardiomyocyte is essential. As reviewed by Woodier et al. (2015), in the resting cardiomyocyte, the intracellular Zn^{2+} concentration is ~100 picomoles, while during EC coupling, Zn^{2+} concentration undergoes transients similar to Ca^{2+}. It has also been suggested that Zn^{2+} enters through the L-type channel with greater affinity than Ca^{2+}, but with lower permeability, reducing the inward current and the sarcoplasmic Ca^{2+} load (Yi et al., 2013). The net result would be improved cardiomyocyte relaxation with a reduction in diastolic intracellular Ca^{2+}.

5.7 TARGETS FOR MYOCARDIAL PROTECTION: CARDIAC REMODELING

5.7.1 Retinoids

Retinoid structure, absorption, transport, and homeostasis have been described elsewhere (refer to Essentials II: Heavy Metals, Retinoids and Precursors: Retinoids). The importance of retinoid signaling in heart development (Pan and Baker, 2007), maintenance of the heart structure, and inhibition of ventricular remodeling has been reviewed (Kotake et al., 2014). Briefly, retinoids prevented cardiomyocyte hypertrophy following exposure to an α-adrenergic receptor agonist and to endothelin in an in vitro culture system. Similarly in an in vivo model of rat myocardial infarction, retinoids reduced the number of myocytes in a cross-section of myocardium. A trial with a vitamin A-deficient diet in these animals increased the left ventricle size, but did not change the size of the infarct. It is of interest that vitamin A increased in the heart following infarction, together with activation of RAR, suggesting that endogenous retinoids are essential for cardiomyocyte repair.

While activation of the renin-angiotensin-aldosterone system maintains hemodynamic stability under normal conditions, pathological activation stimulates cardiomyocyte hypertrophy. All-trans retinoic acid (ATRA) prevented the angiotensin II (ATII)-induced accumulation of contractile proteins in neonatal rat cardiomyocytes without affecting the ATII-induced reorganization of the sarcomeric unit. ATRA also attenuated the ATII-induced increase in intracellular Ca^{2+} (Wang et al., 2002). In a later study, pressure-loaded rats exhibited systolic and diastolic dysfunction with decreased ejection fraction and stroke volume. ATRA treatment prevented these changes in cardiac structure and function, restored the ratio of Bcl2 to Bax, inhibited caspase cleavage, and prevented the reduction of SOD 1 and 2 levels. The pressure overload-induced production of ATII was inhibited in ATRA-treated animals via upregulation of expression of angiotensin-converting enzyme 2 (ACE2). ACE2 mediates the conversion of ATII to AT (1—7) and may act as an endogenous inhibitor of ATII; ACE2 has been shown to be lost in the infarcted area of the myocardium (Keidar et al., 2007). ATRA-treated rats had reduced expression of cardiac and renal renin, angiotensinogen, ACE, and angiotensin type 1 receptor. (Choudhary et al., 2008). The authors concluded that ATRA has a significant inhibitory effect on pressure overload-induced cardiac remodeling through inhibition of the expression of renin—angiotensin system components. The availability of vitamin A in a rat model revealed that following a myocardial infarction, vitamin A is mobilized from liver to heart (Palace et al., 1999). Since no change in plasma vitamin A concentrations was observed, the authors suggested that plasma levels were perhaps not the appropriate tissue to monitor vitamin A distribution during myocardial injury. The transfer from the liver was mediated by an increase in retinyl ester hydrolase, a liver enzyme that hydrolyzes vitamin A storage forms.

5.7.2 Vitamin D

The classic regulation of vitamin D by PTH and fibroblast growth factor 23 (FGF23) has been discussed elsewhere (refer to Essentials III: Nutrients for Bone Structure and Calcification: Vitamin D). Cardiomyocytes can also convert 25D to 1,25D and express vitamin D receptor (VDR) as reviewed by Mizobuchi et al. (2010) and Norman and Powell (2014). Briefly, studies in VDR knockout mice demonstrated ventricular hypertrophy and increased myocardial matrix turnover, but these results were confounded by hypertension with increased plasma renin and PTH levels, also known to increase hypertrophy. In another mouse model, selective deletion of the VDR gene produced hypertrophy with little effect on fibrosis. Several mechanisms have been proposed for the observed amelioration of cardiomyocyte hypertrophy with supplements of vitamin D, but evidence is unclear as to which, if any, of these actions of 1,25D are clinically important in humans.

5.7.3 Phosphorus

The cardiomyopathy in chronic kidney disease (CKD) is often manifest as left ventricular hypertrophy (LVH); ~75% of CKD patients develop LVH by the time they reach end-stage disease (Faul et al., 2011). These authors also reported that fibroblast growth factor 23 (FGF23) (refer to Essentials III: Nutrients for Bone Structure and Calcification: Integrated Regulation of Calcium, Phosphorus and Vitamin D Homeostasis and:Phosphorus) acted independently of CKD. In isolated rat cardiomyocytes FGF23 caused hypertrophy via FGF receptor-dependent activation of the calcineurin-nuclear factor of activated T-cell signaling (NFAT); this effect was independent of α klotho (refer to Essentials III: Nutrients for Bone Structure and Calcification: Vitamin D), the FGF23 receptor found on the kidney and parathyroid glands. Furthermore, study revealed that intramyocardial or intravenous FGF23 injection in wildtype mice without kidney disease produced LVH. Recently, the cardiomyocyte receptor was identified as FGF receptor 4 (Grabner et al., 2015). The mechanism of FGF23-induced hypertrophy has been identified as illustrated in Fig. 5.12. In the absence of α klotho, FGF23 binding to FGFR4 induces interaction with phospholipase Cγ (PLCγ) and stimulates the calcineurin/NFAT downstream cascade that regulates expression of genes controlling pathological, but not physiological, cardiac hypertrophy,

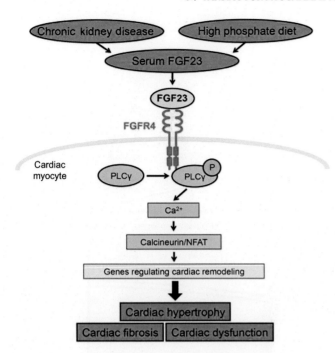

FIGURE 5.12 **FGF23 binding in the cardiomyocyte.** FGF23 is produced by osteocytes in response to high serum phosphorus. Its main function is to increase urinary phosphorous excretion and decrease phosphorous absorption from the enterocyte and resorption from the bone by reducing the production of active 1,25D as described elsewhere (refer to Essentials III: Nutrients for Bone Structure and Calcification: Vitamin D). *Source: Reproduced from Grabner, A., Amaral, A.P., et al., 2015. Activation of cardiac fibroblast growth factor receptor 4 causes left ventricular hypertrophy. Cell Metab. 22 (6), 1020–1032 with permission from Science Direct.*

cardiac fibrosis, and cardiac dysfunction (Wilkins et al., 2004). LVH promotes heart failure and ventricular arrhythmias, and involves ventricular pressure and volume overload.

Since FGF23 is secreted in response to dietary phosphorus load, targeted diet modulation by limiting well-absorbed dietary phosphorus is a powerful tool in prevention and control of cardiac hypertrophy, remodeling, and cardiomyopathy.

5.7.4 Copper

Regression of cardiac hypertrophy with copper supplements was first demonstrated in copper-deficient rats almost 50 years ago (Dallman, 1967), but it was not until 2007 that the mechanisms that underlie the effects of copper on cardiac hypertrophy following pressure overload in mice were proposed (Jiang et al., 2007). The most current understanding is that pressure overload increases ROS production and disrupts the methionine cycle as discussed elsewhere (refer to Essentials I: Life in an Aerobic World: Nutrients and Transcriptional Control of Antioxidant Protection and

Chapter 6: Atherosclerosis and Arterial Calcification: Metabolic Targets for Diet Modification). The cystathione β synthase (CBS) pathway that converts homocysteine to cysteine is blocked by ROS (Zuo et al., 2013). Intramyocyte homocysteine accumulates and complexes with copper to form toxic complexes (Cu−Hcy); this complex is extruded from the cell. The net result is intracardiomyocyte copper deficiency and increased extracellular copper complexes with potential to exert oxidative damage.

Copper normally enters the human cell on the human copper transporter (hCTR1). Upon uptake, Cu is first transferred to reduced glutathione (GSH), presumably responsible for reducing oxidized Cu^{2+} to Cu^+ (Maryon et al., 2013). As depicted in Fig. 5.13, Cu is then transferred to chaperones (COX 17; Atox 1) with greater Cu affinity and ultimately to Cu requiring proteins such as cytochrome c oxidase (COX), the transport proteins (ATP7A; ATP7B), and MT for storage have been described elsewhere (refer to Essentials II: Heavy Metals, Retinoids and Precursors: Copper). hCTR1 is responsive to the extracellular copper concentration. Thus, when human embryonic kidney (HEK293) cells were cultured with high concentrations of $CaCl_2$, hCTR1 immediately translocated to cytoplasmic sites (Molloy and Kaplan, 2009); when extracellular Cu concentration was reduced, hCTR1 returned to its plasma membrane location and was fully functional.

Intracellular copper deficiency reduces assembly and activation of cytochrome oxidase (COX), the last complex in the ETC. Without copper, ETC cannot produce ATP and high-energy electrons escape from the ETC, increasing superoxide production as described above. Cu repletion restores cytochrome oxidase activity, increases ATP production, and improves contractile function. Cu is also important for hypoxia-inducible factor (HIF-1) promotion of VEGF activity (Jiang et al., 2007). In fact, Cu is required for most aspects of compensatory angiogenesis, often reduced in cardiomyopathy. Cu is necessary for migration of endothelial cells to the hypoxic areas by regulating VEGF and fibroblast growth factor 1 (FGF-1); it also modifies the ECM so that the cells can anchor. Cu also increases endothelial cell proliferation through activation of eNOS and modulates lumen formation by regulating VEGF and elastin as reviewed by Zheng et al. (2015).

Copper reduced hypertrophy in phenylephrine (PE)-stimulated neonatal rat myocytes and also in a pressure-overloaded rat model by a VEGF-dependent pathway. The mechanism responsible for these changes related to the VEGF receptors 1 and 2. VEGFR-2 is critical for the hypertrophic growth in cardiomyocytes, while VEGFR-1 causes hypertrophy regression. Copper promoted VEGF production and suppressed VEGFR-2, switching the VEGF signaling to

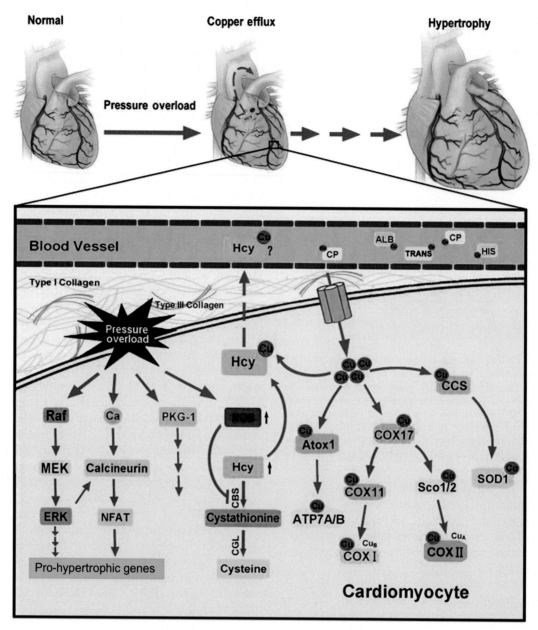

FIGURE 5.13 **Postulated mechanism of Cu efflux with pressure overload.** Cu^+, the primary intracellular Cu ion, makes up ~95% of total body copper; Cu^{2+} is found in the extracellular space (Zhang et al., 2014). The copper transporter protein (CTR1) appears to form a transport pore to mediate oxidized Cu^{2+} uptake. Not shown is glutathione, possibly required to form reduced Cu^+. In the cell, $Cu^{\pm}$ binds to several copper chaperones (eg, COX 17, Atox 1) and is delivered to copper-dependent proteins such as COX, ATP7A/B, and superoxide dismutase (SOD). Hypertrophic signal pathways, such as calcineurin and MAPK, as well as antihypertrophic pathways (PKG-1), are activated in pressure overload, but hypertrophy dominates. Pressure overload-induced ROS inhibits the CBS pathway, allowing homocysteine (Hcy) to accumulate and form toxic Cu—Hcy complexes that leave the cardiomyocyte (Zuo et al., 2013). The presence of increased extracellular copper rapidly internalizes hCTL-1 with a rapid reduction of CTR1-mediated transport of ceruloplasmin (CP)-bound copper. (Molloy and Kaplan, 2009). The authors demonstrated that CTR1 was localized inside the cytoplasm and when extracellular copper was removed, it promptly recycled to the plasma membrane and copper uptake was restored. *Source: Reproduced from Zheng, L., Han, P., et al., 2015. Role of copper in regression of cardiac hypertrophy. Pharmacol. Ther. 148, 66—84 with permission from Science Direct.*

VEGFR-1, thereby inducing regression of the cardio-myocyte hypertrophy (Zhou et al., 2009).

The primary neonatal rat cardiomyocyte culture with PE was also used to explore the relationship among VEGF, Cu, and COX. Atomic absorption spectrometry revealed that VEGF restored PE-depleted Cu concentrations in hypertrophic cardiomyocytes and restored COX activity. In addition, protein contents of COX-IV and Cu chaperones for COX were restored after VEGF treatment. Since PE treatment elevated cardiomyocyte ROS and homocysteine accumulation, it is likely that VEGF-suppressed oxidative stress lessened the possibility of homocysteine—copper complex formation (Sun et al., 2014).

In summary, cardiomyocyte dependence on intracellular copper is multifaceted. Not only is Cu^+ required for cytochrome oxidase-mediated ATP generation, but for HIF-1 and VEGF response for compensatory angiogenesis in the damaged myocardium. Cu also suppresses VEGFR-2 pathways that promote hypertrophy and orchestrate the switch to VEGFR-1 pathways that induce regression.

5.8 CHRONIC HEART FAILURE AND CARDIAC CACHEXIA

Cardiac cachexia is a complex syndrome characterized by progressive weight loss in all body compartments, fat as well as muscle and bone. Multiple processes feed into cachexia, including activation of neuroendocrine and inflammatory systems, increased lipolysis, muscle wasting, anorexia, and malabsorption (von Haehling et al., 2009). While diagnostic criteria are debated, a consensus document has proposed that cachexia must be considered in adult patients with chronic illness who experience unexplained weight loss of 5% in 12 months or less (or BMI < 20) with three of five clinical or laboratory criteria (Evans et al., 2008). To meet increased energy needs, HF patients require 3–7 kcal/kg per day more than healthy adults and may require ~20% more protein (Jurgens et al., 2015). Although feeding alone does not reverse cachexia, low serum levels of micronutrients (thiamin, selenium, zinc, and copper) and amino acids (eg, branched chain amino acids and taurine) have been reported in patients with cardiac cachexia. For some of these nutrients, supplements have shown beneficial results.

Intestinal function is altered in patients with CHF. The inflammatory state that characterizes cardiac cachexia may be secondary to intestinal translocation of luminal bacteria, lipopolysaccharide (LPS), or other bacterial metabolites. Gut morphology and function studied in 22 patients with CHF revealed that patients, compared with controls, had increased bowel wall thickness, a 35% increase in small intestinal permeability with a decrease in D-xylose absorption, suggesting bowel ischemia. The permeability was 210% greater in the colon and higher concentrations of adherent bacteria were found in the mucus throughout the intestine (Sandek et al., 2007). These observations suggest that bacterial biofilm and altered absorption, as well as chronic inflammation, can contribute to the malnutrition in the patient with cardiac cachexia (refer to Essentials IV: Diet Impact on Microbiota Diversity and Gut Integrity).

5.9 DIETARY GUIDELINES TO PREVENT CARDIOMYOPATHY AND MANAGE CHRONIC HEART FAILURE

The American Heart Association recommends a "heart healthy" diet pattern to reduce risk for cardiovascular disease in the general population (Lichtenstein et al., 2006). The overall pattern is a balanced caloric intake with adequate physical activity. The diet should be rich in vegetables and fruits, whole grains, and high-fiber foods. Fish, especially oily fish, should be consumed at least twice per week; saturated fat should be limited to <7% of energy. Trans fats should be avoided and cholesterol limited to <300 mg/day by selecting lean meats or vegetable protein, fat-free or low-fat dairy products. The AHA also recommends limiting sugar and salt and consuming alcohol in moderation. Recent clinical trials suggest that diets containing either plant or animal protein sources can improve metabolic syndrome criteria, especially if the diet results in and maintains a moderate weight loss (~5%). Sodium intake in healthy individuals should be ~2400 mg/day; reduction to 1500 mg/day will reduce blood pressure to a greater degree (Hill et al., 2015).

The American Heart Association and the Heart Failure Society of America (Jurgens et al., 2015) have published recommendations for managing chronic heart failure. Because heart failure (HF) results in fluid retention, dietary sodium restriction has been considered critical to reduce edema and fatigue, decrease extracellular water, and improve the quality of life. However HF patients, especially those in skilled nursing facilities may not comply with these recommendations, or may not have access to low-sodium diets. For this reason, a moderate sodium restriction <3000 mg is recommended to avoid a rapid increase in extracellular volume. Fluid restriction of HF patients is not supported in the literature and not recommended, especially for HF patients receiving laxatives for constipation. Loop diuretic agents increase excretion not only of sodium, but of potassium, calcium, magnesium, and some vitamins, particularly thiamin.

Despite these recommendations, diet analysis of both HF patients (60+ years) and age-matched controls indicated inadequate intakes of potassium, calcium, magnesium, folate, vitamin E, vitamin D, and zinc in over 50% of participants (Arcand et al., 2009). Daily multivitamin and mineral supplementation to prevent deficiencies should be considered for nursing home residents taking loop diuretics as well as for free-living patients who have decreased intake or consume a poor diet (Jurgens et al., 2015).

5.10 TARGETED DIET MODIFICATION TO REDUCE RISK FOR CARDIOMYOPATHY

While the general dietary guidelines provide a structure for constructing a "heart healthy" diet, it is useful to examine the evidence that micronutrients and specific food types can target specific disease processes and potentially prevent cardiomyopathy and reduce its progression to heart failure. Several micronutrient deficits, including vitamin A, vitamin C, vitamin E, thiamine, other B vitamins, vitamin D, selenium, coenzyme Q10, carnitine, magnesium, zinc, and copper have been reported in the context of cardiomyopathy (Allard et al., 2006). Current evidence on the utility of repleting these nutrients has been reviewed (Soukoulis et al., 2009; McKeag et al., 2012). The cardioprotective actions of bioactive components have been discussed throughout this text. Polyphenols in tea, coffee, and red wine (Mukherjee et al., 2009), as well as curcumin from spices that upregulates Nrf2-mediated transcription of cardioprotective enzymes (Zeng et al., 2015), define a diet with potential to reduce risk for cardiomyopathy. Selected nutrients are discussed below.

5.10.1 Vitamins, Amino Acids, and Cardiomyopathy

5.10.1.1 Thiamin (Vitamin B1)

As indicated in Fig. 5.1 above and described elsewhere (refer to Chapter 4: Type II Diabetes, Peripheral Neuropathy and Gout: Thiamin), thiamin plays a major role in regulating mitochondrial fuel substrate, especially carbohydrate, for ATP synthesis. Thiamin is required for pyruvate dehydrogenase, the multiprotein mitochondrial complex that converts pyruvate to acetyl CoA. Thiamin is also required for transketolase, the hexose monophosphate shunt enzyme that controls NADPH production required for biosynthetic activities, especially for de novo production of the endogenous antioxidant, glutathione, now known to be required for intramyocardial copper transport (Maryon et al., 2013). Severe thiamin deficiency results in the deficiency disease, beriberi, characterized by right-sided heart failure, metabolic acidosis, and massive peripheral edema (Dinicolantonio et al., 2013). It should be noted that patients with beriberi cannot convert pyruvate to acetyl CoA for mitochondrial oxidation and rely heavily on glycolysis for ATP production; the product of this pathway is lactic acid, a likely contributor to metabolic acidosis.

A systematic review of the literature identified 20 articles pertaining to thiamin deficiency in patients with heart failure (Dinicolantonio et al., 2013). The authors concluded that patients with heart failure, especially impaired left ventricular ejection fraction, may be thiamin-deficient and that supplementing them with thiamin had potential to improve cardiac function and left ventricular ejection fraction, urine output, weight loss, and other signs and symptoms of heart failure. The authors propose that thiamin status be assessed in all CHF patients (refer to Chapter 7: Diabetic Nephropathy and Chronic Kidney Disease: Thiamin) and until large clinical trials are conducted, moderate supplementation be provided, especially to diabetic patients.

5.10.1.2 Vitamin A

Metabolism of the retinoids and their precursors has been extensively reviewed (D'Ambrosio et al., 2011) and discussed elsewhere (refer to Essentials II: Heavy Metals, Retinoids and Precursors: Vitamin A (Retinoids)). As described above, mechanisms have been identified through which retinoid signaling plays a major role in preventing cardiomyocyte hypertrophy. However, retinoid signaling is highly complex and plays extensive roles in developmental and other pathways. Several clinical reports and experimental trials have demonstrated deleterious effects of high retinoid intake (Albanes et al., 1996; Alkan et al., 2015). Until clinical trials have more clearly defined the safe therapeutic window, it is prudent to obtain retinoids largely from plant sources containing retinoid precursors, for example, carotenoids in green and yellow vegetables, rather than from vitamin A supplements (Institute of Medicine, 2001).

5.10.1.3 Vitamin D (Cholecalciferol)

As described elsewhere (refer to Essentials III: Nutrients for Bone Structure and Calcification: Vitamin D), although humans obtain vitamin D from the diet and by ultraviolet photolysis of 7-dehydrocholesterol in the skin, many individuals have low circulating levels of 25D. Vitamin D hydroxylated in the liver (25D) is the only currently available assessment tool for vitamin D status, however it does not actually represent vitamin D stores, nor is it the active form (1,25D) that binds the VDR. Classically, 1,25D hormone production by the kidney is tightly regulated by PTH and fibroblast growth factor 23 (FGF23). This system is complemented by paracrine/autocrine systems in most cells, including cardiomyocytes that express genes for the 25D activating enzyme (CYP27B1) and for VDR. Regulation of extrarenal conversion of 25D to 1,25D is not yet been fully characterized. Several mechanisms through which vitamin D modifies cardiovascular function have been postulated (Fig. 5.14).

In the United States, vitamin D is widely prescribed for patients with serum 25D levels below the adequate

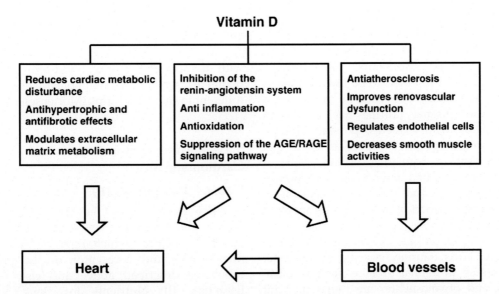

FIGURE 5.14 **Vitamin D actions.** Postulated mechanisms through which 1,25D and/or VDR modifies heart and blood vessel structure and function. Much of this work is preliminary and shown only in animal models or cell culture. *Source: Reproduced from Lee, T.-W., Lee, T.-I., et al., 2015. Potential of vitamin D in treating diabetic cardiomyopathy. Nutr. Res. 35 (4), 269–279 (Lee et al., 2015) with permission Science Direct.*

thresholds as detailed elsewhere (refer to Essentials III: Nutrients for Bone Structure and Calcification: Vitamin D). Since dietary sources of vitamin D include fatty fish (sardines, mackerel), also recommended for their content of omega-3 FAs, it is prudent to consume at least 1–2 fish meals in addition to regular sun exposure when possible. However, a recent meta-analysis demonstrated that consumption of oily fish did raise 25D levels, but was not sufficient to optimize vitamin D status (Lehmann et al., 2015). It should also be noted that since vitamin D is stored in the liver, cod liver oil is a good source of vitamin D, whereas fish oil, taken to increase omega-3 fatty acid levels is made from the whole fish, thus the vitamin D content is minimal.

5.10.2 Taurine, Cysteine, and Sulfur

The dietary sources and metabolism of sulfur compounds and amino acids have been described elsewhere (refer to Essentials I: Life in an Aerobic World: Single Unit Antioxidant Nutrients: Hydrogen Sulfide and to Chapter 6: Atherosclerosis and Arterial Calcification: Organosulfates) and the importance of cysteine as the rate-limiting amino acid in de novo glutathione biosynthesis is well known. As depicted in Fig. 5.15, taurine, cysteine, and hydrogen sulfide are all obtained from dietary sources and generated through vitamin B6 (pyridoxine)-dependent pathways (CBS and downstream enzymes). The availability and metabolism of extracellular, oxidized cystine and the reduced, intracellular bioactive form, cysteine, have recently been reviewed (Yin et al., 2016).

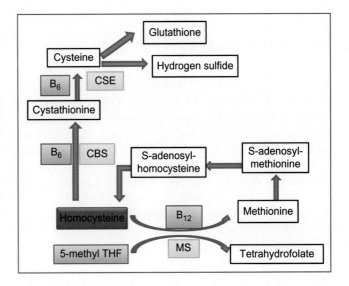

FIGURE 5.15 **Sulfur metabolic pathways.** Dietary sulfur-containing amino acids, methionine and cysteine, are regulated to provide adequate methyl donors and substrate for sulfur-containing metabolites of cysteine including glutathione and taurine (not shown). Nutrients required for these pathways are in pink boxes; enzymes are in green boxes. Homocysteine concentration (red box) is normally maintained at a constant level by these pathways; dysregulated pathways can result in homocysteine efflux from the cell and accumulation in the serum. *Source: Credit: S. Ettinger.*

The protective roles for taurine and cysteine in diabetes and cardiomyopathy have been explored (Tappia et al., 2013; Yang et al., 2013). These authors focused primarily on the ability of these metabolites to prevent cardiomyocyte oxidative stress and myocardial remodeling. A recent review has examined the possibility that taurine, by improving heart function, can prolong lifespan (Schaffer et al., 2015).

Similarly, a relationship between obesity and taurine has been suggested (Murakami, 2015). Briefly, plasma taurine has been reported to be lower in obese humans and animals and decreased in mice fed a high-fat diet. Low plasma taurine has also been reported in patients with diabetes, possibly due to impaired renal reabsorption. Evidence is cited that taurine is important in regulating osmotic balance, modulating the immune and inflammatory responses and regulating lipid and glucose metabolism. Furthermore, higher excretion of taurine (as an index of taurine intake) was inversely related to risk factors for heart disease (Sagara et al., 2015). Since sulfur-containing amino acids (taurine, cysteine, methionine) are well represented in animal proteins such as meat, fish, and eggs, and organosulfates are found in plant sources (cabbage, kale, garlic, onion), these foods would be a prudent addition to the diet of an individual at risk for obesity and its comorbid complications.

5.10.3 Coenzyme Q-10

The utility of supplemental CoQ10 (refer to Essentials I: Life in an Aerobic World: Single Unit Antioxidant Nutrients: Coenzyme Q10) in patients with CHF is still far from resolved. While a meta-analysis of studies testing the influence of statins on plasma CoQ10 levels reported that statins did, indeed, reduce serum CoQ10 levels (Banach et al., 2015), controversy remains (Stocker and Macdonald, 2013). It should be remembered that serum CoQ 10 levels may not reflect tissue concentrations (Bentinger et al., 2007). However, a recent review of intervention trials (DiNicolantonio et al., 2015) noted that myocardial CoQ10 content also tended to decline as the degree of heart failure worsened.

A number of controlled pilot trials with supplemental CoQ10 in heart failure found improvements in functional parameters such as ejection fraction, stroke volume, and cardiac output, without side effects. However, a recent Cochrane review concluded that evidence for use of CoQ10 in heart disease was inconclusive and based on only a small number of adequately controlled trials (Madmani Mohammed et al., 2014). Since then, the impact of supplemental CoQ10 in patients in chronic heart failure was tested in a multicenter randomized placebo-controlled trial (Q-SYMBIO trial). A total of 420 patients received either CoQ10 (100 mg three times daily) or placebo and were followed for 2 years. CoQ10 significantly reduced development of a major adverse cardiovascular event, taken as a hard endpoint. Secondary endpoints

including cardiovascular mortality, all-cause mortality and incidence of hospital stays for heart failure were also significantly lower in the CoQ10 group compared with the placebo group. Short-term functional endpoints (eg, walking for 6 minutes) were not statistically different in the two groups (Mortensen et al., 2014).

5.10.4 Mineral Homeostasis and Cardiomyopathy

5.10.4.1 Phosphorus (P)

When osteocytes in the bone sense elevations in serum phosphorous they secrete fibroblast growth factor 23 (FGF23) which interacts with α klotho, a membrane-bound coreceptor in the proximal renal tubule as described above and elsewhere (refer to Essentials III: Nutrients for Bone Structure and Calcification: Phosphorus). Phosphate is excreted in the urine by upregulation of phosphate transporters while 25D conversion to 1,25D is suppressed to reduce 1,25D-dependent actions at the enterocyte and the bone. In this way, phosphorus concentrations return to normal, reducing risk for soft-tissue calcification. FGF23 is also bound to the cardiomyocyte by non-Klotho-dependent binding to FGFR-4 and stimulates pathways leading to cardiomyocyte hypertrophy as illustrated in Fig. 5.12 above.

Dietary phosphorus intake has risen sharply in recent years due to increased consumption of processed foods containing highly absorbable (~100% available) phosphate salts and cola beverages (diet or sweetened) containing phosphoric acid. Increased intake of bioavailable phosphate, especially in beverages consumed without foods, dramatically increases the risk for elevated serum phosphorus and FGF23 secretion. The association of high FGF23 plasma levels and cardiac hypertrophy in patients with CKD is well described; high FGF23 levels have been associated with endothelial dysfunction, arterial stiffness, LVH, and cardiovascular mortality (Faul et al., 2011). Recently, attention has been focused on risk for LVH and cardiomyopathy from increased phosphorus intake in the population at large (Calvo et al., 2014).

Prudent dietary advice to prevent deleterious effects of high phosphorus intake on the heart would be to avoid processed foods, especially those that list phosphate salts on the label and avoid cola beverages containing phosphoric acid. Although foods in their natural state do contain phosphorus, this element is usually bound to phytates (grains and seeds) and phosphoproteins (animal protein sources) resulting in slowed release and reduced absorption as discussed elsewhere (refer to Essentials III: Nutrients for Bone Structure and Calcification: Phosphorus).

5.10.4.2 Zinc (Zn)

Zinc deficiency is common in diabetic patients due to reduced absorption and hyperzincuria (Kinlaw et al., 1983). Zinc is known to induce cardiomyocyte MT, thus enhancing protection and function as described above. Zinc is widely available in the food supply, especially in protein sources that do not contain absorption inhibitors as discussed elsewhere (refer to Essentials II: Heavy Metals, Retinoids and Precursors: Zinc). Zinc supplementation has been used clinically in many trials as reviewed by Wang et al. (2006). While significant toxic effects have not been reported, long-term supplements (eg, 100 mg/day for 5 years) have been associated with copper-deficiency anemia. This anemia, secondary to copper sequestration in the enterocyte by zinc-induced MT, is readily reversible by lowering the zinc dose and adding dietary copper. Copper deficiency also increases risk for cardiomyopathy (Elsherif et al., 2004b) as discussed above. The authors proposed that a small zinc and copper supplement (30 mg/day zinc with 3 mg/day Cu) shown to have no apparent deleterious effect on copper status (Bonham et al., 2003) could replete zinc status and induce protective myocardial MT. Further trials are needed to titrate the supplemental dose to the extent of zinc and copper losses.

5.10.4.3 Copper (Cu)

Copper-deficient rat hearts consistently exhibit connective tissue abnormalities, ventricular aneurysms, hemothorax, pleural effusion, cardiac hypertrophy, abnormal electrocardiograms, and disturbed response to glucose, FAs, and norepinephrine (Kopp et al., 1983). The dysregulation of copper distribution in cardiomyopathy has been outlined above. As discussed elsewhere (refer to Essentials II: Heavy Metals, Retinoids and Precursors: Copper), copper deficiency is difficult to assess due to lack of appropriate biomarkers. The primary circulating copper protein, ceruloplasmin, is an acute-phase protein and is elevated in inflammation, aging, and chronic disease. Furthermore, serum nonceruloplasmin copper may be primarily bound to homocysteine in a toxic complex and elevated during oxidative stress. Until better biomarkers are available, dietary assessment may be the best indicator of copper status.

In view of its importance in maintaining energy balance (cytochrome oxidase) and antioxidant protection (Zn-Cu superoxide dismutase) as well as other functions, copper sources such as nuts, seeds, green leafy vegetables, and chocolate should be well represented in the diet. While the RDA for copper is only 0.9 mg/day, several investigators have questioned this recommendation, some suggesting an increase to ~3 mg/day in patients with cardiomyopathy (Bashyam, 2007).

5.10.4.4 Magnesium (Mg)

Magnesium plays multiple protective roles in the heart as described above and elsewhere (de Baaij et al., 2015) (refer to Essentials III: Nutrients for Bone Structure and Calcification: Magnesium). Briefly Mg^{2+} regulates ion channels in the heart, improves myocardial contractility and has an antiinflammatory and vasodilatory effect. While most (~99%) Mg is intracellular, serum levels are often assessed as predictive of total-body Mg. Serum and dietary Mg levels have been found to be inversely associated with risk factors for cardiomyopathy including hypertension, type II DM, and metabolic syndrome. The Atherosclerosis Risk in Communities (ARIC) study assessed serum Mg in a cohort (n=14,232) of subjects aged 45–64 years. At 12 years of follow-up, 264 cases of sudden cardiac death (SCD) were observed (Peacock et al., 2010). After adjusting for potential confounders (eg, diuretic use), individuals in the highest quartile of serum Mg had a significantly lower risk for SCD. It is of interest that no relation to dietary Mg intake was reported.

Magnesium is the central atom in chlorophyll, thus dark green leafy vegetables are rich in this nutrient. Mg is released from chlorophyll by heat and enters the cooking liquid. Cooking processes (steaming, canning, blanching) in which the liquid is discarded, will cause Mg losses in the food as eaten. Seeds, legumes, and chocolate are also rich sources of Mg.

5.10.4.5 Selenium (Se)

Selenium was identified as essential for humans over 60 years ago (Schwarz and Foltz, 1957); since then over 25 selenoproteins with active selenocysteine centers have been identified. Se sources, absorption, homeostasis, and utilization are discussed elsewhere (refer to Essentials I: Life in an Aerobic World: Single Unit Nutrients: Selenium). Selenium was first identified as a toxic metal (Fairweather-Tait et al., 2010), and indeed, the therapeutic window between deficiency (<55 mg/day) and the upper limit (<400 µg/day) is narrow (Institute of Medicine, 2000). Because selenoproteins mediate the function of the metal, and selenomethionine is stored in protein, animal products are the major source of organic selenium in the diet (Holben and Smith, 1999). Grains and seeds also contain selenium in multiple forms, but since the content of Se in the soil varies regionally, selenium intake from the diet is difficult to quantitate. Worldwide regions have been characterized by different selenium soil levels and have been shown to correspond with whole-blood selenium levels and the selenium content of hair samples taken from healthy volunteers.

The first evidence linking chronic selenium deficiency with cardiomyopathy was reported about 80 years ago in the Keshan province of north-western China (Yang et al., 1988). (Keshan disease (named after the Chinese region) was used to characterize the fibrotic and degenerative changes observed in patients.) The condition included widespread myocyte necrosis and replacement fibrosis (Benstoem et al., 2015) The soil in this region was found to be selenium-depleted, thus the cardiac manifestations were assumed to be due to selenium depletion. More recent studies have suggested that the influence of selenium depletion is indirect and other factors are involved in the etiology of Keshan cardiomyopathy; studies have proposed mycotoxins such as moniliformin, known to be cardiotoxic, as well as infection with enteroviruses (eg, Coxsackie virus) may be the direct causes. It is also possible that selenium deficiency increased susceptibility to viruses that target the heart muscle (Sun, 2010).

In view of the difficulty in determining selenium content in specific foods, the most prudent course would be to consume animal and plant foods from widely varied sources. Since animal protein contains selenomethionine, adequate protein intake will also improve selenium status.

5.10.4.6 Iron (Fe)

The regulation of iron, both essential and deadly to the human organism, is under tight control as described above and elsewhere (refer to Essentials II: Heavy Metals, Retinoids and Precursors: Iron). In health, iron absorption and iron excretion are in balance such that sufficient iron is available to provide oxygen-carrying capacity as hemoglobin and to serve as cofactor and electron transporter in critical enzymes and systems. To date, over 14 gene mutations have been identified that dysregulate iron homeostasis and increase risk for iron overload (Roy, 2013). *HFE*, the first mutated gene identified as a cause of late-onset hereditary hemochromatosis (type 1) is very common in the US population; approximately 1 million people are affected, mostly of Northern European descent. This gene mutation, together with accessory genes, upregulates hepcidin and potentiates iron deposition in organs such as the heart, liver, and pancreas; if not treated early, excess iron can cause severe tissue damage and lead to cardiomyopathy, cirrhosis, diabetes, and death. Men typically develop symptoms between age 40 and 60. Since blood loss results in iron loss, women do not develop the disease until after menopause. Iron overload is diagnosed by an elevation in serum ferritin and in the transferrin saturation, indicating that more iron is available than can be appropriately utilized.

Since many mutations have been described that disrupt the normal ability to regulate iron absorption and prevent iron overload, it is likely that individuals at risk for iron overload will be underdiagnosed at early stages. It is therefore prudent to avoid intake of dietary iron excess of need. Several strategies are suggested to prevent excess iron availability:

- Unless there is a medical reason (eg, blood loss), no man or postmenopausal woman should take iron supplements. Diagnosed iron deficiency should be promptly corrected and supplements discontinued.
- Check labels as, in the United States and many other countries, iron can be added to fortify cereals and flour, processed snacks, and other prepared products, infant foods, and formulas, as well as milk drinks and malt beverages, meal replacers, slimming beverages, sports beverages, fruit juice drinks, and others (Barrett and Ranum, 1985).
- Red meat contains well-absorbed heme iron derived largely from myoglobin in the muscle. Red meat should be limited and portions consumed should be small. For example, a grilled 16-oz. T-bone steak contains 15.2 mg iron. The RDA for men and postmenopausal women is 8 mg iron.
- Processed red meat usually contains the same amount of iron as its fresh counterpart. The iron content is lower if the meat is mixed with filler (bologna) or brined (corned beef) or processed in a way that dilutes the iron content.
- Other animal proteins (chicken, fish) have iron content similar to red meat per unit weight.
- Legumes contain ferritin-bound iron that appears to be well absorbed (Theil et al., 2012). It is not clear whether high legume intake can contribute to iron overload.
- Patients with anemia should be treated with iron only after iron deficiency has been diagnosed. Other anemias resulting from ineffective erythropoiesis or hemolysis contribute to iron retention.
- "Tonics" to boost energy are available commercially or can be made at home. Many contain iron and should be taken with caution.
- Blood donations significantly reduce an iron load and are recommended if the patient is in good health and not anemic.

5.11 FUTURE RESEARCH NEEDS

Results from intervention trials with single-unit antioxidants (vitamin E, β carotene) have repeatedly demonstrated that these supplements are not uniformly effective in reducing cardiomyopathy, do not reduce premature mortality, and may produce deleterious consequences. However, preliminary mechanistic studies suggest oxidative stress, electrolyte imbalance, and cardiac hypertrophy can be modified by targeted diet modulation that enhances endogenous protective systems. Experimental and clinical trials are urgently needed to test the efficacy of such dietary components to prevent cardiomyocyte damage and to control symptoms in the patient with CHF. Furthermore, studies designed to confirm the observation that dietary phosphorus can increase ventricular remodeling even in normal individuals are required. These studies are urgently needed in view of the widespread practice of adding highly bioavailable phosphate salts to the food supply.

Since genetic polymorphisms in addition to the *HFE* gene have potential to increase the body iron load, strategies for early identification of patients at risk for cardiomyocyte damage due to iron overload and micronutrient depletion should be developed. This is especially important in the obese patient with comorbidities such as diabetes, since these conditions are associated with acquired hepcidin dysregulation due to inflammatory signals. Finally, the inflammatory state associated with obesity can alter nutrient distribution, for example, intracellular copper, leading to deleterious effects on ATP production and antioxidant protection. Further studies are needed to characterize the dysregulation of copper status in the obese human and to develop strategies for restoring homeostasis.

References

Albanes, D., Heinonen, O.P., et al., 1996. α-tocopherol and β-carotene supplements and lung cancer incidence in the alpha-tocopherol, beta-carotene cancer prevention study: effects of base-line characteristics and study compliance. J. Natl. Cancer. Inst. 88 (21), 1560–1570.

Alkan, S., Kayiran, N., et al., 2015. Isotretinoin-induced Spondyloarthropathy-related symptoms: a prospective study. J. Rheumatol. 42 (11), 2106–2109.

Allard, M.L., Jeejeebhoy, K.N., et al., 2006. The management of conditioned nutritional requirements in heart failure. Heart Fail. Rev. 11 (1), 75–82.

Alpert, M.A., Lavie, C.J., et al., 2014. Obesity and heart failure: epidemiology, pathophysiology, clinical manifestations, and management. Transl. Res. 164 (4), 345–356.

Arcand, J., Floras, V., et al., 2009. Nutritional inadequacies in patients with stable heart failure. J. Am. Diet. Assoc. 109 (11), 1909–1913.

Ashoori, M., Saedisomeolia, A., 2014. Riboflavin (vitamin B2) and oxidative stress: a review. Br. J. Nutr. 111 (11), 1985–1991.

Banach, M., Serban, C., et al., 2015. Statin therapy and plasma coenzyme Q10 concentrations—a systematic review and meta-analysis of placebo-controlled trials. Pharmacol. Res. 99, 329–336.

Bani, D., Nistri, S., 2014. New insights into the morphogenic role of stromal cells and their relevance for regenerative medicine. Lessons from the heart. J. Cell. Mol. Med. 18 (3), 363–370.

Barrett, F., Ranum, P., 1985. 6—Wheat and blended cereal foods A2—wiemer. In: Clydesdale, F.M., Wiemer, K.L. (Eds.), Iron Fortification of Foods. Academic Press, Orlando, FL, pp. 75–109.

Bashyam, H., 2007. Heavy metal for a troubled heart. J. Exp. Med. 204 (3), 455a.

Bay, J., Kohlhaas, M., et al., 2013. Intracellular Na^+ and cardiac metabolism. J. Mol. Cell. Cardiol. 61, 20–27.

Benstoem, C., Goetzenich, A., et al., 2015. Selenium and its supplementation in cardiovascular disease—what do we know? Nutrients. 7 (5), 3094–3118.

Bentinger, M., Brismar, K., et al., 2007. The antioxidant role of coenzyme Q. Mitochondrion. 7 (Supplement), S41–S50.

Bernardi, P., Di Lisa, F., 2015. The mitochondrial permeability transition pore: molecular nature and role as a target in cardioprotection. J. Mol. Cell. Cardiol. 78, 100–106.

Blankenhorn, M.A., Vilter, C.F., et al., 1946. Occidental beriberi heart disease. J. Am. Med. Assoc. 131 (9), 717–726.

Bobbert, P., Jenke, A., et al., 2012. High leptin and resistin expression in chronic heart failure: adverse outcome in patients with dilated and inflammatory cardiomyopathy. Eur. J. Heart Fail. 14 (11), 1265–1275.

Boettler, U., Sommerfeld, K., et al., 2011. Coffee constituents as modulators of Nrf2 nuclear translocation and ARE (EpRE)-dependent gene expression. J. Nutr. Biochem. 22 (5), 426–440.

Bonham, M., O'Connor, J.M., et al., 2003. Zinc supplementation has no effect on lipoprotein metabolism, hemostasis, and putative indices of copper status in healthy men. Biol. Trace Elem. Res. 93 (1), 75–86.

Borkowski, B.J., Cheema, Y., et al., 2011. Cation dyshomeostasis and cardiomyocyte necrosis: the Fleckenstein hypothesis revisited. Eur. Heart J. 32 (15), 1846–1853.

Calvo, M.S., Moshfegh, A.J., et al., 2014. Assessing the health impact of phosphorus in the food supply: issues and considerations. Adv. Nutr. Int. Rev. J. 5 (1), 104–113.

Choudhary, R., Palm-Leis, A., et al., 2008. All-trans retinoic acid prevents development of cardiac remodeling in aortic banded rats by inhibiting the renin-angiotensin system. Am. J. Physiol. Heart Circ. Physiol. 294 (2), H633–H644.

Clusin, W.T., 2008. Mechanisms of calcium transient and action potential alternans in cardiac cells and tissues. Am. J. Physiol. Heart Circ. Physiol. 294 (1), H1–H10.

Cooper, L.T., 2009. Myocarditis. N Eng. J. Med. 360 (15), 1526–1538.

Corvisart des Marets, J.N., 1812. Of the degeneration of the muscular tissue of the heart into fat. An Essay on the Organic Diseases and Lesions of the Heart and Great Vessels. Bradford & Read, Boston, MA, pp. 153–156.

Crawford, J., Haudek, S., et al., 2012. Origin of developmental precursors dictates the pathophysiologic role of cardiac fibroblasts. J. Cardiovasc. Transl. Res. 5 (6), 749–759.

Dallman, P.R., 1967. Cytochrome oxidase repair during treatment of copper deficiency: relation to mitochondrial turnover. J. Clin. Invest. 46 (11), 1819–1827.

D'Ambrosio, D.N., Clugston, R.D., et al., 2011. Vitamin A metabolism: an update. Nutrients. 3 (1), 63.

Das, S.K., Wang, W., et al., 2015. Iron-overload injury and cardiomyopathy in acquired and genetic models is attenuated by resveratrol therapy. Sci. Rep. 5, 18132.

de Baaij, J.H.F., Hoenderop, J.G.J., et al., 2015. Magnesium in man: implications for health and disease. Physiol. Rev. 95 (1), 1–46.

de Brito, O.M., Scorrano, L., 2008. Mitofusin 2 tethers endoplasmic reticulum to mitochondria. Nature. 456 (7222), 605–610.

Dinicolantonio, J.J., Niazi, A.K., et al., 2013. Thiamine supplementation for the treatment of heart failure: a review of the literature. Congest. Heart Fail. 19 (4), 214–222.

Dinicolantonio, J.J., Bhutani, J., et al., 2015. Coenzyme Q10 for the treatment of heart failure: a review of the literature. Open Heart. 2 (1), e000326.

Dorn, G.W., Song, M., et al., 2015. Functional implications of mitofusin 2-mediated mitochondrial-SR tethering. J. Mol. Cell. Cardiol. 78, 123–128.

Duncan, J.G., Fong, J.L., et al., 2007. Insulin-resistant heart exhibits a mitochondrial biogenic response driven by the peroxisome proliferator-activated receptor-α/PGC-1α gene regulatory pathway. Circulation. 115 (7), 909–917.

Eisner, D., Bode, E., et al., 2013. Calcium flux balance in the heart. J. Mol. Cell. Cardiol. 58, 110–117.

Elsherif, L., Jiang, Y., et al., 2004a. Dietary copper restriction-induced changes in myocardial gene expression and the effect of copper repletion. Exp. Biol. Med. 229 (7), 616−622.

Elsherif, L., Wang, L., et al., 2004b. Regression of dietary copper restriction-induced cardiomyopathy by copper repletion in mice. J. Nutr. 134 (4), 855−860.

Epelman, S., Liu, P.P., et al., 2015. Role of innate and adaptive immune mechanisms in cardiac injury and repair. Nat. Rev. Immunol. 15 (2), 117−129.

Evans, W.J., Morley, J.E., et al., 2008. Cachexia: a new definition. Clin. Nutr. 27 (6), 793−799.

Fairweather-Tait, S.J., Bao, Y., et al., 2010. Selenium in human health and disease. Antioxid. Redox Signal. 14 (7), 1337−1383.

Fang, K.-M., Lee, A.-S., et al., 2008. Free fatty acids act as endogenous ionophores, resulting in Na+ and Ca2+ influx and myocyte apoptosis. Cardiovasc. Res. 78 (3), 533−545.

Faul, C., Amaral, A.P., et al., 2011. FGF23 induces left ventricular hypertrophy. J. Clin. Invest. 121 (11), 4393−4408.

Firth, B.G., Yancy, C.W., 1990. Survival in congestive heart failure: have we made a difference? Am. J. Med. 88 (Suppl. 1N), 3N−8N.

Flanagan, J.L., Simmons, P.A., et al., 2010. Role of carnitine in disease. Nutr. Metab. (Lond). 7 (1), 1−14.

Folkers, K., Vadhanavikit, S., et al., 1985. Biochemical rationale and myocardial tissue data on the effective therapy of cardiomyopathy with coenzyme Q10. Proc. Natl. Acad. Sci. 82 (3), 901−904.

Folkers, K., Langsjoen, P., et al., 1990. Lovastatin decreases coenzyme Q levels in humans. Proc. Natl. Acad. Sci. USA. 87 (22), 8931−8934.

Frey, N., Olson, E.N., 2003. Cardiac hypertrophy: the good, the bad, and the ugly. Annu. Rev. Physiol. 65 (1), 45−79.

Gore, J.M., Fallon, J.T., 1994. Case 31-1994. A 25-year-old man was admitted to the hospital because of the recent onset of congestive heart failure and atrial fibrillation. N Eng. J. Med. 331 (7), 460−466.

Grabner, A., Amaral, A.P., et al., 2015. Activation of cardiac fibroblast growth factor receptor 4 causes left ventricular hypertrophy. Cell Metab. 22 (6), 1020−1032.

Green, E.D., Maffei, M., et al., 1995. The human obese (OB) gene: RNA expression pattern and mapping on the physical, cytogenetic, and genetic maps of chromosome 7. Genome Res. 5 (1), 5−12.

Groenendyk, J., Agellon, L.B., et al., 2013. Coping with endoplasmic reticulum stress in the cardiovascular system. Annu. Rev. Physiol. 75 (1), 49−67.

Hafstad, A.D., Nabeebaccus, A.A., et al., 2013. Novel aspects of ROS signalling in heart failure. Basic Res. Cardiol. 108 (4), 1−11.

Hall, M.E., Harmancey, R., et al., 2015. Lean heart: role of leptin in cardiac hypertrophy and metabolism. World J. Cardiol. 7 (9), 511−524.

Hill, A.M., Harris Jackson, K.A., et al., 2015. Type and amount of dietary protein in the treatment of metabolic syndrome: a randomized controlled trial. Am. J. Clin. Nutr. 102 (4), 757−770.

Hoffman, W.H., Passmore, G.G., et al., 2013. Increased systemic Th17 cytokines are associated with diastolic dysfunction in children and adolescents with diabetic ketoacidosis. PLoS ONE. 8 (8), e71905.

Holben, D.H., Smith, A.M., 1999. The diverse role of selenium within selenoproteins: a review. J. Am. Diet. Assoc. 99 (7), 836−843.

Hunter, D.R., Haworth, R.A., 1979. The Ca2+-induced membrane transition in mitochondria. Arch. Biochem. Biophys. 195 (2), 453−459.

Huynh, K., Bernardo, B.C., et al., 2014. Diabetic cardiomyopathy: mechanisms and new treatment strategies targeting antioxidant signaling pathways. Pharmacol. Ther. 142 (3), 375−415.

Institute of Medicine, 2000. Dietary Reference Intakes for Vitamin C, Vitamin E, Selenium, and Carotenoids. The National Academies Press, Washington, DC.

Institute of Medicine (US) Panel on Micronutrients, 2001. Dietary Reference Intakes for Vitamin A, Vitamin K, Arsenic, Boron, Chromium, Copper, Iodine, Iron, Manganese, Molybdenum, Nickel, Silicon, Vanadium, and Zinc. National Academy Press, Washington, DC, PANEL ON MICRONUTRIENTS. Available from: http://www.ncbi.nlm.nih.gov/books/NBK222316/.

Iseri, L.T., French, J.H., 1984. Magnesium: nature's physiologic calcium blocker. Am. Heart J. 108 (1), 188−193.

Jankovic, A., Korac, A., et al., 2015. Redox implications in adipose tissue (dys)function—a new look at old acquaintances. Redox Biol. 6, 19−32.

Jiang, Y., Reynolds, C., et al., 2007. Dietary copper supplementation reverses hypertrophic cardiomyopathy induced by chronic pressure overload in mice. J. Exp. Med. 204 (3), 657−666.

Jurgens, C.Y., Goodlin, S., et al., 2015. Heart failure management in skilled nursing facilities: a scientific statement from the American Heart Association and the Heart Failure Society of America. Circ.: Heart Fail. 8 (3), 655−687.

Katz, A.M., Katz, P.B., 1962. Disease of the heart in the works of Hippocrates. Br. Heart J. 24 (3), 257−264.

Keidar, S., Kaplan, M., et al., 2007. ACE2 of the heart: from angiotensin I to angiotensin (1−7). Cardiovasc. Res. 73 (3), 463−469.

Kinlaw, W.B., Levine, A.S., et al., 1983. Abnormal zinc metabolism in type II diabetes mellitus. Am. J. Med. 75 (2), 273−277.

Kopp, S.J., Klevay, L.M., et al., 1983. Physiological and metabolic characterization of a cardiomyopathy induced by chronic copper deficiency. Am. J. Physiol. Heart Circ. Physiol. 14 (5), H855−H866.

Kotake, D., Sato, T., et al., 2014. Retinoid signaling in pathological remodeling related to cardiovascular disease. Eur. J. Pharmacol. 729, 144−147.

Krizkova, S., Kepinska, M., et al., 2016. Microarray analysis of metallothioneins in human diseases—a review. J. Pharm. Biomed. Anal. 117, 464−473.

Kuramoto, K., Okamura, T., et al., 2012. Perilipin 5, a lipid droplet-binding protein, protects heart from oxidative burden by sequestering fatty acid from excessive oxidation. J. Biol. Chem. 287 (28), 23852−23863.

Laflamme, M.A., Murry, C.E., 2011. Heart regeneration. Nature. 473 (7347), 326−335.

Lee, T.-W., Lee, T.-I., et al., 2015. Potential of vitamin D in treating diabetic cardiomyopathy. Nutr. Res. 35 (4), 269−279.

Lehmann, U., Gjessing, H.R., et al., 2015. Efficacy of fish intake on vitamin D status: a meta-analysis of randomized controlled trials. Am. J. Clin. Nutr. 102 (4), 837−847.

Leri, A., Rota, M., et al., 2015. Origin of cardiomyocytes in the adult heart. Circ. Res. 116 (1), 150−166.

Lichtenstein, A.H., Appel, L.J., et al., 2006. Diet and lifestyle recommendations revision 2006: a scientific statement from the American Heart Association Nutrition Committee. Circulation. 114 (1), 82−96.

Lopaschuk, G.D., Ussher, J.R., et al., 2010. Myocardial fatty acid metabolism in health and disease. Physiol. Rev. 90 (1), 207−258.

MacLennan, D.H., Kranias, E.G., 2003. Phospholamban: a crucial regulator of cardiac contractility. Nat. Rev. Mol. Cell Biol. 4 (7), 566−577.

Madmani Mohammed, E., Yusuf Solaiman, A., et al., 2014. Coenzyme Q10 for heart failure. Cochrane Database Syst. Rev. Available from: http://dx.doi.org/10.1002/14651858.CD008684.pub2.

Mandavia, C.H., Aroor, A.R., et al., 2013. Molecular and metabolic mechanisms of cardiac dysfunction in diabetes. Life Sci. 92 (11), 601−608.

Maron, B.J., Towbin, J.A., et al., 2006. Contemporary definitions and classification of the cardiomyopathies: an American Heart Association Scientific Statement from the Council on Clinical

Cardiology, Heart Failure and Transplantation Committee; Quality of Care and Outcomes Research and Functional Genomics and Translational Biology Interdisciplinary Working Groups; and Council on Epidemiology and Prevention. Circulation. 113 (14), 1807–1816.

Maryon, E.B., Molloy, S.A., et al., 2013. Cellular glutathione plays a key role in copper uptake mediated by human copper transporter 1. Am. J. Physiol. Cell Physiol. 304 (8), C768–C769.

Matsushima, S., Sadoshima, J., 2015. The role of sirtuins in cardiac disease. Am. J. Physiol. Heart Circ. Physiol. 309 (9), H1375–H1389.

McBride, H.M., Neuspiel, M., et al., 2006. Mitochondria: more than just a powerhouse. Curr. Biol. 16 (14), R551–R560.

McKeag, N.A., McKinley, M.C., et al., 2012. The role of micronutrients in heart failure. J. Acad. Nutr. Diet. 112 (6), 870–886.

Mitchell, P., 1961. Coupling of phosphorylation to electron and hydrogen transfer by a chemi-osmotic type of mechanism. Nature. 191 (4784), 144–148.

Mizobuchi, M., Nakamura, H., et al., 2010. Myocardial effects of VDR activators in renal failure. J. Steroid Biochem. Mol. Biol. 121 (1–2), 188–192.

Molkentin, J.D., 2004. Calcineurin–NFAT signaling regulates the cardiac hypertrophic response in coordination with the MAPKs. Cardiovasc. Res. 63 (3), 467–475.

Molloy, S.A., Kaplan, J.H., 2009. Copper-dependent recycling of hCTR1, the human high affinity copper transporter. J. Biol. Chem. 284 (43), 29704–29713.

Moore-Morris, T., Guimar, et al., 2014. Resident fibroblast lineages mediate pressure overload–induced cardiac fibrosis. J. Clin. Invest. 124 (7), 2921–2934.

Mortensen, S.A., Rosenfeldt, F., et al., 2014. The effect of coenzyme Q10 on morbidity and mortality in chronic heart failure: results from Q-SYMBIO: a randomized double-blind trial. JACC: Heart Fail. 2 (6), 641–649.

Mukherjee, S., Lekli, I., et al., 2009. Expression of the longevity proteins by both red and white wines and their cardioprotective components, resveratrol, tyrosol, and hydroxytyrosol. Free Radic. Biol. Med. 46 (5), 573–578.

Murakami, S., 2015. Role of taurine in the pathogenesis of obesity. Mol. Nutr. Food Res. 59 (7), 1353–1363.

Murphy, M.P., 2009. How mitochondria produce reactive oxygen species. Biochem. J. 417 (1), 1–13.

Murphy, M.P., 2015. Redox modulation by reversal of the mitochondrial nicotinamide nucleotide transhydrogenase. Cell Metab. 22 (3), 363–365.

Murphy, C.J., Oudit, G.Y., 2010. Iron-overload cardiomyopathy: pathophysiology, diagnosis, and treatment. J. Card. Fail. 16 (11), 888–900.

Nakayama, H., Chen, X., et al., 2007. Ca(2+)- and mitochondrial-dependent cardiomyocyte necrosis as a primary mediator of heart failure. J. Clin. Invest. 117 (9), 2431–2444.

Negrette-Guzmán, M., Huerta-Yepez, S., et al., 2013. Modulation of mitochondrial functions by the indirect antioxidant sulforaphane: a seemingly contradictory dual role and an integrative hypothesis. Free Radic. Biol. Med. 65, 1078–1089.

Nickel, A.G., von Hardenberg, A., et al., 2015. Reversal of mitochondrial transhydrogenase causes oxidative stress in heart failure. Cell Metab. 22 (3), 472–484.

Nickel, A., Löffler, J., et al., 2013. Myocardial energetics in heart failure. Basic Res. Cardiol. 108 (4), 358.

Nickel, A., Kohlhaas, M., et al., 2014. Mitochondrial reactive oxygen species production and elimination. J. Mol. Cell Cardiol. 73, 26–33.

Noland, R.C., Koves, T.R., et al., 2009. Carnitine insufficiency caused by aging and overnutrition compromises mitochondrial performance and metabolic control. J. Biol. Chem. 284 (34), 22840–22852.

Norman, P.E., Powell, J.T., 2014. Vitamin D and cardiovascular disease. Circ. Res. 114 (2), 379–393.

Okada, K.-I, Minamino, T., et al., 2004. Prolonged endoplasmic reticulum stress in hypertrophic and failing heart after aortic constriction: possible contribution of endoplasmic reticulum stress to cardiac myocyte apoptosis. Circulation. 110 (6), 705–712.

Otsuka, M., Matsuzawa, M., et al., 1999. Contribution of a high dose of L-ascorbic acid to carnitine synthesis in guinea pigs fed high-fat diets. J. Nutr. Sci. Vitaminol. (Tokyo). 45 (2), 163–171.

Palace, V.P., Hill, M.F., et al., 1999. Metabolism of vitamin A in the heart increases after a myocardial infarction. Free Radic. Biol. Med. 26 (11), 1501–1507.

Palomer, X., Salvadó, L., et al., 2013. An overview of the crosstalk between inflammatory processes and metabolic dysregulation during diabetic cardiomyopathy. Int. J. Cardiol. 168 (4), 3160–3172.

Pan, J., Baker, K.M., 2007. Retinoic acid and the heart. Vitam. Horm. 75, 257–283.

Peacock, J.M., Ohira, T., et al., 2010. Serum magnesium and risk of sudden cardiac death in the Atherosclerosis Risk in Communities (ARIC) Study. Am. Heart J. 160 (3), 464–470.

Piper, H.M., 2000. The calcium paradox revisited. An artefact of great heuristic value. Cardiovasc. Res. 45 (1), 123–127.

Pollak, N.M., Jaeger, D., et al., 2015. The interplay of protein kinase A and perilipin 5 regulates cardiac lipolysis. J. Biol. Chem. 290 (3), 1295–1306.

Posadino, A.M., Cossu, A., et al., 2015. Resveratrol alters human endothelial cells redox state and causes mitochondrial-dependent cell death. Food Chem. Toxicol. 78, 10–16.

Punjabi, N.M., 2008. The epidemiology of adult obstructive sleep apnea. Proc. Am. Thorac. Soc. 5 (2), 136–143.

Rebouche, C.J., 1991. Ascorbic acid and carnitine biosynthesis. Am. J. Clin. Nutr. 54 (6), 1147S–1152S.

Reid, I., 2013. Cardiovascular effects of calcium supplements. Nutrients. 5 (7), 2522.

Rey, S., Semenza, G.L., 2010. Hypoxia-inducible factor-1-dependent mechanisms of vascularization and vascular remodelling. Cardiovasc. Res. 86 (2), 236–242.

Roger, V.L., Go, A.S., et al., 2012. Heart disease and stroke statistics—2012 update: a report from the American Heart Association. Circulation. 125 (1), e2–e220.

Roy, C.N., 2013. An update on iron homeostasis: make new friends, but keep the old. Am. J. Med. Sci. 346 (5), 413–419.

Sagara, M., Murakami, S., et al., 2015. Taurine in 24-h urine samples is inversely related to cardiovascular risks of middle aged subjects in 50 populations of the world. In: Marcinkiewicz, J., Schaffer, W.S. (Eds.), Taurine, 9. Springer International Publishing, Cham, pp. 623–636.

Sandek, A., Bauditz, J., et al., 2007. Altered intestinal function in patients with chronic heart failure. J. Am. Coll. Cardiol. 50 (16), 1561–1569.

Schaffer, S.W., Ramila, K.C., et al., 2015. Does taurine prolong lifespan by improving heart function? In: Marcinkiewicz, J., Schaffer, W.S. (Eds.), Taurine 9. Springer International Publishing, Cham, pp. 555–570.

Schilling, J.D., 2015. The mitochondria in diabetic heart failure: from pathogenesis to therapeutic promise. Antioxid. Redox Signal. 22 (17), 1515–1526.

Schwarz, K., Foltz, C.M., 1957. Selenium as an integral part of factor 3 against dietary necrotic liver degeneration. J. Am. Chem. Soc. 79 (12), 3292–3293.

Sica, D.A., 2007. Loop diuretic therapy, thiamine balance, and heart failure. Congest. Heart Fail. (Greenwich, Conn.). 13 (4), 244–247.

Singh, P., Sharma, R., et al., 2015. Sulforaphane protects the heart from doxorubicin-induced toxicity. Free Radic. Biol. Med. 86, 90–101.

Soukoulis, V., Dihu, J.B., et al., 2009. Micronutrient deficiencies: an unmet need in heart failure. J. Am. Coll. Cardiol. 54 (18), 1660–1673.

Stark, R., Reichenbach, A., et al., 2015. Hypothalamic carnitine metabolism integrates nutrient and hormonal feedback to regulate energy homeostasis. Mol. Cell Endocrinol. 418 (Part 1), 9–16.

Stefanson, A., Bakovic, M., 2014. Dietary Regulation of Keap1/Nrf2/ARE pathway: focus on plant-derived compounds and trace minerals. Nutrients. 6 (9), 3777–3801.

Stocker, R., Macdonald, P., 2013. The benefit of coenzyme Q10 supplements in the management of chronic heart failure: a long tale of promise in the continued absence of clear evidence. Am. J. Clin. Nutr. 97 (2), 233–234.

Sun, S., 2010. Chronic exposure to cereal mycotoxin likely citreoviridin may be a trigger for Keshan disease mainly through oxidative stress mechanism. Med. Hypotheses. 74 (5), 841–842.

Sun, M., Zuo, X., et al., 2014. Vascular endothelial growth factor recovers suppressed cytochrome c oxidase activity by restoring copper availability in hypertrophic cardiomyocytes. Exp. Biol. Med. 239 (12), 1671–1677.

Sweeney, G., 2010. Cardiovascular effects of leptin. Nat. Rev. Cardiol. 7 (1), 22–29.

Tappia, P.S., Xu, Y.J., et al., 2013. Cardioprotective effects of cysteine alone or in combination with taurine in diabetes. Physiol. Res. 62 (2), 171–178.

Tashiro, M., Inoue, H., et al., 2013. Magnesium homeostasis in cardiac myocytes of Mg-deficient rats. PLoS ONE. 8 (9), e73171.

Tham, Y., Bernardo, B., et al., 2015. Pathophysiology of cardiac hypertrophy and heart failure: signaling pathways and novel therapeutic targets. Arch. Toxicol. 89 (9), 1401–1438.

Theil, E.C., Chen, H., et al., 2012. Absorption of iron from ferritin is independent of heme iron and ferrous salts in women and rat intestinal segments. J. Nutr. 142 (3), 478–483.

Towbin, J.A., Bowles, N.E., 2002. The failing heart. Nature. 415 (6868), 227–233.

Unger, R.H., 2005. Hyperleptinemia: protecting the heart from lipid overload. Hypertension. 45 (6), 1031–1034.

Vaz, F.M., Wanders, R.J.A., 2002. Carnitine biosynthesis in mammals. Biochem. J. 361 (3), 417–429.

von Haehling, S., Lainscak, M., et al., 2009. Cardiac cachexia: a systematic overview. Pharmacol. Ther. 121 (3), 227–252.

Wagner, S., Maier, L.S., et al., 2015. Role of sodium and calcium dysregulation in tachyarrhythmias in sudden cardiac death. Circ. Res. 116 (12), 1956–1970.

Wang, H.-J., Zhu, Y.-C., et al., 2002. Effects of all-trans retinoic acid on angiotensin II-induced myocyte hypertrophy. J. Appl. Physiol. 92 (5), 2162–2168.

Wang, J., Song, Y., et al., 2006. Cardiac metallothionein induction plays the major role in the prevention of diabetic cardiomyopathy by zinc supplementation. Circulation. 113 (4), 544–554.

Wang, H., Sreenivasan, U., et al., 2013. Cardiomyocyte-specific perilipin 5 overexpression leads to myocardial steatosis and modest cardiac dysfunction. J. Lipid Res. 54 (4), 953–965.

Wang, Y., Zhou, S., et al., 2014. Inhibition of JNK by novel curcumin analog C66 prevents diabetic cardiomyopathy with a preservation of cardiac metallothionein expression. Am. J. Physiol. Endocrinol. Metab. 306 (11), E1239–E1247.

Wilkins, B.J., Dai, Y.-S., et al., 2004. Calcineurin/NFAT coupling participates in pathological, but not physiological, cardiac hypertrophy. Circ. Res. 94 (1), 110–118.

Wongjaikam, S., Kumfu, S., et al., 2015. Current and future treatment strategies for iron overload cardiomyopathy. Eur. J. Pharmacol. 765, 86–93.

Woodier, J., Rainbow, R.D., et al., 2015. Intracellular zinc modulates cardiac ryanodine receptor-mediated calcium release. J. Biol. Chem. 290 (28), 17599–17610.

Wu, H., Kong, L., et al., 2015. Metallothionein plays a prominent role in the prevention of diabetic nephropathy by sulforaphane via up-regulation of Nrf2. Free Radic. Biol. Med. 89, 431–442.

Xiao, Q., Murphy, R.A., et al., 2013. Dietary and supplemental calcium intake and cardiovascular disease mortality: the National Institutes of Health–AARP diet and health study. JAMA Intern. Med. 173 (8), 639–646.

Yan, L., Vatner, D.E., et al., 2005. Autophagy in chronically ischemic myocardium. Proc. Natl. Acad. Sci. U.S.A. 102 (39), 13807–13812.

Yang, G., Ge, K., et al., 1988. Selenium-related endemic diseases and the daily selenium requirement of humans. World Rev. Nutr. Diet. 55, 98–152.

Yang, Q., Yang, J., et al., 2013. Effects of taurine on myocardial cGMP/cAMP ratio, antioxidant ability, and ultrastructure in cardiac hypertrophy rats induced by isoproterenol. Adv. Exp. Med. Biol. 776, 217–229.

Yi, T., Vick, J.S., et al., 2013. Identifying cellular mechanisms of zinc-induced relaxation in isolated cardiomyocytes. Am. J. Physiol. Heart Circ. Physiol. 305 (5), H706–H715.

Yin, J., Ren, W., et al., 2016. l-Cysteine metabolism and its nutritional implications. Mol. Nutr. Food Res. 60 (1), 134–146.

Yusuf, J., Khan, M.U., et al., 2012. Disturbances in calcium metabolism and cardiomyocyte necrosis: the role of calcitropic hormones. Prog. Cardiovasc. Dis. 55 (1), 77–86.

Zeng, C., Zhong, P., et al., 2015. Curcumin protects hearts from FFA-induced injury by activating Nrf2 and inactivating NF-κB both in vitro and in vivo. J. Mol. Cell Cardiol. 79, 1–12.

Zhang, Y., Xia, Z., et al., 2011. Activation of Akt rescues endoplasmic reticulum stress-impaired murine cardiac contractile function via glycogen synthase kinase-3β-mediated suppression of mitochondrial permeation pore opening. Antioxid. Redox Signal. 15 (9), 2407–2424.

Zhang, S., Liu, H., et al., 2014. Diabetic cardiomyopathy is associated with defective myocellular copper regulation and both defects are rectified by divalent copper chelation. Cardiovasc. Diabetol. 13 (1), 1–18.

Zheng, L., Han, P., et al., 2015. Role of copper in regression of cardiac hypertrophy. Pharmacol. Ther. 148, 66–84.

Zhou, Y., Bourcy, K., et al., 2009. Copper-induced regression of cardiomyocyte hypertrophy is associated with enhanced vascular endothelial growth factor receptor-1 signalling pathway. Cardiovasc. Res. 84 (1), 54–63.

Zhou, S., Yin, X., et al., 2014. Metallothionein prevents intermittent hypoxia-induced cardiac endoplasmic reticulum stress and cell death likely via activation of Akt signaling pathway in mice. Toxicol. Lett. 227 (2), 113–123.

Zimmerman, A. N.E., Daems, W., et al., 1967. Morphological changes of heart muscle caused by successive perfusion with calcium-free and calcium-containing solutions (calcium paradox). Cardiovasc. Res. 1 (3), 201–209.

Zuo, X., Dong, D., et al., 2013. Homocysteine restricts copper availability leading to suppression of cytochrome C oxidase activity in phenylephrine-treated cardiomyocytes. PLoS ONE. 8 (6), e67549.

Further Reading

Goldman, L., Schafer, A.I., 2012. Chapter 58. Heart failure: pathophysiology and diagnosis. Goldman Cecil Medicine. Elsevier Saunders, Philadelphia, PA.

Gore, J.M., Fallon, J.T., 1994. Case 31-1994—a 25 year-old man with the recent onset of diabetes mellitus and congestive heart failure. N Eng. J. Med. 331, 460–466. Available from: http://dx.doi.org/10.1056/NEJM199408183310708.

Kumar V. et al. 2014. Robbins and Cotran Pathologic Basis of Disease, ninth Ed. Chapter 12. The heart: cardiomyopathy. Elsevier Saunders, Philadelphia, PA, London.

6

Atherosclerosis and Arterial Calcification

CHIEF COMPLAINT (TYPICAL PATIENT)

Severe, unrelenting anterior chest pain for the past 1.5 hours.

PATIENT HISTORY

Mrs. K provided the history because Mr. K was in severe pain. GK, 61-year-old white male, has been a Lutheran minister in the same parish for the past 35 years. He enjoys his work, but sometimes becomes quite agitated when confronted with difficult problems in his church. Today he arrived home complaining of nausea and chest pain. He did not complain of palpitations or shortness of breath, nor has he complained of chest pain or fainting in the past. He has smoked (1 pack per day) for over 40 years. Both he and his wife abstain from alcohol. His father succumbed to a sudden fatal myocardial infarction (MI) at age 59 years. His mother still lives alone and is in good health at 85 years.

MEDICAL HISTORY

The medical record indicated that the patient has visited a private physician annually for many years for routine check-ups. He has no history of hypertension, diabetes, or hyperlipidemia. His total cholesterol levels have been steadily rising over the years. He is on no medication at present; Mrs. K. indicated that he has a history of allergy to sulfa drugs. He underwent surgery for a cholecystectomy 10 years ago and an appendectomy 30 years ago.

DIET AND LIFESTYLE HISTORY

Mrs. K advised that her husband has been cutting down on fat intake in an attempt to manage his rising cholesterol numbers. He is also rather sedentary and recently he has become concerned about his increasing girth. In the last year he has been trying to walk a bit more when he finds the time. He does not take vitamin supplements except 1000 μg folic acid daily. His wife found information on the internet that this vitamin was protective against peripheral vascular disease (PVD). Because their milk intake is low, both he and his wife have been taking 1000 mg calcium citrate daily.

PHYSICAL EXAMINATION

General appearance: Overweight male with significant abdominal obesity in acute distress from chest pain
Vital signs: Temperature 37°C; BP 118/78 mmHg; HR 92 bpm; RR 20 bpm
Heart: PMI 5 ICS MCL focal. SI normal intensity; S2 normal intensity and split. S4 gallop at the apex. No murmurs, clicks or rubs
HEENT: Head, ears, nose, and Throat WNL
Eyes: EOMI, fundoscopic exam WNL, no evidence of atherosclerosis, diabetic retinopathy, or early hypertensive changes
Genitalia: Grossly normal
Neurologic: No focal localizing abnormalities. DTR symmetric bilaterally
Skin: Diaphoretic and pale
Chest/lungs: Lungs clear to auscultation and percussion
Peripheral vascular: PPP normal
Abdomen: RLQ scar and midline suprapubic scar. No hepatomegaly, splenomegaly, masses, inguinal lymph nodes, or abdominal bruits
Anthropometric indices: Height 70 inches; weight 220 lbs; BMI 31.6; waist circumference 45″.

DIET HISTORY

Mr. K has a good appetite. Recently, he has tried to cut down on fat and switched to safflower oil margarine instead of butter.

129

24-Hour recall:	(Provided by Mrs. K)
Breakfast:	Coffee with ¼ cup low-fat milk and two sugars
Mid-morning snack:	One large cinnamon raisin bagel with 2 tbsp raspberry jam, 8 oz apple juice, coffee with ¼ cup low-fat milk and two sugars
Lunch:	Sandwich with four slices roast beef, two slices Genoa salami, and two slices low-fat Swiss cheese with lettuce, tomato, and low fat-mayonnaise on a roll. Half sour pickle, 8 oz apple juice
Dinner:	Two lean grilled pork chops, baked potato with substitute (corn oil margarine and milk) sour cream. One cup sweetened apple sauce, ½ cup green beans with corn oil margarine, one slice angel-food cake with ½ cup low-fat ice cream
Before-bed snack:	8 oz tea with milk and sugar with six low-fat chocolate chip cookies

Pertinent Laboratory	Normal	Patient Admission
Glucose (mg/dL)	70–110	136
Cholesterol (mg/dL)	120–199	235
HDL-C (mg/dL)	>45	30
VLDL (mg/dL)	7–32	45
LDL (mg/dL)	<130	160
LDL/HDL	<3.55	5.3
APO A (mg/dL)	94–178	72
APO B (mg/dL)	63–133	143
Homocysteine (μmol/L)	5.9–11.4	13
Troponin I (mg/dL)	<0.2	2.4
Troponin T (mg/dL)	<0.03)	2.1
Triglyceride (mg/dL)	40–160	201
HbA$_{1C}$	3.9–5.2	7.2
WBC ($\times 10^3$/mm^3)	4.8–11.8	11
Hct (%)	40–54	54
MCV (μm^3)	80–96	91
Differential		
%GRANs	34.6–79.2	86
SEGS	50–62	84
Ferritin (mg/mL)	20–300	305

HOSPITAL COURSE

Patient's chest pain resolved after two sublingual nitroglycerine (NTG) at 3-minute intervals. Pain resolution maintained with 2 mg morphine IV. Patient taken to cath. lab. A soft plaque in the left anterior descending (LAD) coronary artery causing a 70% stenosis was noted. Angioplasty of the LAD resulted in a patent infarct-related artery with near normal flow. A stent was placed to stabilize the patient and limit infarct size.

Patient was referred for nutrition counseling and for cardiac rehabilitation. Patient discharged with the following medications:

Lopressor 50 mg/day; lisinopril 10 mg/day; nitroBid 9 mg twice daily; NTG 0.4 mg sublingually prn for chest pain; Aspirin 81 mg/day; Lipitor 10 mg/day at bedtime.

RESOURCES

6.1 THE ATHEROSCLEROTIC PLAQUE

Vascular disorders typically involve muscular arteries, but venous diseases also occur. The lesions are of two major types: (1) *Narrowing (stenosis)* or *complete obstruction* of vessel lumens, either progressively (eg, by atherosclerosis) or precipitously (eg, by thrombosis or embolism) and (2) *weakening* of the vessel walls, leading to dilation or rupture (Fig. 6.1).

The normal vessels are covered by a continuous layer of multifunctional endothelial cells (intima). These cells maintain blood flow, create a barrier to prevent entrance of circulating immune cells/monocytes, regulate growth and contraction of smooth muscles in the underlying media, maintain the tone of the vessel wall and control synthesis of extracellular matrix (ECM) (collagen, elastin, and proteoglycans) that provides support for the vessel.

6.1.1 Atherosclerosis (AS)

Diseases of the vasculature and heart, including coronary heart disease, cerebrovascular disease, peripheral artery disease, cardiac hypertrophy, hypertension, congenital heart disease, deep vein thrombosis, and pulmonary embolism have been classified by the WHO under the general term, cardiovascular disease (CVD) (http://www.who.int/mediacentre/factsheets/fs317/en/index.html provides a summary of CVD risk factors, symptoms, and treatment). CVD is accompanied by chronic injury to the vessels, most commonly the muscular arteries, resulting in the formation of an atherosclerotic plaque. Behavioral risk factors such as smoking, lack of physical inactivity, and an unhealthy diet, leading to intermediate metabolic risk factors including obesity, hypertension, dyslipidemia, and glucose intolerance, account for ~80% of CVD worldwide. The WHO has issued a fact sheet on CVD useful for the clinician.

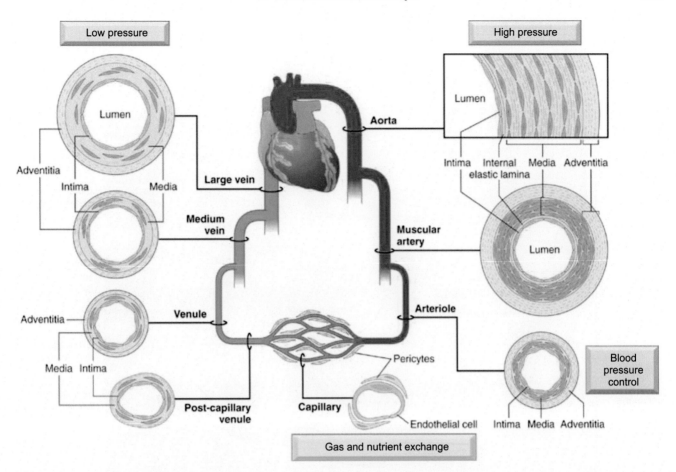

FIGURE 6.1 Low- and high-pressure vessel structure. The low-pressure venous system and the high-pressure arterial system have similar structures, but the wall thickness and composition varies with hemodynamic forces and functional requirements. The systems meet at the thin-walled capillaries that facilitate gas and nutrient exchange. *Source: Reproduced from Robbins and Cotran Pathologic Basis of Disease, 9th Ed. With permission from Elsevier Saunders.*

Arteriosclerosis is the generic term for thickened and stiffened arteries of all sizes. *Atherosclerosis* is more specific and describes thickened and hardened lesions of the medium and large muscular and elastic arteries. Atherosclerotic plaques protrude into the vascular lumen and mechanically reduce blood flow. Advanced plaques can also rupture, leading to catastrophic vessel thrombosis (Fig. 6.2). In addition, plaques weaken the underlying vascular media and contribute to formation of aneurysms. Plaques in the coronary arteries obstruct blood flow to the heart, resulting in a myocardial infarction (MI), responsible for almost a quarter of all deaths in the United States. Significant morbidity and mortality are also caused by aortic and carotid atherosclerotic disease resulting in ischemic stroke.

The American Heart Association has separated atherosclerotic plaque development into six phases:

- *Phase 1* lesions are small and commonly found in persons younger than 30. Type I lesions contain macrophages with lipid inclusions (foam cells) while type II lesions contain both macrophages and smooth muscle cells with mild extracellular lipid deposits. Type III lesions contain smooth muscle cells surrounded by extracellular connective tissue, fibrils, and lipid deposits.
- *Phase 2* lesions have high lipid content with increased inflammation and a thin, fibrous cap. They are described as "vulnerable or high-risk" because they may be susceptible to disruption. These plaques are categorized as type IV lesions: extracellular lipid mixed with normal intima as an outer layer or cap, and type V lesions that possess an extracellular lipid core covered by a thin fibrous cap.
- *Phases 3, 4, 5, and 6* are more complex lesions which can contain calcium precipitates, accumulate more necrotic debris and thrombotic detritus, erode into the vessel wall, or become stenotic (Fig. 6.3).

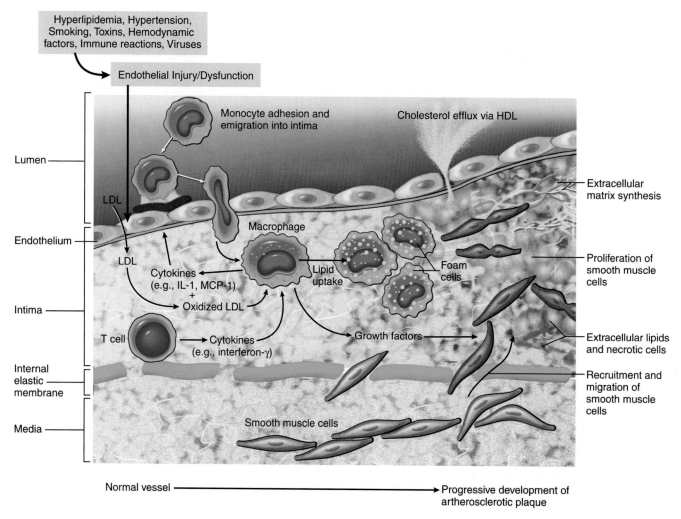

FIGURE 6.2 The lipid-filled AS plaque. The atherosclerotic plaque consists of a central area filled with necrotic debris, covered with a fibrous cap produced by resident fibroblasts and myointimal cells derived from underlying smooth muscle. In response to injury and endothelial cell dysfunction, macrophages enter the lesion, phagocytize oxidized lipoproteins and become activated foam cells that secrete inflammatory mediators. Smooth muscle cells proliferate and migrate to the intima and platelets aggregate in the prothrombotic environment. Cholesterol is released to high density lipoproteins (HDL) and returned to the liver. *Source: Reproduced from Robbins and Cotran Pathologic Basis of Disease, 9th Ed. With permission from Elsevier Saunders.*

6.1.2 Lipid Content of the Atheroma

Atheromas are raised lesions that protrude into the vessel lumen and contain a soft, yellow, grumous (thick and lumpy) core consisting mainly of cholesterol and cholesterol esters, covered by a white, fibrous cap. Extracellular lipid enters the atheroma by bulk transport from the vessel lumen, especially when dyslipidemia is present. Lipid is also released into the necrotic core from degenerating foam cells. Cholesterol is frequently present in the atheroma center as crystalline aggregates seen as cholesterol clefts in histological preparation. (Routine tissue processing washes out the cholesterol and leaves behind only empty "clefts.") Cholesterol accumulation in the

plaque reflects an imbalance between lipid influx and efflux. High-density lipoproteins (HDLs) facilitate cholesterol efflux from these accumulations by reverse cholesterol transport (see below). Under the influence of inflammatory mediators, smooth muscle cells migrate to the intima, differentiate into fibroblast-like cells, proliferate, and produce ECM elements, including collagen and proteoglycans as shown below.

6.1.3 Cardiovascular Calcification

Complex atherosclerotic lesions contain radio-opaque calcium hydroxyapatite deposits. The degree of

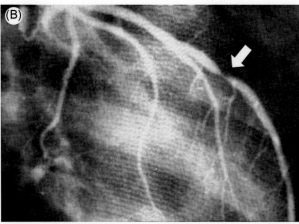

FIGURE 6.3   (A) The lumen of this coronary artery is partially occluded with lipid-laden atherosclerotic plaque. (B) Arterial narrowing is visible in coronary artery angiography (white arrow). *Source: Reproduced from Robbins and Cotran. Pathologic Basis of Disease, ninth ed. Pathology Cases: Burns et al. Cardiovascular Diseases Case 1 (figure A) and Kumar et al. Hemodynamic Disorders, Thrombosis and Shock Case 1 (figure B). With permission from Elsevier Saunders.*

calcification correlates with the extent of atherosclerosis (Allison et al., 2004). The significance of cardiovascular calcification (CVC) was illustrated in a meta-analysis of prospective studies reporting calcification and CVD endpoints. The presence of calcification in the arteries increased risk 3–4-fold for mortality and cardiovascular events (Rennenberg et al., 2009). Three main types of vascular calcification have been reported; Mönckeberg's arterial calcification; intimal calcification, and infantile calcification. Intimal calcification has been classified as calcification within an atherosclerotic plaque, while Mönckeberg's medial calcific sclerosis is characterized by calcific deposits within the media of small- and medium-sized muscular arteries. Medial calcification is more common in patients with chronic kidney disease (CKD) or diabetes, and involves the differentiation of vascular smooth muscle cells (VSMCs) into osteoblast-like cells. In contrast, intimal calcification is found in vessels with atherosclerotic plaque and has been related to aging and endothelial injury.

6.2 PATHOGENESIS OF THE COMPLEX ATHEROSCLEROTIC PLAQUE

Atherosclerotic lesions are found primarily within the intima, but can involve the media and adventitia. Vascular lesions generally occur randomly, often at vascular branch points, despite uniform endothelial exposure to the luminal contents. As lesions mature, they can become complicated by mural or occlusive thrombosis which may cause ischemia. Complicated AS lesions can have a thinning fibrous cap, making them vulnerable to rupture. These lesions predispose

to clinical angina, MI with tissue necrosis and other sequelae, cerebral infarction (stroke), or gangrene of the extremities. Complicated lesions are described by the term *atherothrombosis.*

The current mechanistic hypothesis holds that atherosclerosis is a *tissue response to injury.* The model, proposed over 40 years ago (Ross and Glomset, 1973), holds that atherosclerosis occurs in the context of endothelial injury followed by migration of VSMCs into the intima and their subsequent lipid uptake. More recent work has explored the functional capacities and variable morphology of the endothelial cell barrier in response to mechanical stress (Tabas et al., 2015), the generation, differentiation, activation, and efflux of infiltrating blood monocytes (Dutta et al., 2012; Moore et al., 2013), and exploration of the signals that trigger pluripotent stem cells to differentiate into endothelial precursors competent to replace injured and dying endothelial cells (Ditadi et al., 2015).

Lipid accumulation in the intima was initially thought to be due to monocyte-derived macrophage uptake and conversion to lipid-laden "foam" cells. It is now recognized that at least 40% of foam cells have VSMC characteristics (Huff and Pickering, 2015). VSMCs normally secrete collagen I that provides stability to the vessel wall. Lipid-laden VSMCs are unable to effectively synthesize collagen I, resulting in thinning of the fibrous cap and plaque instability (Frontini et al., 2009). Macrophages normally phagocytize dead cells and remove them by efferocytosis. (Efferocytosis is from the Latin, *efferre,* meaning "to bury." It refers to phagocytic uptake and degradation of dead and dying cells.) When they become filled with lipid, macrophages cannot effectively perform this function, thus the necrotic core of the AS lesion is studded with dead and fragmented cells, stimulating additional inflammatory infiltration. Signals from the dysfunctional endothelial cells contribute to dysregulated mineralization pathways. VSMCs transdifferate to osteoblast-like cells competent to release calcium-phosphate containing matrix vesicles (MVs) into the ECM (Kapustin et al., 2015). Apoptotic vesicles containing calcified matrix remain in the ECM after VSMC death. In an inflammatory environment, iron retained in macrophages is released as unbound iron upon macrophage death. Unbound iron can stimulate reactive oxygen species (ROS) production and exacerbate inflammation (Sullivan, 2007). Endothelial injury stimulates simultaneous processes of repair. Thus, repair and remodeling are also ongoing. In fact, intimal thickening is a normal component of tissue repair that resolves with healing.

These insights allow construction of a pathogenic model in which injury is persistent, resolution is delayed and/or prevented, and the lesions progress with the following pathogenic events:

- Endothelial injury from oxidative stress, mechanical stress, toxins, or other agents increases vascular permeability, reduces endothelial anticoagulant activities, and upregulates expression of leukocyte adhesion molecules.
- Oxidative damage stimulates monocyte precursors to proliferate and adhere to adhesion molecules on the damaged endothelium. They migrate into the intima where they differentiate further into activated innate immune cells and secrete proinflammatory mediators.
- Signals from endothelial injury stimulate VSMCs to migrate from the media to the intima and dedifferentiate into myointimal cells. Modified VSMCs lose their contractile capacity and assume fibroblast capability to secrete ECM elements. These fibers stabilize the lesion and produce a fibrous cap that covers the necrotic center of the atherosclerotic plaque.
- Circulating lipoproteins, especially low-density lipoproteins (LDLs) and oxidized LDLs (oxLDL), enter the intima and populate the ECM where the oxidative environment further modifies their structure. Structurally modified lipids generate "oxidation-specific" epitopes recognized by macrophage pattern recognition receptors (PRRs) (refer to Chapter 2: The Obese Gunshot Patient: Injury and Septic Shock: Innate Host Defense) and by a myriad of scavenger receptors, the most prominent of which are CD36 and scavenger receptor A (SRA).
- Macrophages avidly internalize modified lipoproteins and hydrolyze the cholesterol esters

into free cholesterol and fatty acids. Free cholesterol is re-esterified in the endoplasmic reticulum by acyl-CoA:cholesterol ester transferase (ACAT) and the cholesterol ester "foam" is stored in vesicles.
- Hypercholesterolemia overwhelms ACAT esterification capacity, resulting in accumulation of toxic free cholesterol. Lipid-loaded macrophages are unable to effectively remove dead cells by efferocytosis; continued exposure to free cholesterol can result in macrophage apoptosis, increasing the lipid-rich necrotic core that fuels chronic inflammatory reactions.
- In a hyperlipidemic environment, VSMCs also take up lipids. While they are more resistant to free cholesterol than macrophages, lipid-laden VSMCs lose their ability to synthesize collagen I and ECM, resulting in thinning of the fibrous cap and plaque instability.
- The oxidative environment reduces the endothelial progenitor cell (EPC) pool, thus injured endothelial cells die, but are inadequately replaced.
- Platelets adhere to collagen underlying the injured endothelial monolayer, change shape, activate and aggregate to form a thrombus. Activated platelets, macrophages, and endothelial cells secrete mediators that augment the inflammatory process.
- Lipid-laden macrophages take up iron from red blood cells trapped in the plaque. Upon macrophage death, free iron is released into the necrotic center, stimulates ROS, and causes further oxidative damage.
- Vulnerable plaques with thinned fibrous caps and large necrotic cores can rupture. Sudden stress from adrenergic stimulation that increases systemic blood pressure or induces local vasoconstriction can potentiate a rupture, spilling noxious debris into the circulation and dislodging thrombi. Migrating clots can cause catastrophic events including MI and stroke.

6.3 TARGETS FOR ATHEROSCLEROTIC PREVENTION AND CONTROL

6.3.1 Triggers for Oxidative Damage

As discussed elsewhere (refer to Essentials I: Life in an Aerobic World and Chapter 5: Cardiomyopathy and Congestive Heart Failure: Mitochondrial Production and Management of Reactive Oxygen Species) mitochondria use oxygen to generate ATP. This process necessarily produces ROS. Normally, sublethal mitochondrial stress and superoxide production rapidly induce mitochondrial antioxidant systems that maintain ROS in balance and provide low levels of H_2O_2 for redox signaling. Exposure to toxic environmental agents and metabolic stress can overwhelm these systems and generate higher ROS production and contribute to an oxidative milieu. For example, phagocytes use ROS for oxygen-dependent microbial killing and high-energy radicals are generated from proinflammatory arachidonic acid metabolism through lipoxygenase and cyclooxygenases. Activation of oxidases (NADPH oxidase, xanthine oxidase, and others) as well as high concentrations of unbound metals such as iron and copper can produce ROS. As described below, dysregulated nitric oxide synthases (NOS) can produce superoxide rather than nitric oxide (NO). Classic risk factors for AS, such as tobacco smoke, metabolic syndrome with obesity, sedentary lifestyle, insulin resistance, hyperglycemia, hyperlipidemia, and hyperhomocysteinemia are also associated with redox imbalance and oxidative stress.

The endothelial cell (EC) monolayer is susceptible to damage and death from oxidative stress from NO, a labile (half-life ~5 second) reactive free radical produced by activated macrophages and inflammatory cells upon stimulation of inducible nitric oxide synthase (iNOS). The forkhead transcription factor (FOXO1) responds to stimulation by cytokines and microbial products and activates iNOS, generating high amounts of NO. At the same time, the activity of the constitutive endothelial enzyme (eNOS), that produces small amounts of NO required for normal vascular function, is reduced. As reviewed by Tanaka et al. (2009), FOXO1 activation is also essential for expression of cell adhesion molecules (CAMs), inflammatory mediator activation, and promotes EC senescence. NADPH oxidase-generated ROS was shown to stimulate transforming growth factor beta (TGFβ), thereby activating the p38 mitogen-activated protein kinase (MAPK)/ c-Jun N-terminal kinase (JNK) pathway leading to EC apoptosis (Yan et al., 2014). Thus oxidative stress and EC injury are a major factor in AS pathogenesis. As discussed elsewhere (refer to Essentials I: Life in an Aerobic World: Nutrients, Mitochondrial Oxidants and Antioxidant Protection) and reviewed by Saad et al. (2015), the human organism has evolved many pathways that can counter oxidative stress, reducing EC death and dysfunction and thereby modulating AS risk.

6.3.2 Hyperlipidemia and Reverse Cholesterol Transport

Normally, dietary long-chain fatty acids are incorporated into apoprotein B (ApoB)-rich chylomicrons and transported through lymphatic channels to the right atrium of the heart. In the systemic circulation, the triglyceride-rich lipoproteins encounter lipoprotein lipase (LPL) expressed on the endothelium of capillaries that sustain peripheral tissues such as skeletal muscles and adipose tissue. LPL hydrolyzes triglycerides to fatty acid and monoglyceride products that are taken up by the tissues for utilization and/or storage. The partially lipid-depleted chylomicron remnants acquire ApoE, return to the liver where they bind to Apo E receptors, and are recycled into smaller, but still triglyceride-rich VLDL. Nascent VLDL particles mature in the bloodstream by also acquiring ApoE and ApoC proteins from mature HDL2 lipoproteins. VLDL triglycerides enter the systemic circulation and are again hydrolyzed by capillary LPL and their products taken up by muscles and adipocytes. Hydrolyzed VLDL remnants, known as intermediate-density lipoproteins, can be removed by the liver or undergo further hydrolysis in the plasma to become cholesterol-rich LDLs. VLDL can also be taken up intact by the VLDL receptor (VLDLr), widely distributed on all tissues except the liver. Cholesterol-rich LDL binds its receptor found on all nucleated cells and is endocytosed, allowing the transported cholesterol to be utilized for diverse biosynthetic and structural purposes (Fig. 6.4).

Macrophages and VSMC in the vessel wall maintain their cholesterol balance by facilitating uptake and efflux. Cholesterol esters can be off-loaded from the "foam" cells by several transporters that hydrolyze the esters and transfer free cholesterol to HDL via reverse cholesterol transport. The ABCA1 cassette transfers cholesterol to lipid-poor immature HDL, while another transporter, ABCG1, donates cholesterol to mature HDL particles (Moore et al., 2013). Genes for these transporters are transcriptionally upregulated by the nuclear liver X receptors (LXRs) that bind oxidized cholesterol (oxysterols) and act as sterol sensors. Experimental upregulation of LXRα by oxysterols has revealed that LXRα augments several antiinflammatory mechanisms (refer to Essentials I: Life in an Aerobic World: Nutrient Transcriptional Control of Antioxidant Protection). These include repression of proinflammatory nuclear factor κB (NFκB) signaling pathways, induction of ABCA1 and ABCG1 transporters and upregulation of the antiinflammatory nuclear factor erythroid 2-related factor 2 (Nrf2) pathways as discussed below (Hasty and Yvan-Charvet, 2013). Increased cholesterol efflux and plaque regression are associated with upregulation of the chemokine receptor CCR7 that decreases myeloid-derived expression of retention factors and/ or macrophage reverse transmigration to the lumen or lymphatics. Macrophage transmigration is regulated by modulation of macrophage migration factors (Fig. 6.5).

Hypercholesterolemia increases circulating monocytes in experimental models by enriching their pool of precursor cells, the hematopoietic stem and progenitor cells (HSPCs), as reviewed by Moore et al. (2013). Cholesterol enrichment upregulates expansion signals, including granulocyte/macrophage colony-stimulating factor (GM-CSF), thereby enhancing HSPC proliferation. In the mouse model, cholesterol-enhanced HSPC

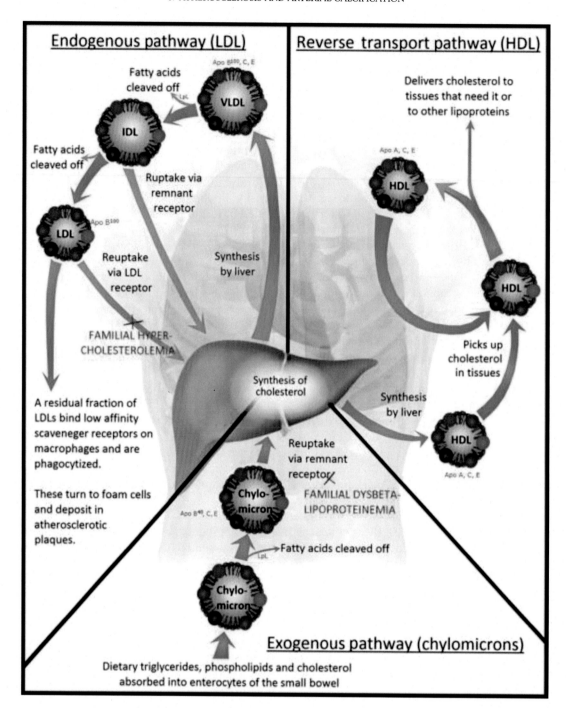

FIGURE 6.4   Lipid transport pathways. Exogenous dietary lipid is delivered to the peripheral tissues by chylomicrons and triglycerides and hydrolyzed by LDL. Chylomicron remnants return to the liver where they are recycled into VLDL and returned to the peripheral tissues. Residual cholesterol-rich LDL are taken up by tissues via receptor-mediated endocytosis. Reverse cholesterol transport from peripheral cells to the liver is facilitated by ApoA-rich phospholipid disks secreted by the liver (HDL). Immature HDL collect free cholesterol from plasma membrane via ATP binding cassette (ABC) transporters. Free cholesterol is esterified in the HDL and transported to the liver for excretion as bile acids or redistribution in newly formed VLDL. As noted in the figure, apoproteins can be transferred among different lipoprotein forms. *Source: By Npatchett (own work) (GFDL (http://www.gnu.org/copyleft/fdl.html) or CC BY-SA 3.0 (http://creativecommons.org/licenses/by-sa/3.0)), via Wikimedia Commons.*

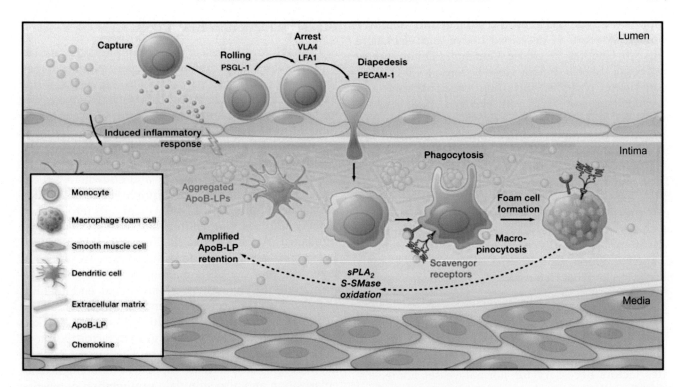

FIGURE 6.5 ApoB-LP retention promotes monocyte recruitment and foam cell formation. Lipoproteins that contain ApoB (APOB-LP) enter the intima where they bind to the extracellular matrix and undergo modifications including oxidation. They also incite an inflammatory response, possibly by binding to pattern recognition receptors (PRRs) on macrophages and dendritic cells. Inflammatory mediators alter expression of endothelial adhesion molecules and contribute to aggregation and retention of lipoproteins; mediators also recruit additional monocytes and potentiate uptake of native and modified oxLPL, resulting in foam cell formation. Foam cells further amplify lipoprotein modifications and retention. *Source: Reproduced from Moore, K.J., Tabas, I., 2011. The cellular biology of macrophages in atherosclerosis. Cell 145 (3), 341–355 (Moore and Tabas, 2011) with permission from Science Direct.*

proliferation was associated with an increased fraction of the inflammatory LY6C^{hi} monocyte subset, believed to be the source of the M1 (activated) macrophages found in plaques. In the plaque, both macrophages and VSMCs take up lipids in proportion to high circulating lipid levels.

6.3.3 Inflammation, Monocyte Binding, and Macrophage Activation

The intact endothelium does not normally bind inflammatory cells. Several inflammatory mediators can stimulate dysfunctional endothelial cells to express adhesion molecules such as vascular cell adhesion molecule 1 (VCAM-1) that specifically binds monocytes and T lymphocytes (T cells). Locally produced chemokines facilitate migration of inflammatory leukocytes into the intima. Current understanding of monocyte action in the intima is presented in the following model:

- Monocytes transform into macrophages and avidly take up lipoproteins, especially oxLDL. This process

could be considered protective, since structurally altered lipid particles are cytotoxic. However, proinflammatory M1 macrophages stimulated by aging, obesity, and other inflammatory states secrete tumor necrosis factor (TNFα), monocyte chemotactic protein, and other cytokines that increase leukocyte adhesion and translocation. In this milieu, additional blood monocytes are recruited to the area. Activated macrophages produce ROS as well as secrete factors that stimulate smooth muscle cell migration and proliferation.

- Cross-talk between macrophages and T cells generates a chronic inflammatory state. Activated T cells produce inflammatory cytokines such as interferon gamma (IFNγ), which stimulate resident cells, including macrophages, endothelial cells, and smooth muscle cells.

Evidence is accumulating suggesting that only areas in the vessel wall containing atherosclerotic plaques are inflamed as reviewed by Curtiss and Tobias (2009). Hallmarks of inflammation include secretion of inflammatory mediators and infiltration of the lesion sites with cells of the innate and adaptive immune systems as described above and elsewhere (refer to Chapter 2: Obese Gunshot Patient: Injury and Septic Shock: Innate Host Defense). Circulating acute-phase proteins, including ceruloplasmin and C-reactive protein (CRP), are elevated. PRRs including CD36, toll-like receptors (eg, TLR-4) and others recognize oxidation-modified patterns on oxidized phospholipids and oxidized cholesterol esters, as well as products of bacteria such as lipopolysaccharide (LPS) and mediate immune responses (Miller et al., 2011). Increased TLR activity amplifies the inflammatory processes and when combined with formation of cholesterol crystals may stimulate inflammasome activation (Tall and Yvan-Charvet, 2015) (refer to Chapter 2: Obese Gunshot Patient: Injury and Septic Shock: Innate Host Defense).

6.3.4 Endothelial Dysfunction

The vascular endothelium constitutively elaborates factors necessary for normal vessel homeostasis. Endothelial cells modulate anticoagulant, antithrombotic and fibrinolytic processes, ECM secretion, blood flow and vascular reactivity, inflammatory and immune processes, and cell proliferation. These endothelial-derived homeostatic functions can be compromised by gross damage, but also by injury at the cellular or biochemical level. Factors that can negatively modulate endothelial cell activities include:

- Hypertension and turbulent flow can upregulate expression of CAMs, facilitating monocyte adherence to the vessel wall.
- Chronic hyperlipidemia, particularly hypercholesterolemia, can directly impair endothelial cell function by increasing local ROS production; increased ROS exposure damages the endothelium, in part by reducing endothelial NO synthesis and/or accelerating its decay.
- Hyperlipidemia increases endothelial uptake of lipoproteins that bind to the ECM and are susceptible to oxidation by ROS produced locally by activated endothelial cells and macrophages. oxLDL are recognized by PRR and ingested by macrophages and VSMCs.
- Uptake of oxidized lipids activates endothelial cells and macrophages to produce chemokines and cytokines that increase monocyte recruitment and retention in the intima. Oxidized lipids such as free cholesterol are cytotoxic to endothelial and smooth muscle cells and can induce endothelial cell dysfunction and death.

6.3.5 Reduction in EPC Pools

Damage to the vessels from multiple CVD risk factors (smoking, hyperlipidemia, hyperglycemia, hypertension) is known to precipitate oxidative stress and endothelial apoptotic suicide pathways (Dimmeler and Zeiher, 2004). While endothelial repair was originally thought to be mediated by migration and proliferation of existing endothelial cells, it was suggested almost 50 years ago that most replacement cells came from stem cell niches in the bone marrow and elsewhere. Several reviews defining the signals required for cells to differentiate from human pluripotent stem cells into hematopoietic and vascular endothelial lineages have been published (Ackermann et al., 2015; Ditadi et al., 2015). In general, cells that reside in the bone marrow stromal layer and bear the CD34 marker are stimulated by signals from the injured or dysfunctional endothelium. Mobilized $CD34^+$ cells differentiate in the circulation into EPCs and acquire additional surface markers including vascular endothelial growth factor receptor 2 (CD309;VEGFR2) as reviewed by Dickinson-Copeland et al. (2015). These cells are capable of inducing neovascularization and also stimulate proliferation of existing endothelial cells.

Endothelial cell precursor (ECP) pools can be reduced for several reasons (Dimmeler and Zeiher, 2004; Saad et al., 2015). These include exhaustion of the ECP pool, reduction or modification of mobilization and/or differentiation signals, and compromised ECP survival at the target. Current consensus is that a major contributor to reduction of the ECP pool is oxidative stress. High levels of H_2O_2 decrease EPC viability, increase apoptosis, and impair tube formation required for angiogenesis. Damage is mediated by increased expression of the forkhead transcription factor (FOXO3) that stimulates downstream proapoptotic targets. Oxidized-LDL induces the Akt/p53/p21 signaling pathway that accelerates EPC senescence. Furthermore, H_2O_2 oxidizes sulfhydryl groups on antioxidant proteins (refer to Essentials I: Life in an Aerobic World: Nutrients that Modify Transcriptional Control of Antioxidant Protection), reducing their protective ability and permitting EPC apoptosis. EPC availability is reduced under conditions of insulin resistance and hyperglycemia, especially in obese patients and those with types I and II DM (Thum et al., 2007; Saad et al., 2015). Inflammation and infection may also deplete ECP through apoptosis mediated by induction of the proinflammatory chemokine, CXCL10, and toll-like receptor (TLR) activation (Dickinson-Copeland et al., 2015) (refer to Essentials IV: Diet, Microbial Diversity and Gut Integrity for discussion on CXCL10 and TLR upregulation in peripheral blood monocytes from celiac patients following in vitro exposure to gluten fractions).

6.3.6 VSMC Proliferation, Remodeling, and Neointima Formation

In response to inflammatory and hemodynamic stresses associated with injury, disease, and aging, vessels remodel (van Varik et al., 2012). The type of remodeling differentiates *atherosclerosis*, characterized by a focal inflammatory process in the intima initiated by accumulation of lipids in plaques, from *arteriosclerosis*, diffusely localized alterations of the medial arterial vascular wall. Remodeling involves both hyperplasia of the arterial intimal and medial components and changes in the ECM components with deposition of calcium (described below). Maintenance of the vessel wall integrity is of utmost importance to control blood pressure.

VSMCs in the media constrict or dilate (vasoreactivity) in response to vasoactive stimuli. Under physiological conditions, the turnover rate of endothelial and smooth muscle cells is low. When vascular injury occurs, whether from environmental agents, angioplasty, or surgery, mature vascular stem cells proliferate to replace lost cells. If the injury is severe or chronic, recent evidence suggests that stem cells, found in all three vessel layers, are responsible for the formations of neointimal lesions. These resident progenitor cells have the capacity to differentiate into a variety of cell types (Torsney and Xu, 2011) (Fig. 6.6).

Vascular endothelium responds to gross or biochemical insults by elaborating mediators that recruit smooth muscle cells (SMC) to the intima from the vascular media. (SMC proliferation and ECM synthesis are stimulated by several mediators, including PDGF (released by locally adherent platelets, as well as macrophages, endothelial cells, and smooth muscle cells), FGF, and TGF-α.) Upon entry into the intima, SMCs lose their contractile properties and dedifferentiate into proliferative, fibroblast-like

cells (myointimal cells) competent to secrete ECM. The intima thickens as a result of accumulation of cells and matrix, including collagen, elastin, proteoglycans, and other secreted ECM components that initially stabilize the atherosclerotic plaque. SMC activities are regulated by a spectrum of growth promoters, inhibitors, and regulators such as products of the renin—angiotensin—aldosterone system, especially angiotensin II, catecholamines, the estrogen receptor, and osteopontin in the ECM. Endothelial cells migrate from adjacent uninjured areas or are recruited from the EPC pool. The resulting neointima is typically completely covered by endothelial cells. *Thus, intimal thickening is the stereotypical response of the vessel wall to any insult.*

6.3.7 Platelets and Hypercoagulability

Platelets are derived from megakaryocytes in the bone marrow and circulate as small biconvex disks, ~3 μm in size. While they have no nucleus, they contain granules that can rapidly discharge their mediators through an open canalicular system. When platelets traveling in the periphery of the vessel lumen detect endothelial injury they undergo structural and biochemical changes that allow them to bind underlying connective tissue elements, limit injury, and facilitate vessel wall repair (Van Varik et al., 2012). Thus, although intact, endothelial cells secrete factors that inhibit platelet adhesion and clot formation at the endothelial surface, endothelial injury, inflammation, or activation results in a procoagulant phenotype. Platelets interact with the vascular endothelium in the formation of thrombin clots using the following general mechanisms:

- *Platelet binding*: Endothelial-derived prostacyclin (PGI$_2$) and NO produced by endothelial cell eNOS

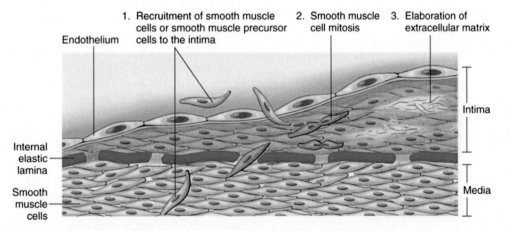

FIGURE 6.6 Tissue response to injury. Healing involves elaboration of ECM for vascular support and cross-talk with other tissue components. Removal of the injurious stimulus sometimes, but not always, results in resolution and return to normal morphology. *Source: Reproduced from Robbins and Cotran Pathologic Basis of Disease, 9th Ed. With permission from Elsevier Saunders.*

normally prevent platelet adherence to the endothelium. Endothelial cells also produce adenosine diphosphatase, which degrades adenosine diphosphate (ADP), further inhibiting platelet aggregation. Endothelial injury allows platelet contact with the underlying ECM through von Willebrand factor, an essential cofactor for platelet binding to matrix elements.

- *Platelet activation*: Platelet activation is enhanced and amplified by production of thromboxane A2 (TxA2). This prostaglandin is derived from arachidonic acid in the platelet membrane via the cyclooxygenase (COX) pathway (refer to Chapter 2: The Obese Gunshot Patient: Injury and Septic Shock: Coagulopathy of Trauma). Dietary omega-3 fatty acids from marine sources, metabolized through the same pathway, produce the far less vasoconstrictive product, TxA3.

- *Platelet recruitment*: Proaggregating factors include superoxide produced by NADPH oxidase system and ADP that scavenges NO. NO is critical for late disaggregation of the thrombus.

Hypercoagulability and thrombotic disorders increase the risk for clot formation and stroke under conditions commonly associated with AS, including hyperlipidemia, hypertension, diabetes, and smoking. Indeed, risk for stroke is significantly elevated in diabetic patients with hypertension and peripheral vascular disease. (Jover et al., 2013). The pathogenesis for both conditions includes vascular damage, inflammatory processes, and formation of a thrombus. Additionally, activated platelets increase hypercoagulability and incorporate higher amounts of thrombin, creating a positive feedback. Coagulation factors require vitamin K-dependent calcium binding for activation; calcium also regulates platelet function by a calcium-sensing receptor (House et al., 1997). The relationship of calcium to coagulation raises concern over the practice of taking large doses of calcium supplements. In postmenopausal women, calcium supplements increased blood coagulability at 4 hours post-administration (Bristow et al., 2015a,b).

Drugs that inhibit platelet function (aspirin and clopidogrel (Plavix)) reduce risk for coagulopathy; it must be noted that there is evidence that certain diets (Mediterranean, Eskimo) and the bioactive components in the food that they contain (fish oil, organosulfur compounds, flavonoids) can also inhibit platelet activation and clot formation as described below (Violi et al., 2010).

6.3.8 Cardiovascular Calcification

As reviewed by Reynolds et al. (2004) and New and Aikawa (2011), while VSMCs in the normal artery wall constitutively express potent inhibitors of calcification, these inhibitors are downregulated in atherosclerotic calcification while markers of both osteoblast and chondrocyte differentiation are simultaneously upregulated (Tyson et al., 2003). Vascular calcification occurs in a manner analogous to that in bone (refer to Chapter 9: Osteoporosis and Fracture Risk: Bone Cells and Normal Mechanisms of Remodeling). In bone, osteoblasts respond to mineralization signals by secreting small membrane-bound matrix vessicles (MVs) containing supraphysiologic concentrations of Ca^{2+} and HPO_4^{2-}. $CaHPO_4$ crystals precipitate (refer to Essentials III: Nutrients for Bone Structure and Calcification: Regulation of Calcium, Phosphorus and Vitamin D) in the MV and form a nidus for the irreversible formation of larger hydroxyapatite crystals and subsequent secretion onto the extracellular surface. In the absence of calcification inhibitors, the VSMC, transdifferentiated to an osteoblast phenotype, secretes both mineral crystals and proteins such as osteocalcin (OCN) into the ECM. OCN has high affinity for hydroxyapatite crystals where it appears to fine-tune crystal formation (Murshed et al., 2004).

VSMCs are derived from the same lineage as osteoblasts and chondrocytes and, when stimulated, can switch to an osteogenic phenotype and migrate into the intima. Stimuli such as high extracellular levels of calcium and phosphate ions combined with lack of vascular calcification inhibitors can facilitate this change. Other facilitators of calcification include downregulation of mineralization-inhibitory proteins and upregulation of osteogenic markers such as alkaline phosphatase. Bone morphogenic protein (BMP), known to induce osteogenic differentiation, has been localized to human atherosclerotic lesions. Other factors that promote calcification include elastic fiber degradation, exposure to high glucose and insulin levels, inflammatory mediators and dysregulation of parathyroid hormone—calcium—phosphorus and vitamin D metabolism as described elsewhere (refer to Chapter 9: Osteoporosis and Fracture Risk: Bone Cells and Normal Mechanisms of Remodeling).

In the vessels, VSMCs express matrix gla protein (Mgp) that requires vitamin K-dependent gamma carboxylation. Mgp is expressed by other cells in the atherosclerotic plaque (macrophages, endothelial cells, platelets) and in chondroblasts, but not in osteoblasts. By virtue of its four carboxyl groups, Mgp has a high affinity for calcium and acts locally to inhibit Ca/P mineralization in VSMCs. Like osteoblasts, VSMCs concentrate calcium and phosphorus in the MVs. In contrast to osteoblasts, VSMCs take up a circulating liver-derived protein, fetuin-A, which they also transfer to the MVs. The MVs are extruded from the VSMCs and localize to overlying collagen and connective tissue (Kapustin et al., 2015) where fetuin A forms calciprotein particles with the calcium and facilitates Ca/P particle removal as waste. These processes normally prevent Ca/P mineralization in the arterial wall. Additionally, the multifunctional fetuin A protein prevents VSMC transdifferentiation into an osteogenic cell under hypercalcemic conditions as reviewed by Jahnen-Dechent et al. (2011). Hyperphosphatemia as well as injury due to inflammation, oxidative, and mechanical shear stress (hypertension) or hyperglycemia increase arterial calcification. Fetuin A has been called a systemic regulator of mineralization because it can also alter cytotoxicity of Ca/P particles and bind directly to BMP to block osteogenic transdifferation of VSMCs (Schlieper et al., 2015).

The impact of calcium supplements on increased risk for cardiovascular-related death has been the subject of several meta-analyses as reviewed by Reid et al. (2016). Calcium supplements (500 mg Ca-citrate) raised serum ionized calcium by about one standard deviation in postmenopausal woman ($n = 10$); the elevation persisted for 6−8 hours after dosing. A similar dose of calcium as a meal caused a much lower rise in serum ionized calcium (Bristow et al., 2015a,b). The mechanism through which elevated serum calcium increases AS risk is unclear, however it is likely to involve increased vascular calcification. VSMCs exposed to elevated extracellular calcium produced MVs that were depleted in the primary calcification inhibitor, Mgp, and also expressed calcium-binding annexins. High media calcium exposed phosphatidylserine on the MV surface, providing nucleation sites for hydroxyapatite formation (Kapustin et al., 2011).

Magnesium (refer to Essentials III: Nutrients for Bone Structure and Calcification), on the other hand, appears to prevent formation and vascular deposition of Ca/P crystals. Although not well studied, magnesium appears to exert its actions in two ways. Mg^{2+} promotes the formation of an unstable amorphous calcium magnesium phosphate crystal and inhibits Ca/P apatite formation (Cheng et al., 1988). Additionally, Mg^{2+} was shown to downregulate genes that promoted bovine VSMC transdifferentiation to an osteoblastogenic phenotype (Kircelli et al., 2012). Higher magnesium levels also prevented VSMC apoptosis in this in vitro study.

6.3.9 Macrophage Uptake and Efflux of Iron

The risk from iron overload in the pathogenesis of AS was first proposed over 30 years ago (Sullivan, 1981). The author proposed that premenopausal women were protected from AS due to monthly blood loses, whereas postmenopausal women and men who did not lose blood (and iron) and were at increased AS risk. Whole-body homeostasis of this redox-active metal is reviewed elsewhere (refer to Essentials II: Heavy Metals, Retinoids and Precursors: Iron). Red blood cells enter the atherosclerotic plaque by hemorrhage or through the porous neovasculature in the vasa vasorum underlying the plaque. Once exposed to the oxidized plaque environment, erythrocytes are lysed, the liberated hemoglobin is oxidized, and heme dissociates from ferrihemoglobin. Free heme can be oxidatively cleaved by oxidized plaque components, releasing iron atoms which can promote the oxidation of plaque lipids through the Fenton and other oxidant-generating reactions as illustrated in Fig. 6.7.

In the past, macrophages have been viewed as a safe repository of stored iron, especially in the liver and bone marrow. However, the inflammatory environment of the AS plaque stimulates hepcidin production that increases macrophage uptake and retention of iron. Red cell entry into the intima and subsequent intraplaque hemorrhage allow macrophages to acquire a load of iron sufficient to exert toxic injury both to the macrophage and surrounding cellular elements. Human carotid atherosclerotic lesions contained 3- to 17-fold more iron than healthy control arteries. Increased availability of redox-active iron results in the formation of highly toxic material including iron-laden ceroid and other cytotoxic lipid materials. As iron-laden macrophages die, the plaque interior becomes increasingly cytotoxic both for existing cellular elements and for new monocytes recruited to the destabilizing lesion (Nagy et al., 2010).

A subset of macrophages have been identified that can participate in iron efflux from the plaque (Bories et al., 2013). After characterizing multiple macrophage phenotypes in the atherosclerotic lesions, the authors identified a specific macrophage phenotype that colocalizes with iron deposits in AS plaques. These cells are CD68 and mannose receptor-positive ($CD68^+MR^+$) alternative M2 macrophages with a gene expression profile that favors iron transport and accumulation. These MR^+ cells have the capacity to take up redox-active, unbound iron from lysed red cells, thereby reducing the ability of unbound iron to generate ROS. Unexpectedly, the authors found that $CD68^+MR^+$ macrophages change character when loaded with iron and acquire a phenotype that favors iron release by upregulating expression of the iron efflux protein, ferroportin. Iron released by ferroportin is safely bound to transferrin or ferritin and is much less dangerous. Iron-bound to transferrin can be returned to the blood circulation. Iron contained in the $CD68^+MR^+$ macrophages enhances oxidation of surrounding lipids, promoting their subsequent uptake into the macrophages. The oxidized lipids (oxysterols) upregulate liver X

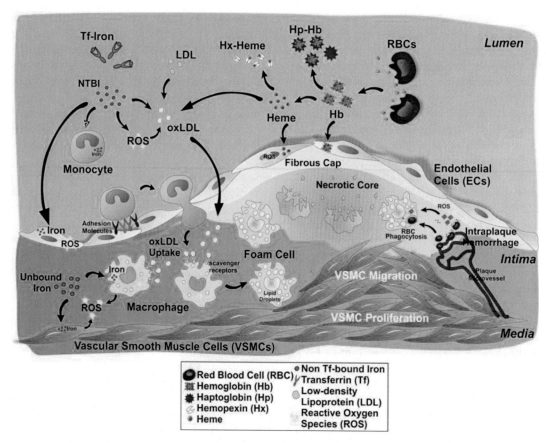

FIGURE 6.7 Iron overload and atherogenesis. Iron is accessible to the growing plaque as inorganic nontransferrin-bound iron (NTBI), as hemoglobin, or as heme from degraded red cells. NTBI is present in conditions of chronic iron overload and assessed as elevated serum ferritin. Intravascular hemolysis and hemorrhage can also provide iron. NTBI enhances oxidative damage, adhesion cell molecule activation, and vascular smooth muscle cell transformation, proliferation, and migration, thereby stimulating LDL oxidation and facilitating monocyte recruitment. *Source: Reproduced from Vinchi, F., Muckenthaler, M.U., et al., 2014. Atherogenesis and iron: from epidemiology to cellular level. Front. Pharmacol. 5, 94 (Vinchi et al., 2014), under the terms of the Creative Commons Attribution License (CC BY).*

receptor α (LXRα), a transcription factor that induces ABCA1 and ABCG1 cassette transporters as well as ApoE expression in association with a downregulation of cellular hepcidin. LXRα also activates nuclear factor erythroid 2-related factor 2 (Nrf2) that stimulates ferroportin expression. Taken together, CD68⁺MR⁺ macrophages exert atheroprotective effects in that they facilitate both iron and lipid efflux from the atherosclerotic plaque.

6.4 INCREASED AS RISK RELATED TO OBESITY AND ITS COMORBITIES

6.4.1 Hyperglycemia and Insulin Resistance

Obesity is associated with insulin resistance and hyperinsulinemia. Insulin signaling through the phosphatidyl-3-kinase (PI3-K) pathway was reduced in obese nondiabetic subjects, and essentially absent in type 2 DM patients (Cusi et al., 2000). Insulin resistance may play a major role in reduced availability of EPCs to proliferate, differentiate, and migrate to areas denuded by injury, or to form collateral circulation by angiogenesis (Hamed et al., 2011). That EPCs are essential for vascular repair and neovascularization is well supported by experimental evidence. Indeed EPC proliferation from diabetic patients was inversely correlated with their plasma glycated hemoglobin levels (HbA1c), supporting the hypothesis that glycemic control influences EPC number and proliferation.

The mechanistic plausibility for the impaired EPC pool currently proposed (Saad et al., 2015) is that hyperglycemia and insulin resistance inhibit eNOS, thereby uncoupling NO formation and producing peroxynitrite and other ROS that oxidize LDL (oxLDL). (The term ROS is used to refer to all dangerous reactive species including nitrogen radicals and lipid peroxides.) Together, ROS and oxLDL severely impair the EPC count. At the same time, symmetric dimethylarginine (ADMA), a naturally occurring endogenous l-arginine metabolite normally degraded and excreted by the kidney, competes with arginine as eNOS substrate as discussed below. Hyperglycemia inhibits ADMA degradation, thus ADMA accumulates in plasma of diabetic patients. ADMA and ROS both impair eNOS production of NO and severely reduce EPC number and function.

Mechanisms that regulate EPC release and migration to the endothelium include the PI 3-K/Akt (Akt) pathway and eNOS. Akt upregulation by exercise, hydroxymethylglutaryl-coenzyme A (HMG) reductase inhibitors (statins), erythropoietin, estrogens, and VEGF increase EPC proliferation and migration. Compounds that stimulate Akt also activate eNOS, known to be essential for the mobilization of stem and progenitor cells. Thus, it is likely that eNOS activation is the final common pathway regulating ECP number and function (Förstermann and Münzel, 2006). While hyperglycemia can reduce the number and function of EPC, the combination of associated metabolic dysfunctions, such as hyperlipidemia and advanced glycation end-products, appears to be required for the severe reduction of EPC counts and vascular disease observed in diabetic patients.

6.5 INFLUENCE OF THE MICROBIOME ON AS RISK

The gut microbiome is comprised of thousands of microbial species that interact with host metabolism in regulating energy utilization and immune system function and modulating intermediary metabolism as discussed elsewhere (refer to Essentials IV: Diet, Microbial Diversity and Gut Integrity). Recently, a comprehensive study was conducted to determine whether variations in circulating lipids could be explained by microbial diversity and/or function and could partly explain the link between obesity and AS (Fu et al., 2015). Samples from the LifeLines-DEEP cohort (1500 samples) were genotyped and blood lipids analyzed. Fecal samples were sequenced to identify intestinal bacteria taxa whose proportions correlated with host body mass index (BMI) and lipid levels. Blood lipid variance was related to microbiota relative to age, sex, and host genetics. Analysis demonstrated that microbial richness and diversity were inversely correlated with BMI and triglyceride levels. Diversity was also positively associated with HDL cholesterol levels but correlations between microbial taxa and lipids did not appear to be modified by BMI or host genetic factors. It is of interest that no

relationships between total cholesterol and LDLs with microbial diversity were observed.

Several mechanisms for the impact of microbial diversity on blood lipids are possible (Allayee and Hazen, 2015). Certain facultative and anaerobic bacteria produce secondary bile acids from bile salts secreted into the gut lumen. Some of these bile acids are reabsorbed and modulate hepatic as well as systemic fuel metabolism through binding of the Takeda G-protein-coupled receptor 5 (TGR5) as illustrated in Fig. 6.8 (Thomas et al., 2008). While all bile acids bound TGR5, the most powerful agonists tested in vitro were the secondary (bacterially modified) bile acids, lithocholic and deoxycholic acids (Duboc et al., 2014). TGR5 binding increased the expression of several genes in brown fat (mice) and human skeletal muscle myoblasts. One of these genes encodes type 2 deiodinase enzyme, a selenocysteine-requiring enzyme that converts relatively inactive thyroxine (T_4) to the potent thyroid hormone, triiodothyronine (T_3) (Watanabe et al., 2006). TGR5 binding has been shown to regulate energy expenditure and secretion of glucagon-like peptide 1 (GLP1). More recent work has demonstrated that TGR5 signaling downregulates the inflammatory response in macrophages by reducing NFκB activity (Pols, 2014). The author reviews evidence that TGR5 binding may be an important factor in the prevention of cardiovascular and inflammatory disease.

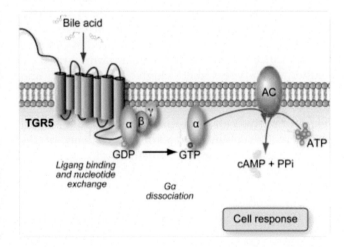

FIGURE 6.8 Bile acid receptor, TGR5. TGR5 is expressed in many organs and cells. Expression is high in gallbladder epithelium and intestine, as well as human monocytes. In the liver, Kupffer cells express TGR5; human skeletal muscle and central nervous system elements also express the gene. In response to bile acids, GDP on the G protein is replaced by GTP, leading to dissociation of protein α that activates adenyl cyclase, leading to the induction of cAMP which leads to downstream signaling. *Source: Reproduced from Pols, T. W.H., 2014. TGR5 in inflammation and cardiovascular disease. Biochem. Soc. Trans. 42 (2), 244−249 with permission from Science Direct.*

6.5.1 Omnivore Intake and Trimethylamine N-Oxide (TMAO)

Dietary components such as choline and betaine (Wang et al., 2011) and carnitine (Koeth et al., 2013) are converted to trimethylamine (TMA) by intestinal microbes and subsequently to trimethylamine N-oxide (TMAO) by hepatic flavin monooxygenases. In the mouse model, TMAO has been shown to promote foam cell formation in cholesterol-laden macrophages and to reduce reverse cholesterol transport. Both in rodents (Gregory et al., 2015) and humans (Tang et al., 2013), choline and carnitine, in the presence of gut microbiota can generate TMAO. The multiple mechanisms through which TMAO potentiates AS and inhibits reverse cholesterol transport have been studied (Koeth et al., 2013). Although not entirely clear, data suggest that TMAO alters the hepatic bile acid pathway for cholesterol elimination. These findings were confirmed by the observation that mice supplemented with TMAO had significantly decreased synthetic bile acid pool sizes. However, it is not yet clear whether the observed reduction of reverse cholesterol transport can be attributed entirely to changes in bile acid-induced cholesterol excretion.

The authors noted that the microbiota of vegans and vegetarians produced little or no TMA when challenged with carnitine (Koeth et al., 2013). This observation was supported by results of the PREDIMED study that showed a 30% reduction in cardiovascular events in 7447 experimental subjects observed over a 4.8-year timeframe. The PREDIMED study provided two experimental diets high in fat in the form of olive oil or nuts, compared with the control, low-fat diet (Estruch et al., 2013). Both experimental arms used a Mediterranean-type diet high in fruits and vegetables and low in red meat. Thus, the high-fat content of the diets did not increase AS risk. The PREDIMED study did not measure TMAO, nor did it quantitate the choline and carnitine contents of the diet. However the authors (Koeth et al., 2013) suggested that some component(s) of the vegetarian/vegan and Mediterranean diets changed the character of the microbiota and reduced the proportion of microbes competent to convert choline and carnitine to TMA.

It is thus of interest that a recent study in mice has demonstrated that allicin, an organosulfate component in garlic, reduced TMAO production from carnitine to baseline in mice supplemented with allicin (Wu et al., 2015). The authors proposed that the antimicrobial activities of allicin and other organosulfates modulated microbiota diversity in these animals, reducing the microbial species capable of forming TMA.

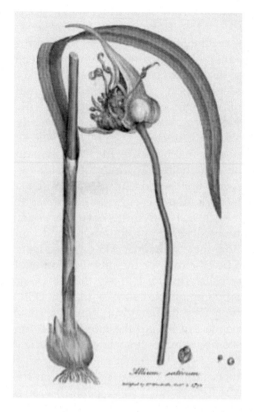

FIGURE 6.9 *Allium sativum*. William Woodville: "Medical botany," London, James Phillips, 1793, Vol. 3, Plate 168: *Allium sativum* (Garlic). *Source: This figure is in the public domain.*

Allicin (diallylthiosulfinate) is synthesized in garlic (*Allium sativum* L.) and other members of the allium family (onion, shallots, scallions, leeks, chives, and other wild varieties). It is produced from the amino acid alliin (S-allylcysteine sulfoxide) by alliinase, an enzyme released by crushing or otherwise injuring the plant, as reviewed by Amagase (2006) and Borlinghaus et al. (2014). Allicin is volatile and rapidly degrades following extraction. It is possible that its sulfur-containing metabolites, diallyl disulfide (DADS), diallyl sulfide (DAS), diallyl trisulfide, and sulfur dioxide exert the observed biological actions; for example, sulfur donors contribute to the formation of vasodilatory H_2S gas (refer to Essentials I: Life in an Aerobic World: Nutrient Transcriptional Control of Antioxidant Protection). Allicin can also inhibit the proliferation of bacteria and fungi; it can also kill antibiotic-resistant bacterial strains such as meticillin-resistant *Staphyloccus aureus* (MRSA) (Fig. 6.9).

Examination of the PREDIMED dietary recommendations revealed that the experimental diets, but not the control diet, included sofrito, a sauce made from tomato and onion, often garlic and peppers, simmered in olive oil with aromatic herbs. Garlic, onion and aromatic spices and herbs contain volatile and fragrant substances (essential oils; EOs), many with antimicrobial actions. The medicinal properties of EO were described centuries ago in the Ebers papyrus and have been widely recommended through the years as reviewed by Bassolé and Juliani (2012). Abundant in vitro work has shown that EOs containing aldehydes or phenols (cinnamaldehyde, citral, eugenol, thymol) show greatest antibacterial activity. Recently, a "colonic model" was used to evaluate the influence of garlic powder on mixed fecal bacterial species. This study revealed a differential effect on the microbiota; commensal species were resistant, while pathogenic species were temporarily more susceptible to antimicrobial action of EOs (Filocamo et al., 2012). Since carnitine and choline are nutrients essential for optimal intermediary metabolism, more research is urgently needed to clarify the potential for dietary EO to modify human microbiota and reduce TMA production and AS risk in the vast omnivore population.

6.6 METABOLIC TARGETS FOR DIET MODIFICATION

Vasoreactivity is controlled by the endothelial cell (EC). This cell produces mediators that both relax (eg, NO and prostacyclin) and contract (eg, endothelin and thromboxane) the vessel. In the normal endothelium, these processes are tightly regulated. Because endothelial dysfunction and reduced NO production are a major contributor to atherosclerosis (Davignon and Ganz, 2004), the ability of EC to secrete the vasodilatory gas, NO formed by the constitutive action of endothelial nitric oxide synthase (eNOS) is considered as the first line of EC vasoactive defense.

NO is a biologically active, unstable radical synthesized by three mammalian nitric oxide synthase isoforms: neuronal (nNOS), endothelial (eNOS), and inducible or inflammatory (iNOS). Vascular eNOS is regulated at the transcriptional, posttranscriptional, and posttranslational levels (Rafikov et al., 2011). As has been reviewed (Saad et al., 2015), in the unstimulated EC, eNOS is inhibited by binding to caveolin-1, a major protein in lipid rafts; caveolin compartmentalizes and concentrates signaling molecules such as G-protein subunits, tyrosine kinases, and eNOS (Williams and Lisanti, 2004). EC stimulation by sheer stress or vasoactive agents induces an increase in intracellular Ca^{2+} that binds to calmodulin and releases eNOS from caveolar binding. The NO gas diffuses to surrounding cells, activating guanylate cyclase that converts guanosine triphosphate (GTP) to cGMP in the underlying

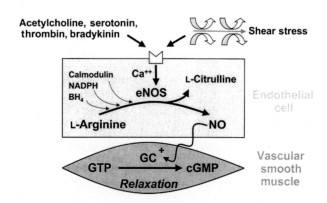

FIGURE 6.10 Biofactors required for activation of eNOS. Adrenergic stimulation and shear stress increase Ca^{2+} influx into the EC and activate eNOS. L-arginine substrate, in a reaction that requires tetrahydrobiopterin (BH_4), NADPH, and calmodulin, is converted to NO and L-citrulline. NO relaxes VSMC contraction. If substrate or cofactors are limiting, eNOS is "uncoupled," producing superoxide anion (O_2^-) and hydrogen peroxide (H_2O_2) with peroxynitrite anion ($OONO^-$) formation. *Source: Reproduced from Behrendt, D., Ganz, P., 2002. Endothelial function: from vascular biology to clinical applications. Am. J. Cardiol. 90 (10, Supplement 3), L40–L48 (Behrendt and Ganz, 2002) with permission from Science Direct.*

TABLE 6.1   Anti-AS Mechanisms for NO

- Decreases endothelial permeability
- Inhibits low-density lipoprotein oxidation
- Reduces influx of lipoproteins into the vascular wall
- Inhibits proliferation of vascular smooth muscle cells
- Prevents leukocyte adhesion to vascular endothelium
- Inhibits leukocyte migration into the vascular wall

VSMCs. cGMP enhances cGMP-dependent protein kinase, decreasing intramyocyte Ca^{2+}, reducing VSMC contraction, and dilating the vessel (Fig. 6.10). Persuasive evidence (Zhang et al., 2011) suggests that modulators of eNOS activity, including substrate and cofactor bioavailability, as well as stimuli that enhance its transcription and interfere with inhibitors, can be effective targets to prevent and treat CVDs as shown in Table 6.1.

6.6.1 Nutrient Requirements for NO Synthesis: Arginine (a Nonessential Amino Acid)

Plasma L-arginine levels are regulated by dietary arginine intake, global protein turnover, arginine synthesis, and metabolic factors (Boger and Bode-Boger, 2001; Förstermann and Münzel, 2006). On an average

US diet, adults absorb approximately 5.4 g of L-arginine. Asymmetric dimethylarginine (ADMA) is an endogenous arginine analog often elevated in AS, that inhibits eNOS activity. Inhibition can be overcome by supplemental arginine, indicating that AMDA competes with L-arginine for enzyme binding. Arginine destruction by arginase II has been reported in both animal and human models of CVD. Thus while a diet with adequate protein should provide sufficient arginine for NO biosynthesis, it may not be sufficient to ameliorate arginine "starvation" in the dysfunction endothelial cell.

Several clinical trials (Böger, 2014) have reported improved vascular function following L-arginine supplements in patients with hypercholesterolemia and/or hypertension, although some larger doses were associated with adverse effects; supplemental L-arginine continues to be investigated as an adjuvant therapy for endothelial dysfunction. The authors noted that the effects of L-arginine supplementation on human physiology appear related to baseline ADMA and are multicausal and dose-related. Doses of 3–8 g/day appear to be safe and do not cause acute pharmacologic effects in humans. Additionally, plasma ADMA was elevated both in prediabetic and diabetic subjects at baseline and increased in tandem with adhesion molecules and CRP upon glucose loading (Konukoglu et al., 2008). Thus, in the presence of glucose intolerance, presumably adequate arginine availability may be compromised. While this could explain, in part, the improved vasoreactivity observed with arginine supplementation (the "arginine paradox"), mechanisms are still not entirely clarified (Fig. 6.10).

6.6.2 Tetrahydrobiopterin

All NOS species require reduced tetrahydrobiopterin (BH_4) as a cofactor for catalytic activity. BH_4 is essential for NOS structure, allowing the enzyme to shift its heme iron to a high spin state, increase arginine binding, and stabilize the active NOS dimeric form (Katusic, 2001). Suboptimal concentrations of BH_4 reduce the generation of NO and favor NOS "uncoupling," leading to NOS-mediated reduction of molecular oxygen and the generation of ROS. Reduced availability of BH_4 has been reported in vessels of diabetic and hypertensive patients (Rubio-Guerra et al., 2010). Insufficient reduced biopterin (BH_4) limits L-arginine oxidation. Under these conditions, insufficient endothelial NO is produced for its vasodilatory, antiatherogenic, antithrombotic, antiinflammatory, and antiproliferative effects (Table 6.1).

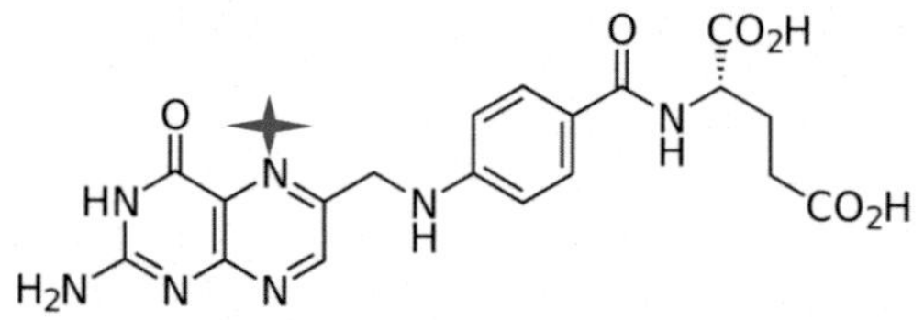

FIGURE 6.11 Folic acid (FA). Folic acid is a synthetic, nonreduced form of folate. To be used as a cofactor, folic acid must be methylated in the 5-position (star) and reduced in four positions to form 5-methyl-tetrahydrofolate (5-meTHF). *Source: This figure is in the public domain.*

6.6.3 $BH_2 \rightarrow BH_4$ Reduction: Role of Folate Metabolism

Folate deficiency has been linked to CVD through multiple mechanisms (Moens et al., 2008). As illustrated in Fig. 6.12, reduced and methylated folate, as 5-methyl tetrahydrofolate (5-meTHF), donates its methyl group to homocysteine to form methionine. If 5-meTHF availability is inadequate, serum homocysteine is not remethylated and homocysteine levels rise. Other theoretical roles for folate interaction with endothelial function have been postulated, including direct or indirect interaction with eNOS and NO biosynthesis. Strong hypothetical associations between folate status and AS have led to multiple clinical trials using the stable, nonreduced analog, folic acid (FA). Unexpectedly, outcome data from these trials have been inconclusive (Cholesterol Homocysteine Collaborative Group, 2010; Clarke et al., 2010) (Fig. 6.11).

A possible explanation for inconsistency in outcome data in trials using folic acid relates to the availability of tetrahydrobiopterin (BH_4) for eNOS activity. Tetrahydrobiopterin is formed de novo from GTP, and when exposed to oxidative stress, it is oxidized to BH_2. BH_2 is reduced back to BH_4 by an enzyme in the folate pathway, dihydrofolate reductase (DHFR). While the total amount of biopterin in the cell is regulated by the biosynthetic pathway, the activity of DHFR determines the BH_4:BH_2 ratio critical to eNOS activity and NO production (Crabtree et al., 2009). DHFR activity is variable and is known to be inhibited by the 4-amino derivative of FA, aminopterin (methotrexate; MTX), widely used as an antineoplastic drug (Huennekens, 1994). In a manner analogous to MTX, nonreduced synthetic FA is also reduced by DHFR to function as the cofactor, tetrahydrofolate.

FA is not found in the food supply, but is used as a dietary supplement because this form is stabile and does not degrade during storage or heat. Upon ingestion, FA is first reduced to dihydrofolate and subsequently to tetrahydrofolate by DHFR. In humans, in contrast to rats, the activity of hepatic DHFR is extremely slow (<2% of the activity in rats) and varies widely between individuals (Bailey and Ayling, 2009). As is the case with MTX, large supplements of FA can rapidly saturate and inactivate the DHFR enzyme (Huennekens, 1994). Thus, large supplements of FA could theoretically prevent BH_2 reduction to active BH_4, thereby uncoupling eNOS, reducing NO biosynthesis and generating oxidative damage.

A very recent study suggests that cutaneous vascular contraction in old, but not young, subjects is improved by local perfusion or oral supplementation of folate as the cofactor (5-meTHF) or synthetic FA (Stanhewicz et al., 2015). The authors also reported that the improvement was independent of homocysteine. The fact that only a fraction of individuals receiving oral FA supplements have high levels of unmetabolized FA in their serum (Mason, 2009) reaffirms the variability of DHFR activity in humans. Thus, the presence of unmetabolized FA in the serum following an FA supplement could reflect lower activity of DHFR and be an index of concomitant reduction of reduced BH_4. This question should be clarified since some investigators have postulated that high levels of unmetabolized FA can have other deleterious effects as well (Smith et al., 2008).

6.6.4 Redox Control of eNOS Function

If eNOS function depends on the reduced cofactors, GSH and BH_4, it is likely that the redox state in the EC is important, as discussed below. Two clinical trials were conducted to test whether a reducing agent, such as ascorbate, could improve eNOS function. While supraphysiologic doses of parenteral or oral ascorbate were reported to increase NO bioactivity in atherosclerotic patients, analysis of ascorbate kinetics suggested that it was unlikely that extracellular ascorbate acts as a radical scavenger (Huang et al., 2000). Subsequent in vitro trials revealed that ascorbate, but not Zn-Cu SOD increased eNOS activity by ~73%. Additionally, ascorbate, but not glutathione, increased BH_4 by 226% compared with vehicle-treated cells. Ascorbate did not alter arginine uptake by the cells. These observations suggest that ascorbate may stabilize the enzyme or act by other, as yet not fully characterized, mechanisms (De Tullio, 2012) to increase endothelial NO availability (refer to Essentials I: Life in an Aerobic World: Non-enzymatic Antioxidant Nutrients: Ascorbate).

As discussed below and elsewhere (refer to Essentials I: Life in an Aerobic World: Nutrient Transcriptional Control of Antioxidant Protection) eNOS activity in a mouse model was increased by the gasotransmitter, H_2S, produced by reverse trans-sulfuration from cysteine. The authors proposed that the cardioprotective actions of H_2S are mediated by crosstalk and augmentation of eNOS action (King et al., 2014).

Recent work (Crabtree et al., 2013) suggests that superoxide production triggered by BH_4 deficiency also oxidizes glutathione, decreasing the GSH:GSSG ratio and initiating glutathionylation. S-glutathionylation is the posttranslational modification of a protein by glutathione that modifies protein activity (Dalle-Donne et al., 2009); S-glutathionylation of eNOS results in eNOS uncoupling. These investigators found that both BH_4 levels and GSH levels are important in the regulation of eNOS function and that BH_4-dependent and 5-glutathionylation-induced eNOS uncoupling are mechanistically independent but functionally linked.

6.6.5 Sulfur Metabolism, Homocysteine, and AS Risk

Several reports have demonstrated an association of hyperhomocysteinemia (HHcy) with risk for peripheral vascular disease (PVD) and CVD (McCully, 1969, 1996; Loscalzo, 1996; Jakubowski, 2007). As described elsewhere (refer to Essentials I: Life in an Aerobic World: Single Unit Antioxidant Nutrients: Vitamin C: Ascorbate and to Figure EI-6 Homocysteine Homeostasis), cellular levels of homocysteine are tightly regulated by dietary sulfur containing amino acids (SAA; methionine and cysteine) and by the folate and trans-sulfuration pathways as illustrated in Fig. 6.12. HHcy has been well described in patients with mutations in the trans-sulfuration pathway, especially the enzyme cystathionine β synthase (CBS). Similarly, altered folate availability due to polymorphisms in methylene tetrahydrofolate reductase (MTHFR) or other enzymes in the methionine pathway can elevate serum homocysteine. Alternatively, failure to methylate homocysteine to methionine as a result of dietary folate and vitamin B12 deficiency, or failure to convert homocysteine to cystathionine and other sulfur compounds by deficiency of vitamin B6 can also result in HHcy. Choline and betaine contribute methyl groups to the methionine cycle and may indirectly modify the trans-sulfuration cycle by modifying homocysteine availability.

Constancy of the methionine cycle is essential for the maintenance of methyl balance and for the provision of methyl groups donated by S-adenosyl-methionine (SAM) (Mudd et al., 2007). This necessity predicts that in the face of reduced availability of sulfur-containing amino acids (SAA) from diet as methionine and cysteine, the body adapts by breaking-down lean tissue to maintain a constant supply of methionine to form SAM.

A recent study demonstrated that a diet inadequate in SAA raised serum homocysteine (Ingenbleek and McCully, 2012). SAA are limiting amino acids in vegetable protein sources. This is especially true when

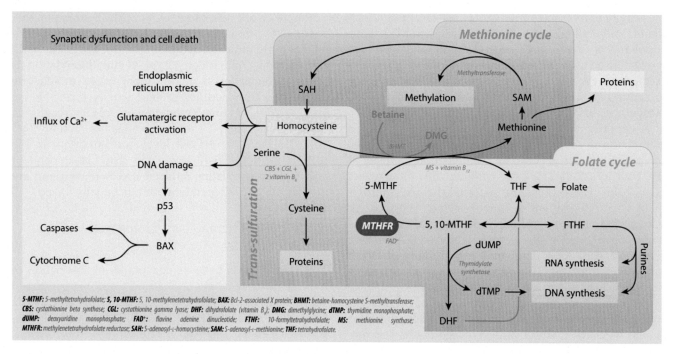

FIGURE 6.12　Homocysteine homeostasis. Homocysteine concentrations in the cell are tightly regulated by the methionine, folate, and *trans*-sulfuration cycles. The left panel indicates some of the deleterious changes in the cell associated with hyperhomocysteinemia. Not shown are the pathways through which cysteine is the rate-limiting amino acid for de novo glutathione biosynthesis and is a precursor for hydrogen sulfide (H_2S) production. (Refer to Essentials I: Life in an Aerobic World: Figure EI-6 for more information.) *Source: By Epgui (own work) (CC BY-SA 3.0 (http://creativecommons.org/licenses/by-sa/3.0)), via Wikimedia Commons.*

plants are grown in soil depleted of sulfur (Young and Pellett, 1994), as is the case in several areas of the globe. Diets depleted of SSA can cause subclinical protein malnutrition and reduced lean body mass (Jahoor et al., 2006). Reduced lean mass cannot provide adequate SAA to maintain the trans-sulfuration pathway, thus serum homocysteine rises (Ingenbleek and McCully, 2012). Under these conditions, the *trans*-sulfuration pathway is inhibited at CBS, limiting the formation of downstream metabolites including cysteine, glutathione, and H_2S. As described below, these metabolites are important in maintaining redox homeostasis and reducing AS risk.

The observation that reduced availability of SAA increases CVD risk by raising homocysteine levels suggests that patients with age- and/or obesity-associated sarcopenia characterized by reduction in lean body mass and increased fat deposition may also be at risk for SAA malnutrition. Indeed, a recent paper demonstrated increased risk for CVD in patients with sarcopenic obesity (Bahat and İlhan, 2016).

Experimental evidence implicating elevated homocysteine concentrations as contributors to the pathogenesis of atherosclerosis has been comprehensively reviewed (McCully, 1996; Gurda et al., 2015). Homocysteine thiolactone (tHcy) is especially toxic in that it can form complexes with enzymes and alter

their function (Jakubowski, 2007). For example, tHcy binds to fibrinogen, decreasing clot permeability and susceptibility to lysis (Undas et al., 2006). Preliminary evidence also suggests that homocysteine forms complexes with intracellular copper, contributing to its efflux from the cell as described elsewhere (refer to Chapter 5: Cardiomyopathy and Congestive Heart Failure: Targets for Myocardial Protection: Cardiac Remodeling). The resulting intracellular copper deficiency impairs essential copper-dependent enzymes, such as mitochondrial cytochrome oxidase (Dong et al., 2013). These authors demonstrated that exposure of human umbilical vein endothelial cells to increasing concentrations of homocysteine resulted in a concentration-dependent decrease in intracellular copper, formation of copper-homocysteine complexes and necrotic cell death; pretreatment with copper prevented these effects.

Other mechanisms for homocysteine-dependent vascular damage have been reviewed (McCully, 1996). HHcy increases risk for AS by increasing production of sulfated proteoglycans in the intima, production of oxygen radicals, stimulation of VSMC hyperplasia and migration, and mitochondrial damage that interferes with electron transport. In addition, homocysteine stimulates activity of elastase with degradation of vascular elastin, and exerts a thromboogenic effect by enhancing platelet aggregation and inhibiting endothelial anticoagulant functions.

Vascular lesions associated with elevated homocysteine levels do not necessarily involve dyslipidemia. In fact, a large fraction of patients with vascular lesions both in the United States and globally have normal lipid levels (Ingenbleek and McCully, 2012). It is of interest that while clinical trials using vitamin supplementation (FA, vitamins B12 and B6) did not modulate AS risk, despite reducing homocysteine levels (Cholesterol Homocysteine Collaborative, 2010), folate supplementation and reduction of homocysteine levels did increase clot permeability and shorten lysis time (Undas et al., 2006).

6.6.6 NO Recycling and Exogenous Nitrates

The L-arginine-dependent eNOS pathway for NO synthesis described above is a highly complex system that consumes energy and requires many substrates and cofactors. Once formed, some NO makes its way to the vascular smooth muscle where it stimulates vasodilation, some is taken up by platelets and leukocytes. Any NO that remains is rapidly terminated by oxidation to nitrite and nitrate by NADPH oxidases and circulating oxidants such as hemoglobin as reviewed by Lundberg and Weitzberg (2005). The nitrate and nitrite oxidation products of NO were long considered inert, but recent work suggests that the nitrite ion can be recycled back to NO (Kevil et al., 2011; McNally et al., 2016) and is cardioprotective. Thus, the EC can reutilize both endogenous nitrite formed by NO oxidation and exogenous nitrite available from the diet.

Nitrite as (KNO_3) in its natural mineral form (saltpeter) or as pure sodium or potassium salts has been added to salted meats since the Middle Ages and has been used as a food additive in the United States since 1925. Nitrite is antibacterial and has been used to prevent growth of Clostridium botulinum in the anaerobic center of large cuts of meat such as ham. Nitrite also oxidizes the ferrous iron in myoglobin, preventing oxygen binding, and maintaining the pleasing pink color change in cooked processed meat. It contributes to the taste of processed meats, by an unknown mechanism. In large doses, nitrite salts can oxidize ferrous iron on hemoglobin, reducing oxygen-carrying capacity and resulting in potentially lethal methemoglobinemia. Nitrates in plants (beets, spinach) can be converted to nitrite through a gut–salivary gland cycle as shown in Fig. 6.13.

While dietary nitrites have been implicated in the etiology of multiple cancers (Kuhnle and Bingham, 2007), an increased risk for cancer has not been confirmed in recent reports (Lundberg and Weitzberg, 2010; Bryan et al., 2012); however nitrite is toxic in large doses. In fact, dietary nitrate consumption and

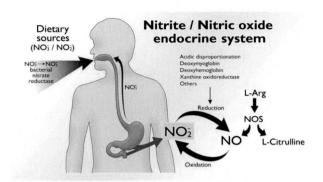

FIGURE 6.13 Nitrate, nitrite, nitric oxide pathway. Dietary sources of nitrate and nitrite can be converted to NO and augment the pool of NO available for vasodilation. Green leafy vegetables are rich in nitrates. Dietary nitrates are recycled by an enterosalivary circulation that delivers ingested nitrate to discrete regions in the posterior areas of the tongue that contain nitrate-reducing bacteria. Nitrite that enters the plasma is reduced to NO and rapidly oxidized back to nitrite (Kevil et al., 2011). *Source: Reproduced from Kevil, C.G., Kolluru, G.K., et al., 2011. Inorganic nitrite therapy: historical perspective and future directions. Free. Radic. Biol. Med. 51 (3), 576–593 with permission from Science Direct.*

subsequent increase in NO availability have been associated with reduced risk for obesity and its comorbidities (Kobayashi et al., 2015). It is of interest that use of chlorhexidine-containing antimicrobial mouthwash decreased the absorption of dietary nitrates as nitrites in human volunteers, likely by inactivating bacterial conversion in the mouth (Govoni et al., 2008).

6.6.7 Red Wine, Resveratrol, Anthocyanins, and EPC Pools

Red wine contains several polyphenols as well as resveratrol (3,5,4′-trihydroxy-*trans*-stilbene). These compounds belong to a group of phytoalexins produced in plants as a defense against disease and predators. (Phytoalexins are produced in plants in response to injury and stimulate genes that upregulate plant defenses.) Resveratrol has been extensively studied since it was shown to increase cell survival in budding yeast (*Saccharomyces cerevisiae*) by stimulating sirtuin 1-dependent deacetylation of p53 (Howitz et al., 2003). SIRT molecules 1–7 are found in humans and regulate various metabolic processes that allow the cell to adapt to nutrient stress. As reviewed by Nakagawa and Guarente (2011) and Matsushima and Sadoshima (2015), SIRT molecules positively regulate liver receptor X (LRXα), increasing cholesterol efflux from foam cells. They also increase mitochondrial biogenesis and coordinate glucose and fatty acid oxidation. Inflammation downregulation is facilitated in part by inhibiting the proinflammatory NFκB cascade and reducing ROS by upregulating superoxide dismutase (SOD) and other

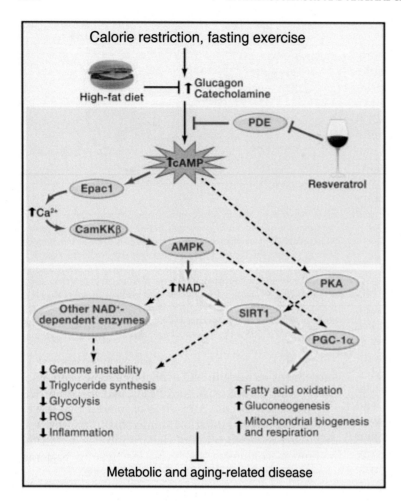

FIGURE 6.14 Proposed actions of resveratrol. Resveratrol inhibits phosphodiesterases and increases cAMP levels. Subsequent activation of signaling pathways, together with increased calcium influx activates AMPK. Downstream, an increase in NAD$^+$ leads to SIRT1 activation, promoting beneficial changes, primarily through deacetylation. *Blue arrows* indicate recent work (Park et al., 2012), while dashed lines indicate molecular functions previously identified. *PDE,* phosphodiesterase; *cAMP,* cyclic AMP; *Epac1,* cAMP-regulated guanine nucleotide exchange factor 1; *CamKKβ,* calcium/calmodulin-dependent kinase kinase beta; *AMPK,* AMP-activated protein kinase; *PKA,* protein kinase A; *NAD$^+$,* nicotinamide adenine dinucleotide; *PGC-1α,* peroxisome proliferator-activated receptor gamma coactivator-1 alpha; *ROS,* reactive oxygen species. *Source: Reproduced from Tennen, R.I., Michishita-Kioi, E., et al., 2012. Finding a target for resveratrol. Cell 148 (3), 387–389 with permission from Science Direct.*

transcriptionally regulated systems. Sirtuins may not be the direct target for resveratrol. A recent review of the actual functions of resveratrol revealed that it may act upstream of sirtuins at the level of phosphodiesterases that activate adenyl cyclase and increase cAMP, thereby stimulating AMPK and mimicking the effect of calorie restriction to mitigate the deleterious effects of a high-fat diet (Park et al., 2012; Tennen et al., 2012).

While resveratrol isolated from red wine and other sources (grape skin, purple, and blue berries) has been the subject of extensive mechanistic studies, red wine actually contains a complex mixture of flavonoids, including anthocyanins and flavan-3-ols (and polymeric procyanidins, eg, condensed tannins). Nonflavonoids, such as resveratrol and gallic acids, make up only a small fraction of the total phenolic compounds in wine; analysis of phenolics in an array of fruit juices revealed that resveratrol and its glucoside represent 0.4 – 6.6% of the total phenolics in red wine compared to other potentially protective components such as flavonols, anthocyanins, flavan-3-ols, gallic acid, and hydroxycinnamates (Mullen et al., 2007) (Fig. 6.14).

A number of factors determine the level of phenolic constituents in red wine. Different grape varietals, the elevation of the vineyard, and the duration of the alcoholic maceration with grape seeds and skin in the formation the must determine the character of the wine. (Must is from the Latin word (*vinum mustum*) meaning young wine. It is the freshly pressed fruit juice that contains seeds, skins, and stems of the fruit.) Aging in oak barrels adds phenolics from the wood to the finished wine (Corder et al., 2006). The authors identified oligomeric procyanidins (OPC) in red wine to be most effective in suppressing synthesis of endothelin 1, a vasoconstricting peptide, in cultured endothelial cells. Additional work to clarify the action of OPC on endothelial cells suggested that OPC stimulate a pseudolaminar shear stress response, suppressing endothelin 1 and upregulating eNOS, thereby facilitating vasodilation and return to laminar flow (Khan et al., 2015).

Preliminary observations have suggested that ingested OPC and/or its metabolites can transit the intestinal tract and reach the human vasculature to elicit protective effects. The possibility that consumption of wine with increased OPC improves human longevity stems from the observation that higher OPC levels measured in wines from certain regions (eg, southern France and Sardinia) were associated with the longevity of the native population in the region in which the wine was made. Additionally, a human trial provided 80 young healthy subjects with red wine (100 ml/day), vodka, beer, or water and quantitated the availability of epithelial precursor cells (EPCs) in the blood. Even at this physiologically relevant dose, circulating EPCs, defined by CD34 and other markers, as well as EPC colony-forming units significantly increased, but only in the RW group (Huang et al., 2010). The authors reported an increase in plasma levels of NO associated with a decrease in the arginine analog, ASMA.

Several questions remain to be answered. While cell culture and ex vivo studies with red wine and/or OPC have demonstrated atheroprotective responses at doses likely to be consumed by humans, OPC molecules are very large and their bioavailability is not well studied, nor has their metabolism by intestinal microbiota been clarified. In view of the variability of phenolic compounds in the food supply, not only in wine but in fruits and vegetables as well as tea, coffee, and chocolate, it is important to develop methods to more accurately quantitate total dietary content of these potentially atheroprotective components.

6.6.8 Flavonoids, Flavanols, and Cardiovascular Risk Reduction

Polyphenols are also abundant in fruits and plant-derived beverages such as fruit juices, tea, coffee, red wine, cereals, chocolate, and dry legumes. The total dietary intake of dietary polyphenols can be very high (~ 10 times higher than vitamin C and ~ 1000 times higher than vitamin E and carotenoids). While polyphenols were initially thought to act as antioxidants due to their conjugated aromatic structure, it now appears likely that they modulate signal transduction through direct interactions with receptors (Heiss et al., 2010). They are also thought to trigger a series of redox-dependent reactions (Ignarro et al., 2007).

Flavonoids and polyphenols in dark chocolate, tea, and other dietary sources must be available at the target of action. Thus, the following processes modulate flavonoid efficacy in vivo: (1) rate and extent of absorption in the intestine as well as of enterohepatic circulation; (2) metabolic conversions by gut commensal microbes, enterocytes, hepatocytes, and target cells; (3) degree of binding to serum albumin or cellular proteins; (4) plasma levels of aglycones and conjugates; (5) accumulation in target tissues such as vascular endothelium; (6) urinary elimination rate; and (7) the

molecular mode of actions. For example, dietary polyphenols and their metabolites may contribute to the control of NADPH oxidase activity, thus lowering ROS generation, which, in turn, leads to elevation of the steady-state level of NO in the cells. These complex pathways require additional substrates, cofactors, and energy which can also modify the action of bioactive food components.

Despite the fact that most evidence comes from experimental models and small-scale intervention studies, these studies have been important for elucidating potential mechanisms of action through use of surrogate end points such as endothelium-dependent vasodilation, blood pressure, platelet function, and glucose tolerance. Flavanol intake has been linked to short-term increases in NO bioactivity both in healthy subjects and those with increased CVD risk. Mechanistically focused clinical research is needed to assess the impact of food phytochemicals on AS risk in humans (Traka and Mithen, 2011). Thus, while flavanol data are promising, in many trials flavanol content is poorly defined, bioavailability is unquantified and the mechanisms of action unknown. Longer-term, randomized controlled interventions to better define the protective vascular effects of these nutrients are required. A sample of polyphenol-rich dietary sources that have been shown to reduced risk for CVD includes the following.

6.6.9 Pomegranate

Pomegranate fruit (*Punica granatum* L.) contains a mixture of flavonoids including anthocyanins, catechins, ellagic tannins, and gallic and ellagic acids. Pomegranate juice can inhibit proinflammatory cyclo-oxygenases and lipo-oxygenases. Polyphenolics contained in the juice have been shown to target genes upregulated in oxidative stress developed in arterial segments exposed to disturbed blood flow, normalizing gene expression, and enhancing eNOS activity (de Nigris et al., 2007) both in vitro in cultured human coronary artery EC and in vivo in hypercholesterolemic mice (Fig. 6.15).

6.6.10 Chocolate

The scientific name for the cocoa tree, *Theobroma cacao*, is taken from the Greek words: *theo* (god) and *broma* (drink). The "drink of the Gods" reflects the high esteem in which this fruit was held throughout antiquity (Henderson et al., 2007). (Anthropological evidence of cacao use in Honduras dates from 1000 BC. Early concoctions were probably made from fermentation of the fruity pulp surrounding the seeds. Evidence suggests that around 1600 BC, Aztecs and contemporaries fermented and ground the seeds to make a "drink of the gods.") The bitter drink made

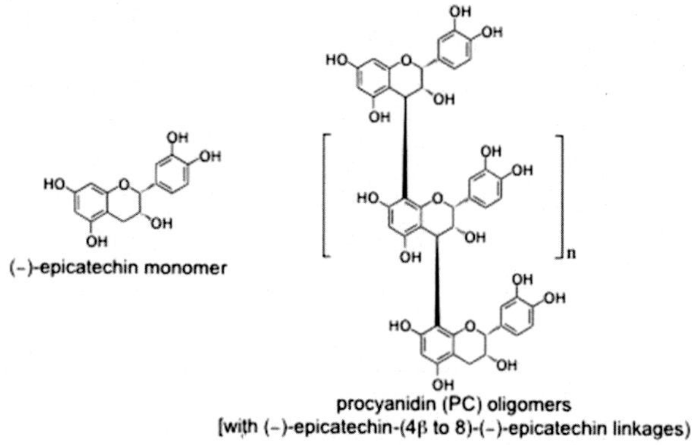

FIGURE 6.15 Selected flavan-3-ols in cocoa. High-molecular-weight oligomers differ from low-molecular-weight structures with respect to their ability to preserve gut integrity and reduce inflammation (Bitzer et al., 2015). *Source: Reproduced from Bitzer, Z.T., Glisan, S.L., et al., 2015. Cocoa procyanidins with different degrees of polymerization possess distinct activities in models of colonic inflammation. J. Nutr. Biochem. 26 (8), 827–831 with permission from Science Direct.*

from the fermented seeds was used by ruling Aztecs for its health benefits and to reduce fatigue. The natural product *cacao* is different from the processed product, *chocolate*, which refers to the combination of cocoa, sugar, milk, and other ingredients in the solid food product. Further studies are underway to separate the effects of cocoa mass, concomitant ingredients, and the impact of cacao processing on the biochemical activities of the final product.

Epidemiological data suggest that cocoa intake exerts strong protective effects on cardiovascular and all-cause use mortality (Corti et al., 2009). The authors reported a 50% reduction in risk for cardiovascular mortality for men in the highest tertile of cocoa intake (RR for all-cause mortality: 0.53; $p \leq 0.001$). Cocoa polyphenols have several possible actions to enhance endothelial function. For example, they increase Ca^{2+} or eNOS phosphorylation via the PI3-kinase/Akt pathway. Cocoa also lowered vascular arginase activity in human endothelial cells in vitro, allowing greater local L-arginine availability. As reviewed by Corti et al. (2009), cocoa polyphenols may also activate endothelium-derived hyperpolarizing factor (EDHF), increase endothelial prostacyclin release, or inhibit the synthesis of endothelin-1 (ET). Moreover, polyphenols may directly inhibit angiotensin-converting enzyme (ACE) and cyclo-oxygenase (COX-1), contributing to their observed effects on blood pressure and inflammation.

6.6.11 Organosulfur Donors

6.6.11.1 *Garlic* (**Allium sativum L.**)

Garlic and its close relatives (onion, leek, shallot) have been used as flavoring and medicinal agents for hundreds of years. Epidemiological and experimental

TABLE 6.2 Possible Mechanisms by Which Garlic May Inhibit Atherosclerosis

- Inhibition of stenosis caused by damage induced by balloon catheterization (in vivo)
- Inhibition of cell transformation and cell growth in the smooth muscle cells (in vitro)
- Inhibition of lipid accumulation into macrophage (foam cells) (in vitro)
- Inhibition of LDL oxidation-caused endothelial cell damage in artery (in vitro)
- Inhibition of LDL oxidation-induced free radical generation from damaged endothelial cells in artery (in vitro)
- Inhibition of glutathione depletion from the endothelial cells (in vitro)
- Activation of cNOS (in vitro)
- Increase of nitrous oxide metabolites; eNOS activation (in vivo)
- Lowering of cholesterol, raising of HDL cholesterol
- Lowering blood pressure
- Reduction of homocysteine
- Improvement of endothelium function (in vivo)

studies have repeatedly shown an inverse correlation between garlic consumption and AS. Garlic is known to inhibit key enzymes involved in cholesterol and lipid synthesis, reduce platelet aggregation and thrombus formation, enhance fibrinolysis, and maintain vascular tone (as reviewed by Vazquez-Prieto and Miatello (2010)). Clinical trials have been less conclusive; the most consistent effect shown in clinical trials was platelet aggregation reduction in seven of seven studies conducted. It is likely that inconsistency in clinical trials was partly due to varied methods of garlic processing, the presence of unknown active constituents, inadequate randomization, inadequate subject selection, and other methodologic issues.

Diallyldisulfide and diallyltrisulfide, two of three major organosulfur compounds in garlic oil, suppressed LDL_{ox} induction of HL60 (human promyelocytic leukemia cell line) adhesion to human vascular endothelial cells (Lei et al., 2008); the authors attributed this reduction to downregulation of ICAM-1 and VCAM-1 mRNA and protein expression. Other mechanisms through which garlic may inhibit atherosclerosis are indicated in Table 6.2 (Budoff, 2006).

6.6.12 Cabbage and Cruciferous Vegetables

Isothiocyanates are organosulfur compounds found in cruciferous vegetables such as cabbage, kale, and broccoli. They also have cardioprotective actions, but the mechanisms are far from clear. Sulforaphane activates transcriptional factor Nrf2 that regulates gene expression through the promoter antioxidant response element (ARE). Nrf2 regulates the transcription of a battery of protective and metabolic enzymes as discussed elsewhere (refer to Essentials I: Life in an Aerobic World: Nutrient Transcriptional Control of Antioxidant Protection).

6.7 NUTRIENTS THAT POTENTIALLY INCREASE AS RISK

6.7.1 The Iron-Overload Hypothesis

Dietary iron availability is a strong contributor to iron homeostasis: the estimated iron content of a typical mixed diet (6 mg/1000 kcal) dictates that young women of child-bearing age (RDA = 18 mg/day) will have difficulty obtaining sufficient iron from food, while postmenopausal women and men (RDA = 8 mg/day) can easily consume iron in excess of their needs (Micronutrients, 2001). Regulation of iron homeostasis is discussed elsewhere (refer to Essentials II: Heavy Metals, Retinoids and Precursors: Iron). An imbalance of homeostatic mechanisms that control the risk from unbound iron can result in systemic and parenchymal siderosis and contribute to organ damage such as β-cell dysfunction, fibrosis in liver diseases, and atherosclerotic plaque growth and instability as reviewed by Fernández-Real and Manco (2014).

Dietary iron availability can be reduced by limiting consumption of heme in red meat, avoiding iron in fortified foods or as dietary iron supplements and reducing intake of other readily absorbable iron sources. It should be noted that soybeans and other legumes contain iron as ferritin, a large, stable, phytate-resistant nanocage that surrounds hundreds of iron and oxygen atoms. Iron absorbed as ferritin is taken up by receptor-mediated endocytosis and absorption is much more efficient than iron presented as single inorganic iron atoms (Theil, 2011). It is not yet certain whether consumption of legume ferritin contributes to elevated iron status in men and postmenopausal women.

6.8 NUTRIENT DYSREGULATIONS THAT LEAD TO ARTERIAL CALCIFICATION

Several nutrients have been implicated in the regulation of arterial calcification, as recently reviewed by Nicoll et al. (2015). VSMC calcification shares mechanisms in common with those of bone mineralization, discussed elsewhere (refer to Essentials III: Nutrients for Bone Structure and Calcification: Vitamin K, also refer to Chapter 9: Osteoporosis and Fracture Risk). Two areas of particular importance in regulating calcification are as follows.

6.8.1 Vitamin K Deficiency and Loss of Calcification Inhibitor

Pathological calcification of the ECM can occur in the arteries unless actively inhibited. Mgp, a mineral-binding ECM protein (Hauschka et al., 1989) synthesized by VSMC and identified in vivo as an active inhibitor of vascular smooth muscle calcification (Luo et al., 1997).

Vitamin K is required for posttranslational carboxylation of several vitamin K-dependent Mgps in the coagulation cascade (refer to Essentials III: Nutrients for Bone Structure and Calcification: Vitamin K). Vitamin K requirements have been tied to optimal coagulation ability (Micronutrients, 2001) and patients with hypercoagulability have been treated with the antivitamin K drug Coumadin (Warfarin).

It is now appreciated that Mgp, the vascular inhibitor of calcification, requires vitamin K-dependent carboxylation for activity. Inadequate availability of vitamin K due to low dietary intake, fat malabsorption, antibiotics, and especially vitamin K antagonists can inhibit Mgp carboxylation and lead to subsequent VSMC injury and apoptosis (Danziger, 2008). Because Mgp also contributes to inhibiting the VSMC switch to an osteogenic phenotype (Schurgers et al., 2013), vitamin K deficiency promotes vascular calcification. Vitamin K is found primarily in dark green leafy vegetables and fermented dairy products. It is also synthesized by the gut bacteria, but since it is a fat-soluble vitamin it requires micelles for absorption. The degree to which vitamin K is absorbed from the colon has not yet been determined (Shearer et al., 2012).

6.8.2 Calcium and Phosphorus Excess

While arterial calcification is increasingly recognized as an early indication of arterial disease, the influence of dietary calcium from diets or supplements has not been carefully studied. Serum calcium concentrations have been positively associated with carotid artery plaque thickness, aortic calcification, incidence of MI, and mortality, as recently reviewed by Bolland et al. (2013). Coronary arterial calcification (CAC) scores are estimates of calcium deposition obtained by cardiac computed tomographic estimation (CT scan). While CAC scores in 23,652 Korean participants without CKD or clinically overt CVD revealed no relationship between CAC scores and either dietary or supplemental calcium, combined serum calcium and phosphorus levels were significantly related to CAC scores (Kwak et al., 2014). These results were supported in a miniature swine model that revealed no increased calcification with high dietary calcium (Phillips-Eakley et al., 2015), suggesting that the calcium-phosphorus product may be a risk factor.

On the other hand, as described above, following consumption of calcium supplements, ionized serum calcium levels were elevated to high normal range for as long as 8 hours. Food calcium did not produce this elevation (Bristow, Gamble et al., 2015a,b). While the ionized calcium levels are still in the normal range, there is preliminary evidence that this degree of elevation can predispose VSMC to produce MV and promote vascular calcification (Kapustin et al., 2011).

Phosphorus and calcium are sparingly soluble in the blood and their concentrations must be tightly regulated to prevent crystal precipitation in the soft tissue. Major regulators of these minerals are fibroblast growth factor 23 (FGF23) and parathyroid hormone (PTH); this regulation is described elsewhere (refer to Essentials III: Nutrients for Bone Structure and Calcification). Both hormones increase urinary phosphate excretion by reducing proximal tubular expression of the sodium-dependent phosphate cotransporters, thereby lowering serum phosphate levels. They differ functionally in that FGF23 decreases serum levels of activated vitamin D (1,25D) by inhibiting renal 1α-hydroxylase and stimulating 24-hydroxylase (Emmett, 2008; Scialla and Wolf, 2014) while PTH increases renal 1α-hydroxylase production of 1,25D, which consequently enhances the absorption of phosphate (and calcium) from the intestinal tract.

Evidence is accumulating that excess dietary phosphorus can increase FGF23 secretion not only in patients with renal failure, but in normal individuals as well (Gutiérrez, 2013), increasing risk for arterial calcification. Pathologic calcification in the vascular media and heart valves has been amply demonstrated in renal patients, and reports are emerging that this may be occurring in patients with normal kidney function as well. High dietary phosphorus is also thought to induce cardiomyocyte hypertrophy, impair vascular reactivity by inhibiting NO synthesis, and cause endothelial dysfunction in animals and humans (Shuto et al., 2009). Finally, high serum phosphate has been associated with a procoagulant milieu and inflammatory cytokines, linking excess phosphorous intake with inflammatory processes which themselves are mediators of CVD (Abbasian et al., 2015).

Dietary phosphorus from whole foods, such as grains and legumes, is primarily in the form of phytate and poorly absorbable; phosphorus in meat and dairy products is bound to proteins and is only about 50–70% absorbable. In contrast, phosphate salts, used in processed foods to improve shelf-life and maintain freshness, are 100% absorbable. Foods containing phosphate salts include baked goods, processed cheeses, soft drinks, cereals, and meats that have been bathed or injected with a phosphate cocktail to improve moisture and tenderness. The impact of processed foods on the dietary phosphate is uncertain, but has been estimated to more than double the dietary phosphorous intake.

6.8.3 Magnesium Deficiency

Magnesium and possibly other components of a whole-food diet appear to somewhat ameliorate the effects of high-phosphorus intake on cardiovascular risk (Kircelli et al., 2012). Magnesium homeostasis and food sources are discussed elsewhere (refer to Essentials III: Nutrients for Bone Structure and Calcification: Magnesium. Further information can be accessed at https://ods.od.nih.gov/factsheets/Magnesium-HealthProfessional/). Observational studies suggested an inverse relationship between serum magnesium concentrations and arterial calcification as well as with cardiovascular and all-cause mortality, not only in renal patients (de Roij van Zuijdewijn et al., 2015), but nonuremic populations as well (Peacock et al., 2010).

Magnesium intake below the recommended intake (320 mg/day female; 420 mg/day male) was associated with coronary artery calcification in 2695 participants of the Framingham Heart Study who were free of CVD and underwent multidetector computed tomography (MDCT) of the heart and abdomen (Hruby et al., 2014). Intracellular Mg^{2+} deficiency alters vascular function in multiple ways including promotion of inflammation, oxidative stress, and calcification. Mg^{2+} increases eNOS producing vasodilation with NO and suppresses the secretion of vasoconstrictor endothelin 1 as reviewed by de Baaij et al. (2015). Mg^{2+} deficiency in the endothelium promotes a state of permanent inflammation, associated with increased NFκB activity. An in vitro model using vascular bovine smooth muscle cells revealed that magnesium prevented phosphate-induced calcium deposition and inhibited alkaline phosphatase activity (Kircelli et al., 2012). It also reduced differentiation of smooth muscle cells into osteoblast-like cells and prevented cell damage (apoptosis). This model suggests that magnesium reduces arterial calcification by multiple molecular mechanisms. In addition, dietary magnesium as well as calcium sources are known to bind ingested phosphorous, thereby reducing phosphorous absorption from the gut lumen.

6.9 PUBLISHED DIET RECOMMENDATIONS FOR CARDIOVASCULAR RISK REDUCTION

6.9.1 2013 AHA/ACC Guideline on Lifestyle Management to Reduce Cardiovascular Risk

The American College of Cardiology/American Heart Association Task Force on Practice Guideline reviewed evidence on lifestyle management to reduce

TABLE 6.3   Summary: 2013 AHA/ACC Lifestyle Management to Reduce Cardiovascular Risk

Recommendations to Lower LDL Cholesterol	Strength of Evidence
1. Consume a diet pattern that includes vegetables, fruits, whole grains, low-fat dairy products, poultry, fish, legumes, nontropical vegetable oils	Strong (A)
2. Aim for a dietary pattern that achieves 5–6% of calories from saturated fats	Strong (A)
3. Reduce percent of calories from *trans*-fats	Strong (A)
RECOMMENDATIONS TO LOWER BP	
1. Consume a dietary pattern that emphasizes intake of vegetables, fruits, and whole grains; includes low-fat dairy products, poultry, fish, legumes, nontropical vegetable oils, and nuts	Strong (A)
2. Limit intake of sweets, sugar-sweetened beverages, and red meats	
3. Adapt this dietary pattern to appropriate calorie requirements, personal and cultural food preferences, and nutrition therapy for other medical conditions (including diabetes)	
4. Achieve this pattern by following plans such as the DASH dietary pattern, the USDA Food Pattern, or the AHA Diet	
5. Lower sodium intake	Strong (A)
6. a. Consume no more than 2400 mg of sodium/day	Moderate (B)
b. Further reduction of sodium intake to 1500 mg/day can result in even greater reduction in BP	
c. Even without achieving these goals, reducing sodium intake by at least 1000 mg/day lowers BP	
7. Combine DASH diet pattern with lowered sodium intake	Strong (A)
RECOMMENDATIONS FOR PHYSICAL ACTIVITY	
1. Lipid lowering	Strong (A)
In general, advise adults to engage in aerobic physical activity to reduce LDL-C and non-HDL-C: 3–4 sessions per week, lasting on average 40 minute per session, and involving moderate- to vigorous-intensity physical activity	
2. Blood Pressure Lowering	Strong (A)
In general, advise adults to engage in aerobic physical activity to lower BP: 3–4 sessions per week, lasting on average 40 minute per session, and involving moderate- to vigorous-intensity physical activity	

ACC, American College of Cardiology; AHA, American Heart Association; BP, blood pressure; DASH, Dietary Approaches to Stop Hypertension; HDL-C, high-density lipoprotein cholesterol; LDL-C, low-density lipoprotein cholesterol; NHLBI, National Heart, Lung, and Blood Institute; USDA, US Department of Agriculture.

cardiovascular risk (Eckel et al., 2014). A summary of the recommendations are listed in Table 6.3. In addition to the recommendations, the task force identified the following areas in need of further study:

- Interaction between dietary modification and statin treatment;
- Relative effects of saturated fats, monounsaturated fatty acids, polyunsaturated fatty acids, trans fatty acids, omega-3 fatty acids, and types of carbohydrates on lipids, inflammation, microbiome, and other newer potential CVD risk factors;
- Relative effects of naturally occurring fiber (cereal (whole grains) and vegetable/fruit) and supplemental fiber on lipids, inflammation, microbiome, and other newer potential CVD risk factors;
- Effects of dietary cholesterol on LDL-C and HDL-C over the current ranges of cholesterol and saturated-fat intakes (5th and 95th percentiles);
- Effects of minerals in combination (other than sodium) on BP;

- HDL function in studies that modify HDL-C by changes in diet;
- Is the minimal effect of dietary carbohydrate on plasma triglycerides harmful?
- The effect of sodium reduction in patients with diabetes, heart failure, and CKD;
- Effect of dietary pattern and sodium intake in adults taking BP-lowering or lipid-lowering medications (eg, effects on BP/lipids, achieving BP/lipid goals, medication needs/costs, outcomes);
- Effect of dietary pattern and sodium intake in adults with CVD (eg, after MI; after stroke; with coronary artery disease, heart failure, or CKD);
- Strategies for effectively (and cost-effectively) implementing these evidence-based recommendations:
  - How can primary care providers, health systems, public health agencies, local and federal government, community organizations, and other stakeholders help patients adopt these diet and sodium intake recommendations?

- Increased understanding of racial/ethnic/
socioeconomic factors that may influence:
  - effect of dietary pattern and sodium on BP and
lipids
  - adoption of diet/sodium recommendations
  - method of diet assessment.

Evidence from major clinical trials has demonstrated
the efficacy of nutrition practices in "curing" diseases,
slowing disease progression, and markedly decreasing
risk of chronic diseases to a similar extent as pharmacologic therapy (Kris-Etherton, 2009). Several areas of
controversy as to the ideal recommendations remain,
and required continued investigation. For example:

- The ideal dietary macronutrient composition to
reduce atherogenic dyslipidemia has not been
clarified. Results from the randomized OMNIHeart
trial revealed that partial substitution of the
low-fat/high-carbohydrate diet with either protein
or monounsaturated fat could lower blood pressure,
improve lipid levels, and reduce cardiovascular risk
(Appel et al., 2005).
- A recent review (Siri-Tarino et al., 2015) suggests
that the recommendation to reduce saturated fat
(SF) intake must be evaluated in the context of
replacement by other macronutrients. The authors
cite evidence from clinical trials that replaced SF
with polyunsaturated or monounsaturated fat and
reported lower LDL but also lower HDL and
generally show a reduction or no change in CVD
events. Replacement of SF with carbohydrate,
especially refined carbohydrate, exacerbated
atherogenic dyslipidemia (increased triglycerides
and small LDL particles; reduced HDL cholesterol),
and potentiated insulin resistance and obesity
(Mangravite et al., 2011). Since high triglycerides and
low HDL have been implicated as major risk factors
for atherosclerosis, these authors suggested that
dietary modification should emphasize limitation of
refined carbohydrate intakes; this change is
mediated at least in part through effects on
atherogenic dyslipidemia, a cluster of traits
including elevation of small, dense LDL particles.
- The effects of dietary SFs on insulin sensitivity,
inflammation, vascular function, and thrombosis are
less clear. It is also likely that SF from different
protein sources modify AS differently. A clinical
trial comparing high-fat diets from cheese or meat
sources to a low-fat control diet in which the fat
calories were isocalorically substituted with
carbohydrate was conducted in 14 healthy
postmenopausal women (Thorning et al., 2015).
Both high-SF diets resulted in higher HDL
cholesterol and Apo A-1 levels, suggesting the
high-SFA diets might be less atherogenic than the

low-fat, high-carbohydrate control diet. On the other
hand, it is possible that other components of the
diets (eg, fiber content) may have biased the results.
- The poor compliance with the low-fat DASH diet
prompted a larger study to test the effect of
substituting full-fat dairy products for non/low-fat
dairy products (HF-DASH) in a randomized
crossover trial with 36 participants (Chiu et al.,
2016). The HF-DASH diet with 14% SFA and a
reduction of 12% CHO, largely from sugars, was
tested against the standard DASH diet and a control
diet. Compared with the control diet, the HF-DASH
diet lowered blood pressure to the same extent as
the DASH diet but also reduced plasma triglyceride
and VLDL concentrations without significantly
increasing LDL cholesterol. The DASH diet lowered
HDL cholesterol and apo A-1 compared with the
control diet; the HF-DASH diet reduced HDL to
a lesser extent but did not reach significance.

A recent review of clinical trials concluded that
current evidence does not clearly support cardiovascular guidelines that encourage high consumption
of polyunsaturated fatty acids and low consumption of
total saturated fats (Chowdhury et al., 2014). In fact,
there is growing evidence that SF in the context of
dairy foods, particularly fermented dairy products,
have neutral or inverse associations with CVD. Overall
dietary patterns emphasizing vegetables, fish, nuts,
and whole versus processed grains form the basis of
heart-healthy eating and should supersede a focus on
macronutrient composition. Furthermore, differences
in metabolic response to refined carbohydrates are also
well documented. Fructose is known to raise triglyceride levels in animal studies. A recent clinical trial
(Stanhope et al., 2011) revealed that fructose, but not
glucose, raised triglycerides, LDL, and Apo B in healthy
young human subjects. The authors suggested that fructose, alone or in combination with glucose as sucrose,
could aggravate dyslipidemia in susceptible patients.

6.10 FUTURE RESEARCH NEEDS

The optimal macronutrient content of a diet that not
only optimizes blood lipid levels but achieves ideal
lipid particle composition is yet unclear. It will also
be important to determine whether macronutrient
recommendations should change with lifecycle phases;
for example, is the recommendation for the growing
child the same as for the elderly individual?

Additionally, since whole-food contains nutrients
(magnesium, copper, folate) as well as non-nutritive
bioactive components (flavanols, procyanidins, terpenes,
organosulfates) with potential to modify risk for

vascular disease, more research is needed to identify the utility of including specific foods in dietary recommendations. In addition, dietary components are acted upon, and modify the extensive gut microbiota and influence the permeability of the intestinal barrier. Translocation of bacterial products can have deleterious influences on cardiovascular risk. For example, the potential for dietary iron and inorganic phosphorus to cause vascular injury and calcification in normal patients, as well as those with kidney disease, should be clarified. Studies to confirm the influence of nutrients on endothelial integrity and dysfunction are needed, specifically the utility of increasing food sources of nitrate and organosulfates to enhance NO and H_2S availability for vasodilation. Similarly, mechanistically focused research on the efficacy and appropriate form of folate supplementation to reduce HHcy are required, as is an examination of the influence of age- and obesity-related sarcopenia on SAA malnutrition and dysregulation of methionine homeostasis. The connection between homocysteine elevation and intracellular copper efflux is also important, since intracellular copper is essential for ATP production as well as antioxidant protection as Zn-Cu SOD.

Of particular interest is the possibility that antimicrobial EOs could specifically alter the bacterial populations that convert dietary carnitine and choline to the toxic trimethylamine (TMA) metabolite, implicated in risk for atherosclerosis. Since these molecules are essential nutrients, it is important to study means through which microbial production of TMA can be reduced.

Until strong clinical evidence has been generated that provides evidence for targeted dietary guidelines, it would appear prudent to substitute dark green leafy and root vegetables for less nutrient-dense, processed foods and to increase consumption of sulfur-containing garlic, onions, and cruciferous vegetables. Avoidance of processed foods and adherence to a Mediterranean or DASH diet pattern that emphasizes fatty fish and olive oil, minimizes hemoglobin-containing red meat and complements the meal with a glass of red wine has been associated with primary prevention and reduced disease progression following a cardiovascular event.

References

Abbasian, N., Burton, J.O., et al., 2015. Hyperphosphatemia, phosphoprotein phosphatases, and microparticle release in vascular endothelial cells. J. Am. Soc. Nephrol. 26 (9), 2152–2162.

Ackermann, M., Liebhaber, S., et al., 2015. Lost in translation: pluripotent stem cell-derived hematopoiesis. EMBO Mol. Med. 7 (11), 1388–1402.

Allayee, H., Hazen, S.L., 2015. Contribution of gut bacteria to lipid levels: another metabolic role for microbes? Circ. Res. 117 (9), 750–754.

Allison, M.A., Criqui, M.H., et al., 2004. Patterns and risk factors for systemic calcified atherosclerosis. Arterioscler. Thromb. Vasc. Biol. 24 (2), 331–336.

Amagase, H., 2006. Clarifying the real bioactive constituents of garlic. J. Nutr. 136 (3), 716S–725S.

Appel, L.J., Sacks, F.M., et al., 2005. Effects of protein, monounsaturated fat, and carbohydrate intake on blood pressure and serum lipids: results of the OmniHeart randomized trial. JAMA. 294 (19), 2455–2464.

Bahat, G., İlhan, B., 2016. Sarcopenia and the cardiometabolic syndrome: a narrative review. Eur. Geriatr. Med. 7 (3), 220–223.

Bailey, S.W., Ayling, J.E., 2009. The extremely slow and variable activity of dihydrofolate reductase in human liver and its implications for high folic acid intake. Proc. Natl. Acad. Sci. 106 (36), 15424–15429.

Bassolé, I.H.N., Juliani, H.R., 2012. Essential oils in combination and their antimicrobial properties. Molecules. 17 (4), 3989.

Behrendt, D., Ganz, P., 2002. Endothelial function: from vascular biology to clinical applications. Am. J. Cardiol. 90 (10, Supplement 3), L40–L48.

Bitzer, Z.T., Glisan, S.L., et al., 2015. Cocoa procyanidins with different degrees of polymerization possess distinct activities in models of colonic inflammation. J. Nutr. Biochem. 26 (8), 827–831.

Böger, R.H., 2014. The pharmacodynamics of L-arginine. Altern. Ther. Health. Med. 20 (3), 48–54.

Boger, R.H., Bode-Boger, S.M., 2001. The clinical pharmacology of L-arginine. Annu. Rev. Pharmacol. Toxicol. 41 (03621642), 79–99.

Bolland, M.J., Grey, A., et al., 2013. Calcium supplements and cardiovascular risk: 5 years on. Ther. Adv. Drug Saf. 4 (5), 199–210.

Bories, G., Colin, S., et al., 2013. Liver X receptor activation stimulates iron export in human alternative macrophages. Circ. Res. 113 (11), 1196–1205.

Borlinghaus, J., Albrecht, F., et al., 2014. Allicin: chemistry and biological properties. Molecules. 19 (8), 12591.

Bristow, S.M., Gamble, G.D., et al., 2015a. Acute effects of calcium supplements on blood pressure and blood coagulation: secondary analysis of a randomised controlled trial in post-menopausal women. Br. J. Nutr. 114 (11), 1868–1874.

Bristow, S.M., Gamble, G.D., et al., 2015b. Acute effects of calcium citrate with or without a meal, calcium-fortified juice and a dairy product meal on serum calcium and phosphate: a randomised cross-over trial. Br. J. Nutr. 113 (10), 1585–1594.

Bryan, N.S., Alexander, D.D., et al., 2012. Ingested nitrate and nitrite and stomach cancer risk: an updated review. Food Chem. Toxicol. 50 (10), 3646–3665.

Budoff, M., 2006. Aged garlic extract retards progression of coronary artery calcification. J. Nutr. 136 (3), 741S–744S.

Cheng, P.T., Grabher, J.J., et al., 1988. Effects of magnesium on calcium-phosphate formation. Magnesium. 7 (3), 123–132.

Chiu, S., Bergeron, N., et al., 2016. Comparison of the DASH (Dietary Approaches to Stop Hypertension) diet and a higher-fat DASH diet on blood pressure and lipids and lipoproteins: a randomized controlled trial. Am. J. Clin. Nutr. 103 (2), 341–347.

Cholesterol Homocysteine Collaborative, Group, 2010. Effects of homocysteine-lowering with folic acid plus vitamin b12 vs placebo on mortality and major morbidity in myocardial infarction survivors: a randomized trial. JAMA. 303 (24), 2486–2494.

Chowdhury, R., Warnakula, S., et al., 2014. Association of dietary, circulating, and supplement fatty acids with coronary risk a systematic review and meta-analysis. Ann. Intern. Med. 160 (6), 398–406.

Clarke, R., Halsey, J., et al., 2010. Effects of lowering homocysteine levels with b vitamins on cardiovascular disease, cancer, and cause-specific mortality: meta-analysis of 8 randomized trials involving 37 485 individuals. Arch. Intern. Med. 170 (18), 1622–1631.

Corder, R., Mullen, W., et al., 2006. Oenology: red wine procyanidins and vascular health. Nature. 444 (7119), 566.

Corti, R., Flammer, A.J., et al., 2009. Cocoa and cardiovascular health. Circulation. 119 (10), 1433–1441.

Crabtree, M.J., Tatham, A.L., et al., 2009. Critical role for tetrahydrobiopterin recycling by dihydrofolate reductase in regulation of endothelial nitric-oxide synthase coupling: relative importance of the de novo biopterin synthesis versus salvage pathways. J. Biol. Chem. 284 (41), 28128–28136.

Crabtree, M.J., Brixey, R., et al., 2013. Integrated redox sensor and effector functions for tetrahydrobiopterin- and glutathionylation-dependent endothelial nitric-oxide synthase uncoupling. J. Biol. Chem. 288 (1), 561–569.

Curtiss, L.K., Tobias, P.S., 2009. Emerging role of toll-like receptors in atherosclerosis. J. Lipid. Res. 50 (Suppl), S340–S345.

Cusi, K., Maezono, K., et al., 2000. Insulin resistance differentially affects the PI 3-kinase– and MAP kinase–mediated signaling in human muscle. J. Clin. Invest. 105 (3), 311–320.

Dalle-Donne, I., Rossi, R., et al., 2009. Protein S-glutathionylation: a regulatory device from bacteria to humans. Trends Biochem. Sci. 34 (2), 85–96.

Danziger, J., 2008. Vitamin K-dependent proteins, warfarin, and vascular calcification. Clin. J. Am. Soc. Nephrol. 3 (5), 1504–1510.

Davignon, J., Ganz, P., 2004. Role of endothelial dysfunction in atherosclerosis. Circulation. 109 (23 suppl 1), III-27–III-32.

de Baaij, J.H.F., Hoenderop, J.G.J., et al., 2015. Magnesium in man: implications for health and disease. Physiol. Rev. 95 (1), 1–46.

de Nigris, F., Williams-Ignarro, S., et al., 2007. Effects of a pomegranate fruit extract rich in punicalagin on oxidation-sensitive genes and eNOS activity at sites of perturbed shear stress and atherogenesis. Cardiovasc. Res. 73 (2), 414–423.

de Roij van Zuijdewijn, C.L.M., Grooteman, M.P.C., et al., 2015. Serum magnesium and sudden death in European hemodialysis patients. PLoS ONE. 10 (11), e0143104.

De Tullio, M.C., 2012. Beyond the antioxidant: the double life of vitamin C. Subcell Biochem. 56, 49–65.

Dickinson-Copeland, C.M., Wilson, N.O., et al., 2015. Heme-mediated induction of CXCL10 and depletion of CD34 + progenitor cells is toll-like receptor 4 dependent. PLoS ONE. 10 (11), e0142328.

Dimmeler, S., Zeiher, A.M., 2004. Vascular repair by circulating endothelial progenitor cells: the missing link in atherosclerosis? J. Mol. Med. 82 (10), 671–677.

Ditadi, A., Sturgeon, C.M., et al., 2015. Human definitive haemogenic endothelium and arterial vascular endothelium represent distinct lineages. Nat. Cell Biol. 17 (5), 580–591.

Dong, D., Wang, B., et al., 2013. Disturbance of copper homeostasis is a mechanism for homocysteine-induced vascular endothelial cell injury. PLoS ONE. 8 (10), e76209.

Duboc, H., Taché, Y., et al., 2014. The bile acid TGR5 membrane receptor: from basic research to clinical application. Dig. Liver Dis. 46 (4), 302–312.

Dutta, P., Courties, G., et al., 2012. Myocardial infarction accelerates atherosclerosis. Nature. 487 (7407), 325–329.

Eckel, R.H., Jakicic, J.M., et al., 2014. 2013 AHA/ACC guideline on lifestyle management to reduce cardiovascular risk: a report of the American College of Cardiology/American Heart Association task force on practice guidelines. J. Am. Coll. Cardiol. 63 (25), 2960–2984.

Emmett, M., 2008. What does serum fibroblast growth factor 23 do in hemodialysis patients? Kidney. Int. 73 (1), 3–5.

Estruch, R., Ros, E., et al., 2013. Primary prevention of cardiovascular disease with a Mediterranean diet. N Engl. J. Med. 368 (14), 1279–1290.

Fernández-Real, J.M., Manco, M., 2014. Effects of iron overload on chronic metabolic diseases. Lancet Diabetes Endocrinol. 2 (6), 513–526.

Filocamo, A., Nueno-Palop, C., et al., 2012. Effect of garlic powder on the growth of commensal bacteria from the gastrointestinal tract. Phytomedicine. 19 (8–9), 707–711.

Förstermann, U., Münzel, T., 2006. Endothelial nitric oxide synthase in vascular disease: from marvel to menace. Circulation. 113 (13), 1708–1714.

Frontini, M.J., O'Neil, C., et al., 2009. Lipid incorporation inhibits Src-dependent assembly of fibronectin and type i collagen by vascular smooth muscle cells. Circ. Res. 104 (7), 832–841.

Fu, J., Bonder, M.J., et al., 2015. The gut microbiome contributes to a substantial proportion of the variation in blood lipids. Circ. Res. 117 (9), 817–824.

Govoni, M., Jansson, E.Å., et al., 2008. The increase in plasma nitrite after a dietary nitrate load is markedly attenuated by an antibacterial mouthwash. Nitric Oxide. 19 (4), 333–337.

Gregory, J.C., Buffa, J.A., et al., 2015. Transmission of atherosclerosis susceptibility with gut microbial transplantation. J. Biol. Chem. 290 (9), 5647–5660.

Gurda, D., Handschuh, L., et al., 2015. Homocysteine thiolactone and N-homocysteinylated protein induce pro-atherogenic changes in gene expression in human vascular endothelial cells. Amino Acids. 47 (7), 1319–1339.

Gutiérrez, O.M., 2013. The connection between dietary phosphorus, cardiovascular disease, and mortality: where we stand and what we need to know. Adv. Nutr.: Int. Rev. J. 4 (6), 723–729.

Hamed, S., Brenner, B., et al., 2011. Nitric oxide: a key factor behind the dysfunctionality of endothelial progenitor cells in diabetes mellitus type-2. Cardiovasc. Res. 91 (1), 9–15.

Hasty, A.H., Yvan-Charvet, L., 2013. Liver X receptor α-dependent iron handling in M2 macrophages: the missing link between cholesterol and intraplaque hemorrhage?. Circ. Res. 113 (11), 1182–1185.

Hauschka, P.V., Lian, J.B., et al., 1989. Osteocalcin and matrix Gla protein: vitamin K-dependent proteins in bone. Physiol. Rev. 69 (3), 990–1047.

Heiss, C., Keen, C.L., et al., 2010. Flavanols and cardiovascular disease prevention. Eur. Heart. J. 31 (21), 2583–2592.

Henderson, J.S., Joyce, R.A., et al., 2007. Chemical and archaeological evidence for the earliest cacao beverages. Proc. Natl. Acad. Sci. 104 (48), 18937–18940.

House, M.G., Kohlmeier, L., et al., 1997. Expression of an extracellular calcium-sensing receptor in human and mouse bone marrow cells. J. Bone Miner. Res. 12 (12), 1959–1970.

Howitz, K.T., Bitterman, K.J., et al., 2003. Small molecule activators of sirtuins extend Saccharomyces cerevisiae lifespan. Nature. 425 (6954), 191–196.

Hruby, A., O'Donnell, C.J., et al., 2014. Magnesium intake is inversely associated with coronary artery calcification: the Framingham Heart Study. JACC Cardiovasc. Imaging. 7 (1), 59–69.

Huang, A., Vita, J.A., et al., 2000. Ascorbic acid enhances endothelial nitric-oxide synthase activity by increasing intracellular tetrahydrobiopterin. J. Biol. Chem. 275 (23), 17399–17406.

Huang, P.-H., Chen, Y.-H., et al., 2010. Intake of red wine increases the number and functional capacity of circulating endothelial progenitor cells by enhancing nitric oxide bioavailability. Arterioscler. Thromb. Vasc. Biol. 30 (4), 869–877.

Huennekens, F.M., 1994. The methotrexate story: a paradigm for development of cancer chemotherapeutic agents. Adv. Enzyme Regul. 34 (0), 397–419.

Huff, M.W., Pickering, J.G., 2015. Can a vascular smooth muscle–derived foam-cell really change its spots?. Arterioscler. Thromb. Vasc. Biol. 35 (3), 492–495.

Ignarro, L.J., Balestrieri, M.L., et al., 2007. Nutrition, physical activity, and cardiovascular disease: an update. Cardiovasc. Res. 73 (2), 326–340.

Ingenbleek, Y., McCully, K.S., 2012. Vegetarianism produces subclinical malnutrition, hyperhomocysteinemia and atherogenesis. Nutrition. 28 (2), 148–153.

Jahnen-Dechent, W., Heiss, A., et al., 2011. Fetuin-A regulation of calcified matrix metabolism. Circ. Res. 108 (12), 1494−1509.

Jahoor, F., Badaloo, A., et al., 2006. Sulfur amino acid metabolism in children with severe childhood undernutrition: cysteine kinetics. Am. J. Clin. Nutr. 84 (6), 1393−1399.

Jakubowski, H., 2007. The molecular basis of homocysteine thiolactone-mediated vascular disease. Clin. Chem. Lab. Med. 45 (12), 1704.

Jover, E., Marín, F., et al., 2013. Atherosclerosis and thromboembolic risk in atrial fibrillation: focus on peripheral vascular disease. Ann. Med. 45 (3), 274−290.

Kapustin, A.N., Davies, J.D., et al., 2011. Calcium regulates key components of vascular smooth muscle cell−derived matrix vesicles to enhance mineralization. Circ. Res. 109 (1), e1−e12.

Kapustin, A.N., Chatrou, M.L.L., et al., 2015. Vascular smooth muscle cell calcification is mediated by regulated exosome secretion. Circ. Res. 116 (8), 1312−1323.

Katusic, Z.S., 2001. Vascular endothelial dysfunction: does tetrahydrobiopterin play a role? Am. J. Physiol. Heart Circ. Physiol. 281 (3), H981−H986.

Kevil, C.G., Kolluru, G.K., et al., 2011. Inorganic nitrite therapy: historical perspective and future directions. Free Radic. Biol. Med. 51 (3), 576−593.

Khan, N.Q., Patel, B., et al., 2015. Regulation of vascular endothelial function by red wine procyanidins: implications for cardiovascular health. Tetrahedron. 71 (20), 3059−3065.

King, A.L., Polhemus, D.J., et al., 2014. Hydrogen sulfide cytoprotective signaling is endothelial nitric oxide synthase-nitric oxide dependent. Proc. Natl. Acad. Sci. 111 (8), 3182−3187.

Kircelli, F., Peter, M.E., et al., 2012. Magnesium reduces calcification in bovine vascular smooth muscle cells in a dose-dependent manner. Nephrol. Dial. Transplant. 27 (2), 514−521.

Kobayashi, J., Ohtake, K., et al., 2015. NO-rich diet for lifestyle-related diseases. Nutrients. 7 (6), 4911.

Koeth, R.A., Wang, Z.Z., et al., 2013. Intestinal microbiota metabolism of l-carnitine, a nutrient in red meat, promotes atherosclerosis. Nat. Med. 19 (5), 576−585.

Konukoglu, D., Fırtına, S., et al., 2008. The relationship between plasma asymmetrical dimethyl-l-arginine and inflammation and adhesion molecule levels in subjects with normal, impaired, and diabetic glucose tolerance. Metabolism. 57 (1), 110−115.

Kris-Etherton, P.M., 2009. Adherence to dietary guidelines: benefits on atherosclerosis progression. Am. J. Clin. Nutr. 90 (1), 13−14.

Kuhnle, G., Bingham, S., 2007. Dietary meat, endogenous nitrosation and colorectal cancer. Biochem. Soc. Trans. 35 (5), 1355−1357.

Kwak, S.M., Kim, J.S., et al., 2014. Dietary intake of calcium and phosphorus and serum concentration in relation to the risk of coronary artery calcification in asymptomatic adults. Arterioscler. Thromb. Vasc. Biol. 34 (8), 1763−1769.

Lei, Y.-P., Chen, H.-W., et al., 2008. Diallyl disulfide and diallyl trisulfide suppress oxidized LDL−induced vascular cell adhesion molecule and E-selectin expression through protein kinase A− and B−dependent signaling pathways. J. Nutr. 138 (6), 996−1003.

Loscalzo, J., 1996. The oxidant stress of hyperhomocyst(e)inemia. J. Clin. Invest. 98 (1), 5−7.

Lundberg, J.O., Weitzberg, E., 2005. NO generation from nitrite and its role in vascular control. Arterioscler. Thromb. Vasc. Biol. 25 (5), 915−922.

Lundberg, J.O., Weitzberg, E., 2010. The biological role of nitrate and nitrite: the times they are a-changing'. Nitric Oxide. 22 (2), 61−63.

Luo, G., Ducy, P., et al., 1997. Spontaneous calcification of arteries and cartilage in mice lacking matrix GLA protein. Nature. 386 (6620), 78−81.

Mangravite, L.M., Chiu, S., et al., 2011. Changes in atherogenic dyslipidemia induced by carbohydrate restriction in men are dependent on dietary protein source. J. Nutr. 141 (12), 2180−2185.

Mason, J.B., 2009. Folate, cancer risk, and the Greek god, Proteus: a tale of two chameleons. Nutr. Rev. 67 (4), 206−212.

Matsushima, S., Sadoshima, J., 2015. The role of sirtuins in cardiac disease. Am. J. Physiol. Heart Circ. Physiol. 309 (9), H1375−H1389.

McCully, K., 1969. Vascular pathology of homocysteinemia: implications for the pathogenesis of arteriosclerosis. Am. J. Pathol. 56, 111−128.

McCully, K.S., 1996. Homocysteine and vascular disease. Nat. Med. 2 (4), 386−389.

McNally, B., Griffin, J.L., et al., 2016. Dietary inorganic nitrate: from villain to hero in metabolic disease? Mol. Nutr. Food. Res. 60 (1), 67−78.

Micronutrients, 2001. Dietary Reference Intakes for Vitamin A, Vitamin K, Arsenic, Boron, Chromium, Copper, Iodine, Iron, Manganese, Molybdenum, Nickel, Silicon, Vanadium, and Zinc. I. Medicine. National Academy Press, Washington, DC.

Miller, Y.I., Choi, S.-H., et al., 2011. Oxidation-specific epitopes are danger associated molecular patterns recognized by pattern recognition receptors of innate immunity. Circ. Res. 108 (2), 235−248.

Moens, A.L., Vrints, C.J., et al., 2008. Mechanisms and potential therapeutic targets for folic acid in cardiovascular disease. Am. J. Physiol. Heart Circ. Physiol. 294 (5), H1971−H1977.

Moore, K.J., Tabas, I., 2011. The cellular biology of macrophages in atherosclerosis. Cell. 145 (3), 341−355.

Moore, K., Sheedy, F., et al., 2013. Macrophages in atherosclerosis: a dynamic balance. Nat. Rev. Immunol. 13 (10), 709−721.

Mudd, S.H., Brosnan, J.T., et al., 2007. Methyl balance and transmethylation fluxes in humans. Am. J. Clin. Nutr. 85 (1), 19−25.

Mullen, W., Marks, S.C., et al., 2007. Evaluation of phenolic compounds in commercial fruit juices and fruit drinks. J. Agric. Food Chem. 55 (8), 3148−3157.

Murshed, M., Schinke, T., et al., 2004. Extracellular matrix mineralization is regulated locally; different roles of two gla-containing proteins. J. Cell Biol. 165 (5), 625−630.

Nagy, E., Eaton, J.W., et al., 2010. Red cells, hemoglobin, heme, iron, and atherogenesis. Arterioscler. Thromb. Vasc. Biol. 30 (7), 1347−1353.

Nakagawa, T., Guarente, L., 2011. Sirtuins at a glance. J. Cell Sci. 124 (6), 833−838.

New, S.E.P., Aikawa, E., 2011. Cardiovascular calcification; an inflammatory disease. Circ. J. 75 (6), 1305−1313.

Nicoll, R., Howard, J., et al., 2015. A review of the effect of diet on cardiovascular calcification. Int. J. Mol. Sci. 16 (4), 8861.

Park, S.-J., Ahmad, F., et al., 2012. Resveratrol ameliorates aging-related metabolic phenotypes by inhibiting cAMP phosphodiesterases. Cell. 148 (3), 421−433.

Peacock, J.M., Ohira, T., et al., 2010. Serum magnesium and risk of sudden cardiac death in the Atherosclerosis Risk in Communities (ARIC) Study. Am. Heart J. 160 (3), 464−470.

Phillips-Eakley, A.K., McKenney-Drake, M.L., et al., 2015. Effect of high-calcium diet on coronary artery disease in Ossabaw miniature swine with metabolic syndrome. J. Am. Heart Assoc. 4 (8), e001620.

Pols, T.W.H., 2014. TGR5 in inflammation and cardiovascular disease. Biochem. Soc. Trans. 42 (2), 244−249.

Pols, T.W.H., Noriega, L.G., et al., 2011. The bile acid membrane receptor TGR5 as an emerging target in metabolism and inflammation. J. Hepatol. 54 (6), 1263−1272.

Rafikov, R., Fonseca, F.V., et al., 2011. eNOS activation and NO function: structural motifs responsible for the posttranslational control of endothelial nitric oxide synthase activity. J. Endocrinol. 210 (3), 271−284.

Reid, I.R., Gamble, G.D., et al., 2016. Circulating calcium concentrations, vascular disease and mortality: a systematic review. J. Intern. Med. 279 (6), 524−540.

Rennenberg, R.J.M.W., Kessels, A.G.H., et al., 2009. Vascular calcifications as a marker of increased cardiovascular risk: a meta-analysis. Vasc. Health Risk Manag. 5, 185–197.

Reynolds, J.L., Joannides, A.J., et al., 2004. Human vascular smooth muscle cells undergo vesicle-mediated calcification in response to changes in extracellular calcium and phosphate concentrations: a potential mechanism for accelerated vascular calcification in ESRD. J. Am. Soc. Nephrol. 15 (11), 2857–2867.

Ross, R., Glomset, J.A., 1973. Atherosclerosis and the arterial smooth muscle cell. Science. 180 (4093), 1332–1339.

Rubio-Guerra, A.F., Vargas-Robles, H., et al., 2010. Is tetrahydro-biopterin a therapeutic option in diabetic hypertensive patients? Integr. Blood Press. Control. 3, 125–132.

Saad, M., Abdelkhalek, T., et al., 2015. Insights into the molecular mechanisms of diabetes-induced endothelial dysfunction: focus on oxidative stress and endothelial progenitor cells. Endocrine. 50 (3), 537–567.

Schlieper, G., Schurgers, L., et al., 2015. Vascular calcification in chronic kidney disease: an update. Nephrol. Dial. Transplant. 31 (1), 31–39.

Schurgers, L.J., Uitto, J., et al., 2013. Vitamin K-dependent carboxylation of matrix Gla-protein: a crucial switch to control ectopic mineralization. Trends Mol. Med. 19 (4), 217–226.

Scialla, J.J., Wolf, M., 2014. Roles of phosphate and fibroblast growth factor 23 in cardiovascular disease. Nat. Rev. Nephrol. 10 (5), 268–278.

Shearer, M.J., Fu, X., et al., 2012. Vitamin K nutrition, metabolism, and requirements: current concepts and future research. Adv. Nutr. Int. Rev. J. 3 (2), 182–195.

Shuto, E., Taketani, Y., et al., 2009. Dietary phosphorus acutely impairs endothelial function. J. Am. Soc. Nephrol. 20 (7), 1504–1512.

Siri-Tarino, P.W., Chiu, S., et al., 2015. Saturated fats versus polyunsaturated fats versus carbohydrates for cardiovascular disease prevention and treatment. Annu. Rev. Nutr. 35, 517–543.

Smith, A.D., Kim, Y.-I., et al., 2008. Is folic acid good for everyone? Am. J. Clin. Nutr. 87 (3), 517–533.

Stanhewicz, A.E., Alexander, L.M., et al., 2015. Folic acid supplementation improves microvascular function in older adults through nitric oxide-dependent mechanisms. Clin. Sci. 129 (2), 159–167.

Stanhope, K.L., Bremer, A.A., et al., 2011. Consumption of fructose and high fructose corn syrup increase postprandial triglycerides, LDL-cholesterol, and apolipoprotein-B in young men and women. J. Clin. Endocrinol. Metab. 96 (10), E1596–E1605.

Sullivan, J.L., 1981. Iron and the sex difference in heart disease risk. Lancet. 8233, 1293–1294.

Sullivan, J.L., 2007. Macrophage iron, hepcidin, and atherosclerotic plaque stability. Exp. Biol. Med. 232 (8), 1014–1020.

Tabas, I., García-Cardeña, G., et al., 2015. Recent insights into the cellular biology of atherosclerosis. J. Cell Biol. 209 (1), 13–22.

Tall, A.R., Yvan-Charvet, L., 2015. Cholesterol, inflammation and innate immunity. Nat. Rev. Immunol. 15 (2), 104–116.

Tanaka, J., Qiang, L., et al., 2009. Foxo1 links hyperglycemia to LDL oxidation and endothelial nitric oxide synthase dysfunction in vascular endothelial cells. Diabetes. 58 (10), 2344–2354.

Tang, W.H.W., Wang, Z., et al., 2013. Intestinal microbial metabolism of phosphatidylcholine and cardiovascular risk. N Eng. J. Med. 368 (17), 1575–1584.

Tennen, R.I., Michishita-Kioi, E., et al., 2012. Finding a Target for resveratrol. Cell. 148 (3), 387–389.

Theil, E.C., 2011. Iron homeostasis and nutritional iron deficiency. J. Nutr. 141 (4), 724S–728S.

Thomas, C., Pellicciari, R., et al., 2008. Targeting bile-acid signalling for metabolic diseases. Nat. Rev. Drug Discov. 7 (8), 678–693.

Thorning, T.K., Raziani, F., et al., 2015. Diets with high-fat cheese, high-fat meat, or carbohydrate on cardiovascular risk markers in overweight postmenopausal women: a randomized crossover trial. Am. J. Clin. Nutr. 102 (3), 573–581.

Thum, T., Fraccarollo, D., et al., 2007. Endothelial nitric oxide synthase uncoupling impairs endothelial progenitor cell mobilization and function in diabetes. Diabetes. 56 (3), 666–674.

Torsney, E., Xu, Q., 2011. Resident vascular progenitor cells. J. Mol. Cell Cardiol. 50 (2), 304–311.

Traka, M.H., Mithen, R.F., 2011. Plant science and human nutrition: challenges in assessing health-promoting properties of phytochemicals. Plant Cell Online. 23 (7), 2483–2497.

Tyson, K.L., Reynolds, J.L., et al., 2003. Osteo/chondrocytic transcription factors and their target genes exhibit distinct patterns of expression in human arterial calcification. Arterioscler. Thromb. Vasc. Biol. 23 (3), 489–494.

Undas, A., Brożek, J., et al., 2006. Plasma homocysteine affects fibrin clot permeability and resistance to lysis in human subjects. Arterioscler. Thromb. Vasc. Biol. 26 (6), 1397–1404.

Van Varik, B.J., Rennenberg, R.J., et al., 2012. Mechanisms of arterial remodeling: lessons from genetic diseases. Front. Genet. 3, 290.

Vazquez-Prieto, M.A., Miatello, R.M., 2010. Organosulfur compounds and cardiovascular disease. Mol. Aspects. Med. 31 (6), 540–545.

Vinchi, F., Muckenthaler, M.U., et al., 2014. Atherogenesis and iron: from epidemiology to cellular level. Front. Pharmacol. 5, 94.

Violi, F., Pignatelli, P., et al., 2010. Nutrition, supplements, and vitamins in platelet function and bleeding. Circulation. 121 (8), 1033–1044.

Wang, Z., Klipfell, E., et al., 2011. Gut flora metabolism of phosphatidylcholine promotes cardiovascular disease. Nature. 472 (7341), 57–63.

Watanabe, M., Houten, S.M., et al., 2006. Bile acids induce energy expenditure by promoting intracellular thyroid hormone activation. Nature. 439 (7075), 484–489.

Williams, T.M., Lisanti, M.P., 2004. The caveolin proteins. Genome. Biol. 5 (3), 214.

Wu, W.-K., Panyod, S., et al., 2015. Dietary allicin reduces transformation of L-carnitine to TMAO through impact on gut microbiota. J. Funct. Foods. 15, 408–417.

Yan, F., Wang, Y., et al., 2014. Nox4 and redox signaling mediate TGF-β-induced endothelial cell apoptosis and phenotypic switch. Cell Death Dis. 5 (1), e1010.

Young, V.R., Pellett, P.L., 1994. Plant proteins in relation to human protein and amino acid nutrition. Am. J. Clin. Nutr. 59 (5), 1203S–1212S.

Zhang, Y., Janssens, S.P., et al., 2011. Modulating endothelial nitric oxide synthase: a new cardiovascular therapeutic strategy. Am. J. Physiol. Heart Circ. Physiol. 301 (3), H634–H646.

Further Reading

Albert, et al., 2014. Chapter 22. Stem cells and tissue renewal: blood vessels, lymphatics and endothelial cells, Molecular Biology of the Cell, sixth ed. Garland Science, New York, NY.

Goldman, L., Schafer, A.I., 2012. Chapter 70. Atherosclerosis, thrombosis and vascular biology; Chapter 420. Peripheral neuropathy. Goldman Cecil Medicine. Elsevier Saunders, Philadelphia, PA.

Kumar, V., et al., 2014. Chapter 4. Hemodynamic disorders, thromboembolic disease and shock; Chapter 11. Blood vessels, Robbins and Cotran Pathologic Basis of Disease, ninth ed. Elsevier Saunders, Philadelphia, PA.

7

Diabetic Nephropathy, Chronic Kidney Disease

CHIEF COMPLAINT (TYPICAL PATIENT)

Patient complains of increasing anorexia and nausea over the past month, coupled with a 4 kg weight gain. She complains that her rings are too tight and she cannot wear her shoes because her feet have swollen. Her sleep is disturbed because she cannot catch her breath without three pillows, and she suffers from severe itchiness and muscle cramps.

MEDICAL HISTORY

A 48-year-old Hispanic woman had been followed in the clinic for more than 20 years. At age 22 she was diagnosed with type II diabetes, hypertension, and hyperlipidemia, and was initially treated with metformin, atorvastatin, and hydrochlorothiazide. On this regimen, she had maintained HbA1c <7.0%, low-density lipoprotein (LDL) cholesterol <100 mg/dL, and blood pressure <135/80. At age 40, her urine albumin increased to 54 mg/g creatinine and lisinopril was added to the regimen. At age 45 her serum creatinine increased to 4.56 mg/dL and her glomerular filtration rate (GFR) was estimated at 11 mL/minutes/1.73 m^2. Her HbA1c increased to 7.2% and she began to complain of bone pain and parasthesias in her feet. At this time, metformin was discontinued and she was started on glargine insulin. Blood sugars checked two to four times each day at home ranged from 150 to 250 mg/dL. She tries to walk 30 minutes each day, but finds she tires easily.

CURRENT MEDICATIONS

Lisinopril, hydrochlorothiazide, diltiazem, glargine, atorvastatin, kayexalate, and aspirin. She does not take nutrient supplements (too expensive), but has frequently resorted to additional over-the-counter pain medications for aches and pains.

PATIENT HISTORY

Mrs. S has worked in a local factory, but lost her job 2 years ago when the factory closed. She has been chronically tired and has not been able to find employment. She lives with her husband, his mother and her two children, aged 25 and 23. Without medical insurance, she has reduced her medications to make them go farther.

PHYSICAL EXAMINATION

- *Anthropometrics:* weight 115 kg (253 lb); BMI 42.8
- *Vital signs:* BP 220/80 mmHg; pulse 84
- *Retinal exam:* significant for panretinal photocoagulation changes with scattered dot hemorrhages and macular edema in both eyes
- *Heart, lung, and abdominal exams:* unremarkable
- *Skin:* dry and yellowish brown
- *Extremities:* muscle weakness; 3 + pitting edema to the knees, no cyanosis. Pulses in her feet were reduced, but her feet were warm and without ulcers
- *Neurologic:* pupils equal and reactive; intermittent deep tendon reflexes and plantar response absent
- *Renal:* estimated glomerular filtration rate (eGFR) 3–4 mL/minutes/1.73 m^2; 24-hour urine output <300 mL.

DIET HISTORY

- *General*
  - In the past, patient has been fond of "comfort foods" such as macaroni and cheese and chicken nuggets. She dislikes vegetables but usually drinks several glasses of fruit juice (apple) daily.
  - Recently, she has been unable to eat and has mainly been subsisting on colas and ginger ale to control the nausea.

161

- *Usual intake*
  - Breakfast: juice, cold cereal and toast. Occasionally a fried egg and fried potatoes
  - Lunch: American cheese sandwich, chips and sweetened cola (meat is unpleasant to her).
  - Dinner: fried eggs, fried potatoes or plantains and dessert—cake/pudding.
  - Snack: chips or crackers with cheese dip or peanut butter.
- *Diet analysis*
  - Not done. Patient's intake has been minimal for the last few weeks.

LABORATORY DATA: BLOOD

Albumin	3.0 g/dL	3.5–5 g/dL
Total protein	5.8 g/dL	6–8 g/dL
Sodium	130 mEq/L	136–145 mEq/L
Potassium	5.8 mEq/L	3.5–5.5 mEq/L
PO4	9.5 mg/dL	2.3–4.7 mEq/L
Osmolality	317 mmol/kg/H_2O	285–295 mM/kg/H_2O
Glucose	282 mg/dL	70–110 mg/dL
BUN	69 mg/dL	8–18 mg/dL
Creatinine	12 mg/dL	0.6–1.2 mg/dL
Calcium	8.2 mg/dL	9–11 mg/dL
Phosphorus	9.5 mg/dL	2.3–4.7 mg/dL
Total cholesterol	220 mg/dL	120–199 mg/dL
Triglycerides	200 mg/dL	35–135 mg/dL
HDL	27 mg/dL	33–96 mg/dL
HbA1c	8.9 %	3.8–5.2%
Hct	29 %	37–47%
MCV	76 μm^3	80–96 μm^3
HCO$_3$	20 mEq/L	22–28 mEq/L

LABORATORY DATA: URINE

pH	7.9	5–7
Protein	2+	neg
WBC	20/HPF	1–5/HPF

RESOURCES

7.1 DIABETIC KIDNEY DISEASE

Approximately 40% of patients with either type I or II diabetes mellitus (DM) will develop diabetic nephropathy and progress to end-stage renal disease (ESRD), characterized by glomerular disease, nonnephrotic proteinuria, and nephritic syndrome. Hyalinizing arteriolar sclerosis is seen in the renal arterioles, and patients have increased susceptibility to pyelonephritis and a variety of tubular lesions. Morphologic changes in diabetic nephropathy include capillary basement membrane thickening, diffuse mesangial sclerosis, and nodular glomerulosclerosis (Figs. 7.1 and 7.2).

Diabetic kidney disease (DKD) is thought to arise both from an underlying metabolic defect that results in hyperglycemia and from hemodynamic changes including increased GFR and glomerular capillary pressure. Glomerular hypertrophy and an increased glomerular filtration area contribute to glomerulosclerosis. Both processes are associated with loss of podocyte filtration and result in protein wasting. A well-accepted model holds that high circulating glucose levels activate protein kinase C (PKC) and induce transforming growth factor-beta (TGF-β), leading to expression of p27^{Kip1} protein with subsequent binding to cyclin–CdK complexes and cell cycle arrest in the G_1 phase. Cell cycle arrest results

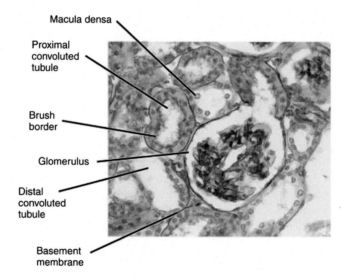

FIGURE 7.1 **Normal glomerulus.** This stained preparation of normal glomerulus illustrates thin and delicate capillary loops, tubules, and basement membrane. The mesangial region is of normal size and unencumbered by inclusions. *Source: By OpenStax College (CC BY 3.0 (http://creativecommons.org/licenses/by/3.0)), via Wikimedia Commons.*

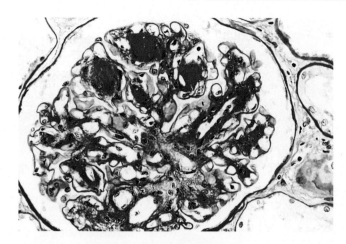

FIGURE 7.2 **Diabetic glomerular disease.** The arterial walls of the diabetic glomerulus contain hyaline thickening and the mesangial area contains pink, amorphous inclusions. In DKD, there is also evidence of tubular destruction with an increase in the interstitial fibrous tissue. *Source: Reproduced from Robbins and Cotran Pathologic Basis of Disease, ninth ed. With permission from Elsevier Saunders.*

in cell hypertrophy. Additional factors such as activation of the renin−angiotensin−aldosterone system (RAAS) pathway with vasoactive angiotensin II (Ag II) production, hemodynamic stretch and creation of advanced glycation end products (AGEs) may further stimulate p27^{Kip1} expression and accelerate hypertrophy and matrix accumulation (Wolf and Ziyadeh, 1999).

Deleterious changes occur even with moderate hyperglycemia insufficient to exceed the renal threshold for glucose (~180 mg/dL). Experimental work suggests that glucose reaching the kidney through the vascular system stimulates initial, self-limited proliferation followed by cell cycle arrest and hypertrophy. Thus total kidney expansion with mesangial and proximal tubular cell hypertrophy are early hallmarks of diabetic nephropathy. Cell cycle arrest in the kidney is often accompanied by increased deposition of extracellular matrix (ECM) proteins.

7.2 TARGETS FOR SLOWING PROGRESSION OF DIABETIC NEPHROPATHY

7.2.1 Renin−Angiotensin−Aldosterone System (RAAS) Dysregulation

Glomerular filtration is mediated by podocytes and the endothelial lining of the glomerular capillary.

Podocyte injury and a reduction of podocyte number are prominent observations in diabetic nephropathy. Several studies have linked elevated blood glucose levels to podocyte toxicity and subsequent activation of maladaptive signaling events as reviewed by Durvasula and Shankland (2008) and Márquez et al. (2015).

AgII is emerging as a critical mediator of podocyte injury in diabetic renal disease. Thus therapies that block AgII production or utilization have become cornerstones in the management of diabetic nephropathy; these strategies afford renal protection beyond blood pressure lowering effects alone. AgII is the end product of the RAAS signaling pathway. This pathway is normally activated in response to low renal perfusion signals from the distal tubule and the renal arterioles. Renin is secreted by the macula densa cells of the juxtaglomerular apparatus and regulates blood pressure by multiple actions as shown in Fig. 7.3.

However, plasma renin, which mediates the rate-limiting step in the classical pathway of AG II generation, is typically low in patients with diabetic nephropathy. Attempts to resolve the apparent paradox of the low renin state of diabetic nephropathy resulted in growing awareness that Ag II formed in peripheral tissues is not under central RAAS regulation. Recent work (Rüster and Wolf, 2011) has shown that non-kidney tissues contain all necessary RAAS components for AgII biosynthesis. In contrast to the classic pathway, tissue RAAS is not regulated by renal perfusion and blood volume, rather it is stimulated by multiple factors, including hyperglycemia. In diabetic nephropathy, classic RAAS is suppressed, but tissue RAAS, including the system in adipose tissue, is activated. Hyperglycemia induces oxidative stress and endothelial damage, and also stimulates downstream mediators including Ag II and TGF-β. Through a series of nonhemodynamic actions including apoptosis, hypertrophy, matrix accumulation, and fibrosis, AgII can damage the kidney. Accordingly, delineation of mechanisms by which local angiotensin production is regulated is of utmost important, especially in diabetic nephropathy.

7.2.2 Nonclassic RAS Production of AgII in Obesity

As reviewed by Luther and Brown (2011), adipocytes express angiotensinogen and are competent to form AgII. Elevated AgII isoforms generated through Nonclassic RAS (NC-RAS) systems appear to mediate

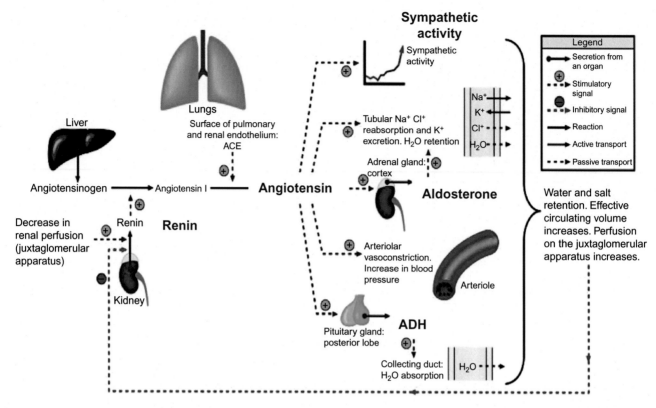

FIGURE 7.3   **Classic (circulatory) RAAS regulates blood pressure and volume.** RAAS secretion is initiated when macula densa cells of the juxtaglomerular apparatus detect low renal blood perfusion and secrete renin. Renin is a protease that cleaves hepatic angiotensinogen to angiotensin I (Ag I). Angiotensin-converting enzyme (ACE) produced by the epithelial cells in the lung cleaves AgI to AgII; this product normalizes blood pressure by multiple actions including raised sympathetic activity, enhanced aldosterone secretion and ADH secretion and vasoconstriction. *Source: By A. Rad (own work) (GFDL (http://www.gnu.org/copyleft/fdl.html) or CC-BY-SA-3.0 (http://creativecommons.org/licenses/by-sa/3.0/)), via Wikimedia Commons.*

different responses depending on the receptors to which they bind. The possibility of dual responses, one protective and one destructive, has been comprehensively reviewed (Favre et al., 2015). Briefly, angiotensin-converting enzyme inhibitors target the AgII type 1 receptor (AT1R) and are used to prevent diabetes and insulin resistance in patients with hypertensive or ischemic cardiopathy. Mechanistically, these drugs inhibit potentially destructive AgII activity in multiple tissues. In adipocytes, AT1R binding inhibits the PI3K cascade directly and indirectly via proinflammatory nuclear factor kappa B (NFκB) activation resulting in reactive oxygen species (ROS) generation and secretion of cytokines, IL-6 and monocyte-chemoattractant protein 1 (MCP-1). Together these changes lead to an insulin-resistant state and to recruitment of macrophages into the adipose tissue. AT1R binding also changes the polarity of macrophages from M2 to M1 macrophages that secrete proinflammatory mediators. At the same time, AT1R activation inhibits adipogenesis by decreasing fatty acid synthase and inhibiting the differentiation promotor, peroxisome proliferator-activated receptor gamma, reducing lipid storage capacity in the adipose tissue. In

the skeletal muscle, glucose transporter (GLUT 4) translocation is inhibited, lowering insulin-stimulated glucose uptake, while in the liver, AT1R binding stimulates gluconeogenesis through upregulation of phosphoenolpyruvate carboxykinase (PEPCK). AT1R binding in the pancreas produces ROS, promoting β-cell apoptosis and reducing glucose sensing by suppressing GLUT 2 action. In summary, AgII—AT1R binding increases insulin resistance, elevates blood glucose, reduces the capacity for lipid storage in hypertrophic adipocytes, and mediates ectopic fat deposition.

AgII also acts as a morphogen (Rüster and Wolf, 2011). It binds to downstream AT1R on the kidney and stimulates pathways that facilitate vasoconstriction, thrombosis, inflammation, fibrosis, and vascular remodeling. Strong clinical data are emerging that RAAS antagonists (angiotensin-converting enzyme (ACE) inhibitors or AgII receptor blocking agents) preserve kidney function not only by lowering blood pressure but also by reducing tissue AgII-associated renal damage, such as fibrosis and inflammation, that can occur with diabetes even in the absence of blood pressure elevation.

Ag II is not the only peptide produced by NC-RAS. As reviewed by Favre et al. (2015) other members of the family include Ag III, Ag IV, and Ag 1−7. The latter member, Ag 1−7 binds the G-protein-coupled receptor, *Mas*, and appears to be involved in the protective actions of NC-RAS, including adipogenesis and lipogenesis. Interestingly, agonists for the AT2R receptor did not lower blood pressure in experimental models. Ag 1−7−Mas receptor binding inhibited the production of ROS in adipocytes and increased insulin actions. It also increased adipocyte production of adiponectin, thereby sensitizing cells to insulin action. In skeletal muscle, MasR binding increased translocation of GLUT 4, facilitating glucose uptake. In the liver, MasR promoted glycogen synthesis and blocked gluconeogenesis (downregulating PEPCK). Finally, MasR binding reduced NADPH oxidase activation in β cells, protecting them from apoptosis. Protective effects have also been seen when AgII binds to its alternate receptor, AT2R. Binding improved insulin sensitivity by stimulating preadipocyte differentiation and protected pancreatic β cells by reducing ROS generation.

ACE inhibitors and AT1R receptor blockers, although clinically effective in lowering blood pressure, do not appear to reduce rates of mortality and progression to renal failure. It has been suggested that this dichotomy is due to high renin secretion that results when classic RAAS is blocked (Mende, 2010). In the classic RAAS system, renin, stored in secretory granules of the juxtaglomerular apparatus, is released in response to decreases in renal perfusion pressure, decreases in Cl⁻ in the distal tubule fluid, or increased sympathetic nerve stimulation via β1 adrenoceptors. Renin release and/or expression are inhibited by a direct action of Ag II, as well as by uric acid, TGFβ, and TNFα. ACE inhibitors, by blocking production of AgII, prevent normal feedback inhibition and allow unrestricted renin secretion. Treatment strategies with RAAS inhibitors as well as renin inhibitors are now being explored. It is of interest that active vitamin D (1,25D) inhibits RAS by blocking the activity of the cAMP response element in the renin gene promoter, thereby suppressing renin gene transcription (Li et al., 2002; Yuan et al., 2007; Lai and Fang, 2013). An additional factor that reduces clinical effectiveness of RAAS inhibitors is that AgII appears be converted from AgI through non-ACE-dependent pathways, thus circumventing the effectiveness of ACE inhibition.

AgII binding can also cause deleterious changes in the kidney by inducing TGF-β. TGF-β is a member of a large family (>40 proteins) of signaling molecules that regulate cell and matrix homeostasis (Catherwood et al., 2002; Ruiz-Ortega et al., 2007). TGF-β is synthesized as an inactive protein that interacts with latent TGF-β binding proteins (LTBP) anchored in the extracellular matrix (ECM). When bound, TGF-β is activated by proteolytic cleavage and migrates to the plasma membrane where it facilitates complex pathways in TGF-β-mediated fibrosis. By reducing collagenase production and upregulating expression of tissue inhibitor of metalloproteinases (TIMP), TGF-β causes an overall inhibition of ECM degradation that leads to excessive matrix accumulation. Several factors regulate TGF-β synthesis and activation, including hyperglycemia, AgII, mechanical stress, and advanced glycation endproducts (AGEs). High blood levels of oxidized LDLs and high dietary sodium (Sanders, 2009) have also been reported to increase TGF-β activity in mesangial cells.

7.2.3 Oxidative Damage

An important factor contributing to chronic kidney disease (CKD) progression is the imbalance between generation of ROS and antioxidant protection (refer to Essentials I: Life in an Aerobic World). Overactivation of RAAS produces ROS that signals downstream cascades such as TGF-β1 and the proinflammatory cascade, NFκB, leading to cell injury, fibrosis, and end-stage kidney disease (ESKD) (Liu et al., 2012; Rojas-Rivera et al., 2012). Hyperglycemia stimulates oxidative stress in mesangial cells with depletion of glutathione (Catherwood et al., 2002). Exposure to high levels of glucose also activates several other pathways associated with oxidative damage and diabetic complications (Son, 2012), including activation of the polyol pathway, de novo synthesis of diacylglycerol, increased PKC activity, alterations in the hexosamine pathway, and production of nonenzymatic AGEs (Pácal et al., 2011) as shown (Fig. 7.4).

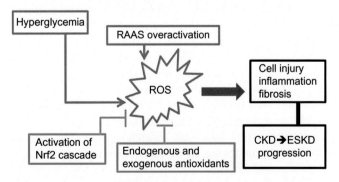

FIGURE 7.4 **CKD progression.** Both hyperglycemia and RAAS activation stimulate ROS production with predictable sequelae including cell injury, increased TGFβ and NFkβ cascades (red arrows). Antioxidant cascades such as nuclear factor (erythroid-derived 2)-related factor 2 (Nrf2) as well as endogenous and exogenous antioxidants (blue blunt arrows) have been shown to reduce progression to ESRD. *Source: Adapted from Son, S.M., 2012. Reactive oxygen and nitrogen species in pathogenesis of vascular complications of diabetes. Diabetes Metab. J. 36 (3), 190−198, redrawn by Susan Ettinger.*

Adequate concentrations of reduced glutathione are critical for endogenous antioxidant protection and require a spectrum of micronutrients. Glutathione is a three-amino-acid peptide that reduces oxidized molecules by a selenium-dependent reaction (*glutathione peroxidase*). Regeneration of reduced glutathione (*glutathione reductase*) requires NADPH and riboflavin. NADPH depletion reported in diabetic nephropathy can be due to several factors. While upregulation of nonphagocytic NADPH oxidase has been implicated in NADPH destruction, NADPH biosynthesis can also be compromised by diabetes-associated loss of thiamin (see below). Glutathione biosynthesis de novo requires the availability of cysteine, which can be limiting in oxidative stress, especially if folate, vitamin B12- and vitamin B6-dependent pathways are compromised (refer to Chapter 6: Atherosclerosis and Arterial Calcification: Strategies to Repair Endothelial Dysfunction, Fig. 6.11).

Folate depletion in the CKD patient can compromise homocysteine remethylation and increase serum levels of the toxic metabolite, homocysteine thiolactone (Jakubowski, 2007), implicated in vascular damage, a major risk for the CKD patient (McCully, 1996). It should also be noted that homocysteine forms complexes with copper, reducing intracellular copper concentration, thereby compromising oxidative phosphorylation and adenosine triphosphate (ATP) production (Zuo et al., 2013) (refer to Chapter 5: Cardiomyopathy and Congestive Heart Failure: Targets for Myocardial Protection: Cardiac Remodeling). Interference with the electron transport chain (ETC) increases the probability that oxygen radicals will escape and initiate oxidative damage as discussed above and elsewhere (refer to Chapter 5: Cardiomyopathy and Congestive Heart Failure: Mitochondrial Production and Management of Reactive Oxygen Species).

Micronutrient depletion, although likely in patients with CKD, especially those with type II DM (Pastore et al., 2014), has not been well studied (Clase et al., 2013). Because tissue distribution of many micronutrients is tightly controlled, serum levels may not be representative of the physiological tissue status of the nutrient. For example, measurement of red cell folate is difficult and highly inaccurate (Golding, 2014), thus serum folate levels are often taken as evidence of folate status. However tissue (brain, liver, heart, intestine) expressions of folate and thiamin transporters were shown to be reduced in a rodent model of CKD, despite normal serum levels of the nutrients (Bukhari et al., 2011), making accurate assessment of nutrient status uncertain.

Interestingly, CKD is also associated with altered copper homeostasis. Hypercupremia and elevated chelatable (nonceruloplasmin) copper reported in diabetic CKD patients have been implicated in oxidative damage and risk for cardiovascular disease (Sondheimer et al., 1988; Kasama, 2010; Shaban et al., 2014). CKD patients have been observed to transport a greater fraction of their serum copper in a form that could be chelated and thus was not bound to the primary circulating copper protein, ceruloplasmin. Factors that regulate intracellular copper concentrations have been discussed elsewhere (refer to Essentials II: Heavy Metals, Retinoids and Precursors: Copper). It must be remembered that ceruloplasmin is an acute-phase protein responsive to upregulation by oxidative damage, and that unbound copper can participate in ROS generation. Further, by downregulating copper transporters unbound extracellular copper can compromise intracellular copper concentrations and limit function of the electron transport chain (refer to Chapter 5: Cardiomyopathy and Congestive Heart Failure). These observations may have important implications for oxidative damage and cardiovascular risk seen in CKD patients.

Finally, nuclear factor (erythroid-derived 2)-like 2 (Nrf2) is a transcription factor that regulates hundreds of genes involved in cellular protection against ischemia−reperfusion, oxidative and xenobiotic damage as reviewed by Leonard et al. (2006) and Noel et al. (2015). As described elsewhere (refer to Essentials I: Life in an Aerobic World: Nutrient Transcriptional Control of Antioxidant Protection), Nrf2 is bound to its inhibitor, kelch-like ECH-associated protein (Keap1), and maintained at homeostatic levels by ubiquitination and proteasomal degradation. Upon exposure to damaging stimuli, Nrf2 is released and translocates to the nucleus where it binds to its antioxidant response element and stimulates expression of an array of protective and/or detoxification genes. Crosstalk between Nrf2 and other systems such as proinflammatory NFκB create an integrated response to stress, thus enhancing the protective effect of Nrf2 (Wakabayashi et al., 2010). Actions of the Nrf2 system were downregulated in the progression of experimental diabetic nephropathy (Zoja et al., 2015). Additionally, Nrf2 upregulates several target genes that could provide protection against kidney injury leading to end-stage disease (Noel et al., 2015). While drugs that upregulate Nrf2 have been developed, they are associated with significant side effects. In contrast, several nonnutritive dietary components in cruciferous vegetables and other plant products (refer to Chapter 5: Cardiomyopathy and Congestive Heart Failure: Mitochondrial Production and Management of Reactive Oxygen Species) have also been shown to upregulate Nrf2 activity in experimental models of diabetic nephropathy (Zheng et al., 2011).

7.2.4 Vascular Tone, Diuresis, and Hypoxia

Vascular tone is regulated in part by the production of the two gasotransmitters, nitric oxide (NO) and hydrogen sulfide (H₂S) (refer to Chapter 6: Atherosclerosis and Arterial Calcification: Metabolic Targets for Diet Modification). In the kidney, H₂S stimulates natriuresis and diuresis by increasing glomerular filtration and inhibiting tubular sodium reabsorption (Bełtowski, 2010). Data suggest that baseline renal H₂S production may also control the classic RAAS by reducing renin production, limiting hyperglycemia and ROS production, and rescuing mesangial cells from glucose-induced overproliferation and ECM secretion (Lobb et al., 2015).

H₂S is oxidized in the mitochondria in a pO₂-dependent manner. Since the renal medulla has a lower pO₂ than the cortical region, it is likely that H₂S accumulation in the medulla acts as an O₂ sensor, restoring O₂ balance by increasing medullary blood flow, reducing energy requirements for tubular transport, and directly modulating mitochondrial respiration. Tubulointerstitial hypoxia has been suggested as the major event that leads to progression of ESKD (Mimura and Nangaku, 2010). This vicious cycle can be modulated by many of the biological functions of H₂S as illustrated in Fig. 7.5. Deficiency of H₂S in the kidney could contribute to the pathogenesis of CKD by compromising medullary oxygenation, potentiating damage from ischemia, hyperglycemia, hypertension, and other injuries (Wollesen et al., 1999; Koning et al., 2015).

Endogenous H₂S biosynthesis is compromised in obesity and several of its chronic comorbidities, including diabetes, cardiomyopathy, cardiovascular disease, and asthma (Wang, 2012). Pharmaceutical agents that provide H₂S or its sulfur-containing precursors are currently under investigation. However, sulfur-containing precursors are well represented in dietary sources. For example, sulfur-containing amino acids (SAA), methionine and cysteine, are abundant in animal-derived protein sources, especially egg yolk. Vegetables of the cruciferous family, broccoli, kale, and cabbage, as well as the allium family, onions, garlic, shallots, and leeks, contain a variety of organosulfur compounds. Garlic may actually provide an "active H₂S pool" (Zhao et al., 2015). The allicin (diallyl thiosulfinate) contained in ingested garlic decomposes to reactive sulfur-containing compounds (diallyl sulfide, diallyl disulfide, diallyl trisulfide) during digestion. H₂S release from these metabolites is thought to be initiated by the nucleophilic substitution of a thiol (eg, glutathione) with subsequent rearrangement and release of H₂S.

Finally, hemodynamic changes that occur even in the early stages of diabetes contribute to diabetic nephropathy. These include increased GFR (hyperfiltration) mediated by proportionately greater relaxation of the afferent arteriole than the efferent arteriole. Hyperfiltration leads to increased glomerular blood flow and elevated glomerular capillary pressure. If diabetes is poorly controlled, hyperglycemia results in glomerular hypertrophy with

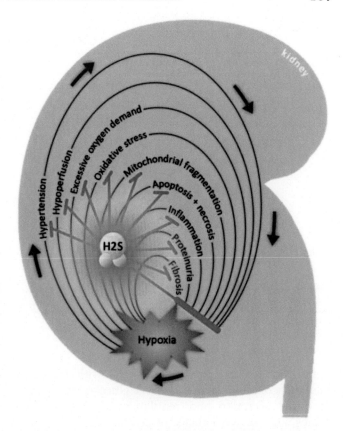

FIGURE 7.5 **The chronic hypoxia hypothesis in ESKD.** Medullary hypoxia can potentiate many of the disease processes leading to ESKD. Cystathionine γ lyase (CSE) requires a vitamin B6 cofactor to generate H₂S from a cysteine substrate. Hypoxia increases CSE translocation into the mitochondria where the H₂S produced supports ATP production by donating electrons to the ETC. This reaction is believed to reflect the evolution of mitochondria from the ancient bacterial precursor, competent to use H₂S as an energy source (Fu et al., 2012). *Source: Reproduced from Koning, A.M., Frenay, A.-R.S., et al., 2015. Hydrogen sulfide in renal physiology, disease and transplantation—the smell of renal protection. Nitric Oxide 46, 37–49 with permission from Science Direct.*

an increased glomerular capillary surface area and fibrosis. These intraglomerular hemodynamic and structural alterations may contribute to the development or progression (or both) of diabetic renal injury.

Once begun, injurious processes crosstalk and amplify, such that progression of CKD to ESKD is inexorable and predictable. However, there is evidence that diet modification, physical exercise, and other lifestyle changes have potential to intervene and thus slow the rate of functional decline (Table 7.1).

7.3 MACRONUTRIENT IMPLICATIONS IN DIABETIC NEPHROPATHY PROGRESSION

Metabolic dysregulations observed in diabetes include insulin resistance in muscle and adipose tissue,

TABLE 7.1   Stages of Diabetic Nephropathy

		Chronology of diabetic nephropathy[a]				
	Designation	Clinical	GFR	Albumin loss	Blood pressure	Time line
Stage 1	Hyperfunction/ hypertrophy	Glomerular hyperfiltration	Increased	May be increased	Normal/ hypertension	At diagnosis
		Some microalbuminuria				
Stage 2	Silent stage	Thickened BM	Normal	May be <30–300 mg/day	Normal/ hypertension	First 5 years
		Obstructed mesangium				
Stage 3	Incipient stage	Fixed microalbuminuria	GFR reduced	30–300 mg/day	Normal/ hypertension	6–10 years
		Proteinuria >500 mg/24 h				
Stage 4	Overt nephropathy	Proteinuria 3.5 g/24 h	GFR below normal	>300 mg/day	Hypertension	15–25 years
Stage 5	Uremic	ESRD	GFR 0–10	Decreasing	Hypertension	25–39 years

[a]Adapted from http://www.medisuite.ir/medscape/a238946-business.html
GFR, glomerular filtration rate.

dysfunction of the pancreatic β cells with impaired insulin secretion, increased hepatic glucose production, abnormal secretion and regulation of incretin hormones, and altered balance of central nervous system pathways that control food intake and energy expenditure. Mechanisms under investigation to modulate insulin resistance and/or diabetes progression include (refer to Chapter 4: Type II Diabetes, Peripheral Neuropathy and Gout):

- Altered insulin signaling due to cellular lipid accumulation, proinflammatory signals, and endoplasmic reticulum stress;
- Altered insulin production due to pancreatic injury and inadequate micronutrient availability;
- Reduced incretin-dependent and -independent β-cell insulin secretion;
- Impaired mitochondrial capacity and/or function.

Medical nutrition therapy currently focuses on changing macronutrient availability to ameliorate these processes. In doing so, macronutrient modification also has the potential to slow progression to ESKD.

7.3.1 Carbohydrate Intake and Metabolism

Diabetic ESKD has steadily increased over the past three decades, reaching epidemic proportions. While published treatment guidelines target the underlying diabetic disease processes and resultant hyperglycemia, these remedies have not been successful in reducing the incidence of diabetic ESKD. This dichotomy could be due, in part, to qualitative changes in macronutrient

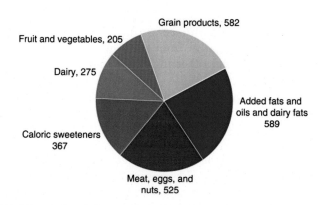

Added fats and oils and added sugars are added to foods during processing or preparation. They do not include naturally-occurring fats and sugars in food (eg, fats in meat or sugars in fruits).

FIGURE 7.6   **Daily calories per capita by food group 2010.** Total calorie consumption has increased from 2039 calories per person in 1970 to 2544 calories in 2010. Refined grains with added fats and oils contributed 398 calories to this increase. Vegetables contributed fewer calories in 2010 than in 1970 while protein sources made up a greater percent of total calories in 1970 than in 2010. *Source: Reproduced from "U.S. Department of Agriculture" website. The information is in the public domain.*

consumption that have occurred over the past 3–4 decades as shown in Fig. 7.6. Increased consumption of fats and sugar has been linked to the increased prevalence of obesity and diabetes (Fagherazzi et al., 2013), as well as a threefold increased risk of CKD over a 5-year period (Gopinath et al., 2011).

Daily per capita calorie intake data compiled by the USDA Economic Research Service from 1970 through

2006 suggests that the increase in calorie intake is primarily from added fats, grains, and sugars and that in the US the average per capita fructose intake made up more than 10% of calories ingested (Vos et al., 2008). These data suggested that total intake of meat, eggs, nuts, and dairy increased slightly, while fruits and vegetables remained relatively constant (data source: http://www.ers.usda.gov). These data are consistent with data from 2010 presented in Fig. 7.6.

7.3.2 Carbohydrate Intake and CKD

Approximately 50% of US sweetener consumption is provided by high-fructose corn syrup (HFCS) (https://fnic.nal.usda.gov/food-composition/nutritive-and-nonnutritive-sweetener-resources). HFCS is made from corn treated with *glucose isomerase* to convert some of the glucose to fructose. Because HFCS is less expensive than sucrose, it is widely used for sweetened beverages. Its ability to retain moisture better than other sweeteners makes HFCS an excellent additive to baked and processed foods to improve quality and texture, and to maintain a longer shelf-life. While HFCS is purported to contain 45–55% fructose, preparations that approach 65% fructose have been reported for a number of commercially available soft drinks.

Sucrose, common table sugar, is a disaccharide made up of fructose linked by a glycosidic bond to glucose. Ingested sucrose is cleaved to its monosaccharide components by the intestinal brush-border enzyme, *sucrose*. Whether the sweetener is sucrose or HFCS, the glucose is absorbed into the enterocyte by a sodium-dependent active transporter. In contrast, liberated fructose is rapidly taken up by facilitated diffusion, mediated by the non-energy requiring GLUT-5 enterocyte transporter. Fractional absorption of fructose is enhanced in the presence of luminal glucose, suggesting the presence of an additional, possibly glucose-specific, transporter.

Absorbed monosaccharides also differ in their metabolic pathways (Tappy and Lê, 2010). While glucose oxidation through the glycolysis pathway is highly regulated, fructose oxidation has no negative feedback system. Most absorbed fructose is phosphorylated to fructose-1-phosphate by the hepatic enzyme, *fructokinase*. Since this action requires ATP, excess dietary fructose maximizes *fructokinase* activity, leading to transient ATP depletion. The AMP produced is metabolized by *AMP deaminase* to inosine monophosphate and eventually to uric acid (Kretowicz et al., 2011). Emerging evidence suggests that elevated uric acid predicts kidney damage through endothelial dysfunction, increased RAAS activity as well as induction of inflammatory cascades and profibrotic cytokine

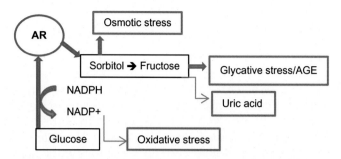

FIGURE 7.7 **Aldose reductase (AR) pathways.** Hyperglycemia upregulates the AR pathway, not only in the eye but in other organs. Biological membranes are impermeable to sorbitol, which can accumulate inside the tissues resulting in osmotic stress. Sorbitol conversion to fructose leads to production of uric acid and AGE complexes, while increased consumption of NADPH reduces glutathione production and results in oxidative stress. These activities are implicated in secondary diabetic complications such as diabetic cataractogenesis, retinopathy, nephropathy, and neuropathy. *Source: Constructed by S. Ettinger using data from Ramana, K.V., 2011. Aldose reductase: new insights for an old enzyme. Biomol. Concepts 2 (1–2), 103–114 (Ramana, 2011).*

activation (Hovind et al., 2011; Jalal et al., 2011). At the same time, under hyperglycemic conditions, glucose is converted to fructose by aldose reductase (AR) through the polyol pathway. This endogenous source of fructose enters the fructose pool. While several classes of drugs are known to inhibit AR, it should also be noted that many flavonoids, including quercetin and luteolin, have AR-inhibitory activities (Xiao et al., 2015) (Fig. 7.7).

Cells respond to transient ATP depletion as they do to ischemia, by arresting protein synthesis with the induction of oxidative stress and inflammation. A small fraction (~20%) of absorbed fructose escapes into the systemic circulation where it is taken up by other cell types, including endothelial cells. Fructose that remains unmetabolized is filtered and reabsorbed by the GLUT 5 transporter in the proximal tubular cells of the kidney. Thus, both dietary and endogenous fructose are metabolized by fructokinase, and have potential to generate transient ATP depletion, leading to an ischemic-like state with potential to precipitate kidney injury, glomerulosclerosis, and tubulointerstitial fibrosis.

Experimental rat models with 5/6 nephrectomy and treated with a high-fructose (60%) diet exhibited increased renal vasoconstriction, proteinuria, glomerular hypertension, and other biomarkers of renal inflammation and subsequent cell injury (Gersch et al., 2007; Johnson et al., 2009). Similarly, large human population studies have implicated high fructose and sucrose intakes with a decline in renal function and microalbuminuria (Lin and Curhan, 2011). High dietary fructose intake is also associated with depletion of essential nutrients (Douard et al., 2010). Marked depletion of active vitamin D and subsequent reduced calcium absorption were seen in rats subjected to 5/6 nephrectomy and fed high-fructose diets (Douard et al., 2013). It is also likely that the high fructose intake increased fructokinase activity in the enterocyte, reducing intracellular calcium transport as described elsewhere (refer to Essentials III: Nutrients for Bone Structure and Calcification: Calcium). Reduced serum calcium levels stimulate compensatory secretion of parathyroid hormone (PTH) and compromised bone integrity.

The high-fructose diet also increased kidney weight, elevated blood urea nitrogen, increased serum levels of creatinine, fibroblast growth factor 23, and phosphate, and increased the calcium—phosphate product. While compromised copper availability has been well described in the fructose-fed rat, human studies are less conclusive, possibly because the fructose dose was lower. The influence of copper deficiency on CKD progression (see below) has not been explored in rat models or in human subjects.

Finally, high fructose intake is associated with production of advanced glycation end-products (AGEs). Sugars can form nonenzymatic complexes with free amino groups on proteins, lipids, and nucleic acids (the Maillard reaction) and mature into irreversible molecular rearrangements (AGEs). (The Maillard reaction is named after the French chemist Louid-Camille Maillard, who described it in 1912. It is a reaction between amino acids and reducing sugars; the carbonyl group of the sugar reacts with the nucleophilic amino group of the amino acid. The Maillard reaction forms nonenzymatic browning of baked goods, meat, and other protein-rich foods.) Depending on the tissue, AGEs can form crosslinks between key molecules in the basement membrane and the ECM, alter intracellular proteins and interfere with intracellular signaling. Podocyte injury observed early in diabetic nephropathy has been attributed to hyperglycemia and RAAS activation. More recently, murine podocytes were shown to avidly bind AGE complexes with resultant hypertrophy and cell damage by a mechanism involving p27^{Kip1} (Rüster et al., 2008). This damage can contribute to the loss of podocytes observed in diabetic nephropathy. While AGE formation has been associated with hyperglycemia, glucose may not be the major contributor because it has a lower chemical reactivity compared with fructose. Fructose was shown to produce 10-fold more AGEs than glucose in vitro due to its more reactive, open chain form (Suárez et al., 1989; Collino, 2011). Thus, it is possible that basement membrane thickening, podocyte necrosis, and other damage associated with AGE formation and diabetic nephropathy result, in part, from exposure to high levels of simple sugars, and especially to fructose.

7.3.3 Hyperinsulinemia and CKD Damage

In addition to the potentially deleterious effects of high fructose intake, there is evidence that diets high in free sugars, such as sucrose and readily available sources of glucose from starch, can damage the kidney by stimulating hyperinsulinemia. Insulin receptors are expressed on energy storage sites such as adipocytes and muscle cells; insulin receptors have also been detected throughout the length of the renal tubule. Exposure to high plasma insulin concentrations has been shown to downregulate insulin receptor abundance on rat adipocytes (Suárez et al., 1989), and may contribute to insulin resistance in adipose tissue.

Insulin receptor expression has been less well studied in the kidney. While insulin has little effect on the normal kidney, in the diabetic patient, hyperinsulinemia selectively increases urinary albumin secretion and reduces both sodium and uric acid excretion (Quiñones-Galvan and Ferrannini, 1997). At the same time, insulin excretion is impaired in the failing kidney, resulting in higher fasting insulin levels. These preliminary observations suggest that tubular insulin receptor compromise may contribute to a mechanistic explanation for the sodium and uric acid retention in the insulin-resistant diabetic patient.

While glucose in the portal vein is the major stimulus for insulin secretion, studies suggest that noncaloric sweeteners can also stimulate insulin secretion and increase risk for type II DM (Brown et al., 2009), possibly by stimulating gut dysbiosis (Suez et al., 2014) as described elsewhere (refer to Chapter 1: Obesity and Metabolic Syndrome: Specific Diet Components that Target Obesity and MS). Analysis of the Nurses' Health study revealed a twofold increased odds ratio for eGFR decline over 11 years in women who consumed >2 servings per day of diet, but not regular, cola (Lin and Curhan, 2011). While it is possible that intestinal sweet receptors increased insulin secretion by stimulating glucagon-like peptide 1, it is also possible that sweet receptor stimulation upregulates sugar transporters, thereby increasing glycemic excursions (Mansour et al., 2013). At this time, the impact, if any, of noncaloric sweeteners on risk for kidney damage or diabetic nephropathy remains unclear.

7.3.4 Glucose Metabolism, Thiamin, and Glutathione

Thiamin is essential for glucose metabolism, yet current literature suggests that thiamin status may be compromised in diabetic patients. Thiamin depletion was not related to reduced thiamin intake, but rather to increased renal clearance and fractional excretion of thiamin in diabetics, compared to healthy controls

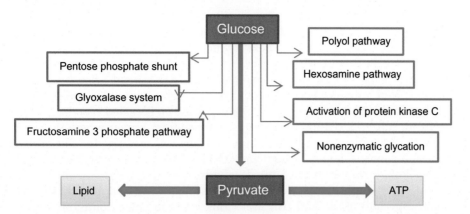

FIGURE 7.8 **Glucose pathways.** In addition to the glycolytic pathway (center), glucose can enter several other pathways. "Harmful" pathways (blue) contribute to oxidative metabolites and diabetic complications. For example, glucose is converted to fructose by aldose reductase in the polyol pathway. "Protective" metabolic pathways (red) convert glucose into harmless or useful metabolites. The pentose phosphate shunt (PPP), also known as the hexose monophosphate shunt (HMS), is especially useful, since it also produces both ribose substrate for nucleotide biosynthesis and NADPH required for antioxidant protection and for oxygen-dependent killing by phagocytic cells. Thiamin is required as cofactor for *transketolase* in PPP. *Source: Constructed by S. Ettinger using data from Pácal, L., Tomandl, J., et al., 2011. Role of thiamine status and genetic variability in transketolase and other pentose phosphate cycle enzymes in the progression of diabetic nephropathy. Nephrol. Dial. Transpl. 26 (4), 1229–1236.*

(Larkin et al., 2012). The positive correlation between increased renal clearance of thiamin and the level of glycated hemoglobin (HbA1C) provides further evidence that the two factors may be related. Diabetic patients had normal serum levels of thiamin and folate, but micronutrient stores in liver and other tissues were compromised (Bukhari et al., 2011). Clinical trials with both oral thiamin and its lipid-soluble analog, benfothiamine, have demonstrated significant blood glucose lowering (Pácal et al., 2014), and several animal and human studies have now demonstrated that oral thiamin supplements can prevent or delay development of diabetic nephropathy in streptozotocin-induced rats and in human subjects with type II DM and microalbuminemia. (It should be noted that benfothiamine is lipid-soluble and compared with hydrophilic thiamin more easily crosses the lipid biomembrane, however, a beneficial effect was also seen with the parent compound, thiamin.) Whether diabetic nephropathy progression results in accelerated thiamin loss has not yet been determined.

Glucose is metabolized through several pathways, as indicated in Fig. 7.8. Glycolysis, leading to glucose oxidation and ATP synthesis, is the major pathway for glucose utilization. Other pathways, such as nonenzymatic glycation and the polyol pathway, are designated as "harmful" because they are implicated in formation of AGE complexes and diabetic complications. Some pathways that utilize glucose are considered "protective" because they divert circulating glucose to other uses. For example, the PPP produces sugar substrates (ribose) for nucleotide biosynthesis,

as well as NADPH, required for biosynthetic pathways and efficient phagocyte killing. Glucose flux through the PPP requires adequate intracellular thiamin as a cofactor for *transketolase*, an enzyme critical for optimal functioning of the PPP. Note that the NADPH produced by the PPP can partly replace NADPH consumed when the "harmful" polyol pathway is activated.

Thiamin is also required for pyruvate dehydrogenase, the mitochondrial enzyme complex that converts pyruvate to acetyl coenzyme A (Pácal et al., 2011). This function was the basis for linking thiamin requirement to caloric intake; early evidence suggested that the dietary thiamin requirement was proportional to the intake of absorbable carbohydrates. Alcohol and folate depletion reduce thiamin absorption via downregulation of transporters on the polarized absorptive epithelial cells. Furthermore, both alcoholic and nonalcoholic liver damage limit thiamin phosphorylation, required for enzyme activity. CKD has been shown to downregulate thiamin and folate transporters in the 5/6 nephrectomy rat model. Although plasma concentrations of these nutrients were normal, plasma membrane and mitochondrial thiamin (and folate) transporters in the intestine, liver, heart, and brain were lower in these animals (Bukhari et al., 2011).

At this time, data on the extent of thiamin status compromise with diabetes, CKD, aging, and with excess alcohol consumption are incompletely described. However, given its requirement for glucose homeostasis, its increased loss with hyperglycemia, and its potential transport compromise in CKD, thiamin intake at recommended dietary allowance (RDA) level may be insufficient under conditions of diabetic nephropathy. A recent randomized, double-blind clinical trial using 3×100 mg thiamin/day for 6 weeks reported a significant decrease in 2-hour plasma glucose relative to baseline; no changes in 2-hour plasma insulin or renal function were observed within or between study arms (Alaei et al., 2013).

The RDA for thiamin for adults is 1.1–1.2 mg/day for healthy adults (Board, 1998). Thiamin is found in the seed coat of grains and seeds; thiamin content in the starch (endosperm) is low, however in the United States, flour, cereals, and other staple foods are fortified with thiamin. Meat, especially pork, and fish are rich sources of thiamin; dairy products contain minimal thiamin. Thiamin content in food is destroyed by heat and because it is water-soluble, cooking processes that discard cooking liquids also reduce its content. For this reason, sautéing is preferable to boiling and pan sauces or soups should be prepared from the fond remaining in the pan after roasting meat.

Thiamin status can be assayed in whole blood, reflecting the thiamin content in the red blood cells; the normal range is 70–180 nM/L. An indirect functional measure of erythrocyte transketolase can also provide useful information. Erythrocyte hemolysates are incubated in excess substrate (ribose-5-phosphate) in the presence and absence of thiamin pyrophosphate. The percent increase in product reflects thiamin adequacy in the erythrocytes; an increase >25% indicates thiamin deficiency. No toxic upper limit for oral thiamin intake has been set. Further information on thiamin requirements and functions is available (https://ods.od.nih.gov/factsheets/Thiamin-HealthProfessional/#en6).

7.3.5 Resistant Starch, Nonstarch Polysaccharide, Dietary Fiber, and Gut Microbiota in CKD

Increased intestinal permeability and altered microbiota have been demonstrated in animal models of CKD (Vaziri et al., 2015). Changes observed in vivo include disruption of tight junctions throughout the gastrointestinal tract (colon, ilium, jejunum, stomach). These observations have been supported by in vitro studies in cultured human colonocytes exposed to uremic human plasma. In addition to the altered barrier function, dysbiosis has also observed in animal models and in humans with CKD. Disrupted barrier function can contribute to inflammation and elevated levels of circulating uremic toxins by allowing translocation of microbes and their products, including proinflammatory uremic toxins such as indoxyl sulfate, P-cresol sulfate, and trimethylamine-N-oxide (refer to Chapter 6: Atherosclerosis and Arterial Calcification: Influence of the Microbiome on AS Risk), as well as protein-bound solutes such as phenols and hippurates generated by microbial fermentation.

These profound changes in the gastrointestinal tract of the CKD patient arise from a variety of factors as reviewed by Vanholder and Glorieux (2015) and Vaziri et al. (2015). Elevated blood urea leads to its influx into the intestinal lumen where it is a substrate for microbial urease leading to the formation of ammonia (NH_3). NH_3 is converted to alkaline ammonium hydroxide (NH_4OH), thereby raising the pH of the intestinal lumen and altering the bacterial milieu; often this leads to enterocolitis. Because urea and uric acid are not excreted by the failing kidney, they are secreted into the intestinal lumen where they provide an energy-rich substrate for the microbiota, reducing the utilization of complex carbohydrates and facilitating production of uremic toxins. Thus the uremic environment shifts the microbial balance to proteolytic microbes that produce toxin precursors. CKD also reduces intestinal motility, contributing to bacterial overgrowth in the small intestine and increasing transit time, allowing greater bacterial toxin absorption (Vanholder and Glorieux, 2015).

A major cause of altered microbial diversity in CKD is the restriction of potassium-rich fruits and vegetables to prevent hyperkalemia, especially in the aneuric patient. As discussed elsewhere (refer to Essentials IV: Diet, Microbial Diversity and Gut Integrity), fruits and vegetables, as well as whole grains, legumes, nuts, and seeds provide a complex carbohydrate substrate for commensal bacteria. Enhanced microbial action that provides short-chain fatty acids as an energy source for colonocytes and for regulatory T-lymphocytes (Kinoshita and Takeda, 2014) (refer to Chapter 3: Type I Diabetes and Celiac Disease: Evidence for Targeted Nutrient Modification in Type I DM), modulates intestinal immune function. The possibility that a potassium-restricted diet could exacerbate dysbiosis in CKD was confirmed in a rat model of CKD. Rats fed a low-fiber diet had reduced creatinine clearance, interstitial fibrosis, inflammation, tubular damage, disruption of colonic tight junctions, increased tissue damage through upregulation of inflammatory, oxidative and fibrotic pathways, and an impaired Nrf2 pathway. In contrast, these damages were significantly ameliorated in rats fed the high-fiber diet (Vaziri et al., 2014). Several clinical trials have tested various types of dietary fiber (resistant starch, nonstarch polysaccharides) but these trials have been small and/or inconclusive. A current review (Vanholder and Glorieux, 2015) suggests that modifying microbial diversity and function by diet holds promise, but that well-designed, robust clinical trials are needed to derive dietary recommendations for the patient with diabetic neuropathy. In the interim, a potassium-poor diet with increased resistant starch and dietary fiber is presented below.

7.3.6 Lipid Dysregulation and Dyslipidemia in CKD

The lipid profiles associated with metabolic syndrome and CKD are phenotypically similar and linked to insulin resistance (Vaziri, 2006, 2013) (refer to Chapter 6: Atherosclerosis and Arterial Calcification: Targets for Atherosclerotic Prevention and Control). Marked reductions in lipoprotein lipase and the very-low-density lipoproteins (VLDL) receptor in skeletal muscle and adipose tissue, as well as downregulation of hepatic lipase (HL) are associated with decreased clearance of triglyceride-rich lipoproteins (chylomicrons and VLDL). Cholesterol transport is impaired by reduced LDL receptor-related protein, hepatic production of Apo-A1 and lecithin-cholesterol acyl transferase (LCAT). At the same time, acyl cholesterol acyltransferase is upregulated in the liver, kidney, and arteries, promoting cholesterol retention and storage in lipid droplets. Plasma lipoproteins are oxidatively modified, contributing to systemic inflammation and vascular disease in CKD patients. In most patients, plasma cholesterol levels are within normal limits, however cholesterol levels can sometimes be elevated in patients with ESKD. These changes result in higher levels of triglyceride-rich lipoproteins (VLDL and chylomicrons) and lower levels of high-density lipoproteins (HDL). Dyslipidemia develops early in CKD progression, presenting as increased concentrations of intact and partially metabolized triglyceride-rich VLDL and IDL, and becomes more abnormal as the disease progresses (Vaziri and Moradi, 2006).

The dyslipidemia observed in a CKD patient reflects reduced triglyceride clearance. Elevated ApoC III concentrations seen in the CKD patient may be a contributory factor (Ooi et al., 2011). ApoCIII has several actions that reduce effective clearance of Apo B containing triglyceride-rich lipoproteins and interfere with HDL-mediated reverse cholesterol transport. In the diabetic patient, insulin-insensitive adipocytes, deprived of the anabolic insulin signal, increase lipolysis, releasing abundant free fatty acid substrate for hepatic VLDL synthesis. Thus delayed clearance of triglyceride-rich lipoproteins, adipocyte lipolysis, and increased production of triglyceride-rich lipoproteins secondary to insulin resistance, may also partially explain the dyslipidemic profile in CKD.

Normal HDL function (Fig. 7.9) is dysregulated in chronic renal failure, with inadequate Apo A availability and reduced activity of (LCAT) as reviewed by Vaziri et al. (2010) and Vaziri (2016). Apo A (I and II) are integral HDL proteins, synthesized in the intestine

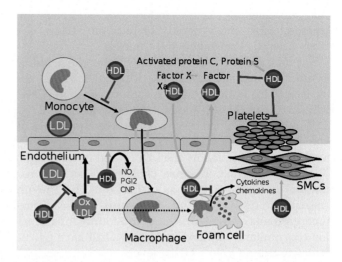

FIGURE 7.9 **HDL function at the vascular endothelium.** In addition to its role in mediating cholesterol efflux from lipid-laden foam cells, HDL enhances endothelial function by activating eNOS to generate vasoactive NO. It exerts antioxidant effects through its paraoxonase and glutathione peroxidase enzymes and by limiting eNOS uncoupling. HDL limits the expression of vascular cell adhesion molecule 1 (VCAM-1) reducing monocyte adhesion to the endothelium and exerts antithrombotic effects through inhibition of tissue factor, P-selectin, protein S, and nitric oxide production. It promotes endothelial growth, prevents apoptosis and increases endothelial progenitor cells (Vaziri, 2016). *Source: By Rfch (own work) (Public domain), via Wikimedia Commons.*

and liver and inserted into chylomicrons and VLDL. When these triglyceride-rich lipoproteins reach the hepatic extracellular space, Apo A dissociates and coalesces to form nascent HDL, a phospholipid-protein disk. Apo AI comprises ~70% of the HDL protein and plays both a structural role and LCAT activator role. Nascent HDL acquires Apo proteins E and C from the plasma or from chylomicron and VLDL remnants and is converted to the small, cholesterol-poor disk known as HDL3. HDL3 circulates in the bloodstream, removing cholesterol from cell membrane caveolae by linking to ATP-binding cassette transporters (ABCA1 and ABCG1).

LCAT is a copper-dependent protein that transfers an acyl group from HDL surface phosphatidylcholine (lecithin) to cholesterol to form a hydrophobic cholesterol ester (CE) central core. HDL2 is formed when the CE core has reached spherical capacity; HDL2 returns to the liver where it is hydrolyzed by HL. Apo AII (20%) activates *HL* and facilitates transfer of lipids to the liver. Cholesterol ester is recycled into lipoproteins for transport, used for biosynthetic pathways, or excreted as bile. Cholesterol can also be removed from cells by passive diffusion of free cholesterol across the caveolae, followed by binding to serum albumin and subsequent transfer to HDL in the bloodstream. HDL also removes oxidized cholesterol from macrophages and endothelial cells (refer to Chapter 6: Atherosclerosis and Arterial Calcification: Targets for Atherosclerotic Prevention and Control).

Several factors could reduce HDL abundance and/or function in diabetic nephropathy. HDL levels are decreased in both insulin resistance and in CKD (Kaysen, 2006). The major cause of low HDL in CKD appears to be the increased fractional catabolic rate of Apo A1; the small, discoid HDL3 is particularly vulnerable to degradation by the kidney and other tissues. LCAT downregulation limits cholesterol ester formation and prevents maturation of HDL3 to HDL2, with subsequent destruction of immature HDL3. Thus hyperglycemia, proteinuria, and profound inflammation concomitant with progressive CKD all have potential to contribute to low HDL2 levels.

A relatively unexamined possibility is that copper deficiency observed in the CKD patient, especially the ESKD patient on dialysis (Kasama, 2010), could compromise LCAT activity and/or reduce ATP production by compromising cytochrome oxidase activity in the electron transport chain. There is fragmentary evidence in rat models that copper deficiency increases cholesterol and triglyceride pools and elevates LDL and also HDL, but reduces LCAT activity (Harvey and Allen, 1981; Lau and Klevay, 1981), which could compromise reverse cholesterol transport. HDL also functions as an effective antioxidant. Its proteins, especially Apo AI and *paraoxonase*, have the capacity to bind and inactivate oxidants. These protective functions of HDL appear to be easily altered by oxidative stress (Norata et al., 2006). For example, HDL can be converted from an antioxidant to a pro-oxidant in the presence of free metal ions, especially iron, in atherosclerotic plaques as described below.

The possible role of copper as a source of oxidative damage and other risk factors for atherosclerosis have been reviewed (Ferns et al., 1997). It should be noted that copper deficiency with reduced ferroxidase activity as a cause for resistance to erythropoietin therapy for renal anemia in dialysis patients has been reported (Higuchi et al., 2006). It should also be noted that elevated chelatable copper observed in CKD could contribute to oxidative stress and at the same time, downregulate copper transport into the cell as reviewed elsewhere (refer to Chapter 5: Cardiomyopathy and Congestive Heart Failure: Targets for Myocardial Protection: Cardiac Remodeling). In view of its potential role in lipid transport and ETC function, further studies on copper status in the CKD patient appear to be warranted.

7.3.7 Reduction of Cholesterol Biosynthesis by HMG CoA Reductase Inhibitors

While CKD patients do not generally have elevated cholesterol, it is possible that hypercholesterolemia in the diabetic patient could play a role in the initial kidney insult (Amann and Benz, 2011). Hypercholesterolemia has been associated with the development of glomerulosclerosis, mesangial expansion, inflammatory cell infiltration, tubulointerstitial injury, and increased glomerular capillary pressure both in rodent models and in patients with reduced renal mass. These observations support the suggestion that inhibitors of *HMG CoA reductase* (HMG-R), the rate-limiting enzyme in the mevalonate pathway for cholesterol synthesis, could be renoprotective (Liang et al., 2004). Statins inhibit MCP-1 gene expression, thereby decreasing infiltration of macrophages across the vascular endothelium. Statins also increase vasodilative and anticonstrictive mechanisms that foster normalized hemostatic effects and exert an antioxidative effect via increased endothelial nitric oxide synthase (eNOS) activity with subsequent NO production.

While there is evidence that statin actions can protect the kidney, there is concern that adverse effects from statin use can increase risk for CKD (Yuet et al., 2015). Statins can increase risk for diabetes by several mechanisms, including downregulation of the insulin-dependent GLUT4, resulting in insulin resistance. Hyperglycemia is a major risk factor for CKD, as described above. Myopathy secondary to HMG-R inhibitors can initiate kidney injury by inducing muscle degeneration; in rare cases, rhabdomyolysis has been reported. Myoglobin and other products of muscle degeneration are implicated in the pathogenesis of acute renal failure. Statin therapy has also been associated with cognitive impairment, neuropathy, and liver damage; the mechanism that underlies these side effects has not yet been clarified. Since statins block HMG-R, they also limit the biosynthesis of complex isoprenoid molecules including ubiquinone (coenzyme Q10), a major component in the oxidative phosphorylation chain (refer to Chapter 5: Cardiomyopathy and Congestive Heart Failure: Mitochondrial Production and Management of Reactive Oxygen Species; and Essentials I: Life in an Aerobic World: Single Unit Antioxidant Nutrients: Coenzyme Q10). Reduced availability of ubiquinone can compromise mitochondrial function, limiting ATP production required for nerve and muscle function.

It must be noted that food contains many bioactive components that modulate HMG-R by alternate mechanisms (Mo and Elson, 2004). (The mevalonate pathway builds acetyl CoA units into isoprenoid precursors that are further polymerized into more complex isoprenoids required for essential cellular functions. For example, bile, steroid hormones, and ubiquinone are formed by the mevalonate pathway). This pathway is regulated by feedback inhibition and by isoprenoid components in the diet. For example, farnesol, an acyclic sesquiterpene, and the farnesyl homologs, γ-tocotrienol and farnesyl derivatives, inhibit HMG-R synthesis and accelerate HMG-R degradation. Cyclic monoterpenes, *d*-limonene, menthol and perillyl alcohol and β-ionone, a carotenoid fragment, lower HMG-R mass; in part by modulating translational efficiency. Thus, although clinical trials are still fragmentary, the possibility remains that statin effects can be synergistically augmented with bioactive dietary components, thereby lowering the drug dose required. On the other hand, it is unclear to what extent modifying the pathway by food components will alter the mechanisms underlying drug function or side effects.

7.3.8 Protein and CKD

7.3.8.1 Energy Balance and Uremic Sarcopenia

Muscle loss in uremia was described as uremic myopathy over 50 years ago (Serratrice et al., 1967). Muscle wasting begins early in renal failure and has been reported in ~50% of patients on dialysis. Indeed, the loss of lean body mass is multifactorial, significant, and similar to sarcopenia observed with aging. Although skeletal muscle physiology appears normal, reduced muscle force, selective structural changes, and significant muscle wasting have been documented (Fahal, 2013). Causes of these changes include reduced protein intake and sedentary lifestyle in CKD patients as well as endocrine, immunologic, and myocellular changes associated with renal dysfunction. Dramatic alterations in body composition with CKD predict rapid progression to ESKD. Thus, although obesity is considered a risk factor for cardiovascular disease and diabetes, a large body of epidemiologic and clinical evidence supports the "dialysis-risk-paradox." This hypothesis states that high body mass index is protective and that renal patients with a low body mass index have worse survival (Kalantar-Zadeh et al., 2005).

Comorbid changes with CKD that can lead to reduction in body mass include:

- Concurrent chronic illnesses and/or superimposed acute illnesses, both of which increase secretion of inflammatory cytokines;
- Diagnostic or therapeutic procedures that reduce nutrient intake or stimulate net tissue autophagy and protein breakdown;
- Chronic blood loss due to losses from dialysis and/or frequent blood sampling;
- Endocrine disorders, especially resistance to insulin and insulin-like growth factor-I, hyperglucagonemia and hyperparathyroidism;

- Deficiency of active vitamin D ($1,25$-OH_2D_3; $1,25D$) and hyperparathyroid secretion associated with renal osteodystrophy;
- Products of metabolism or from dysfunctional intestinal microbiota such as organic and inorganic acids that accumulate in renal failure and may induce wasting;
- Loss of the metabolic actions of the kidney;
- Accumulation of toxic compounds taken up from the environment (eg, aluminum).

7.3.9 Malnutrition in CKD

The most vulnerable period for the CKD patient occurs late in disease progression, just prior to development of ESKD (Kopple, 1999). Protein-energy wasting (PEW) was coined by the International Society of Renal Nutrition and Metabolism to describe the multiple catabolic alterations that occur in CKD. PEW has been identified in 15–54% of late-stage CKD patients (~40% reported by Kopple) and is a powerful predictor for morbidity and mortality as illustrated in Fig. 7.10. During this period, the CKD patient can experience comorbid states such as infection and inflammation that can independently cause PEW or exacerbate malnutrition by decreasing food intake or increasing loss through inflammation and hypercatabolism. The risk for PEW has been confounded because, for over 70 years, standard diet therapy restricted protein to slow the rate of CKD progression and delay the need for dialysis.

For over 100 years, diets containing minimal amounts of high biological value protein have been proposed and studied in patients with CKD (Thilly, 2013) as discussed below. The effectiveness of this intervention to slow disease progression has been investigated in a number of studies, including four meta-analyses (Campbell et al., 2008). Data were confounded because poor compliance with the diet regimen was observed in several of the studies. Taken together, a clinical consensus has been reached that low protein intakes may have no greater benefit than protein intakes at the level recommended for healthy adults (0.6–0.8 g protein/kg/day) (Johnson, 2006). In fact, a low protein intake can be associated with PEW syndrome.

7.3.10 Detection of PEW in the CKD Patient

Protein calorie malnutrition has been reliably assessed in patients with a variety of diseases by combining multiple components, such as dietary and medical history, amount of weight loss, biochemical variables, and anthropometry using the subjective global assessment tool and its multiple variations. Four

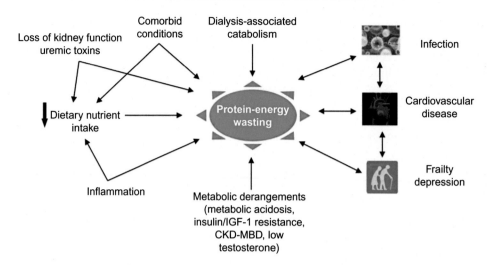

FIGURE 7.10 **Causes and potential sequelae of the protein-energy wasting syndrome.** Multifactorial causes put the CKD patient at risk for severe protein wasting. Nutritional strategies can reduce mortality in malnourished CKD patients, but cannot solely correct protein wasting (Carrero et al., 2013). *Source: Reproduced from Carrero, J.J., Stenvinkel, P., et al., 2013. Etiology of the protein-energy wasting syndrome in chronic kidney disease: a consensus statement from the international society of renal nutrition and metabolism (ISRNM). J. Renal Nutr. 23 (2), 77–90, with permission from Science Direct.*

main and established categories were recognized for the diagnosis of PEW in the CKD patient: biochemical criteria; low body weight, reduced total body fat or weight loss; a decrease in muscle mass; and low protein or energy intakes (Fouque et al., 2007). At least three out of the four categories (with at least one test in each selected category) must be satisfied for a diagnosis of PEW. Each criterion should be documented at least three times (2–4 weeks apart).

- *Biochemical criteria:* serum albumin < 3.8 g/dL, serum prealbumin (transthyretin) <30 mg/dL, and serum cholesterol <100 mg/dL.
- *Body mass criteria:* BMI <23, unintentional weight loss >5% over 3 months or 10% over 6 months, and total body fat percentage <10%.
- *Muscle mass criteria:* reduced muscle mass 5% over 3 months or 10% over 6 months, reduced mid arm muscle circumference area (mid arm muscle mass) >10% lower than 50th percentile of reference population, and creatinine appearance. (Increased creatinine generation reflects muscle wasting. The panel noted that it is not possible to routinely assess the absolute rate of creatinine generation and therefore recommend that the term creatinine appearance be used. This can be estimated by measurement of creatinine in 24-hour urine collection and in the collected spent dialysate.)
- *Dietary intake criteria:* unintentional low protein intake <0.80 g/kg/day for at least 2 months (dialysis patients) or <0.6 g/kg/day for CKD patients in stages 2–5 and unintentional low energy intake <25 kcal/kg/day for at least 2 months.

7.3.11 The Case for a Low-Protein Diet in CKD

Recommendation of the low-protein diet (LPD) to reduce the rate of progression of CKD is not new. It was proposed as early as 1869 and experimentally tested in 1926 as recently reviewed by Thilly (2013). In the 1960s, Giordano and Giovannetti demonstrated that an LPD supplemented with essential amino acids, keto acids or provided with proteins of high biologic value (BV) could achieve zero nitrogen balance and minimize uremic signs and symptoms. More recent studies have shown that by lowering blood urea and nitrogenous wastes, LPD also improved secondary hyperparathyroidism, peripheral resistance to insulin, hyperlipidemia, hypertension, and acid–base disorders, delaying ESKD and dialysis for over a year.

Arguments against a LPD include the risk for predialysis PEW in late-stage CKD patients, associated with increased morbidity and mortality as described above. To test whether an LPD (<0.6 g/kg/day) or a very LPD (<0.3 g/kg/day) supplemented with keto-amino acids reduced progression to ESKD, a large multicenter trial was conducted in 1994 (Klahr et al., 1994). The results were inconclusive. The study authors suggested that although slightly slower progression with the LPD was seen in CKD patients in early stages, patients in late stages showed no benefit. Secondary analysis of the data suggested that the LPD was beneficial for selected patients. Meta-analyses of LPD studies have been similarly confusing because of heterogeneity in diet composition, diverse patient age, variable patient compliance, stage and stage of CKD, and endpoints of the studies.

There is little doubt that the well-constructed LPD or the supplemented VLPD containing adequate calories does not cause PEW or nutrient inadequacy in the predialysis CKD patient and is associated with relatively normal biochemical indices (Mandayam and Mitch, 2006). In contrast, an unrestricted protein intake increases risk for rising serum levels of creatinine, urea nitrogen, and phosphorus, and causes metabolic acidosis, usually treated by dialysis. A recent report (Meek et al., 2013) demonstrated in both animal models and human subjects that high-protein diets, especially in the diabetic patient, increased glomerular cell death-mediated AGE binding to podocyte receptors (RAGE) and was accompanied by inflammation. Thus the LPD appears to reduce kidney damage and uremic symptoms and delay the need for dialysis. However, the success of the LPD is highly dependent on the skill of the dietitian in constructing a diet that is palatable and acceptable to the patient, the ability of the patient and/or caregiver to prepare the diet, and the willingness of the patient to comply with the regimen (Fouque et al., 2011; Ikizler et al., 2013).

7.4 CURRENT PUBLISHED RECOMMENDATIONS TO SLOW PROGRESSION TO ESKD

The Agency for Healthcare Research and Quality and the Effective Health Care has developed methods to assess the overall quality and strength of evidence (AHRQ, 2014): (A) high confidence; (B) moderate confidence suggesting that further research may change the confidence; (C) low confidence, indicating that confidence level may be modified by further research; and (D) insufficient, indicating that evidence was unavailable or did not permit a conclusion. The American Diabetes Association (2010) has developed a similar grading system, however they replace (D) insufficient with (E) expert opinion, when clinical trials have not yet been conducted, or for which there is conflicting evidence.

7.4.1 Fat and Lipid Intake

No consensus exists as to the ideal macronutrient intake for diabetic patients with CKD, as changes in plasma lipids in response to dietary modification are highly dependent on genetic factors. It should be remembered CKD patients have dyslipidemia rather than hyperlipidemia and that CKD is often accompanied by normal serum cholesterol levels.

The following recommendations have been given for patients with diabetes-related diseases and may be applied to the patient with diabetic nephropathy (Mooradian, 2009):

- Reduced saturated fat intake is associated with reduced cholesterol content of various plasma lipoproteins, including LDL and HDL cholesterol.
  - If saturated fat is replaced with carbohydrates, plasma triglycerides are increased, while HDL and LDL cholesterol levels decline.
  - If saturated fat is replaced with mono- or polyunsaturated fat, triglycerides do not rise, and HDL levels do not fall.
- Trans-fatty acids or hydrogenated-fat-enriched diets increase LDL cholesterol level, and either decrease or have no effect on HDL cholesterol and ApoA-I levels.
- Omega-3 fat consumption has a well-described triglyceride-lowering effect. (Note: clinical trials that used a purified omega-3 fatty-acid formulation reported no change, or only a modest increase, in plasma HDL cholesterol levels; the effect on LDL cholesterol levels was variable.)

Guidelines (American Diabetes Association, 2010) for the diabetic patient to reduce dyslipidemia include:

- Saturated fat intake should be <7% of total calories. (A)
- Reducing intake of *trans* fat lowers LDL cholesterol and increases HDL cholesterol (A); therefore, intake of *trans* fat should be minimized. (E)

7.4.2 Protein Intake

- The impact of dietary proteins on plasma lipids is poorly studied
  - Some proteins, such as soy protein, reduce LDL and raise HDL cholesterol.
  - Replacement of carbohydrates with proteins led to modest increases in HDL levels, and a reduction in triglyceride levels.
- Lipid changes could have been the result of limiting glycemic load, rather than a consequence of increasing the protein load.

Recommendations for the CKD patient based on ideal body weight (Imran et al., 2015)

	Nondialysis CKD	**Hemodialysis**
Protein	0.6–0.8 g/kg/day systemic illness 1.0 g/kg[a]	>1.2 g/kg/day
Energy	30–35 kcal/kg/day[b]	30–35 kcal/kg/day[b]
Sodium	80–100 mmol/day	80–100 mmol/day

(Continued)

(Continued)

	Nondialysis CKD	Hemodialysis
Potassium	<1 mmol/kg if elevated	<1 mmol/kg if elevated
Phosphorus	800–1000 mg and phosphate binders if elevated	800–1000 mg and binders if elevated

[a] *> 50% high biological value protein (rich in essential amino acids) is recommended.*
[b] *Based on physical activity level. In sedentary elderly adults, recommended energy intake is 30 kcal/kg/day.*

7.4.3 American Diabetes Strategies to Reduce Hyperglycemia and Insulin Insensitivity (American Diabetes Association, 2010)

Guidelines for maintenance of hyperglycemic control with diet have been published along with their levels of evidence. In general, a structured program emphasizing lifestyle changes is recommended including:

- Moderate weight loss (7% body weight). (A)
- Regular physical activity (150 minutes/week). (A)
- Dietary strategies including reduced calories and reduced intake of dietary fat. (A)
- Dietary fiber (14 g fiber/1000 kcal) and foods containing whole grains (one-half of grain intake). (B)
- Sugar alcohols and nonnutritive sweeteners are safe when consumed within the acceptable daily intake levels established by the Food and Drug Administration. (A)
- Alcohol should be limited to a moderate amount (one drink per day or less for adult women and two drinks per day or less for adult men). (E)
- Routine supplementation with antioxidants, such as vitamins E and C and carotene, is not advised because of lack of evidence of efficacy and concern related to long-term safety. (A)
- Benefit from chromium supplementation in people with diabetes or obesity has not been conclusively demonstrated and, therefore, cannot be recommended. (C)

7.5 TARGETED MACRONUTRIENT MODIFICATION TO SLOW CKD PROGRESSION

7.5.1 Macronutrient Intake, Diet Patterns, and Insulin

In diabetic patients both with and without renal involvement, weight loss and improved control have been achieved by decreasing carbohydrate intake to ~40% of daily kcal, and increasing protein intake to 30% of the diet (Gannon and Nuttall, 2004; Parry-Strong et al., 2013). Improved mitochondrial function and metabolic flexibility have been obtained by physical exercise (Meex et al., 2010). Improved insulin sensitivity has also been achieved by consumption of bioactive components of food, especially resveratrol in red wine and catechins in tea (Lagouge et al., 2006). While it is tempting to speculate that the microalbuminuria and fluid retention in diabetic nephropathy can be ameliorated by lowering plasma insulin levels by dietary reduction of free sugars, simple starches, and noncaloric sweeteners, a causal link has not yet been made between diet and insulin action at the renal tubule. Nonetheless, a recent population study reported that a Western dietary pattern (high red meat, fat, and sugar) was associated with significantly greater risk for microalbuminuria and rapid decline in kidney function, while a dietary approaches to stop hypertension (DASH) diet (high vegetables, fruits, and fiber) appeared to reduce risk for CKD progression (Lin et al., 2011).

7.5.2 Adjuvant Therapy to Minimize Symptoms and Prevent PEW in CKD Patients

Given the difficulty of obtaining patient compliance with a LPD, the protein recommendation for CKD patients with a GFR <25 mL/minutes is 0.6–0.75 g protein/kg/day with ~50% from sources of high BV. (BV is an index of the proportion of dietary protein that can be used for protein synthesis in the cells. Properties of the food that affect BV include its amino acid composition (especially limiting amino acids), its method of preparation, and its nutrient content (required for protein synthesis). In general, egg yolk is considered 100% available, followed by soy and whey proteins. The BV is somewhat less for meat and fish (70–80%) and vegetable protein (60–70%). For further information on BV calculation, see citations in http://en.wikipedia.org/wiki/Biological_value.) Energy recommendations are 35 kcal/kg/day for persons <60 years of age and 30–35 kcal/kg/day for older individuals (recommendations from the Academy of Nutrition and Dietetics). These recommendations are based on the observation that many CKD patients do not maintain neutral or positive nitrogen balance on protein diets providing ~0.6 g protein/kg/day. As indicated above, PEW causes include CKD-induced anorexia and failure to consume sufficient dietary protein and energy to maintain muscle mass, comorbid illness, and/or patient anxiety.

Intensive dietary training is required to assure that the CKD patients ingest adequate amounts of protein and that the protein quality is high enough to provide the spectrum of amino acids needed for biosynthetic use. Ingestion of too few calories for fuel, even with adequate protein intake, diverts protein to energy production under these conditions, amino acids are less available for biosynthetic use. Causes of poor intake, including anorexia and altered taste sensation due to uremic toxins, underlying illness such as advanced diabetes with altered gastric motility, or anxiety and depression, commonly seen in CKD patients should be identified and addressed. Other causes of poor intake including inadequate living conditions, poverty, and food insecurity may require additional public health efforts. Poorly fitting dentures resulting from CKD-induced bone and muscle loss, as well as fatigue-induced inability to shop for food also contribute to malnutrition. Thus, it is clinically astute to recognize that the single most important cause of malnutrition in the CKD patient is inadequate dietary intake.

Weight-loss diets must be carefully considered, since maintenance of adipose tissue stores may be protective in CKD. In contrast to the general medical consensus that obesity is a classical risk factor for cardiovascular disease and diabetes, a large body of epidemiologic and clinical evidence supports the "dialysis-risk-paradox" that states that higher body mass index is protective and that renal patients with a low body mass index have a decreased survival prognosis (Kalantar-Zadeh et al., 2005). It is likely that in the CKD patient, increased calories spare ingested protein for biosynthetic use, thereby minimizing muscle catabolism with attendant production of nitrogenous wastes. Thus, in many cases, calorie restriction for the CKD patient is contraindicated.

7.6 STRATEGIES TO AMELIORATE MICRONUTRIENT ALTERATIONS DUE TO IMPAIRED KIDNEY FUNCTION

The kidney carries out a spectrum of metabolic and physiologic functions outlined in Table 7.2. Nutritional management is complicated because CKD is accompanied by complex regulatory adaptations.

7.6.1 Fluid, Electrolytes, and Acidosis

7.6.1.1 Sodium Load and Fluid Balance in CKD

Exposure to a high dietary sodium load can damage vascular and tubular components of the kidney by both direct and indirect mechanisms (Ritz and Mehls, 2009). Indirect damage can be exerted by a salt-induced increase in blood pressure and/or by withdrawal of drugs (eg, ACE inhibitors) that block RAAS signaling. In CKD patients, dietary salt loading

TABLE 7.2 Functions of the Kidney and Nutritional Problems Due to CKD[a]

Function	Nutritional consequences of dysfunction
Maintains fluid and electrolyte balance	Hyponatremia, hyperkalemia, low-potassium content hypocalcemia, hyperphosphatemia, decreased tolerance to electrolyte, or mineral loading
Regulates blood pressure	Hypertension, cardiovascular disease
Endocrine mediators	Anemia (low erythropoietin), hypertension (renin system activation), bone disease (secondary hyperparathyroidism), low vitamin D activation, prolonged half-lives of peptide hormones (eg, insulin)
Waste product excretion	Acidosis, anorexia, nausea, soft tissue deposition of oxalates and phosphates, neurologic dysfunction, loss of muscle protein

[a]Adapted from Cecil Medicine 23rd Ed. 2008. Chapter 131 (pp. 921–930).

increases renovascular resistance, likely by increasing TGFβ-1, directly involved in vascular and glomerular fibrosis. Decreased renal blood flow and increased glomerular capillary pressure are thus an indirect result of salt loading. A likely mechanism for direct salt-induced damage is oxidative stress due to imbalance in oxidant/antioxidant systems. High salt intake increases ROS generation and upregulates the expression of the mRNA for *NADPH oxidase*, thereby decreasing regeneration of reduced glutathione. Other endogenous antioxidant enzyme systems, for example, *superoxide dismutase*, are also downregulated.

Sodium balance can be impaired early in kidney function decline. As sodium excretion per nephron falls, the remaining nephrons adapt to excrete more sodium. When renal adaptation can no longer maintain sodium balance, excess sodium is retained, leading to volume overload, hypertension, and potential for further kidney damage. Hypertension increases the filtration and excretion of sodium, but the tradeoff for this adaptation is persistent hypertension. Experimental evidence suggests that the kidney adapts by increasing circulating inhibitors of $Na+-K+$ ATPase, thereby restricting sodium resorption at the tubular lumen. Thus, early in renal failure, adaptation can result in sodium wasting in some patients. If dietary sodium is acutely restricted in these "salt-losing" patients, intravascular fluid volume falls and kidney perfusion may be impaired.

Obstructive nephropathy also damages the kidney's ability to reabsorb water and salt. The damage can persist even after the obstruction is removed. Human infants (<1 year) with congenital obstructive hydronephrosis were studied postpyeloplasty. Despite removal of the obstruction, the post-obstructed kidney excreted higher amounts of dilute urine, suggesting that urine-concentrating ability was lower in the postobstructed kidney (Murer et al., 2004). Further work is underway to elucidate the effects of prostaglandin E2, ROS, and other local signaling molecules on downregulation of aquaporins and sodium transporters in the damaged tubules. Nonetheless, it is important to remember that sodium and water reabsorption is compromised in severely damaged tubules. Patients in early stages of CKD, especially those with tubular damage, may also waste salt and should be evaluated for sodium retention. If patients are losing sodium, they should be advised to avoid diets that restrict water and sodium as these diets have potential to cause hypovolemia and accelerate kidney damage.

7.6.1.2 Metabolic Acidosis and Extracellular Potassium

Metabolic processes generate both volatile and non-volatile acids. Volatile acid results from oxidation of carbohydrate and fat; the CO_2 produced is expired through the lungs. In contrast, nonvolatile acids (H^+) derived from sulfur containing amino acids, methionine and cysteine, as well as from organic acids including citrate, urate, and oxalate, are excreted in the urine in the form of ammonium and titratable acid (Scialla and Anderson, 2013). Progressive renal failure is accompanied by a decreased ability to excrete nonvolatile acid, resulting in lower blood pH and reduced plasma bicarbonate levels. As CKD progresses, the acid load is buffered by bone, resulting in CKD-associated bone abnormalities. Experimental metabolic acidosis activates the ubiquitin—proteasome pathway leading to proteolysis and increased protein breakdown (Mitch et al., 1994). Ammonia and nitrogen breakdown products released from body proteins injure the kidney and stimulate chronic tubulointerstitial inflammation through a complement-mediated pathway. Although low plasma bicarbonate levels and frank metabolic acidosis are commonly seen in late-stage CKD, a body of literature suggests that reduced excretion of acid is present even at early stages. Acidosis is not clinically apparent because the acids are buffered by intercellular components and bone (Scialla and Anderson, 2013).

Potassium is critically important for nerve transmission and is the major intracellular cation. As the kidney fails (GFR < 20% of normal), serum potassium concentration rises, increasing risk for potentially fatal cardiac arrhythmias. Normally, potassium is secreted in the proximal tubules on a concentration gradient. If the patient has high-output kidney failure with increased urine flow, secreted potassium will be carried out on the urine flow and hyperkalemia is less likely to occur. As the urine flow declines, secreted potassium is poorly excreted and the concentration gradient is reversed. Potassium is reabsorbed, putting the patient at risk for hyperkalemia. Aldosterone stimulates potassium excretion both in the distal tubule and in the colon. Potassium excretion is reduced by nephrotoxic drugs that block the RAAS system and alter tubular function. These drugs include RAAS inhibitors (ACE inhibitors), aldosterone antagonists, heparin, nonselective beta-blockers, trimethoprim, nonsteroidal antiinflammatory drugs, and calcineurin inhibitors.

As CKD progresses and metabolic acidosis becomes more severe, H^+ enters the cell on a K^+-H^+ exchanger for buffering. Under these conditions, potassium is released from the cell and raises the extracellular potassium pool. Critical conditions such as heart failure, sepsis, and uncontrolled diabetes superimposed on CKD can stimulate additional K^+ efflux, worsening hyperkalemia and sometimes requiring emergency dialysis (http://www.merckmanuals.com/professional/critical_care_medicine/approach_to_the_critically_ill_patient/oliguria.html?qt=&sc=&alt#top. Hyperkalemia can become more clinically apparent as urine output falls below 500 mL/day (oliguria). Severe hyperkalemia is often treated with an ion exchange resin that binds potassium in the gut and facilitates its excretion (Sterns et al., 2010). Kayexalate (sodium polystyrene sulfonate) has been approved for this purpose, however its effectiveness has come into question and concern has been raised by several reports of colonic necrosis with use of the product.

7.6.2 Dietary Strategies to Normalize Sodium, Potassium, and Acid—Base Balance in CKD

7.6.2.1 Sodium

Hypertension is used as an index to reflect sodium retention and subsequent increased blood volume. Although a low-sodium diet is often recommended for the hypertensive, CKD patient, compliance with sodium restriction is difficult to achieve. A major contributing factor is that in developed countries, a large fraction of sodium intake is not under the patient's control. Over 20 years ago, discretionary sources, that is, cooking and table salt use, were found to contribute only 15% of the total salt intake while salt from restaurant and purchased food provided, on average, 85% of total salt intake (Sanchez-Castillo et al., 1987). The Centers for Disease Control and Prevention (CDC; sodium intake data at http://www.cdc.gov/VitalSigns/Sodium/?s_cid=bb-vitalsigns-113) recently estimated that the average daily sodium consumption in the US has risen to >3300 mg/day. Surveys have estimated that 90% of dietary sodium intake comes from prepared foods; 40%

of all sodium consumed comes from the following 10 foods: breads and rolls, cold cuts and cured meats (eg, deli or packaged ham, or turkey), pizza, fresh and processed poultry, soups, sandwiches (eg, cheeseburgers), cheese, pasta dishes, meat-mixed dishes (eg, meat loaf with tomato sauce), and snacks (eg, chips, pretzels, and popcorn). Given the plethora of processed and convenience foods available, CKD patients and their caregivers must be especially vigilant in diet choices, since they can only control table salt and salt added during home cooking.

Compliance with the 2012 KDOQI recommendation of <2.0 g sodium/day for CKD patients requires limitation of processed and heavily salted foods, and special vigilance at restaurants. (On average, fast food restaurant food contains 1848 mg of sodium per 1000 calories and sit-down restaurant food contains 2090 mg of sodium per 1000 calories. http://www.cdc.gov/media/releases/2014/p0123-Reducing-sodium.html/.) A model for achieving this goal is the DASH diet, providing 2–4 g/day of sodium. This diet contains 8–10 servings of fruits and vegetables, low-fat dairy products, whole grains, lean protein, nuts, and seeds and allows only small amounts of red meat, sweets, and sugar-containing beverages, and decreased amounts of total and saturated fat and cholesterol. Compared with a typical mixed diet consumed in the United States, the DASH diet significantly lowers blood pressure in people both with and without hypertension. A subsequent diet (DASH—low-sodium diet) provided ~1.5 g sodium and produced even greater blood pressure lowering. It should be noted that no CKD patients were included in the subject pool for either DASH trial. The 2012 KDOQI guidelines have

been challenged on the grounds that the low-sodium diet has not been adequately tested in CKD patients and potential long-term complications have not been evaluated (Anderson and Ix, 2013). The authors cited the increased GFR and lower vascular resistance observed in patients with early, uncomplicated diabetes following a low-sodium diet. These changes exacerbated existing renal abnormalities and could contribute to hyperfiltration and increased kidney dysfunction as a consequence of severely limiting sodium intake.

7.6.2.2 Acidosis and Potassium

Although neither metabolic acidosis nor hyperkalemia are prominent in early stages of CKD, the net whole-body acid load is increasing during this stage. Undetected acidosis decreases osteoblast-dependent mineralization and increases osteoclast resorption to maximize hydroxyl ions available to buffer elevated serum protons (Brandao-Burch et al., 2005). Several clinical trials have supported the use of sodium bicarbonate to neutralize acid and minimize cell loss of potassium. Sodium bicarbonate was well tolerated and fewer supplemented patients progressed to ESKD (de Brito-Ashurst et al., 2009). Markedly slower progression at early stages (stage 3) of CKD was achieved by aggressive maintenance of serum bicarbonate between 22 and 24 mMol/L by either diet or bicarbonate.

Diet modification can also dramatically lower the potential renal acid load (PRAL) as shown in Fig. 7.11. Diet lowered the PRAL by as much as 30–50%, the equivalent of supplementing 0.5 mEq/kg/day of sodium bicarbonate. Since in clinical trials, each

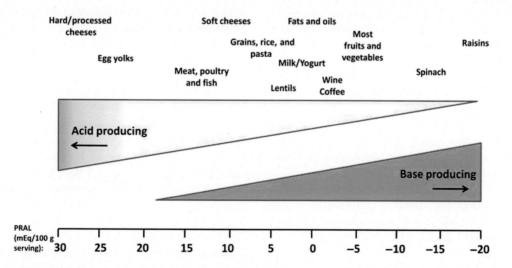

FIGURE 7.11 **Estimated acid-producing potential of selected foods.** Potential renal acid load (PRAL) of selected food items (per 100 g serving). PRAL (mEq/day) = 0.49 × protein (g/day) + 0.037 × P (mg/day) − 0.021 × K (mg/day) − 0.026 × Mg (mg/day) − 0.013 × Ca (mg/day) (Scialla and Anderson, 2013). *Source: Reproduced from Scialla, J.J., Anderson, C.A.M., 2013. Dietary acid load: a novel nutritional target in chronic kidney disease? Adv. Chronic Kidney Dis. 20 (2), 141–149, with permission from Science Direct.*

0.1 mEq/kg/day increase in bicarbonate dose produced 0.33 mEq/L higher serum bicarbonate, it would appear that diet alone has the potential to increase serum bicarbonate by at least 1.5 mEq/L.

While calculations for PRAL vary, in general, cheeses, meat, eggs, milk, and grains produce a greater acid load, while fruits, vegetables, nuts, and seeds are base producers. The typical Western diet with ~15−17% of energy as protein, results in an average dietary acid load of approximately 1 mEq/kg/day. Modifying the diet by increasing fruit and vegetable intake as described in the DASH diet reduces acid load and slows eGFR decline (Goraya et al., 2014). Note that fats, oils, nuts, and seeds are relatively neutral and do not increase the acid load. Since fresh fruits and vegetables do not contain sodium, diet modification of the acid load may be a better choice for the CKD patient than added sodium bicarbonate which does provide sodium. However, if acidosis cannot be controlled by the diet, bicarbonate used as toothpaste is well-tolerated and can be renoprotective even in late-stage CKD.

If hyperkalemia occurs despite control of acid load and dietary fiber, potassium restriction may be required. In CKD, adaptive gastrointestinal responses have been reported to facilitate potassium excretion. Potassium secretion in saliva and across the colonic and rectal mucosa is increased at least threefold. The enhanced colonic K^+ secretion reflects the elevated blood K^+ concentration as well as secondary hyperaldosteronism. Whereas in normal individuals, potassium is reabsorbed in the colon, in the CKD patient, net K^+ secretion occurs. Since, at least in the dog model, fecal excretion of potassium was correlated to the amount of feces, to the fecal dry matter excretion, and to the fecal water excretion, the use of dietary fiber to increase these parameters in the CKD patient and achieve net K + secretion should be tested in clinical trials.

Fruits and vegetables, especially green leafy vegetables, potatoes, and tomatoes, although low in PRAL, tend to be variable and sometimes rich sources of potassium. Food tables (Table 7.3) and published databases make it possible to maintain an alkaline, yet relatively low-potassium diet pattern. In general, vegetables that are less green (eg, cabbage) are lower in potassium that dark green (eg, kale) varieties.

Table 7.3 lists fruits and vegetables lower in potassium, thus minimizing dietary potassium intake. Adapted from the Academy of Nutrition and Dietetics (http://www.eatright.org/Public/content.aspx?id = 4294967541).

While some renal diet manuals suggest soaking and/or boiling potassium-rich vegetables to remove some of the potassium, this practice is relatively ineffective and also results in large losses of other

TABLE 7.3   Fruits and Vegetables Low in Potassium

Apples (best raw)	Blackberries	Blueberries
Cabbage	Carrots	Cauliflower
Corn	Cranberries	Cucumber
Eggplant	Grapes	Green beans
Lemon	Lettuce	Mushrooms
Onions	Peaches	Pear
Pineapple	Plums	Raspberries
Strawberries	Tangerines	Water chestnuts
Watermelon	Fennel	Lima beans

important micronutrients, especially vitamins C, water-soluble B, and calcium. CKD patients are better advised to work with a nutrition professional to structure their diet around lower potassium food choices that also provide a low PRAL. It is also important to counsel CKD patients that most salt substitutes contain potassium chloride and have been reported to precipitate acute hyperkalemia in CKD patients.

7.6.2.3 Added Benefits of a DASH-Type Diet

As indicated above, intestinal barrier function and microbiota are severely compromised in CKD. This increases translocation of proinflammatory bacterial products and toxins. As described elsewhere (refer to Essentials IV: Diet, Microbial Diversity and Gut Integrity) whole grain, seeds, nuts, vegetables, and whole fruits recommended in the DASH diet contain nonstarch polysaccharides, resistant starch, and dietary fiber that support the growth of commensal bacteria. It is also important to determine whether the patient has specific intolerance to substances such as wheat and its proteins and additives in processed foods, as these can compromise gut barrier function and increase permeability.

7.6.2.4 Note on Plant-Induced Nephrotoxicity Due to Oxalates in Foods

Patients with CKD are at special risk for nephrotoxic injury due to salts of oxalic acid absorbed from dietary sources. Certain plants such as spinach, rhubarb, tea, beets (leaves and beet sugar), star fruit, and coconut contain oxalate, sometimes in high quantities, depending on species, growing conditions, and cooking methods. Oxalate can be leached from foods by soaking or boiling, however this practice results in loss of other nutrients as indicated above. Dietary oxalate binds to calcium, thus consumption of a calcium source such as milk or cheese at the same meal can dramatically reduce oxalate bioavailability. Kidney damage from oxalate has been shown to be due both to the obstructive effect of calcium-oxalate crystals and to the induction of apoptosis in renal epithelial cells (Fang et al., 2008).

7.6.2.5 Compromised Bone Homeostasis in CKD

Calcium (Ca), phosphorous (P), and vitamin D are essential for maintenance of bone mineral crystal as discussed elsewhere (refer to Essentials III: Nutrients for Bone Structure and Mineral Metabolism and Chapter 9: Osteoporosis and Fracture Risk). Because Ca^{2+} and inorganic phosphate (P_i) ions are sparingly soluble in body fluids, their concentrations must be tightly controlled to avoid precipitation in the soft tissues, especially in large and small arteries (refer to Chapter 6: Atherosclerosis and Arterial Calcification). Regulation of serum Ca^{2+} by PTH and active vitamin D has been well described (Jones et al., 1998). Briefly, following absorption, dietary P raises the serum P_i concentration and stimulates FGF23 production by osteocytes in bone. FGF23 binds its coreceptor, Klotho, a membrane-bound protein in the kidney; binding downregulates phosphate transporters (Na-Pi type IIa, or NPt2) and facilitates phosphaturia thereby bringing serum P_i into the normal range. FGF23 also downregulates 1,25D biosynthesis, reducing absorption of both Ca and P. Lower serum Ca stimulates PTH, increasing 1,25D biosynthesis and increasing calcium absorption and bone resorption (Emmett, 2008).

In renal disease, blood levels of bone-derived fibroblast growth factor 23 (FGF23) can be up to 1000-fold greater than normal, likely because Klotho is lost as the kidney fails. While FGF23 is secreted in response to a high phosphate levels, elevations in FGF23 are seen early in renal failure, before a rise in serum phosphorus is appreciated, suggesting that FGF23 also responds to self-limited spikes in serum P_i following consumption of highly absorbable phosphate salts. In late CKD, very high levels of FGF23 are associated with high serum phosphate. While it is possible that reduced P_i or FGF23 clearance is the cause of these elevations, the actual mechanism has not been established (Bhattacharyya et al., 2012). The authors postulate that regulation of FGF23 in CKD is different from that in normal individuals and attribute the difference to the presence of a "renal factor" that regulates FGF23 in CKD.

As the number of functioning nephrons declines, less Klotho is available for FGF23 signaling and kidney responsiveness declines. FGF23 rises by orders of magnitude as the patient approaches ESKD, but it cannot facilitate phosphaturia, nor can it suppress PTH. Elevated FGF23 levels are exacerbated by a chronically high phosphate load, active vitamin D therapy, and PTH hypersecretion. High FGF23 may exert extrarenal effects such as ventricular hypertrophy and arterial calcification (refer to Chapter 5: Cardiomyopathy and Congestive Heart Failure: Targets for Myocardial Protection: Cardiac Remodeling). More rapid CKD progression and premature mortality have been positively associated with high FGF23 levels.

7.6.2.6 Calcification and CKD

In CKD, hyperphosphatemia together with hypercalcemia and reduction of inhibitors of mineralization foster the formation and deposition of matrix vesicles containing Ca/P nanocrystals into the vascular wall. While the vascular smooth muscle cells (VSMC) can compensate for high intracellular Ca/P concentrations by forming vesicles, these bodies create a positive feedback loop, inducing expression of genes that promote calcification and repress factors that inhibit calcification. The net result is transdifferentiation of VSMCs to osteoblast-like cells and expansion of the calcification area.

Serum magnesium has been inversely associated with vascular calcification (VC), cardiovascular events, and mortality in patients with CKD (Massy and Drüeke, 2015). Magnesium (Mg) limits VC by forming smaller, Mg-substituted whitlockite crystals that inhibit the transformation of amorphous Ca/P to mature hydroxyapatite. (Whitlockite is an unusual form of calcium phosphate crystal with a formula $Ca_9(MgFe)(PO_4)_6PO_3OH$.) Mg also antagonizes Ca channels, reducing calcium uptake into the cells. Finally, Mg enters the cell via its receptor, TPRM7, and neutralizes the phosphate-induced inhibition of the calcification inhibitors, matrix Gla protein (MGP) and bone matrix protein 7. Mg also downregulates genes that promote VSMC conversion to an osteoblast phenotype and subsequent calcification.

Calciphylaxis is a devastating exacerbation of VC often seen in hemodialysis patients (Brandenburg et al., 2012). While the cause is yet unclear, inadequate vitamin K status induced by drugs (eg, warfarin, coumadin) and antibiotics may play a role. VC is inhibited by MGP, a matrix mineralization inhibitor activated in the presence of vitamin K. While hemodialysis patients are well known to be vitamin K-deficient, deficiency has also been reported at earlier CKD stages. In addition to its role in the clotting cascade, vitamin K is important in bone structure and inhibition of arterial calcification as discussed elsewhere (refer to Chapter 6: Atherosclerosis and Arterial Calcification; Targets for Atherosclerotic Prevention and Control and Essentials III: Nutrients for Bone Structure and Calcification). When dietary vitamin K is limiting or when vitamin K recycling is inhibited by drugs, MGPs in the VSMC remain uncarboxylated and inactive (Danziger, 2008). Under these conditions, and especially with the elevated Ca^{2+} and P_i levels and lowered Mg^{2+} availability seen in CKD, bone crystallization is abnormal and MGPs are unable to prevent Ca:P crystallization in the vessels and soft tissues. The increased risk of CKD patients for coronary arterial calcification and cardiovascular disease has been well described.

7.6.3 Adjuvant Nutrition Modification to Address Abnormal Mineral Homeostasis in CKD

7.6.3.1 Phosphorus

Circulating phosphorus concentrations are maintained within normal levels by the osteocyte that senses phosphate concentration and subsequently secretes FGF23 as described above. Because FGF23 stimulates phosphaturia, serum phosphate in normal individuals remains relatively constant. Randomized trials in CKD patients confirmed that serum FGF23 was strongly influenced by diet and increased as CKD progressed (Sigrist et al., 2013). Given the association of high FGF23 levels with ventricular hypertrophy and more rapid progression to ESKD (see above), these findings suggest that lowering FGF23 secretion by dietary control and/or use of phosphorus binders could be an effective treatment strategy.

While the RDA for phosphorus for adults is 700 mg/day (Board, 1997), dietary phosphorus intakes of ~1033 mg/day did not result in increased mortality in nondialyzed CKD patients (eGFR ~50 mL/minutes/1.73 m^2) compared with patients consuming lower phosphorus intakes. Although these results suggest that the higher levels of phosphorus intake are safe, the CKD stage at which phosphorus intake must be controlled has not yet been determined. Some authors caution that it is not yet known whether aggressive efforts to reduce phosphorus absorption and lower FGF23 levels are wise, since elevated FGF23 levels in CKD are adaptive and may have as yet poorly understood long-term benefit to patients. Long-term intervention trials are needed to address these questions.

However, as dietary modulation of phosphorus intake is a mainstay of therapy as the CKD patient approaches ESKD, it is important to recognize the major sources and bioavailability of this nutrient. Phosphorus in food is variably absorbed: organic phosphorus in meat and fish is ~40–60% absorbed, while phosphorus in grains, legumes, and plants is bound as poorly absorbed phytate and only 10–30% absorbed as described elsewhere (refer to Essentials III: Nutrients for Bone Structure and Mineral Metabolism: Phosphorous. (Phytic acid (inositol hexakisphosphate or phytate as the salt) is the principal storage form of phosphorus in plants, especially in the bran and seed coat. It is not digestible by humans or nonruminant animals and is a poor source of phosphate if eaten directly.) Dairy products do contain phosphorus, but they also contain calcium, known to bind phosphorus and reduce its bioavailability. The rate at which phosphorus is absorbed may be important. Phosphorus from organic sources must be removed from its protein binders in the stomach and proximal small intestine, thus it is slowly absorbed and the rise in serum phosphorus is more gradual and presumably less likely to stimulate robust FGF23 secretion. A 7-day crossover trial testing vegetarian protein sources compared with animal sources in nine patients with a mean eGFR of 32 found that consumption of a vegetarian diet for 1 week led to significantly lower serum phosphate and FGF23 levels (Moe et al., 2011).

In contrast, phosphorus salts widely used in food processing to enhance flavor, improve color and moisture and to extend shelf-life of products are well absorbed; the bioavailability of phosphorus salts has been estimated to approach 100%. Labeling laws do not require food processors to indicate on the label that the product has been treated with phosphates, thus the actual contribution of added salts to the total phosphorus load is impossible to determine. Several analytical studies have reported that phosphorus content in foods containing additives can be as high as 25–70% underestimated and contribute greatly to the phosphorus load. For example, French fries, normally not a rich source of phosphorus, contain modified food starch, sodium acid pyrophosphate, and disodium dihydrogen pyrophosphate, all phosphorous additives that are "Generally Recognized as Safe" (GRAS). ("GRAS" is an acronym for the phrase Generally Recognized As Safe. Under sections 201(s) and 409 of the Federal Food, Drug, and Cosmetic Act, GRAS substances can be added to food for human consumption.) More insidious is the practice by meat processors of using a bath containing phosphorus salts to treat raw meat products both to enhance their color and to decrease microbial growth. In some cases, the composition of the phosphate bath is on the label. However, if the meat is not factory-wrapped, or if the butcher presents the phosphate-enhanced meat in a display case, the presence and content of phosphate salts is usually impossible to detect and quantitate.

Many cola-type beverages, sweetened as well as diet cola, contain phosphoric acid. These inorganic phosphate salts are well absorbed (80–100%), and soda is often consumed between meals. Ingestion of phosphate salts without binders contained in food can cause a rapid spike in serum phosphate and trigger subsequent FGF23 secretion, even in normal individuals. CKD patients at all stages of the disease should avoid sources of inorganic phosphorus salts. Natural foods that have been processed in any way, diet and sugar-sweetened cola-type beverages, enhanced raw and processed meats, baked goods, instant mixes, cereals, and processed cheeses are all sources of well-absorbed phosphorus. On the other hand, unprocessed whole grains, nuts, and seeds contain phosphorus as phytates, which are poorly absorbed. Patients should be taught to read labels and to avoid foods, supplements, and frequently used phosphate containing, over-the-counter medications, such as antacids and laxatives to maintain dietary phosphorus intake close to the recommended intake.

7.6.3.2 Calcium

Because calcium is only available to the body through dietary sources, its recommendation for adults ranges from 1000 to 1500 mg/day, depending on age (Board, 1997). However, balance studies and other determinations of calcium homeostasis were conducted in healthy individuals, and not in CKD patients whose calcium homeostasis is known to be altered. In CKD, the kidney is less able to excrete excess dietary calcium in the urine. To maintain normal extracellular calcium concentrations, net excess calcium can be lost in sweat, incorporated into bone, or deposited in soft tissue. Secondary hyperparathyroidism increases both bone resorption and formation, but in CKD, resorption predominates and calcium is added to the extracellular fluid (ECF). Metabolic acidosis also increases bone mineral dissolution and decreases bone formation, again increasing calcium concentrations in the ECF. Activated vitamin D suppresses serum PTH levels, but also induces bone resorption and increases calcium absorption, further elevating ECF calcium concentrations (Bushinsky, 2012). As kidney function declines, the phosphorus load increases (see above). Phosphate taken into the vascular smooth muscle cells (VSMC) by sodium-dependent phosphate cotransporters stimulates the VSMC to undergo an osteochondrogenic phenotype change and secrete extracellular collagen matrix. Increased ECF calcium concentrations, in the presence of increased phosphorus concentrations overwhelms calcification inhibitors (eg, MGP) and calcium is deposited as Ca:P crystals into the newly formed matrix. Recent studies confirm these findings and suggest that calcium is both a major factor in VSMC apoptosis and also acts synergistically with phosphate to drive accelerated VC.

Since serum ionized calcium levels are highly regulated and do not reflect calcium concentration in bone or tissue, it is difficult to determine the optimal daily calcium intake for CKD patients. In the past, large doses of calcium salts were prescribed to CKD patients as treatment for renal osteodystrophy both to lower parathryroid hormone secretion and to act in the gut as a dietary phosphate binder. These guidelines were modified in the 2002 KDOQI Clinical Practice Guidelines for Bone Metabolism and Disease in CKD. The Working Group reviewed published data and recommended a total daily calcium intake of no more than 2.0 g/day from diet and supplements. In 2009

and 2015 (Ketteler et al., 2015), the Kidney Disease Improving Global Outcome Guideline discussed the calcium-related risk for VC, but provided no additional recommendation for total dietary calcium intake.

Calcium is well absorbed from dairy products ($\sim$ 250 mg/cup), but the similar content of phosphorus may limit dairy as the sole source of calcium. Also, hard cheeses and some dairy products increase the acid load and promote bone buffering and calcium loss. Calcium is also found in legumes and seeds, but as the phytate in these foods binds calcium and prevents its absorption, these foods are not a source of bioavailable calcium.

7.6.3.3 Vitamin K

Vitamin K is found primarily in dark green leafy vegetables and fermented dairy products. Since it is a fat-soluble vitamin, it is better absorbed if fat is also in the meal. Vitamin K can be synthesized by gut bacteria, but the form and degree to which vitamin K is absorbed from the colon have not yet been clarified. CKD patients can be at risk for inadequate dietary vitamin K because of anorexia and/or restricted diet, as well as by use of drugs such as antibiotics and anticoagulants. While high sources of vitamin K tend to also be high in potassium, with care, a diet can be constructed that will provide adequate vitamin K even on a low-potassium diet (Table 7.4).

7.6.4 Nutritional Anemia in CKD

The anemia of CKD is diagnosed when the hemoglobin is <12 g/dL in a man or a postmenopausal woman, or <11 g/dL in a premenopausal woman. The cause of anemia should be investigated in these individuals. Anemia in CKD patients can be secondary to blood loss and anemia of chronic disease associated with inflammation and erythropoietin deficiency. Additionally, anemia can be due to a red cell production defect and deficiencies in nutrients, vitamin B(12), folate, vitamin B6, iron, and copper as well as heavy metal (aluminum, lead) intoxication. After other modifiable causes of anemia have been excluded, anemia of CKD is the most likely etiology, and should be treated with epoetin alfa or darbepoetin alfa, the only two erythropoietins approved for use in the United States.

7.7 FUTURE RESEARCH NEEDS

Although fragmentary, research findings suggest that progressive renal dysfunction is modifiable by dietary strategies. Large population surveys have repeatedly shown that the Western diet pattern (high in animal

TABLE 7.4   Vitamin K Content of Food[a]

Food with low vitamin K content	Foods with moderate vitamin K content	Foods with high vitamin K content
Alfalfa	Asparagus	Broccoli (cooked)
Beans (green)	Avocado	Brussels sprouts
Carrots and cauliflower	Red cabbage	Cabbage (raw)
Celery, corn and potato	Green peas	Endive (raw)
Breads, cereal	Lettuce (iceberg)	Lettuce (gourmet)
Rice	Pickle, dill	Parsley
Fruit and juices (mostly)[b]	Beans (snap)	Silver beet (cooked)
Cheese (cheddar), milk	Cheese (blue)	Spinach (cooked and raw)
Eggs and butter	Margarine	Mayonnaise
Sunflower and sesame oil	Olive oil	Canola and soybean oil
Fish, meat, pork, chicken	Abalone	Liver

[a]Chart taken from http://rehabilitateyourheart.wordpress.com/2013/02/16/vitamin-k-and-coumadin/; USDA database athttp://ndb.nal.usda.gov/ndb/nutrients/report/nutrientsfrm?max=25&offset=0&totCount=0&nutrient1=430&nutrient2=&nutrient3=&subset=1&fg=&sort=f&measureby=m.

[b]Most fruits are low in vitamin K, but some (prune, grape) are higher than others.

For more detailed information, refer to the USDA database, Note that cabbage is high in K and low in potassium. It also contains bioactive food components (sulforaphane) that induce the Nrf2-Keap1 system, lowering the risk for rapid progression to ESKD.

protein, saturated fats, and absorbable sugars) is associated with declining renal function, while a prudent or DASH-type diet (unprocessed plant sources and whole grains) slows progression to CKD. Additional intervention studies using a Mediterranean diet and a diet high in dietary fiber show similar benefits. The grain and especially vegetable fibers in these diets can not only modulate electrolyte balance and bone mineral balance, partly through reducing the renal acid load, but also ameliorate, in part, the dysbiosis and compromised gut barrier function observed in patients with CKD. Diet modulation can also limit excess zinc absorption and provide a good source of plant-derived copper as well as other minerals and vitamins. Further, the bioactive nonnutrient components contained in these foods have potential to reduce oxidative stress and upregulate protective signaling cascades, minimizing the rate of disease progression. Well-designed clinical research is urgently needed to translate these potential dietary mechanisms to targeted patient care.

References

AHRQ, 2014. Methods Guide for Effectiveness and Comparative Effectiveness Reviews. AHRQ, Rockville, MD.

Alaei, S.F., Soares, M.J., et al., 2013. High-dose thiamine supplementation improves glucose tolerance in hyperglycemic individuals: a randomized, double-blind cross-over trial. Eur. J. Nutr. 52 (7), 1821–1824.

Amann, K., Benz, K., 2011. Statins—beyond lipids in CKD. Nephrol. Dial. Transplant. 26 (2), 407–410.

American Diabetes Association, 2010. Executive summary: standards of medical care in diabetes—2010. Diabetes Care. 33 (Suppl. 1), S4–S10.

Anderson, C.A.M., Ix, J.H., 2013. Sodium reduction in CKD: suggestively hazardous or intuitively advantageous? J. Am. Soc. Nephrol. 24 (12), 1931–1933.

Bełtowski, J., 2010. Hypoxia in the renal medulla: implications for hydrogen sulfide signaling. J. Pharmacol. Exp. Ther. 334 (2), 358–363.

Bhattacharyya, N., Chong, W.H., et al., 2012. Fibroblast growth factor 23: state of the field and future directions. Trends Endocrinol. Metabol. 23 (12), 610–618.

Board, F. a N., 1997. Dietary Reference Intakes: Calcium, Phosphorus, Magnesium, Vitamin D and Fluoride. National Academy Press, Washington, DC.

Board, F. a N., 1998. Dietary Reference Intakes: Thiamin, Riboflavin, Niacin, Vitamin B6, Folate, Vitamin B12, Pantothenic Acid, Biotin, and Choline. I. o. Medicine. National Academy Press, Washington, DC.

Brandao-Burch, A., Utting, J.C., et al., 2005. Acidosis inhibits bone formation by osteoblasts in vitro by preventing mineralization. Calcif. Tissue Int. 77 (3), 167–174.

Brandenburg, V.M., Kramann, R., et al., 2012. Calciphylaxis in CKD and beyond. Nephrol. Dial. Transpl. 27 (4), 1314–1318.

Brown, R.J., Walter, M., et al., 2009. Ingestion of diet soda before a glucose load augments glucagon-like peptide-1 secretion. Diabetes Care. 32 (12), 2184–2186.

Bukhari, F.J., Moradi, H., et al., 2011. Effect of chronic kidney disease on the expression of thiamin and folic acid transporters. Nephrol. Dial. Transpl. 26 (7), 2137–2144.

Bushinsky, D.A., 2012. Clinical application of calcium modeling in patients with chronic kidney disease. Nephrol. Dial. Transpl. 27 (1), 10–13.

Campbell, K.L., Ash, S., et al., 2008. Randomized controlled trial of nutritional counseling on body composition and dietary intake in severe CKD. Am. J. Kidney Dis. 51 (5), 748–758.

Carrero, J.J., Stenvinkel, P., et al., 2013. Etiology of the protein-energy wasting syndrome in chronic kidney disease: a consensus statement from the international society of renal nutrition and metabolism (ISRNM). J. Renal Nutr. 23 (2), 77–90.

Catherwood, M.A., Powell, L.A., et al., 2002. Glucose-induced oxidative stress in mesangial cells. Kidney Int. 61 (2), 599–608.

Clase, C.M., Ki, V., et al., 2013. Water-soluble vitamins in people with low glomerular filtration rate or on dialysis: a review. Semin. Dial. 26 (5), 546–567.

Collino, M., 2011. High dietary fructose intake: sweet or bitter life? World J. Diabetes. 2 (6), 77–81.

Danziger, J., 2008. Vitamin K-dependent proteins, warfarin, and vascular calcification. Clin. J. Am. Soc. Nephrol. 3 (5), 1504–1510.

de Brito-Ashurst, I., Varagunam, M., et al., 2009. Bicarbonate supplementation slows progression of CKD and improves nutritional status. J. Am. Soc. Nephrol. 20 (9), 2075–2084.

Douard, V., Asgerally, A., et al., 2010. Dietary fructose inhibits intestinal calcium absorption and induces vitamin D insufficiency in CKD. J. Am. Soc. Nephrol. 21 (2), 261–271.

Douard, V., Sabbagh, Y., et al., 2013. Excessive fructose intake causes 1,25-(OH)2D3-dependent inhibition of intestinal and renal calcium transport in growing rats. Am. J. Physiol. Endocrinol. Metab. 304 (12), E1303–E1313.

Durvasula, R.V., Shankland, S.J., 2008. Activation of a local renin angiotensin system in podocytes by glucose. Am. J. Physiol. Renal Physiol. 294 (4), F830–F839.

Emmett, M., 2008. What does serum fibroblast growth factor 23 do in hemodialysis patients? Kidney Int. 73 (1), 3–5.

Fagherazzi, G., Vilier, A., et al., 2013. Consumption of artificially and sugar-sweetened beverages and incident type 2 diabetes in the Etude Epidémiologique auprès des femmes de la Mutuelle Générale de l'Education Nationale–European Prospective Investigation into Cancer and Nutrition cohort. Am. J. Clin. Nutr. 97 (3), 517–523.

Fahal, I.H., 2013. Uraemic sarcopenia: aetiology and implications. Nephrol. Dial. Transplant. 16 (5), 432–436.

Fang, H.-C., Lee, P.-T., et al., 2008. Mechanisms of star fruit-induced acute renal failure. Food Chem. Toxicol. 46 (5), 1744–1752.

Favre, G.A., Esnault, V.L.M., et al., 2015. Modulation of glucose metabolism by the renin-angiotensin-aldosterone system. Am. J. Physiol. Endocrinol. Metab. 308 (6), E435–E449.

Ferns, G.A.A., Lamb, D.J., et al., 1997. The possible role of copper ions in atherogenesis: the Blue Janus. Atherosclerosis. 133 (2), 139–152.

Fouque, D., Kalantar-Zadeh, K., et al., 2007. A proposed nomenclature and diagnostic criteria for protein-energy wasting in acute and chronic kidney disease. Kidney Int. 73 (4), 391–398.

Fouque, D., Pelletier, S., et al., 2011. Nutrition and chronic kidney disease. Kidney Int. 80 (4), 348–357.

Fu, M., Zhang, W., et al., 2012. Hydrogen sulfide (H2S) metabolism in mitochondria and its regulatory role in energy production. Proc. Natl. Acad. Sci. 109 (8), 2943–2948.

Gannon, M.C., Nuttall, F.Q., 2004. Effect of a high-protein, low-carbohydrate diet on blood glucose control in people with type 2 diabetes. Diabetes. 53 (9), 2375–2382.

Gersch, M.S., Mu, W., et al., 2007. Fructose, but not dextrose, accelerates the progression of chronic kidney disease. Am. J. Physiol. Renal Physiol. 293 (4), F1256–F1261.

Golding, P.H., 2014. Severe experimental folate deficiency in a human subject—a longitudinal investigation of red-cell folate immunoassay errors as megaloblastic anaemia develops. SpringerPlus. 3 (1), 1–13.

Gopinath, B., Harris, D.C., et al., 2011. Carbohydrate nutrition is associated with the 5-year incidence of chronic kidney disease. J. Nutr. 141 (3), 433–439.

Goraya, N., Simoni, J., et al., 2014. Treatment of metabolic acidosis in patients with stage 3 chronic kidney disease with fruits and vegetables or oral bicarbonate reduces urine angiotensinogen and preserves glomerular filtration rate. Kidney Int. 86 (5), 1031–1038.

Harvey, P.W., Allen, K.G.D., 1981. Decreased plasma lecithin:cholesterol acyltransferase activity in copper-deficient rats. J. Nutr. 111 (10), 1855–1858.

Higuchi, T., Matsukawa, Y., et al., 2006. Correction of copper deficiency improves erythropoietin unresponsiveness in hemodialysis patients with anemia. Intern. Med. 45 (5), 271–273.

Hovind, P., Rossing, P., et al., 2011. Serum uric acid as a new player in the development of diabetic nephropathy. J. Renal Nutr. 21 (1), 124–127.

Ikizler, T.A., Cano, N.J., et al., 2013. Prevention and treatment of protein energy wasting in chronic kidney disease patients: a consensus statement by the International Society of Renal Nutrition and Metabolism. Kidney Int. 84 (6), 1096–1107.

Imran, N., Man, S., et al., 2015. Chapter 50—nutritional management of patients with chronic kidney disease A2—Kimmel, Paul L. In: Rosenberg, M.E. (Ed.), Chronic Renal Disease. Academic Press, San Diego, CA, pp. 613–623.

Jakubowski, H., 2007. The molecular basis of homocysteine thiolactone-mediated vascular disease. Clin. Chem. Lab. Med. 45 (12), 1704.

Jalal, D.I., Maahs, D.M., et al., 2011. Uric acid as a mediator of diabetic nephropathy. Semin. Nephrol. 31 (5), 459–465.

Johnson, D., 2006. Dietary protein restriction as a treatment for slowing chronic kidney disease progression: the case against. Nephrology (Carlton). 11 (1), 58–62.

Johnson, R.J., Perez-Pozo, S.E., et al., 2009. Hypothesis: could excessive fructose intake and uric acid cause type 2 diabetes? Endocr. Rev. 30 (1), 96–116.

Jones, G., Strugnell, S.A., et al., 1998. Current understanding of the molecular actions of vitamin D. Physiol. Rev. 78 (4), 1193–1231.

Kalantar-Zadeh, K., Abbott, K.C., et al., 2005. Survival advantages of obesity in dialysis patients. Am. J. Clin. Nutr. 81 (3), 543–554.

Kasama, R.K., 2010. Trace minerals in patients with end-stage renal disease. Semin. Dial. 23 (6), 561–570.

Kaysen, G.A., 2006. Dyslipidemia in chronic kidney disease: causes and consequences. Kidney Int. 70 (S104), S55–S58.

Ketteler, M., Elder, G.J., et al., 2015. Revisiting KDIGO clinical practice guideline on chronic kidney disease—mineral and bone disorder: a commentary from a kidney disease: improving global outcomes controversies conference. Kidney Int. 87 (3), 502–508.

Kinoshita, M., Takeda, K., 2014. Microbial and dietary factors modulating intestinal regulatory T cell homeostasis. FEBS Lett. 588 (22), 4182–4187.

Klahr, S., Levey, A.S., et al., 1994. The effects of dietary protein restriction and blood-pressure control on the progression of chronic renal disease. N Eng. J. Med. 330 (13), 877–884.

Koning, A.M., Frenay, A.-R.S., et al., 2015. Hydrogen sulfide in renal physiology, disease and transplantation—the smell of renal protection. Nitric Oxide. 46, 37–49.

Kopple, J.D., 1999. Pathophysiology of protein-energy wasting in chronic renal failure. J. Nutr. 129 (1), 247S–251S.

Kretowicz, M., Johnson, R.J., et al., 2011. The impact of fructose on renal function and blood pressure. Int. J. Nephrol. 2011, 315879.

Lagouge, M., Argmann, C., et al., 2006. Resveratrol improves mitochondrial function and protects against metabolic disease by activating SIRT1 and PGC-1[alpha]. Cell. 127 (6), 1109–1122.

Lai, Y.-H., Fang, T.-C., 2013. The pleiotropic effect of vitamin D. ISRN Nephrol. 2013, 6.

Larkin, J.R., Zhang, F., et al., 2012. Glucose-induced down regulation of thiamine transporters in the kidney proximal tubular epithelium produces thiamine insufficiency in diabetes. PLoS ONE. 7 (12), e53175.

Lau, B.W.C., Klevay, L.M., 1981. Plasma lecithin:cholesterol acyltransferase in copper-deficient rats. J. Nutr. 111 (10), 1698–1703.

Leonard, M.O., Kieran, N.E., et al., 2006. Reoxygenation-specific activation of the antioxidant transcription factor Nrf2 mediates cytoprotective gene expression in ischemia-reperfusion injury. FASEB J. 20 (14), 2624–2626.

Li, Y.C., Kong, J., et al., 2002. 1,25-Dihydroxyvitamin D3 is a negative endocrine regulator of the renin-angiotensin system. J. Clin. Invest. 110 (2), 229–238.

Liang, K., Kim, C.H., et al., 2004. HMG-CoA reductase inhibition reverses LCAT and LDL receptor deficiencies and improves HDL in rats with chronic renal failure. J. Physiol. Renal Physiol. 288 (3), F539–44.

Lin, J., Curhan, G.C., 2011. Associations of sugar and artificially sweetened soda with albuminuria and kidney function decline in women. Clin. J. Am. Soc. Nephrol. 6 (1), 160–166.

Lin, J., Fung, T.T., et al., 2011. Association of dietary patterns with albuminuria and kidney function decline in older white women: a subgroup analysis from the nurses' health study. Am. J. Kidney Dis. 57 (2), 245–254.

Liu, J., Kennedy, D.J., et al., 2012. Reactive oxygen species modulation of Na/K-ATPase regulates fibrosis and renal proximal tubular sodium handling. Inter. J. Nephrol. 2012, 381320. Available from: http://dx.doi.org/10.1155/2012/381320.

Lobb, I., Sonke, E., et al., 2015. Hydrogen sulphide and the kidney: important roles in renal physiology and pathogenesis and treatment of kidney injury and disease. Nitric Oxide. 46, 55–65.

Luther, J.M., Brown, N.J., 2011. The renin–angiotensin–aldosterone system and glucose homeostasis. Trends Pharmacol. Sci. 32 (12), 734–739.

Mandayam, S., Mitch, W.E., 2006. Dietary protein restriction benefits patients with chronic kidney disease (Review Article). Nephrology. 11 (1), 53–57.

Mansour, A., Hosseini, S., et al., 2013. Nutrients related to GLP1 secretory responses. Nutrition. 29 (6), 813–820.

Márquez, E., Riera, M., et al., 2015. Renin-angiotensin system within the diabetic podocyte. Am. J. Physiol. Renal Physiol. 308 (1), F1–F10.

Massy, Z.A., Drüeke, T.B., 2015. Magnesium and cardiovascular complications of chronic kidney disease. Nat. Rev. Nephrol. 11 (7), 432–442.

McCully, K.S., 1996. Homocysteine and vascular disease. Nat. Med. 2, 386–389.

Meek, R.L., LeBoeuf, R.C., et al., 2013. Glomerular cell death and inflammation with high-protein diet and diabetes. Nephrol. Dial. Transplant. 28 (7), 1711–1720.

Meex, R.C.R., Schrauwen-Hinderling, V.B., et al., 2010. Restoration of muscle mitochondrial function and metabolic flexibility in type 2 diabetes by exercise training is paralleled by increased myocellular fat storage and improved insulin sensitivity. Diabetes. 59 (3), 572–579.

Mende, C., 2010. Application of direct renin inhibition to chronic kidney disease. Cardiovasc. Drugs Ther. 24 (2), 139–149.

Mimura, I., Nangaku, M., 2010. The suffocating kidney: tubulointerstitial hypoxia in end-stage renal disease. Nat. Rev. Nephrol. 6 (11), 667–678.

Mitch, W.E., Medina, R., et al., 1994. Metabolic acidosis stimulates muscle protein degradation by activating the adenosine triphosphate-dependent pathway involving ubiquitin and proteasomes. J. Clin. Invest. 93 (5), 2127–2133.

Mo, H., Elson, C.E., 2004. Studies of the isoprenoid-mediated inhibition of mevalonate synthesis applied to cancer chemotherapy and chemoprevention. Exp. Biol. Med. 229 (7), 567–585.

Moe, S.M., Zidehsarai, M.P., et al., 2011. Vegetarian compared with meat dietary protein source and phosphorus homeostasis in chronic kidney disease. Clin. J. Am. Soc. Nephrol. 6 (2), 257–264.

Mooradian, A.D., 2009. Dyslipidemia in type 2 diabetes mellitus. Nat. Clin. Pract. End. Met. 5 (3), 150–159.

Murer, L., Addabbo, F., et al., 2004. Selective decrease in urinary aquaporin 2 and increase in prostaglandin E2 excretion is associated with postobstructive polyuria in human congenital hydronephrosis. J. Am. Soc. Nephrol. 15 (10), 2705–2712.

Noel, S., Hamad, A.R.A., et al., 2015. Reviving the promise of transcription factor Nrf2-based therapeutics for kidney diseases. Kidney Int. 88 (6), 1217–1218.

Norata, G.D., Pirillo, A., et al., 2006. Modified HDL: biological and physiopathological consequences. Nutr. Metab. Cardiovasc. Dis. 16 (5), 371–386.

Ooi, E.M.M., Chan, D.T., et al., 2011. Plasma apolipoprotein C-III metabolism in patients with chronic kidney disease. J. Lipid. Res. 52 (4), 794–800.

Pácal, L., Tomandl, J., et al., 2011. Role of thiamine status and genetic variability in transketolase and other pentose phosphate cycle enzymes in the progression of diabetic nephropathy. Nephrol. Dial. Transplant. 26 (4), 1229–1236.

Pácal, L., Kuricová, K., et al., 2014. Evidence for altered thiamine metabolism in diabetes: is there a potential to oppose gluco- and lipotoxicity by rational supplementation? World J. Diabetes. 5 (3), 288–295.

Parry-Strong, A., Leikis, M., et al., 2013. High protein diets and renal disease—is there a relationship in people with type 2 diabetes? Br. J. Diabetes Vasc. Dis. 13 (5-6), 238–243.

Pastore, A., Noce, A., et al., 2014. Homocysteine, cysteine, folate and vitamin B12 status in type 2 diabetic patients with chronic kidney disease. J. Nephrol. 28 (5), 571–576.

Quiñones-Galvan, A., Ferrannini, E., 1997. Renal effects of insulin in man. J. Nephrol. 10 (4), 188–191.

Ramana, K.V., 2011. Aldose reductase: new insights for an old enzyme. Biomol. Concepts. 2 (1–2), 103–114.

Ritz, E., Mehls, O., 2009. Salt restriction in kidney disease—a missed therapeutic opportunity? Pediatr. Nephrol. 24 (1), 9–17.

Rojas-Rivera, J., Ortiz, A., et al., 2012. Antioxidants in kidney diseases: the impact of bardoxolone methyl. Inter. J. Nephrol. 2012, 11.

Ruiz-Ortega, M., Rodríguez-Vita, J., et al., 2007. TGF-β signaling in vascular fibrosis. Cardiovasc. Res. 74 (2), 196–206.

Rüster, C., Bondeva, T., et al., 2008. Advanced glycation end-products induce cell cycle arrest and hypertrophy in podocytes. Nephrol. Dial. Transplant. 23 (7), 2179–2191.

Rüster, C., Wolf, G., 2011. Angiotensin II as a morphogenic cytokine stimulating renal fibrogenesis. J. Am. Soc. Nephrol. 22 (7), 1189–1199.

Sanchez-Castillo, C.P., Warrender, S., et al., 1987. An assessment of the sources of dietary salt in a British population. Clin. Sci. (London). 72 (1), 95–102.

Sanders, P.W., 2009. Dietary salt intake, salt sensitivity, and cardiovascular health. Hypertension. 53 (3), 442–445.

Scialla, J.J., Anderson, C.A.M., 2013. Dietary acid load: a novel nutritional target in chronic kidney disease? Adv. Chronic Kidney Dis. 20 (2), 141–149.

Serratrice, G., Toga, M., et al., 1967. Neuropathies, myopathies and neuromyopathies in chronic uremic patients. La Presse Médicale. 75 (37), 1835–1838.

Shaban, H., Ubaid-Ullah, M., et al., 2014. Measuring vitamin, mineral, and trace element levels in dialysis patients. Semin. Dial. 27 (6), 582–586.

Sigrist, M., Tang, M., et al., 2013. Responsiveness of FGF-23 and mineral metabolism to altered dietary phosphate intake in chronic kidney disease (CKD): results of a randomized trial. Nephrol. Dial. Transplant. 28 (1), 161–169.

Son, S.M., 2012. Reactive oxygen and nitrogen species in pathogenesis of vascular complications of diabetes. Diabetes Metab. J. 36 (3), 190–198.

Sondheimer, J.H., Mahajan, S.K., et al., 1988. Elevated plasma copper in chronic renal failure. Am. J. Clin. Nutr. 47 (5), 896–899.

Sterns, R.H., Rojas, M., et al., 2010. Ion-exchange resins for the treatment of hyperkalemia: are they safe and effective? J. Am. Soc. Nephrol. 21 (5), 733–735.

Suárez, G., Rajaram, R., et al., 1989. Nonenzymatic glycation of bovine serum albumin by fructose (fructation). Comparison with the Maillard reaction initiated by glucose. J. Biol. Chem. 264 (7), 3674–3679.

Suez, J., Korem, T., et al., 2014. Artificial sweeteners induce glucose intolerance by altering the gut microbiota. Nature. 514 (7521), 181–186.

Tappy, L., Lê, K.-A., 2010. Metabolic effects of fructose and the worldwide increase in obesity. Physiol. Rev. 90 (1), 23–46.

Thilly, N., 2013. Low-protein diet in chronic kidney disease: from questions of effectiveness to those of feasibility. Nephrol. Dial. Transplant. 28 (9), 2203–2205.

Vanholder, R., Glorieux, G., 2015. The intestine and the kidneys: a bad marriage can be hazardous. Clin. Kidney J. 8 (2), 168–179.

Vaziri, N.D., 2006. Dyslipidemia of chronic renal failure: the nature, mechanisms, and potential consequences. Am. J. Physiol. Renal Physiol. 290 (2), F262–F272.

Vaziri, N.D., 2013. Chapter 3—altered lipid metabolism and serum lipids in kidney disease and kidney failure A2. In: Kopple, J.D., Massry, S.G., Kalantar-Zadeh, K. (Eds.), Nutritional Management of Renal Disease, third ed. Academic Press, New York, NY, pp. 31–48.

Vaziri, N.D., 2016. HDL abnormalities in nephrotic syndrome and chronic kidney disease. Nat. Rev. Nephrol. 12 (1), 37–47.

Vaziri, N.D., Moradi, H., 2006. Mechanisms of dyslipidemia of chronic renal failure. Hemodial. Inter. 10 (1), 1–7.

Vaziri, N.D., Navab, M., et al., 2010. HDL metabolism and activity in chronic kidney disease. Nat. Rev. Nephrol. 6 (5), 287–296.

Vaziri, N.D., Liu, S.-M., et al., 2014. High amylose resistant starch diet ameliorates oxidative stress, inflammation, and progression of chronic kidney disease. PLoS ONE. 9 (12), e114881.

Vaziri, N.D., Zhao, Y.-Y., et al., 2015. Altered intestinal microbial flora and impaired epithelial barrier structure and function in CKD: the nature, mechanisms, consequences and potential treatment. Nephrol. Dial. Transplant. 31 (5), 737–746.

Vos, M.B., Kimmons, J.E., et al., 2008. Dietary fructose consumption among US children and adults: the third national health and nutrition examination survey. Medscape J. Med. 10 (7), 160.

Wakabayashi, N., Slocum, S.L., et al., 2010. When NRF2 talks, who's listening? Antioxid. Redox Signal. 13 (11), 1649–1663.

Wang, R., 2012. Physiological implications of hydrogen sulfide: a whiff exploration that blossomed. Physiol. Rev. 92 (2), 791–896.

Wolf, G., Ziyadeh, F.N., 1999. Molecular mechanisms of diabetic renal hypertrophy. Kidney Int. 56 (2), 393–405.

Wollesen, F., Brattstrom, L., et al., 1999. Plasma total homocysteine and cysteine in relation to glomerular filtration rate in diabetes mellitus. Kidney Int. 55 (3), 1028–1035.

Xiao, J., Ni, X., et al., 2015. Advance in dietary polyphenols as aldose reductases inhibitors: structure-activity relationship aspect. Crit. Rev. Food Sci. Nutr. 55 (1), 16–31.

Yuan, W., Pan, W., et al., 2007. 1,25-Dihydroxyvitamin D3 suppresses renin gene transcription by blocking the activity of the cyclic AMP response element in the renin gene promoter. J. Biol. Chem. 282 (41), 29821–29830.

Yuet, W.C., Khine, H., et al., 2015. Statin-associated adverse events. Clin. Med. Insights: Ther. 7, 17–24.

Zhao, Y., Pacheco, A., et al., 2015. Medicinal chemistry: insights into the development of novel H2S donors. In: Moore, P.K., Whiteman, M. (Eds.), Chemistry, Biochemistry and Pharmacology of Hydrogen Sulfide, 230. Springer International Publishing, pp. 365–388.

Zheng, H., Whitman, S.A., et al., 2011. Therapeutic potential of Nrf2 activators in streptozotocin-induced diabetic nephropathy. Diabetes. 60 (11), 3055–3066.

Zoja, C., Zanchi, C., et al., 2015. Key pathways in renal disease progression of experimental diabetes. Nephrol. Dial. Transpl. 30 (Suppl. 4), iv54–iv59.

Zuo, X., Dong, D., et al., 2013. Homocysteine restricts copper availability leading to suppression of cytochrome c oxidase activity in phenylephrine-treated cardiomyocytes. PLoS ONE. 8 (6), e67549.

Further Reading

Cecil textbook of medicine. 2007. Chapter 125 (Diabetes and the Kidney). 2007. twenty-third ed. Elsevier Saunders, Philadelphia, PA.

KDOQI Clinical Practice Guidelines and Clinical Practice Recommendations for Diabetes and Chronic Kidney Disease updated 2012 <http://www.ajkd.org/article/S0272-6386%2812%2900957-2/pdf>.

Nutritional Management of Renal Disease. 2013. third ed. Elsevier Academic Press, Amsterdam.

Chapters 20 (Kidney) and 24 (Endocrine). 2015. , Robbins and Cotran Pathologic Basis of Disease. 2015. ninth ed. Elsevier, Saunders, Philadelphia, PA.

8

Asthma and Obesity

CHIEF COMPLAINT (ADAPTED FROM WECHSLER, M.E., 2007)

At 3:15 am, the patient, ML (20-year-old black female) was brought to the emergency room (ER) by ambulance with a severe asthma attack. Her mother was awakened by ML's heavy wheezing and by the time her mother reached her, ML was lying on the floor gasping for breath. The mother called emergency medical services who found the patient in status asthmaticus. (Case details derived from Wechsler et al., 2007. Overview, pathogenesis, and treatment strategies for status asthmaticus are presented in Medscape at http://emedicine.medscape.com/article/2129484-overview/). The trachea was intubated, and epinephrine (1 mg) and atropine (1 mg) were administered intravenously; patient responded and was admitted to intensive care for emergency evaluation and treatment.

MEDICAL HISTORY

ML was born to a teenage single mother after a normal pregnancy and delivery. Her mother had smoked three packs of cigarettes/day during her pregnancy. ML developed asthma at the age of 4 years and suffered frequent exacerbations triggered by cold air, hot humid air, physical activity, respiratory infections, anxiety, and exposure to paint, cats, and birds. Attacks occurred at night or early in the morning. The mother noted that ML had been using her inhaler very frequently the evening before.

ML had been followed by multiple pediatricians and pulmonologists. Since she was often noncompliant with outpatient follow-up visits, she was forced to use the ER for her frequent asthma exacerbations. At one visit (age 15) she told the health care staff that she rarely took her medications regularly, relying on bronchodilators when she had symptoms. On several occasions, including an episode of status asmaticus (age 8), oxygen desaturation occurred (80% on ambient air, with respiratory rate of 50 breaths per minute). She was admitted every few months for pneumonia; sometimes she was discharged with antibiotics, but for more serious conditions, she was admitted to pediatric intensive care. She was never intubated; peripheral blood eosinophilia was never documented; immunoglobulin E (IgE) levels were not measured (Table 8.1).

Five months before this admission and 2 days after receiving treatment in an ER for an asthma flare-up, the patient returned to an urgent care facility with shortness of breath, cough productive of white sputum, chills, and wheezing of 2 days' duration, despite around-the-clock inhalational treatments with nebulized albuterol (2.5 mg) and nebulized ipratropium (0.5 mg). Her other asthma medications were salmeterol (one puff twice daily), fluticasone propionate delivered as an aerosol (220 μg per puff, at a dose of two puffs with a spacer adaptor, twice daily), prednisone (5 mg every other day), and albuterol with the use of a metered-dose inhaler (two puffs four times daily as needed). The patient was stabilized and released. During the next weeks, ML made numerous phone calls to her current primary care physician, requesting refills of her medications. The physician also became aware that the patient was receiving asthma medications from several different providers and that ML also visited the ER four or five times a month because of acute symptoms of asthma.

ML has been obese since childhood. Although she has not been previously diagnosed with diabetes, blood glucose levels recorded in her chart have gradually risen since puberty and her recent levels were in the prediabetes range. She has been troubled by eczema for which she has been prescribed corticosteroid salves, and is currently treated for gastroesophageal reflux with esomeprazole. Her history indicates chronic depression, fractures of her wrist after a fall, a parietal skull fracture after a street assault, chickenpox, and pinworms. Her childhood vaccinations were delayed because of missed appointments. No allergies to medications are reported.

191

TABLE 8.1   Data Abstracted From the Patient's Chart Indicated (Wechsler et al., 2007) Her Extensive List of Asthma Therapies Administered in the Hospital and Taken at Home

Age range	Medications taken at home	Medications administered on admission (to emergency department or hospital)
4–10 years	Metaproterenol	Epinephrine (subcutaneous)
	Albuterol by nebulizer and metered-dose inhaler	Aminophylline (intravenous)
	Theophylline	Methylprednisolone (intravenous)
	Isoetharine by nebulizer	
	Cromolyn sodium inhaler	
	Beclomethasone inhaler	
	Tapered prednisone (5–14 days)	
10–15 years	Albuterol by nebulizer and metered-dose inhaler	Epinephrine (subcutaneous)
	Iprathropium bromide by nebulizer	Terbutaline sulfate (subcutaneous or intravenous)
	Salmeterol xinafoate inhaler	
	Fluticasone propionate metered-dose inhaler	
	Budesonide inhaler	
	Tapered prednisone (5–14 days)	
	Montelukast sodium tablets	
	Beclomethasone inhaler	
15–20 years	Albuterol metered-dose inhaler	Epinephrine (subcutaneous or intravenous)
	Albuterol and ipratropium bromide by nebulizer	Dexamethasone
	Theophylline	Terbutaline (subcutaneous or intravenous)
	Salmeterol xinafoate inhaler	Magnesium sulfate (intravenous)
	Fluticasone propionate metered-dose inhaler	
	Montelukast sodium tablets	
	Prednisone (5–60 mg daily)	

FAMILY HISTORY

ML is of South Asian, Nigerian, and Dutch ancestry. She lives with her child (born when ML was 19), her mother, and her three half-siblings (12, 14, and 16 years). The patient's mother and maternal relatives have asthma and eczema; there is also a maternal family history of type II diabetes. The paternal medical history is not known.

ML, her mother, and siblings manage childcare by working at part-time, minimum-wage jobs. The family receives food stamps and ML also receives food for her child through the WIC program. The family has extensive involvement with the Department of Social Services because of their noncompliance with treatment and failure to be available for visiting nurse assessment of the home for possible allergens. The patient's mother underwent unsuccessful counseling to stop smoking. It is not known whether ML smokes; she has given contradictory information to different sources. ML's education has been interrupted for long periods of time because of asthma; she quit school after the eighth grade (15 years).

PHYSICAL EXAMINATION (HOSPITAL DAY 2)

Anthropometric Measures

Height 5′4″; Weight 201 lb.; BMI 34.5
Waist circumference 45 in.
Obese phenotype with cushingoid features

Vital Signs and Lung Function Tests

Blood pressure	152/62 mmHg
Pulse	120 bpm
Temperature	37.4°C
Respirations	28 per min
Oxygen saturation	86−89% on ambient air

Pertinent Laboratory Values

Blood glucose	130 mg/dL
Total cholesterol	198 mg/dL
Triglycerides	175 mg/dL
Eosinophiles	4%

Chest/Lungs

Diffuse wheezes in both lungs, without rales or rhonchi.
A chest X-ray revealed bilateral focal consolidations consistent with pneumonia.

Extremities

No digital clubbing or cyanosis.

Remainder of the Examination

Within normal limits.

DIETARY HISTORY

ML advised that she and her mother take turns cooking, although she is usually the one that cooks. They live in an inner-city neighborhood with few sources of fresh produce. (This type of neighborhood is often characterized as a "food desert." See link for the USDA definition and a map of sites nationwide: http://americannutritionassociation.org/newsletter/usda-defines-food-deserts/.) Meals often consist of meat, beans, and rice. Her favorite meal is collard greens cooked for hours with ham hocks and served over rice.

24-Hour Recall

Breakfast	Two fast-food breakfast burritos with sausage and cheese
	One large coffee with creamer and three sugars
Lunch	One grilled cheese and bacon sandwich (American cheese on white bread with margarine)

(Continued)

(Continued)

	A large bag of Frito corn chips
	20 oz. soda (Coke)
Dinner	Two helpings of leftover rice and beans with collards and ham hock
	Two light beers
Snack	One chocolate-covered ice cream pop (from street vendor)
	One light beer

DISCHARGE INSTRUCTIONS

Patient was discharged home while receiving prednisone (60 mg daily, with instructions to taper the dose), her other medications, and a 5-day course of azithromycin.

RESOURCES

8.1 ANATOMY AND PHYSIOLOGY OF THE NORMAL LUNG

The lung is designed for the exchange of gases, predominately carbon dioxide and oxygen. Gas exchange occurs at the terminal end of the airways, in clusters of tiny acini, each with a diameter of about 7 mm. To reach the acini, air enters the trachea and passes through bronchi and bronchioles of decreasing size and modified structure. The trachea and bronchi (but not bronchioles) are lined with pseudostratified, tall, columnar epithelial cells interspersed with mucus-secreting goblet cells. Cilia on the epithelial cells move mucus and debris toward the oropharynx for expectoration or swallowing. The bronchial mucosa also contains neurosecretory-type granules that secrete calcitonin, serotonin, and gastrin-releasing peptide (bombesin). Smooth muscle and cartilage provide airway support (Figs. 8.1 and 8.2).

Histologically, the thin alveolar walls consist of epithelium supported by a basement membrane. The intertwining network of anastomosing capillaries is separated from the epithelial cells that line the alveolus by interstitial tissue. The interstitial alveolar septum consists of thin sections made up of fused endothelium and epithelium and thick sections in which the two cell types are separated by an interstitial space containing fine elastic fibers, small collagen bundles, a few fibroblast-like interstitial cells, smooth muscle cells,

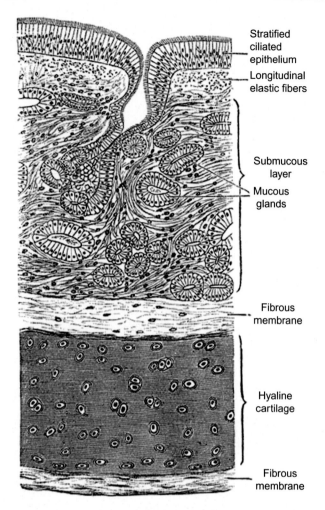

Stratified
ciliated
epithelium

Longitudinal
elastic fibers

Submucous
layer

Mucous
glands

Fibrous
membrane

Hyaline
cartilage

Fibrous
membrane

FIGURE 8.1 **Internal anatomy of the trachea and bronchi.** Pseudostratified columnar epithelial cells line the trachea. Submucosal goblet cells secrete mucus onto the surface of the ciliated epithelial lining; the cilia move the mucus layer with accumulated inhaled debris to the larynx and pharynx where it can be swallowed or expectorated. This process is termed mucociliary clearance. *Figure from Henry Gray (1918) Anatomy of the Human Body Bartleby.com; Plate 946 Henry Vandyke Carter (Public domain), via Wikimedia Commons.*

mast cells, and a rare lymphocyte and monocyte. This space is called the *pulmonary interstitium.*

The continuous alveolar epithelium is made up of two cell types. The type I pneumocytes are flattened, plate-like cells that cover 95% of the alveolar surface. These cells are interspersed with type II pneumocytes with rounded cell bodies. Type II pneumocytes synthesize surfactant and store it in lamellar bodies visualized with osmium in electron micrographs. Type II cells also give rise to type I cells, thus type II cells are essential for alveolar repair. Alveolar macrophages derived from blood monocytes are loosely attached to epithelial cells or lie free within the alveolar space; macrophages ingest inspired carbon and other alveolar debris.

Numerous *pores of Kohn* perforate the alveolar wall and allow passage of bacteria and exudate between adjacent alveoli. A surfactant layer covers the alveolar cell membrane (Fig. 8.3).

8.2 ASTHMA SYMPTOMS, STATISTICS, AND MANAGEMENT

Asthma is associated with chronic inflammation of the airways resulting in recurrent bouts of wheezing, chest tightness, shortness of breath, and coughing. The condition is characterized by increased airway responsiveness to stimuli, inflammation of the bronchial walls, and increased secretion of mucus. Lung epithelium is infiltrated with lymphocytes, eosinophiles, mast cells, macrophages, and neutrophils. Bronchoconstriction contributes to airflow limitation. Severe constriction can remit spontaneously or with treatment, but remodeling is still present.

Currently, about 25 million people in the United States, including 7 million children, have asthma; more than half of these individuals will have at least one episode of worsening symptoms (asthma attacks) each year. These attacks often occur during the night or in the early morning. Attacks lead to more than 1.7 million emergency department visits and about 450,000 hospitalizations annually, at a cost exceeding $20 billion per year. Although asthma deaths are largely preventable, in the United States, over 3000 patients will die from asthma annually (abstracted from the National Asthma Control Initiative (NIHLB) Guidelines for the Diagnosis and Management of Asthma (EPR-3) http://www.nhlbi.nih.gov/guidelines/asthma/index.htm/).

The NIH recommends stratification of patients according to asthma severity, coupled with implementation of treatment algorithms. While it is important to control exposure to environmental triggers, patient management is heavily dependent on controlling airway inflammation (eg, with inhaled corticosteroids) and with therapy escalation as required. Short-acting beta-agonists may be used in the short term to relieve symptoms. If asthma control remains poor despite the use of inhaled corticosteroids, other controller therapies, such as long-acting beta-agonists, leukotriene modifiers, cromolyn (mast cell stabilizers), theophylline, or even systemic corticosteroids should be added to the regimen. Newer treatments such as leukotriene receptor antagonists, omalizumab, a monoclonal antibody against IgE, drugs that stimulate eosinophil apoptosis and tumor necrosis factor inhibitors are also in use, or under investigation. In addition to lung function tests, currently a mainstay of asthma diagnostics, several biomarkers of airway inflammation have been proposed to assist in monitoring treatment. These

Bronchi, bronchial tree, and lungs

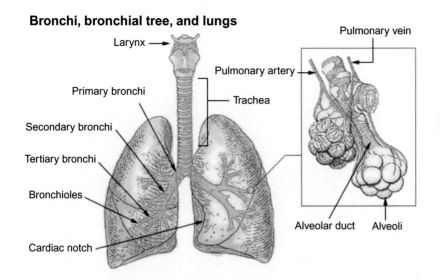

FIGURE 8.2 **Structures of the normal lung.** The trachea divides into bronchi that enter the lung and branch into smaller bronchioles; in this transition, the airways undergo structural changes. The cartilage rings and cilia disappear and submucosal glands are found within the walls. *Terminal bronchioles* are less than 2 mm in diameter. Distal to the terminal bronchioles are the *respiratory bronchioles* that communicate directly with the *alveoli*. Alveolar ducts branch off the respiratory bronchiole wall and terminate in alveolar sacs, the site of gas exchange. A *pulmonary lobule* consists of a cluster of 3–5 terminal bronchioles with their acinus (cluster of alveoli). *Figure in the public domain https://commons.wikimedia. org/w/index.php?curid = 789698.*

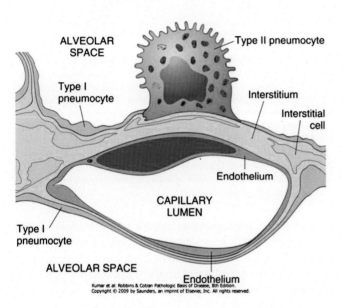

FIGURE 8.3 **A section of the alveolar wall, depicting pneumocytes I and II, interstitial cells, and the endothelial cells lining the capillary lumen.** The basement membrane (yellow) is thin on one side and widened where it is continuous with the interstitial space. *Reproduced from Kumar, V., et al., 2014. Robbins and Cotran Pathologic Basis of Disease, 9th Ed. Chapter 15. The Lung: asthma. Elsevier Saunders, Philadelphia, P.A. With permission from Elsevier Saunders.*

biomarkers can be assayed in several biological samples including sputum, exhaled gases, exhaled breath condensate, and urine (Wadsworth et al., 2011). Exfoliated airway epithelial cells have also been used to improve asthma diagnosis (Zhang et al., 2014).

8.3 PATHOPHYSIOLOGY OF ASTHMA

Bronchial biopsies of asthmatic patients have revealed that asthma is characterized by the accumulation of eosinophils, mast cells, and CD4$^+$ T cells

secreting interleukin 4 (IL-4) and/or IL-5 and IL-13 in patients with atopic hypersensitivity, and genetically mediated predisposition to an excessive IgE reaction (Pawankar et al., 2009). These observations suggested that aberrant T-helper 2 (T$_H$2) cells initiate the pathogenesis of asthma; this mechanism is presented in Fig. 8.4. However, the T$_H$2 response has been shown to be heterogeneous and observed in only about 50% of asthmatic patients (Woodruff et al., 2009), leading to investigations of other mechanisms involved in initiation and progression of asthma. The T$_H$2 hypothesis could not explain the fraction of patients refractory to steroid therapy (Anderson, 2008) and clinical observations revealed that the T$_H$2 response does not adequately fit the clinical data. Thus, the possibility that other T cells might also contribute to asthma has been raised (Lloyd and Hessel, 2010). Regardless of their cause, all forms of asthma induce structural changes in the airway wall (remodeling) with an increase in smooth muscle mass and deposition of extracellular matrix under the epithelial basement membrane and an increase in mucus-secreting goblet cells as comprehensively reviewed by Lambrecht and Hammad (2012).

8.3.1 The Innate Immune Response and Asthma

Epithelial cells that line the airways (AEC) express pattern recognition receptors (PPR), including pathogen-associated molecular patterns (PAMPs) and damage-associated molecular patterns (DAMPs). These sensors allow the AECs to rapidly detect pathogens or cell-derived structures released upon tissue damage, stress, or death. Activation of these PPRs leads to the release of cytokines, chemokines, and other mediators as discussed elsewhere (refer to Chapter 2: The Obese

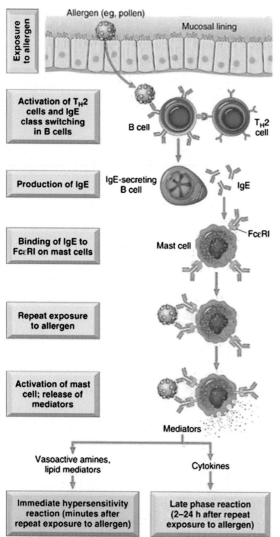

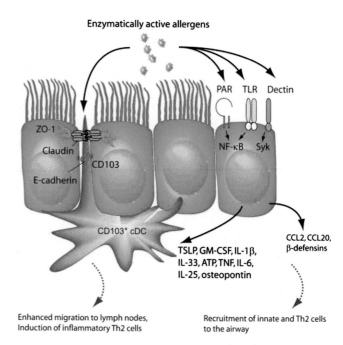

FIGURE 8.5　Dendritic cells and the AEC barrier. Allergens can activate protease-activated receptors (PARs) with subsequent activation of the nuclear factor kappa B (NFκB) cascade and production of chemokines and cytokines. Toll-like receptors (TLRs) and other PPRs also recognize pathogens and cell damage. Allergens can also be recognized by DCs attached to the AEC by tight junctions. Information on AEC attack is communicated to the adaptive immune system, specifically the inflammatory T_H2 cells (Lambrecht and Hammad, 2009). *Reproduced from Lambrecht, B.N., Hammad, H., 2009. Biology of lung dendritic cells at the origin of asthma. Immunity 31 (3), 412–424 with permission from Science Direct.*

FIGURE 8.4　The T_H2 hypothesis of asthma. This hypothesis holds that inhaled allergens bound by dendritic cells in the mucosal lining stimulate induction of T_H2 cells. T_H2 cells secrete cytokines that promote inflammation and stimulate B cells to produce IgE and other antibodies. Major cytokines including IL-4, stimulate the production of IgE. IL-5 activates locally recruited eosinophils, and IL-13 stimulates mucus secretion from bronchial submucosal glands and also promotes IgE production by B cells. Together these mediators result in immediate and late-phase allergic reactions. *Reproduced from Kumar, V., et al., 2014. Robbins and Cotran Pathologic Basis of Disease, 9th Ed. Chapter 15. The Lung: asthma. Elsevier Saunders, Philadelphia, PA. With permission from Elsevier Saunders.*

Gunshot Patient: Injury and Septic Shock: Innate Host Defense). AEC activation through PPRs appears to be a key event in activating dendritic cells (DCs) and coordinating a subsequent immune response.

In a manner similar to intestinal cells, AECs are protected by a mucociliary blanket and closely annexed by tight junctions and zonula occludens proteins; this structure creates an AEC barrier that excludes

antigens and pathogens as reviewed by Hammad and Lambrecht (2008). It must be remembered that the AEC barrier, like the intestinal barrier, contains mucosal DCs situated in the basolateral space. DCs extend processes between epithelial cells directly into the airway lumen allowing them to survey the luminal surface as illustrated in Fig. 8.5. As in the gut barrier, DCs express the tight-junction proteins claudin-1, claudin-7, and zonula-2, and can form tight junctions with airway epithelial cells maintaining an intact AEC barrier.

The lung, similar to the gut, is also colonized by commensal bacteria (Riiser, 2015). Asthmatic lung microbiota differs markedly from normal lung microbiota. These observations have implications for the hygiene and biodiversity hypothesis of asthma prevention as discussed below.

8.3.2 The Gut–Lung Axis

The concept that the gut immune system could participate in the pathogenesis of asthma is now strongly supported by epidemiologic and experimental evidence as comprehensively reviewed by Smith et al. (2013) and Samuelson et al. (2015). Briefly, antigens are processed by DCs in the gut and presented to the adaptive immune system. Immune tolerance is regulated by the gut bacteria through their ability to produce metabolites such as short-chain fatty acids (SCFAs), butyrate, niacin, and other factors needed to facilitate transformation of naïve T cells into immunotolerant T^{reg} cells instead of proinflammatory effector Th17 cells as illustrated in Fig. 8.6.

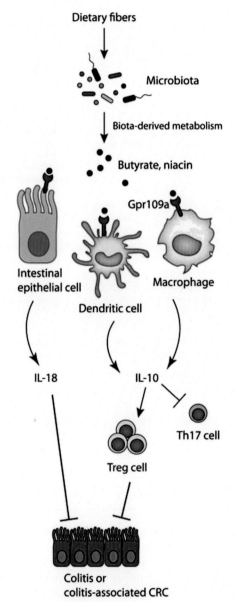

FIGURE 8.6 Dietary fiber, butyrate, and T^{reg} production. The link between bacterial metabolism, SCFA, the innate immune system, and the adaptive immune system is reviewed by Jobin (2014) and discussed elsewhere (refer to Chapter 3: Type I Diabetes and Celiac Disease: Type I DM as Dysregulated Immune Response). Briefly, gut bacteria form SCFA that bind to receptors on the DC and macrophages and in the presence of retinoic acid, produce IL-10 that facilitates differentiation of naïve T cells to immune-tolerant T^{reg} cells. *Reproduced from Jobin, C., 2014. GPR109a: the missing link between microbiome and good health? Immunity 40 (1), 8—10 with permission from Science Direct.*

A gut—lung model has been proposed in which T cells differentiated in the gut home to the sites of infection or antigen exposure. Disruptions in the intestinal microbiota lead to increased expansion of T-cell subsets which are then reflected at the peripheral site, in this case the lung. This link was observed in young and old mice sensitized and challenged with house dust mite antigen. After challenge, fecal pellets excised from the colon and subjected to 16S rRNA analysis were seen to change with age and allergy development. Allergic mice had induced serum levels of IL-17A after challenge; the effect was greater in old than young mice and characterized by an increased pulmonary response in old mice (Vital et al., 2015).

Food allergy, implicated as an initiator of asthma attacks, has also been linked to the microbiota. Several investigators have identified the action of innate lymphoid cells (ILCs) described elsewhere (refer to Chapter 2: The Obese Gunshot Patient: Injury and Septic Shock: Innate Host Defense); in food allergy. ILCs at the intestinal barrier limit inappropriate immune responses to bacteria, regulating interactions between DC and the microbiota. A recent report demonstrated that sensitization to a food allergen was enhanced in mice treated with antibiotics or raised in a germfree environment. The allergy-protective capacity was shown to be provided by clostridia, an anaerobic spore-forming class of *Firmicutes* known to reside near the intestinal epithelium (Stefka et al., 2014). Clostridia were shown to induce IL-22 production by retinoic acid receptor-related orphan receptor gamma $(ROR\gamma\tau)^+$ ILCs and T cells in the lamina propria; this cytokine is reported to reduce dietary antigen entry into the systemic circulation. The authors noted that defects in intestinal permeability have also been implicated in allergic responses to food. The influences of specific dietary components that can damage the intestinal barrier, increase or decrease intestinal permeability are discussed elsewhere (refer to Essentials IV: Diet, Microbial Diversity and Gut Integrity) (Fig. 8.7).

8.4 PATHOLOGY OF THE ASTHMA ATTACK

Asthma is classified as *atopic* and *nonatopic*. Atopic asthma is associated with evidence of allergen sensitization and is a classic IgE-mediated type I hypersensitivity reaction. Nonatopic asthma (also called intrinsic asthma) is brought on by exercise or other triggers but does not have an obvious extrinsic trigger. The atopic allergic response is triggered by an antigen (allergen) and mediated by T_H2 cells, IgE antibodies, and mast cells. Despite its etiologic heterogeneity most types of asthma proceed along a similar clinical course.

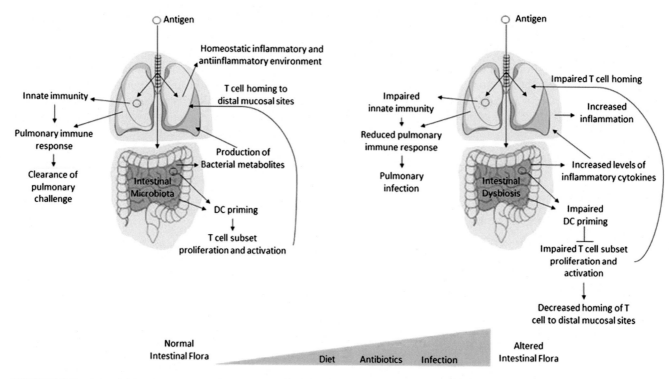

FIGURE 8.7   **A model for the gut—lung axis.** Antigens detected and processed in the gut are presented to diverse T cells that home to the sites of infection or antigen exposure. The immune-modulatory tone of the T cells is dictated by the composition and metabolic activity of the gut microbiota (Samuelson et al., 2015). *Figure is under Copyright © 2015 Samuelson, Welsh and Shellito. This is an open-access article distributed under the terms of the Creative Commons Attribution License (CC BY). The use, distribution or reproduction in other forums is permitted, provided the original author(s) or licensor are credited and that the original publication in this journal is cited, in accordance with accepted academic practice. No use, distribution or reproduction is permitted which does not comply with these terms.*

Mediator release increases capillary permeability and acts on smooth muscles. Proinflammatory mediators recruit inflammatory cells; eotaxin and IL-5 specifically recruit and activate eosinophils. IgE binds to submucosal mast cells; repeated exposures to allergens trigger the mast cells to release granule contents (histamine) and produce additional cytokines and mediators.

The early-phase reaction is dominated by bronchoconstriction, increased mucus production, and variable degrees of vasodilation with increased vascular permeability. Bronchoconstriction is triggered by direct stimulation of subepithelial vagal (parasympathetic) receptors through both central and local reflexes (including those mediated by unmyelinated sensory C fibers). Intense study is underway to understand environmental triggers and the heritable genes that make patients with atopic asthma prone to develop strong T_H2 reactions to environmental allergens to which most individuals do not respond.

8.4.1 Mediators and Outcomes of T-helper Cell (T_H2)-Mediated Inflammation

Cell damage induces AECs to transdifferentiate into goblet cells competent to produce mucus (Fig. 8.8).

Several mediators with the potential to induce airway remodeling are present in the asthmatic lung. These include endothelins, transforming growth factor (TGF)-β, and epidermal growth factor (EGF). Mediators that regulate airway remodeling, such as cysteinyl leukotrienes (CysLTs) and interleukin (IL-13), are also produced.

8.5 TARGETS FOR ASTHMA CONTROL: AIRWAY REMODELING AND MUCUS PRODUCTION

The asthmatic lung undergoes several structural changes as reviewed by Fixman et al. (2007). The epithelial layer becomes fragile and permeable, epithelial cells undergo rapid turnover and frequently transdifferentiate into goblet cells. Bronchial blood vessels increase in size and number possibly due to recurrent hypoxia and angiogenic signals. Bronchial smooth muscle mass increases at the luminal side, bringing it closer to the epithelium. Altered extracellular matrix protein distribution participates in modeling of the airway submucosa and adventitia. Bronchial smooth muscle increases in size and there is evidence that alveolar attachments are disrupted and the elastin content is abnormal. Basement membrane thickening is

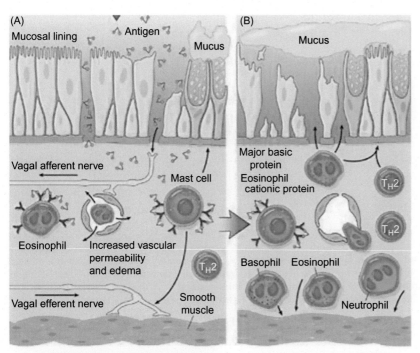

(A) Mucosal lining Antigen Mucus

Vagal afferent nerve

Mast cell

Eosinophil   Increased vascular permeability and edema

T_H2

Vagal efferent nerve   Smooth muscle

(B) Mucus

Major basic protein
Eosinophil cationic protein

T_H2

T_H2

Basophil Eosinophil

Neutrophil

D. IMMEDIATE PHASE (MINUTES)   E. LATE PHASE (HOURS)
Cotran Pathologic Basis of Disease, 8th Edition.

FIGURE 8.8 **Lymphocyte activity in the asthmatic lung.** Atopic allergy is often characterized by mediators that stimulate immature B cells to make the isotopic switch and preferentially secrete IgE, and also to increase the number of eosinophils present. Mast cells bound to IgE release their granule contents in response to allergen binding and produce cytokines and other mediators, which collectively induce the early-phase (A—immediate hypersensitivity) reaction and the late-phase (B) reaction. *Reproduced from Kumar, V., et al., 2014. Robbins and Cotran Pathologic Basis of Disease, 9th Ed. Chapter 15. The Lung: asthma. Elsevier Saunders, Philadelphia, PA. With permission from Elsevier Saunders.*

comprised of connective tissue elements not found in normal basement membrane. Smooth muscle cells proliferate and hypertrophy; the tissue is infiltrated by fibroblasts and myofibroblasts that secrete fibrous matrix into the area.

Smooth muscle cells transdifferentiate into myofibroblasts, maintaining their contractile ability while synthesizing collagen and connective tissue matrix. Asthmatic airways narrow, largely by contraction of smooth muscle. The modified smooth muscle cells also secrete inflammatory cytokines with potential to contribute to these changes. Additionally, molecular changes that influence contractility, including altered myosin light-chain kinase or crosstalk between signaling pathways have been suggested as reviewed by Anderson (2008). Specific molecular changes affecting contractility, including altered myosin light-chain kinase isoforms or crosstalk between signaling G proteins, have been suggested but not confirmed. An increased rate of sarcomere shortening has been shown and could be a functional defect that causes airway narrowing; the mechanism is not yet understood.

In addition to its excess of goblet cells with production of copious mucus, the asthmatic airways are infiltrated with leukocytes (eosinophils, neutrophils, macrophages, and T lymphocytes), especially in the subepithelial layer. Leukocytes are recruited by mediators (chemokines) produced by the mast cells, epithelial cells, and T cells in response to chronic allergenic stimulation. Of special interest is eotaxin, produced by airway epithelial cells. This mediator is a potent chemoattractant and activator of eosinophils. Eosinophils produce proteins (major basic protein, eosinophil cationic protein) that inflict further damage on the epithelial lining and stimulate bronchoconstriction (Fig. 8.9).

8.6 TARGETS FOR ASTHMA CONTROL: OBESITY AND LIFESTYLE RISKS

Several epidemiologic studies have demonstrated associations between asthma and lifestyle choices and complicated by abdominal adiposity, hypertension, dyslipidemia, and insulin resistance (Kim et al., 2009; Agrawal et al., 2011). Obese asthmatics have reduced lung inflammation, yet they have marked respiratory symptoms, poor response to glucocorticoids and peripheral airway dysfunction. Intervention with a very low-calorie diet (~450 kcal) for 8 weeks followed by a low-calorie diet produced weight loss and symptom reduction, with no change in medication (Stenius-Aarniala et al., 2000). It is not clear from these studies whether decreased fat storage and/or associated metabolic changes in inflammatory mediators contributed to the reduction of airway hyperreactivity, or conversely, whether increased activity and reduced stress improved symptoms. It is also possible that the improved composition of the diet (higher fiber, lower fat, and sugar) exerted an antiinflammatory effect and may have improved gut microbial metabolism. In this

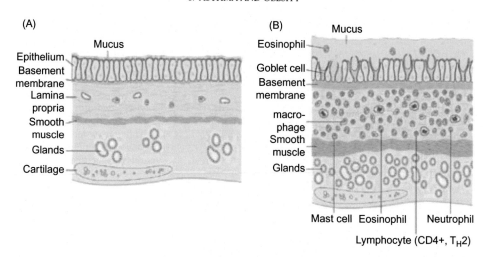

FIGURE 8.9 **Normal airway (A) compared with the asthmatic airway (B).** The normal airway is lined by a barrier of ciliated epithelial cells (AECs); in asthma, the epithelial barrier is compromised and permeable to allergens. Chronic stimulation causes AEC transdifferentiation to goblet cells that produce a thickened mucus layer on the epithelial surface. Not shown is the surfactant layer that lies adjacent to the alveolar cell membrane. Deeper connective tissue and smooth muscle layers are thickened and remodeled. The airways are chronically infiltrated with inflammatory cells and glandular hyperplasia is evident. *Reproduced from Kumar, V., et al., 2014. Robbins and Cotran Pathologic Basis of Disease, 9th Ed. Chapter 15. The Lung: asthma. Elsevier Saunders, Philadelphia, PA. With permission from Elsevier Saunders.*

trial, weight loss may simply have increased the therapeutic efficacy of the subjects' asthma medications. Clearly, more mechanistic studies are needed to identify pathogenic links relating obesity and asthma.

Several possible relationships between obesity and asthma require further investigation (McGinley and Punjabi, 2011):

- Obesity can impose a mechanical load on the diaphragm and decrease expiratory residual volume and functional residual capacity, reducing airway diameter and smooth muscle function.
- Obesity-associated inflammation may increase visceral adipocyte secretion of proinflammatory cytokines such as interleukin-6, leptin, tumor necrosis factor, tumor growth factor-b1, and eotaxin. The altered mediator milieu can modify atopy and exacerbate the T_H1-T_H2 imbalance (common in obesity), increasing airway hyperreactivity.
- Obesity is associated with a cadre of metabolic derangements; there is growing evidence that hyperglycemia, hyperinsulinemia, and insulin-like growth factors exert negative effects on airway structure and function.
- Studies from obese mouse models of asthma have identified abnormalities in nitric oxide—arginine metabolism, associated with oxonitrosative stress in lungs.
- Finally, the microbiota of obese individuals is shifted dramatically away from the "lean phylotype." Obese humans display an increased *Firmicutes* to *Bacteroidetes* balance; this pattern is associated with increased energy extraction from dietary polysaccharides. Dysbiosis of the gut

microbiota has implications for altered sensitivity to food allergens and altered immune function as described above (Sekirov et al., 2010).

Since data are equivocal as to the common pathologic factors that underlie a possible link between obesity and asthma, the search continues.

8.6.1 Influence of Food Allergy on Asthma Development

Suggestive evidence implicates diet quality, nutrient ingestion, and/or hormonal exposure both in utero and throughout life as contributing to the development and severity of asthma. In 2007, the American Academy of Allergy, Asthma and Immunology (see AAAAI website at http://www.aaaai.org/about-the-aaaai/newsroom/allergy-statistics.aspx#Food_Allergy) estimated that approximately 3 million children (under age 18) reported a food or digestive allergy in the previous 12 months. The prevalence of food allergy in this population has increased 18% over the previous decade; children with food allergy were 2—4-fold more likely to also have asthma or other allergic symptoms. Adults (~3 million) also suffer from food allergy. From 2003 to 2006, food allergies resulted in approximately 317,000 visits to hospital emergency departments, outpatient clinics, and physicians' offices. Food allergy-related hospital admissions increased from 2600 per year (1998—2000) to 9500 per year (2004—2006) and caused approximately 150—200 fatalities per year.

The gut—lung axis described above suggests that immune cells activated by ingested food are likely to home to other organs including the lung and stimulate an asthma-like pattern. Such a relationship has recently been demonstrated (Akinbami et al., 2016) in that children with asthma and allergies—particularly food and/or multiple allergies—are at risk for adverse asthma outcomes.

8.6.2 The Hygiene Hypothesis as a Risk Factor for Airway Hypersensitivity

The rapid "Westernization" of the diet is thought to play a key role in the complex genetics and developmental pathophysiology of asthma. The "hygiene hypothesis," first proposed in 1989, holds that bacterial and viral infections early in life shift the maturing immune system away from the proallergic T_H2 responses and toward T_H1 responses. Enhanced understanding of the influence of gut microbes on immune system maturation and tolerance generated the "microflora hypothesis" (Noverr and Huffnagle, 2005). This hypothesis suggests that diet and antibiotic use can reduce exposure of the underdeveloped microbiota to diverse microbes and can delay maturation of the gut immune system. These limitations can reduce normal immunological tolerance and increase allergic hypersensitivity. The hypothesis has been supported by studies in both mice and patients, showing that dysbiosis induced by oral antibiotics elevated serum IgE concentrations, increased steady-state circulating basophil populations and produced exaggerated basophil-mediated T_H2 responses and allergic inflammation (Hill et al., 2012). On the other hand, population studies have been inconsistent and some studies have not found clear-cut immune marker differences in cases and controls, leading to the consensus that the pathogenesis of allergy and airway hypersensitivity is far more complex than originally thought (Figueiredo et al., 2013). In any case, the growing plethora of antibacterial products, heavily advertised in the media for their use in preventing bacterially transmitted disease, may be counterproductive in preventing airway hypersensitivity and asthma.

8.6.3 Maternal and Early-Childhood Feeding Practices and Asthma

The influence of the maternal microbiota, birth type, and early feeding practices are known to play major roles in the diversity and function of the infant's microbiota as discussed elsewere (refer to Essentials IV: Diet, Microbial Diversity and Gut Integrity). Breastfeeding provides immunological protection when the infant's immune system is immature and is also a source of immunomodulary fatty acids. For these reasons, breast milk is recommended by pediatric societies as the sole

infant food at least for the first 4 months. The timing of solid food introduction appears to be more controversial. Gut-associated mucosal immunity develops in early infancy. During this period, genetics, gut flora, antigen dose, and digestive processes influence how antigen-presenting cells process and present antigen to T cells. European and American Societies of Pediatrics differ slightly on their recommendations for cereal grain introduction. One study (Poole et al., 2006) found no evidence that delaying cereal introduction past 6 months was protective; indeed, introduction of cereal grains after 6 months actually increased the risk for wheat allergy. These results support the recommendation by the American Association of Pediatrics that cereal grains should be introduced in small amounts between 4 and 6 months and should not be delayed beyond 6 months.

Epidemiologic patterns as reviewed by Litonjua and Weiss (2007) support the hypothesis that relative maternal vitamin D deficiency can result in aberrant immunomodulation and asthma in the offspring. Hypovitaminosis D is prevalent in developing countries (Arabi et al., 2010) but also in Westernized nations farther away from the equator. Vitamin D deficiency is also concentrated in inner-city populations and in obese individuals. Decreased dietary vitamin D intake, inadequate sun exposure, and presumptive vitamin D sequestration in an expanding fat mass are risk factors for vitamin D deficiency, especially in children (Kumar et al., 2009) and in women of child-bearing age as discussed elsewhere (refer to Essentials III: Nutrients for Bone Structure and Calcification: Vitamin D).

8.6.4 Dietary Patterns Associated With Asthma

Early observational studies focused on the impact of dietary factors such as antioxidants and lipids in the etiology and treatment of asthma as reviewed by Agler et al. (2011), while other investigations have examined related dietary patterns, increased prevalence of obesity, fast foods, and the Mediterranean diet (Hooper et al., 2010). Evidence linking specific nutrients, antioxidant supplementation, food types, or dietary patterns past early childhood on asthma prevalence is inconclusive. However several recent surveys suggest that a diet pattern rich in fruits and vegetables, as well as nuts, is inversely related to asthma symptoms. The influence of diet on airway hypersensitivity has also been linked to the fetal environment. The maternal diet may be a significant factor in the development of the fetal airway and immune system, and may play an epigenetic role in sensitizing fetal airways to respond normally to environmental insults. It must be noted that a diet composition has important influence on the diversity and phylogenetic balance of the microbiota (Claesson et al., 2012).

8.6.5 Diet as a Source of Allergens

The most common allergens in the diet are cow's milk, eggs, peanuts, tree nuts, wheat, soy, fish, and shellfish. Allergens in these foods, or their metabolites can cross the gut mucosa, enter the bloodstream, and bind IgE on mast cells, setting off the atopic allergic cascade as described above. Reactions can be localized (lung, gut, skin, etc.) or result in system-wide anaphylaxis.

Discovery of the specific trigger food(s) is often difficult. An extensive prospective and retrospective diet history is often required to accurately assess the allergenic potential of a specific food. Food-frequency questionnaires and food diaries can be used as tools to correlate foods with allergic symptoms. Food allergies have been classified according to the role of IgE antibodies in precipitating the attack:

- Immediate; IgE-mediated food allergy (eg, peanut anaphylaxis);
- Mixed mechanisms, involving both IgE- and cell-mediated reactions (eg, atopic dermatitis, allergic eosinophilic gastroenteritis);
- Delayed onset, cell-mediated food allergy (eg, food protein-induced enterocolitis syndrome).

Common food patterns associated with food allergy include:

- Inadequate food/oral beverage intake;
- Poor food choices: can be secondary to food insecurity, inadequate cooking facilities, lack of time;
- Disordered eating/meal patterns, reliance on fast food, inadequate knowledge of healthy eating.

No cure currently exists for food allergy. Patients with features suggesting atopic asthma should be assessed for possible food allergens with potential to trigger an attack. As recently reviewed (Albin and Nowak-Wegrzyn, 2015), the standard of care must be strict avoidance of foods to which the patient is allergic. The authors describe protocols for desensitization; this procedure involves administering increasing doses of the allergenic food over time to achieve tolerance. Since this treatment is not successful in all patients, the authors discuss alternative strategies and newer methods for treatment.

Assessment of food triggers is also fraught with difficulty. The double-blind, placebo-controlled food challenge (DBPCF), although labor-intensive, has been the "gold standard" for diagnosis of food allergy for many years (Chinthrajah et al., 2015). Since the DBPCF does carry a risk of anaphylaxis, physician support is required for any formal trial. Several antibody-targeted diagnostic tests have been developed to detect food-specific IgE (skin or serum IgE), but progress in developing tests to detect non-IgE-mediated food allergy has been less rapid. It should be noted that food antigen determinants may be significantly modified during the processes of digestion and gut fermentation. Thus, the true antigen may not be present in the food extracts tested in vitro. A detailed food record with indication of allergic/asthmatic events with respect to food is a useful, noninvasive indication of possible food triggers.

8.7 DIET IN THE MANAGEMENT OF AIRWAY HYPERSENSITIVITY

Accumulating evidence suggests that diet modification may be an important adjuvant therapy in reducing the severity and frequency of asthma attacks (Nowak-Wegrzyn and Groetch, 2012). Recently, several targeted treatment strategies have been proposed to interrupt pathologic processes leading to asthma.

8.7.1 Carbohydrates, Fiber, and Gut Microbiotia

Although positive associations between intake of simple sugars and the prevalence of chronic diseases characterized by inflammatory components have been reported (Thornley et al., 2011), most studies have focused on diet patterns and glycemic index (GI) as independent variables. Some studies have suggested that diet patterns containing simple carbohydrates and sweet desserts are associated with increased severity of wheezing and asthma attacks. A recent epidemiologic study examined the impact of dietary GI, dietary fiber, and carbohydrate-containing food groups on the mortality attributable to noncardiovascular, noncancer inflammatory disease. The investigators concluded that a diet characterized by high GI foods increased mortality attributed to inflammation and oxidative stress. One plausible explanation for this observation is the overproduction of inflammatory cytokines and reactive species seen with recurrent postprandial hyperglycemia (Buyken et al., 2010).

In contrast, dietary fiber has been associated with protection from inflammation. Several plausible mechanisms have been proposed, including the effects of fiber on transit time and maintenance of protective commensal bacterial populations. The microbiota is known to interact with diet and the immune system in a variety of ways as discussed above, including:

- The immune system responds to pathogens by DC sensors and a variety of pattern recognition molecules such as the Toll-like receptors (TLRs), that facilitate recognition of pathogens.

- Commensal bacteria ferment dietary fiber and produce antiinflammatory short chain fatty acids (SCFAs). SCFAs, and other bacterial metabolites have been shown to bind and activate the free fatty acid receptor 2 (FFR2; also known as G-protein-coupled receptor, GPR43) that acts as a nutrient sensor and also regulates immune and inflammatory responses (Covington et al., 2006; Maslowski et al., 2009) and induces tolerance to food antigens.
- Resistant starch (RS) has a three-dimensional organization that prevents digestion by salivary or pancreatic amylase but is fermented by commensal bacteria. Soluble gel-forming fibers include pectins, gums, mucilages, and nonstructural hemicelluloses that are viscous and fermentable. Whole grains, fruits, and vegetables contain fermentable RS and soluble fibers.
- Finally, experimental alteration of the normal bacterial composition by oral antibiotics produces changes in both mice and humans consistent with IgE secretion and increased susceptibility to T_H2 cytokine-dependent inflammation and allergic disease (Hill et al., 2012).

It follows that alteration in microbial species patterns (dysbiosis) has potential to negatively affect the immune response. It has been shown, for example, that the relative proportion of *Bacteroidetes* is lower in obese than lean subjects, and in individuals who consume a high fat and sugar, low fiber "Western"-style diet or take antibiotics frequently. RS and soluble fiber are known to increase the proportion of *Bacteroidetes* in the microbiota. Thus, it is possible that the positive impact of dietary fiber on asthma and allergy could be mediated through this link.

8.7.2 Fatty Acids: Conjugated Linoleic Acid, n-3, and n-6 PUFA

Leukotriene (LT4) and prostaglandin (PG2) products of arachidonic acid (20:4 n-6) participate in early inflammatory processes that lead to asthmatic changes in the airways, sometimes within 2—12 months after diagnosis.

Products of n-3 fatty acids, eicosapentaenoic acid (20:5 n-3), and docosahexaenoic acid (22:6 n-3) induce less severe inflammatory actions than do eicosanoids produced from arachidonic acid. Inflammatory leukotrienes of the 4-series are potent inducers of bronchospasm, airway edema, mucus secretion, and inflammatory cell migration, all of which are important to the asthmatic symptomatology (Simopoulos, 2008; Simopoulos, 2010). (Prostaglandin aromatization uses 2 of the 5 double bonds of eicosapentaenoic acid (20:5) producing eicosanoids of the PG3 and LT5

series; similarly, arachidonic acid (20:4) produces eicosanoids of the PG2 and LT4 series.)

Recently other antiinflammatory, counter-regulatory lipid mediators derived from membrane polyunsaturated FA (PUFA) have been identified. These mediators include lipoxins, resolvins, protectins, cyclopentenones, and presqualene diphosphate (Haworth and Levy, 2007). Some products (resolvins and protectins) are associated with wound healing and are being intensively studied as expediting factors in the resolution of inflammation following trauma (Ariel and Timor, 2013) as discussed elsewhere (refer to Chapter 2: The Obese Gunshot Patient: Injury and Septic Shock: Mediators of the Acute Phase Response). The G-protein-coupled receptor 120 (GPR120) appears to function as an n-3 PUFA receptor and caused antiinflammatory effects in a monocyte cell line (RAW 264.7) and in primary intraperitoneal macrophages (Oh et al., 2010).

PUFAs incorporated into the biomembrane not only modify eicosanoid metabolism, but also modulate lipid—protein lateral organization and inhibit downstream signaling mediated by T-cell receptors, suppressing T-cell activation and proliferation. PUFA can also alter the surface expression of class I and II major histocompatibility complexes, thereby modifying antigen presentation, possibly by altering its conformation, orientation, lateral organization, and trafficking, with consequences for recognition by effector T cells (Shaikh and Edidin, 2006).

Several studies have shown that low dietary availability of n-3 PUFA is associated with increased respiratory distress upon allergen stimulation. High n-3 PUFA ingestion increased urinary 5-series leukotriene excretion and improved respiratory capacity and treatment efficacy as reviewed by Simopoulos (2008). Despite these potential benefits in reducing inflammatory changes associated with allergy and asthma, the most recent Cochrane systemic analysis of studies testing the effect of n-3 PUFA supplementation in pregnant and/or breastfeeding women on allergy outcomes (food allergy, atopic dermatitis or eczema, allergic rhinitis or hay fever, and asthma/wheeze) in their children was inconclusive. There was little long-term difference in the allergy outcome of children of mothers who were supplemented compared with those who were not supplemented (Gunaratne Anoja et al., 2015).

8.7.3 Conjugated Linoleic Acid

Conjugated linoleic acid (CLA) refers to a class of positional and geometric conjugated dienoic isomers of linoleic acid (LA), of which cis-9, trans-11 (c9, t11) and trans-10, cis-12 (t10, c12) CLA predominate. Milk fat contains CLA and full-fat milk consumption from early childhood on has been found to be inversely correlated with allergic sensitization and the onset of bronchial asthma. Similar results were shown with consumption of butter and grass-fed beef, also rich in naturally occurring CLA (O'Shea et al., 2004).

In nature most LA (18:2) is conjugated to c9, t11 (~90%) by bacterial action in the rumen of cows, sheep, goats, and other ruminants. The concentration of CLA in meat and dairy products depends on the availability of LA substrate and rumen bacteria activity. Increasingly in the United States, commercial beef cattle are "finished" on feedlots where they are prevented from exercise and fed a high-grain diet to maximize marbling (fat deposition in the muscle). (Complete breakdown of feedlot management: Iowa State University Extension. "Beef Feedlot Systems Manual." http://www.extension.ias-tate.edu/Publications/PM1867.pdf.) While this regimen optimizes weight gain and produces a tender meat, feedlot finishing dramatically lowers rumen pH, changes the bacterial composition, and can make the animals very sick. Prophylactic antibiotics are routinely given to feedlot animals to prevent overwhelming infections. Meat from animals finished on grass and pasture not only contains more LA substrate, but grass-feeding fosters bacterial species in the rumen that conjugate LA, thus grass-fed meat and dairy products contain substantially more CLA than grain-fed meat (Mir et al., 2004).

Mechanisms that underlie protection from airway hypersensitivity associated with the consumption of CLA are unclear. Recent in vivo studies in mice demonstrated that CLA reduced the T_H2 cytokine IL-5 in the bronchoalveolar lavage fluid, reduced IgE and eosinophil production and modulated allergen-induced airway hyper-responsiveness. Less mucus plugging of segmental bronchi and membrane remodeling were also seen, most likely via a PPARγ-related mechanism and by reducing eicosanoid precursors (Jaudszus et al., 2008).

8.7.4 Magnesium

Magnesium (Mg) sources, absorption, and utilization have been discussed elsewhere (refer to Essentials III: Nutrients for Bone Structure and Calcification: Magnesium). Dietary Mg intake is associated with lung function as comprehensively reviewed by de Baaij et al. (2015). Briefly, inadequate Mg has been associated with impaired lung function as assessed by forced expiratory volume and forced vital capacity, although mechanisms through which Mg supports lung function are not well characterized. Possible explanations include: (1) a strong vasodilator and bronchodilator effect; (2) regulation of acetylcholine and histamine release; and (3) antiinflammatory effects. Several studies have reported low total serum Mg levels in asthmatic patients although others have not. It must be remembered that circulating Mg is ~1% of total body Mg and changes in Mg levels could be due to the severity of asthma, or to comorbid conditions. Since Mg relaxes smooth muscles as discussed elsewhere (refer to Chapter 5: Cardiomyopathy and Congestive Heart Failure: Targets for Myocardial Protection: Calcium and Hypercontracture), if Mg deficiency is present, bronchoconstriction and vasoconstriction could follow, exacerbating asthmatic symptoms.

Mg injections have been used to treat asthmatic paroxysms for almost 80 years (Haury, 1940), however a recent systematic review failed to demonstrate significant improvement in respiratory function or in hospital admissions for patients with acute asthma who had been given nebulized Mg sulfate; however, improvement with both parameters approached significance. For intravenous Mg injection, respiratory function increased in adults and improvement was greater in children. Hospital admissions were reduced (Mohammed and Goodacre, 2007). A Cochrane systematic review of nebulized Mg for acute asthma also found data too heterogeneous for a conclusion (Powell et al., 2012).

8.7.5 Vitamin D

The requirement for vitamin D in calcium metabolism and bone integrity and its regulation by parathyroid hormone and fibroblast growth factor 23 (FGF23) is discussed elsewhere (refer to Essentials III: Nutrients for Bone Structure and Calcification: Vitamin D). Abundant evidence has established vitamin D as having an important role in modulating the immune system (Berridge, 2015). As an immunomodulatory factor, vitamin D regulates the activity of immune cells and production of pro- and antiinflammatory mediators. This role suggests that inadequate vitamin D status could influence the development and severity of asthma and other inflammatory diseases such as infections, cancer, and helper T-cell-related autoimmune diseases, such as type I diabetes.

The role of vitamin D in T-cell responses has been reviewed (Mora et al., 2008). Briefly, active 1,25 $(OH)_2D_3$ (1,25D) blocks induction of helper T-1 (T_H1) cytokines (INFγ) while promoting T_H2 responses including enhancing IL-4 production. 1,25D suppresses DC synthesis of IL-12, also reducing T_H1 responses. T_H17 cell responses are also suppressed, inducing reciprocal differentiation of naïve T cells to T^{reg} cells. B-cell proliferation, plasma cell differentiation, and IgG secretion are also suppressed by 1,25D; this effect might be indirectly mediated through dendritic or macrophage antigen-presenting cells. Innate immune cells are also inhibited by 1,25D; DC differentiation, maturation, and immunostimulatory capacity are inhibited by downregulating their MHC class II molecules and other markers. Finally, 1,25D decreases synthesis of IL-12 and simultaneously increases IL-10 production by DCs. The net result of these changes is a reduction in T_H1 antimicrobial responses and an increase in T^{reg} production.

Observational and clinical studies in humans are suggestive that a causal relationship exists between vitamin D status and asthma development and management (Paul et al., 2012). Not only could vitamin D modulate asthma exacerbations, but it could contribute to host resistance to infection. Analysis of data from the Third National Health and Nutrition Examination Survey (HANES III) (Ginde et al., 2009) revealed an increased susceptibility to upper respiratory infections in patients with pulmonary diseases such as asthma who also had low serum 25(OH) D_3 levels. Several immunomodulatory mechanisms including reduction of atopy and infections and enhancing glucocorticoid action in controlling asthma have been postulated (Finklea et al., 2011).

Six clinical trials are ongoing which can yield, in addition to treatment efficacy, data on dosing, toxicity, and the safety of vitamin D supplementation in different age groups. (Clinical trials are registered at http://clinicaltrials.gov. Two of the trials (NCT00920621 and NCT00856947) are testing the efficacy of supplementation during pregnancy and asthma prevention in the children; three trials (NCT01248065; NCT00978315; NCT01419262) are testing whether vitamin D supplementation added to an inhaled glucocorticoid regimen moderates asthma exacerbation in asthmatic children or adults.) Physicians are cautioned that toxicity of vitamin D is still unknown and until the results of these trials are available, physicians should be guided by the recommendations of the National Academy of Sciences that defined deficiency as serum 25D levels < 20 ng/mL.

8.7.6 Immunomodulatory and Antioxidant Actions of Nonnutritive Components in the Diet

Diets rich in fruits and vegetables have been repeatedly associated with reduced inflammatory lung diseases including asthma (Sharafkhaneh et al., 2007). Although clinical trials with antioxidant nutrients (vitamins C and E) have produced equivocal results, the high flavonoid and polyphenol content of fruits and vegetables containing antioxidant vitamins is likely to play a greater role in their observed biological activity.

The vast (over 5000 members) flavonoid class of bioactive components in fruits and vegetables share a common chemical structure consisting of two phenol rings linked through three carbons. Protective mechanisms may involve scavenging of free radicals by means of their conjugated ring structure and modification of gene transcription via induction or inhibition of specific transcription factors. Antioxidant function is especially important since inhaled oxidants (eg, cigarette smoke) first come in contact with the lung epithelium at the epithelial lining—fluid interface. Endogenous antioxidants such as glutathione and diet-derived antioxidants in the epithelial lining fluid neutralize the oxidants and prevent epithelial damage. Antioxidant cascades such as nuclear factor (erythroid-derived 2)-related factor 2 (Nrf2) and other protective systems enhanced by non-nutritive components in the diet are discussed elsewhere (refer to Essentials I: Life in an Aerobic World).

While flavonoids and other dietary bioactive components are being extensively studied to identify the mechanisms through which they modulate intermediary metabolism, there is limited evidence of their bioavailability, efficacy, and possible toxicity in humans (Finley, 2005). Studies in experimental models support the recommendation that bioactive components from specific fruits, vegetables, and other plant sources may be useful as adjuvant therapy to ameliorate airway hypersensitivity and inflammatory lung disease (Sharafkhaneh et al., 2007). Given their mechanistic potential in reducing the frequency and severity of asthma attacks, it is important that clinical trials be designed to determine dosage, possible toxic complications, and therapeutic efficacy.

8.7.7 Flavonoids, Polyphenols, and Asthma

Resveratrol (3,5,4′-trihydroxystilbene) is a phytoalexin found in the skin and seeds of grapes and produced by grapevines in response to injury or fungal attack. This bioactive component has been reported to have antioxidant, antiinflammatory, and anticarcinogenic properties as described elsewhere (refer to Chapter 6: Atherosclerosis and Arterial Calcification: Red Wine, Resveratrol and Procyanidins). The compound was shown to ameliorate the airway remodeling and hyperreactivity in a murine model of allergic asthma (Royce et al., 2011) and mechanisms through which resveratrol acts to protect the lung from smoke damage are being intensively studied (Kode et al., 2008). Resveratrol has been shown to be a more effective antioxidant than vitamins E and C; it scavenges many species of free radicals including lipid hydroperoxyl, hydroxyl, and superoxide radicals. Resveratrol also inhibits lipopolysaccharide-induced airway neutrophilia and inflammation through an NF-κB-independent mechanism. A recent report demonstrated that oxidants (cigarette smoke) depleted, and resveratrol restored glutathione concentrations in the lung epithelial lining fluid by upregulation of its biosynthetic enzyme (glutamate cysteine ligase) via activation of Nrf2. Resveratrol also quenched Cystathionine γ lyase-induced release of reactive oxygen species and attenuated cytokine release by alveolar macrophages from patients with chronic obstructive pulmonary disease.

On the other hand, although moderate wine consumption is protective in most individuals, it triggers exacerbations in some asthma patients (Guilford and Pezzuto, 2011). Wine allergies and intolerances have been described in a small group of susceptible individuals. Some studies have implicated natural sulfites common to wine and sulfur dioxide added to wine during its fermentation as wine allergy triggers. Other agents including venom from *Hymenoptera* insects (wasps, bees, ants), which may contaminate wines during vinification can also trigger allergies. These symptoms are usually elicited in nondrinkers and are less likely in patients with a history of wine drinking. For patients who tolerate red wine, there appears to be no contraindication, and possibly some benefit, to continued moderate intake.

Curcumin and curcuminoids are derived from the plant, *Curcuma longa*, better known as the powdered yellow spice, turmeric. This plant is a major component in curry, and is used as a medicinal agent in many parts of the world, especially Southeast Asia. The polyphenol curcumin structure has many therapeutic properties as a result of its antioxidant, antiinflammatory, and anticancer effects, mediated in part by inhibition of the cyclooxygenase and lipoxygenase pathways, inhibition of inducible nitric oxide synthase, modulation of the action and secretion of inflammatory cells, and inhibition of the transcription of genes that promote inflammation as reviewed by Meja et al. (2008).

Partly because of its size, the curcumin molecule is poorly bioavailable after oral application, yet its actions are well described. It is currently being studied in at least 40 clinical trials, one of which is testing its efficacy for treatment of asthma (clinical trials are registered at http://clinicaltrials.gov). It should be noted that piperine, a compound in black pepper, has been shown to increase curcumin bioavailability (Sehgal et al., 2012). Curcumin also restores impaired histone deacetylase-2 activity and corticosteroid efficacy in monocytes exposed to oxidative stress. Hence, curcumin has potential to reverse corticosteroid resistance, a common complication in patients with chronic obstructive pulmonary disease and severe asthma.

Kaempferol is a flavonol that has been isolated from tea, broccoli and cabbage, citrus fruits, beans, and other plant sources. Recent reports using an experimental asthma model suggest that kaempferol treatment may inhibit IgE binding and activation of human mast cells and be useful as adjuvant therapy for Th2-driven airway hypersensitivity. A nontoxic dose (1–20 μmol/L) of kaempferol suppressed LPS-induced eotaxin-1 protein expression, possibly mediated via Janus kinase 2 (JAK2) signaling (Gong et al., 2012). Additionally, kaempferol dose-dependently attenuated TNFα-induced expression of epithelial intracellular cell adhesion molecule-1 and eosinophil integrin β2, thus limiting the eosinophil–airway epithelium interaction. Kaempferol blunted TNFα-induced airway inflammation by attenuating monocyte chemoattractant protein-1 transcription, possibly by disturbing NF-κB signaling. Oral administration of kaempferol to mice challenged with an ovalbumin allergen attenuated elevated expression of eotaxin-1 and eosinophil major basic protein via the blockade of NF-κB transactivation, thereby blunting eosinophil accumulation in airway and lung tissue.

Quercetin is the major dietary flavonoid in the human diet (~25 mg/day). It is found in many plants including onions, broccoli, apples, berries, and tea. (Quercetin is also present in extracts from *Gingko biloba* and St. Johns Wort, both popular health supplements.) Quercetin has potent antioxidant effects, combining with free radical species to form considerably less reactive phenoxy radicals. Quercetin also has inhibitory effects on several kinases, mediating many of its potent antiproliferative, proapoptotic, and antiinflammatory effects on biological systems. Quercetin has potential benefit in the treatment of airways disease. It had an inhibitory effect on cytokine and chemokine production in cultured cells (Nanua et al., 2006), attenuated lipopolysaccharide-induced nitric oxide production via inducible nitric oxide synthase expression, inhibited release of TNFα and IL6 from cultured macrophages, and reduced activation of mitogen-activated protein kinases and NFκB. Quercetin has been reported to inhibit mast cell activation and histamine release.

Again, it must be remembered that most studies have been conducted in vitro or in animal models. Until data from clinical trials that assess the safety and efficacy of bioactive components in humans is available, it is prudent to avoid extracts and tonics and to obtain quercetin and other nonnutrient dietary components from whole foods.

8.8 FUTURE RESEARCH NEEDS

Our emerging understanding of the impact of microbes in the gut as well as the lung has shifted the paradigm of asthma from a predominantly T-cell-mediated condition to one much more amenable to targeted diet modification. The link between obesity and asthma is likely to be found in the type of food consumed rather than the magnitude of body fat deposition; the link may be found in the impact of food on microbial diversity and function. The crosstalk among intestinal epithelial cells, DCs, and naïve T cells in the differentiation of tolerant T[regs] under the influence of dietary fiber, resistant starch, and micronutrients such as vitamin D is also of great interest, as is our growing understanding of the influence of maternal diet on respiratory disease in the offspring. Finally, the utility of improving lung function by upregulating protective cascades and possibly by inducing chronic bronchodilation with dietary magnesium and possibly other dietary components will be of great interest.

APPENDIX: DRUGS USED FOR ASTHMA

Ipratropium bromide (trade names Atrovent, λ Apovent, and Aerovent) is an anticholinergic drug—blocks muscarinic receptors.

Salbutamol or *albuterol* is a short-acting β_2-adrenergic receptor agonist used for the relief of bronchospasm. It is marketed as Ventolin among other brand names.

Fluticasone propionate is a synthetic corticosteroid derived from fluticasone used to treat asthma and allergic rhinitis (hayfever). It is also used to treat eosinophilic esophagitis.

- GalaxoSmithKline currently markets fluticasone propionate as Flovent (United States and Canada) and Flixotide (EU) for asthma, and as Flonase (United States and Canada) Flixonase (EU, South Africa, Israel, and Brazil) for allergic rhinitis (hayfever), as well as a combination of fluticasone and salmeterol as Advair (United States and Canada) or Seretide (EU). Teva markets fluticasone propionate as Nasofan in Italy.
- It is also available as a cream (marketed as Cutivate or Flutivate) for the treatment of eczema and psoriasis.

Esomeprazole is a proton pump inhibitor (brand name Nexium).

References

Agler, A.H., Kurth, T., et al., 2011. Randomised vitamin E supplementation and risk of chronic lung disease in the Women's Health Study. Thorax. 66 (4), 320–325.

Agrawal, A., Mabalirajan, U., et al., 2011. Emerging interface between metabolic syndrome and asthma. Am. J. Respir. Cell Mol. Biol. 44 (3), 270–275.

Akinbami, L.J., Simon, A.E., et al., 2016. Trends in allergy prevalence among children aged 0–17 years by asthma status, United States, 2001–2013. J. Asthma. 53, 1–7.

Albin, S., Nowak-Wegrzyn, A., 2015. Potential treatments for food allergy. Immunol. Allergy Clin. North Am. 35 (1), 77–100.

Anderson, G.P., 2008. Endotyping asthma: new insights into key pathogenic mechanisms in a complex, heterogeneous disease. Lancet. 372 (9643), 1107–1119.

Arabi, A., El Rassi, R., et al., 2010. Hypovitaminosis D in developing countries—prevalence, risk factors and outcomes. Nat. Rev. Endocrinol. 6 (10), 550–561.

Ariel, A., Timor, O., 2013. Hanging in the balance: endogenous anti-inflammatory mechanisms in tissue repair and fibrosis. J. Pathol. 229 (2), 250–263.

Berridge, M.J., 2015. Vitamin D cell signalling in health and disease. Biochem. Biophys. Res. Commun. 460 (1), 53–71.

Buyken, A.E., Flood, V., et al., 2010. Carbohydrate nutrition and inflammatory disease mortality in older adults. Am. J. Clin. Nutr. 92 (3), 634–643.

Chinthrajah, R.S., Tupa, D., et al., 2015. Diagnosis of food allergy. Pediatr. Clin. North. Am. 62 (6), 1393–1408.

Claesson, M.J., Jeffery, I.B., et al., 2012. Gut microbiota composition correlates with diet and health in the elderly. Nature. 488 (7410), 178–184.

Covington, D.K., Briscoe, C.A., et al., 2006. The G-protein-coupled receptor 40 family (GPR40–GPR43) and its role in nutrient sensing. Biochem. Soc. Trans. 34 (5), 770–773.

de Baaij, J.H.F., Hoenderop, J.G.J., et al., 2015. Magnesium in man: implications for health and disease. Physiol. Rev. 95 (1), 1–46.

Figueiredo, C.A., Amorim, L.D., et al., 2013. Environmental conditions, immunologic phenotypes, atopy, and asthma: new evidence of how hygiene hypothesis operates in Latin America. J. Allergy Clin. Immunol. 131, 1064–1068.

Finklea, J.D., Grossmann, R.E., et al., 2011. Vitamin D and chronic lung disease: a review of molecular mechanisms and clinical studies. Adv. Nutr. Int. Rev. J. 2 (3), 244–253.

Finley, J.W., 2005. Proposed criteria for assessing the efficacy of cancer reduction by plant foods enriched in carotenoids, glucosinolates, polyphenols and selenocompounds. Ann. Bot. (Lond). 95 (7), 1075–1096.

Fixman, E.D., Stewart, A., et al., 2007. Basic mechanisms of development of airway structural changes in asthma. Eur. Respir. J. 29 (2), 379–389.

Ginde, A.A., Mansbach, J.M., et al., 2009. Association between serum 25-hydroxyvitamin D level and upper respiratory tract infection in the third national health and nutrition examination survey. Arch. Intern. Med. 169 (4), 384–390.

Gong, J.-H., Shin, D., et al., 2012. Kaempferol suppresses eosionphil infiltration and airway inflammation in airway epithelial cells and in mice with allergic asthma. J. Nutr. 142 (1), 47–56.

Guilford, J.M., Pezzuto, J.M., 2011. Wine and health: a review. Am. J. Enol. Viticult. 62 (4), 471–486.

Gunaratne Anoja, W., Makrides, M., et al., 2015. Maternal prenatal and/or postnatal n-3 long chain polyunsaturated fatty acids (LCPUFA) supplementation for preventing allergies in early childhood. Cochrane Database Sys. Rev. 7, CD010085. Available from: http://dx.doi.org/10.1002/14651858.CD010085.pub2.

Hammad, H., Lambrecht, B.N., 2008. Dendritic cells and epithelial cells: linking innate and adaptive immunity in asthma. Nat. Rev. Immunol. 8 (3), 193–204.

Haury, V.G., 1940. Blood serum magnesium in bronchial asthma and its treatment by the administration of magnesium sulfate. J. Lab. Clin. Med. 26, 340–344.

Haworth, O., Levy, B.D., 2007. Endogenous lipid mediators in the resolution of airway inflammation. Eur. Respir. J. 30 (5), 980–992.

Hill, D.A., Siracusa, M.C., et al., 2012. Commensal bacteria-derived signals regulate basophil hematopoiesis and allergic inflammation. Nat. Med. 18 (4), 538–546.

Hooper, R., Heinrich, J., et al., 2010. Dietary patterns and risk of asthma: results from three countries in European Community Respiratory Health Survey-II. Br. J. Nutr. 103 (09), 1354–1365.

Jaudszus, A., Krokowski, M., et al., 2008. Cis-9, trans-11-conjugated linoleic acid inhibits allergic sensitization and airway inflammation via a PPARgamma-related mechanism in mice. J. Nutr. 138 (7), 1336–1342.

Jobin, C., 2014. GPR109a: the missing link between microbiome and good health? Immunity. 40 (1), 8–10.

Kim, J.-H., Ellwood, P., et al., 2009. Diet and asthma: looking back, moving forward. Respir. Res. 10 (1), 49.

Kode, A., Rajendrasozhan, S., et al., 2008. Resveratrol induces glutathione synthesis by activation of Nrf2 and protects against cigarette smoke-mediated oxidative stress in human lung epithelial cells. Am. J. Physiol. Lung Cell. Mol. Physiol. 294 (3), L478–L488.

Kumar, J., Muntner, P., et al., 2009. Prevalence and associations of 25-hydroxyvitamin D deficiency in US children: NHANES 2001–2004. Pediatrics. 124 (3), e362–e370.

Lambrecht, B.N., Hammad, H., 2009. Biology of lung dendritic cells at the origin of asthma. Immunity. 31 (3), 412–424.

Lambrecht, B.N., Hammad, H., 2012. The airway epithelium in asthma. Nat. Med. 18 (5), 684–692.

Litonjua, A.A., Weiss, S.T., 2007. Is vitamin D deficiency to blame for the asthma epidemic? J. Allergy Clin. Immunol. 120 (5), 1031–1035.

Lloyd, C.M., Hessel, E.M., 2010. Functions of T cells in asthma: more than just T(H)2 cells. Nat. Rev. Immunol. 10 (12), 838–848.

Maslowski, K.M., Vieira, A.T., et al., 2009. Regulation of inflammatory responses by gut microbiota and chemoattractant receptor GPR43. Nature. 461 (7268), 1282–1286.

McGinley, B., Punjabi, N.M.M.D.P., 2011. Obesity, metabolic abnormalities, and asthma: establishing causal links. Am. J. Respir. Crit. Care. Med. 183 (4), 424–425.

Meja, K., Rajendrasozhan, S., et al., 2008. Curcumin restores corticosteroid function in monocytes exposed to oxidants by maintaining HDAC2. Am. J. Respir. Cell Mol. Biol. 39, 312–323.

Mir, P.S., McAllister, T.A., et al., 2004. Conjugated linoleic acid–enriched beef production. Am. J. Clin. Nutr. 79 (6), 1207S–1211S.

Mohammed, S., Goodacre, S., 2007. Intravenous and nebulised magnesium sulphate for acute asthma: systematic review and meta-analysis. Emerg. Med. J. 24 (12), 823–830.

Mora, J.R., Iwata, M., et al., 2008. Vitamin effects on the immune system: vitamins A and D take centre stage. Nat. Rev. Immunol. 8 (9), 685–698.

Nanua, S., Zick, S.M., et al., 2006. Quercetin blocks airway epithelial cell chemokine expression. Am. J. Respir. Cell Mol. Biol. 35 (5), 602–610.

Noverr, M.C., Huffnagle, G.B., 2005. The 'microflora hypothesis' of allergic diseases. Clin. Exp. Allergy. 35 (12), 1511–1520.

Nowak-Wegrzyn, A., Groetch, M., 2012. Let them eat cake. Ann. Allergy Asthma Immunol. 109 (5), 287–288.

Oh, D.Y., Talukdar, S., et al., 2010. GPR120 is an omega-3 fatty acid receptor mediating potent anti-inflammatory and insulin-sensitizing effects. Cell. 142 (5), 687–698.

O'Shea, M., Bassaganya-Riera, J., et al., 2004. Immunomodulatory properties of conjugated linoleic acid. Am. J. Clin. Nutr. 79 (6), 1199S–1206S.

Paul, G., Brehm, J.M., et al., 2012. Vitamin D and asthma. Am. J. Respir. Crit. Care Med. 185 (2), 124–132.

Pawankar, R., Holgate, S.T. (Eds.), 2009. Allergy Frontiers: Classification and Pathomechanisms. Springer, Tokyo.

Poole, J.A., Barriga, K., et al., 2006. Timing of initial exposure to cereal grains and the risk of wheat allergy. Pediatrics. 117 (6), 2175–2182.

Powell, C., Dwan, K., et al., 2012. Inhaled magnesium sulfate in the treatment of acute asthma. Cochrane Database Sys. Rev. (Online). 12, CD003898.

Riiser, A., 2015. The human microbiome, asthma, and allergy. Allergy Asthma Clin. Immunol. 11 (1), 1–7.

Royce, S., Dang, W., et al., 2011. Resveratrol has protective effects against airway remodeling and airway hyperreactivity in a murine model of allergic airways disease. Pathobiol. Aging Age Relat. Dis. 1. Available from: http://dx.doi.org/10.3402/PBA.v1i0.7134.

Samuelson, D.R., Welsh, D.A., et al., 2015. Regulation of lung immunity and host defense by the intestinal microbiota. Front. Microbiol. 6, 1085.

Sehgal, A., Kumar, M., et al., 2012. Piperine as an adjuvant increases the efficacy of curcumin in mitigating benzo(a)pyrene toxicity. Hum. Exp. Toxicol. 31 (5), 473–482.

Sekirov, I., Russell, S.L., et al., 2010. Gut microbiota in health and disease. Physiol. Rev. 90 (3), 859–904.

Shaikh, S.R., Edidin, M., 2006. Polyunsaturated fatty acids, membrane organization, T cells, and antigen presentation. Am. J. Clin. Nutr. 84 (6), 1277–1289.

Sharafkhaneh, A., Velamuri, S., et al., 2007. Review: the potential role of natural agents in treatment of airway inflammation. Therap. Adv. Respir. Dis. 1 (2), 105–120.

Simopoulos, A.P., 2008. The importance of the omega-6/omega-3 fatty acid ratio in cardiovascular disease and other chronic diseases. Exp. Biol. Med. 233 (6), 674–688.

Simopoulos, A.P., 2010. Genetic variants in the metabolism of omega-6 and omega-3 fatty acids: their role in the determination of nutritional requirements and chronic disease risk. Exp. Biol. Med. 235 (7), 785–795.

Smith, P.M., Howitt, M.R., et al., 2013. The microbial metabolites, short-chain fatty acids, regulate colonic treg cell homeostasis. Science. 341 (6145), 569–573.

Stefka, A.T., Feehley, T., et al., 2014. Commensal bacteria protect against food allergen sensitization. Proc. Natl. Acad. Sci. 111 (36), 13145–13150.

Stenius-Aarniala, B., Poussa, T., et al., 2000. Immediate and long term effects of weight reduction in obese people with asthma: randomised controlled study. BMJ. 320 (7238), 827–832.

Thornley, S., Stewart, A., et al., 2011. Per capita sugar consumption is associated with severe childhood asthma: an ecological study of 53 countries. Prim. Care Respir. J. 20 (1), 75–78.

Vital, M., Harkema, J.R., et al., 2015. Alterations of the murine gut microbiome with age and allergic airway disease. J. Immunol. Res. 2015, 8.

Wadsworth, S.J., Sin, D.D., et al., 2011. Clinical update on the use of biomarkers of airway inflammation in the management of asthma. J. Asthma Allergy. 4, 77–86.

Wechsler, M.E., et al., 2007. Case 15-2007—a 20-year-old woman with asthma and cardiorespiratory arrest. N. Engl. J. Med. 356, 2083–2091, doi:10.1056/NEJMcpc079006; <http://www.nejm.org/doi/full/10.1056/NEJMcpc079006>.

Woodruff, P.G., Modrek, B., et al., 2009. T-helper type 2–driven inflammation defines major subphenotypes of asthma. Am. J. Respir. Crit. Care Med. 180 (5), 388–395.

Zhang, Y., Zhou, C., et al., 2014. A new index to identify risk of multi-trigger wheezing in infants with first episode of wheezing. J. Asthma. 51 (10), 1043–1048.

Further Reading

Kumar, V., et al., 2014. Robbins and Cotran Pathologic Basis of Disease, 9th Ed. Chapter 15. The Lung: asthma. Elsevier Saunders, Philadelphia, PA.

9

Osteoporosis and Fracture Risk

CHIEF COMPLAINT (ADAPTED FROM EBELING, P.R., 2008)

Mr. A, a 67-year-old Caucasian man, presented for a routine annual physical. He expressed worry about a chronic "bad back" for which he takes OTC NSAIDs (Tylenol; acetaminophen) as needed for pain.

PAST MEDICAL HISTORY AND MEDICATIONS

The patient has been seen annually since age 60. He is on a statin (Lipitor) for high cholesterol and an angiotensin-converting enzyme inhibitor (Lisinopril) and a diuretic (furosemide) for moderately elevated blood pressure. He also uses a high-potency topical steroid cream for psoriasis. Over the past year, he has been troubled by chronic bronchitis, for which he has been prescribed multiple courses of antibiotics. Ten years ago, the patient had a basal cell carcinoma removed from the back of his neck, and since then has protected his pale skin with sun block, hats, and long sleeves. He has no history of fractures. Mr. A does not bother with nutritional supplements, since he considers his diet adequate and balanced.

LIFE STYLE AND FAMILY HISTORY

Patient has been retired for 2 years and lives with his wife of 45 years. Although he has a bad back, he is able to volunteer weekly for the local food bank. They have a large back yard and Mr. A is very proud of his collection of rare roses. He spends as much time as possible on their care, but is concerned that his back problems may further limit this activity since he experiences pain on bending. This is his major regular physical activity. Mr. A does not smoke. He drinks 2–3 glasses of beer daily with his evening meals, and more on weekends when watching the games with his friends.

Mr. A's father died at age 83 of a stroke. His mother died at age 85 after suffering a hip fracture and developing pneumonia. A younger sister, aged 62 years, has been diagnosed with severe osteoporosis for which she is taking Fosamax.

NUTRITION HISTORY

General

- Mr. A describes his appetite as "good." He usually has a cup of black coffee when he arises, followed by a second cup with a roll and jam. He never liked dairy products and 10 years ago, when his doctor told him to cut back on fat because of his high cholesterol levels, he cut out dairy products altogether.
- Patient usually eats a bowl of canned soup and chips for lunch along with some type of regular cola.
- Dinner is the main meal. Mrs. A usually prepares some type of meat with a potato (usually mashed or boiled) and some type of cooked vegetable, usually canned green beans or glazed carrots and a dinner roll (purchased) with margarine. Mrs. A often serves cake or pie (purchased) for dessert.
- The couple treat themselves to Friday and/or Saturday night out, often at a pizza restaurant or the local steakhouse. Mr. A usually has 2–4 beers on these evenings.

Diet Analysis

- The dietitian conducted a 24-hour recall and assessment and obtained the following information:

• Total calories	1903 kcal/day
• Protein	47.6 g/day (10% of calories)
• Fat	74 g/day (35% of calories)
• Carbohydrate	262 g/day (55% of calories)

(Continued)

Nutritional Pathophysiology of Obesity and its Comorbidities
DOI: http://dx.doi.org/10.1016/B978-0-12-803013-4.00009-0

(Continued)

Basal caloric requirements based on Harris Benedict equation[a]	1474 kcal/day
Estimated total energy requirements (activity factor = 1.25)	1843 kcal/day
Estimated protein requirements (kg × 0.8 g/kg)[b]	57 g/day

[a]Please see Cornell site for calculation: http://www-users.med.cornell.edu/~spon/picu/calc/beecalc.htm.
[b]Based on Dietary Reference Intake (DRI) for men >50 = 0.8 m/day.

ANTHROPOMETRICS

General Appearance

Mr. A appears slightly overweight, with fat distribution to the midsection and relatively thin arms and legs.

- Height — 5′7″ (172 cm) (height at exam 3 years ago 5′10″ (179.6 cm))
- Weight — 178 lb (80.7 kg) (±5 lbs from first exam; +12 lbs since exam 3 years ago)
- BMI — 27.3 kg/m^2
- DEXA measurements — T scores of −2.6 at the spine and −2.2 at the femoral neck. (Normal BMD is defined as a T score greater than or equal to −1)

Vitals

- Blood pressure: 135/88
- Heart rate: 72 bpm
- Respiration: 15 bpm

Laboratory data	Patient's values	Normal values
Albumin	3.3 g/dL	3.5–5.8 g/dL
Hemoglobin	11.5 g/dL	11.8–15.5 g/dL
Hematocrit	34%	36–46%
Serum 25-OH D	26 ng/mL	>25 ng/mL (62.5 nmol/L)
Serum PTH	80 pg/mL	10–65 pg/mL
Triglycerides	203 mg/dL	<150 mg/dL
Total cholesterol	215 mg/dL	<200 mg/dL
HDL-C	32 mg/dL	>40–50 mg/dL

RESOURCES

9.1 MORPHOLOGY AND DIAGNOSIS OF OSTEOPOROSIS

Osteoporosis has been defined as a "progressive systemic skeletal disease characterized by low bone mass and microarchitectural deterioration of bone

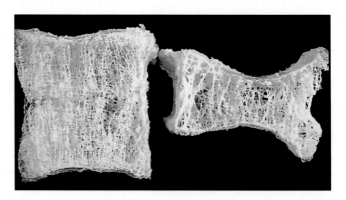

FIGURE 9.1 Osteoporotic vertebral body (right) shortened by compression fractures compared with a normal vertebral body (left). Note that the osteoporotic vertebra has a characteristic loss of horizontal trabeculae and thickened vertical trabeculae. *Source: Reproduced from Robbins and Cotran Pathologic Basis of Disease, ninth ed. With permission from Elsevier Saunders.*

tissue, with a consequent increase in bone fragility and susceptibility to fracture" (Consensus, 1993). A progressive loss of bone mass is observed in the axial (trabecular) as well as the appendicular (cortical) skeleton that leads to bone fragility and fractures. Clinical symptoms include back pain, height loss (stooping), and/or history of fractures. The histological hallmark of age-related bone loss in humans and animals is a decline in mean wall thickness (MWT), an index of the amount of bone made by each team of osteoblasts during bone remodeling (Kanis et al., 2008). Bone mineral density (BMD) is often, but not always, reduced in osteoporotic patients. The wide spectrum of BMD in patients with fractures suggests that factors other than bone mineral contribute to changes in the trabecular and cortical skeleton as shown in Fig. 9.1.

Qualitative abnormalities such as trabecular perforations, microcracks, increased porosity, mineralization defects, reduced bone size, and/or rapid bone turnover have been observed in osteoporotic bone (NIH Consensus Development Panel on Osteoporosis Prevention, Diagnosis, and Therapy, 2001) but are difficult to assess clinically. In view of the difficulty in predicting fracture risk from available clinical methods, the National Guidelines on Diagnosis and Treatment of Osteoporosis (Florence et al., 2013) have defined normal BMD as a T-score greater than or equal to −1. Osteoporosis is defined as a BMD less than or equal to −2.5 standard deviations (SDs) below peak bone mass; T-score is measured as the units below the normative mean of a 35-year-old. BMD with a T-score between −1.0 and −2.5 has been defined as *osteopenia*, but the clinical implications of this term are imprecise (Fig. 9.2).

Bone loss in humans (and other animals) progresses throughout life. In fact, one in two women and one in

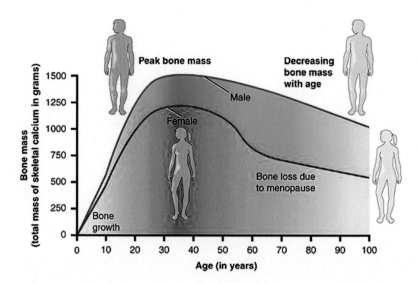

FIGURE 9.2 **Bone mass in men and women.** Both men and women achieve peak bone mass in early adulthood, after which they inexorably lose bone. Women lose bone rapidly during menopause, but thereafter at the same rate as men. *Source: From Anatomy & Physiology, Connexions website. http://cnx.org/content/col11496/1.6/, Jun 19, 2013. (OpenStax College) (CC BY 3.0 (http://creativecommons.org/licenses/by/3.0)), via Wikimedia Commons.*

four men over the age of 50 are predicted to break a bone due to osteoporosis; this amounts to 2 million broken bones every year (National Osteoporosis Foundation statistics, http://www.nof.org/advocacy/resources/prevalencereport). In humans, bone loss begins in the third decade of life and is due mainly to a reduction in the number of osteoblasts due to decreased osteoblastogenesis and/or increased apoptosis rather than to impaired capacity of the mature osteoblast to synthesize bone matrix. If the resorption cavities are not completely filled with each remodeling cycle, net bone loss ensues with an increased risk for fractures (Kearns et al., 2008). Despite this important histological insight, the molecular and cellular mechanisms of this adverse effect of aging on bone have, until fairly recently, remained elusive.

9.2 BONE CELLS AND NORMAL MECHANISMS OF BONE REMODELING

9.2.1 The Bone Cells

Bone is remodeled throughout life in discrete areas termed bone-forming units. Four cell types, bone-lining cells, osteoblasts, osteoclasts, and osteocytes engage in complex crosstalk to remove and replace effete sections of bone, thereby maintaining skeletal integrity and responding to mechanical and environmental changes as recently reviewed by Florencio-Silva et al. (2015). Bone cells are derived from two classes of stem cells located in the bone marrow. The common mesenchymal stem cell (MSC) differentiates into an osteoblast in response to bone morphogenetic proteins (BMP) and Wnt signals that express osteogenic transcription factors such as Runx and Osterix. (Wnt is an abbreviation for "Wingless-related integration site."

Wnt signaling pathways are a group of signal transduction pathways that transmit signals into a cell through protein binding to cell surface receptors.) BMP isoforms are members of the transforming growth factor (TGF) family that are synthesized from local sources and diffuse into surrounding tissue to provide information in a dose-dependent manner (Modica and Wolfrum, 2013). Once differentiated, these cells function as osteoblasts for only a short time. Following this active period, osteoblasts either die by apoptosis or lose their matrix deposition capacity and become flattened bone-lining cells that cover the surface of bone. Alternately, osteoblasts can become embedded in bone matrix as terminally differentiated osteocytes. The MSC from which the osteoblasts derive are competent to respond to other signals and differentiate into chondrocytes (Karsenty, 2008), smooth muscle cells or, when exposed to adipogenic signals such as peroxisome proliferator-activated receptor gamma (PPARγ), become adipocytes (Fig. 9.3).

Single osteoblasts do not, by themselves, form bone. They function as groups of cells connected by tight junctions into a basic multicellular unit (BMU) called an osteon. Cells in the BMU monolayer communicate through gap junctions and create a discrete compartment between the osteon and the bone. Osteoblasts secrete bone matrix into this compartment in two steps. Organic matrix contains at least 20 proteins, the largest fraction is type I collagen. Noncollagen proteins include osteocalcin (OCN), osteopontin (OSP), osteonectin (OSN), bone sialoprotein (BSP); BMP as well as proteoglycans. This secreted matrix adheres to, and is assembled on, the denuded bone. The second step, mineralization, requires formation of matrix vesicles (MVs) that bud off from the osteoblast membrane adjacent to the bone (Anderson, 2003). Osteoblasts secrete enzymes that degrade proteoglycans in the underlying

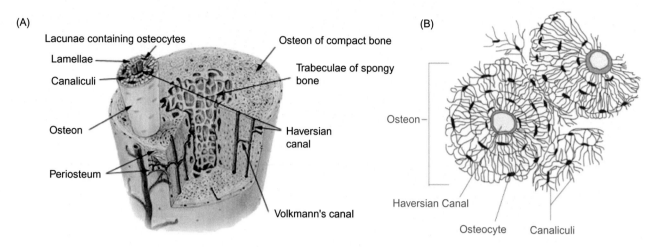

FIGURE 9.3   Diagram of a long bone. (A) Overview of the anatomy of the long bone. Blood vessels and nerves travel through Volkmann's canals perpendicular to the Haversian canals. (B) Detail of the concentric arrangement of osteocytes and the communicating lacunar-canalicular network. *Source: From (A) SEER (Public domain), via Wikimedia Commons; (B) BDB (CC BY-SA 2.5 (http://creativecommons.org/licenses/by-sa/2.5)), via Wikimedia Com.*

organic bone matrix, releasing calcium ions that are taken up by the MVs through calcium-binding channels (annexin 1). MVs contain a supersaturated soup of phosphate ions released by MV phosphatases and calcium ions. Supersaturated ions spontaneously form an intermediate crystal ($CaHPO_4$) that over time assembles into mature hydroxyapatite crystals [$Ca_5(PO_4)_3(OH)$] (refer to Essentials III: Nutrients for Bone Structure and Calcification). Supersaturation of Ca^{2+} and PO_4^{3-} ions in the MVs causes the MV membrane to rupture, spreading hydroxyapatite crystals onto the surrounding matrix. MV-derived hydroxyapatite crystals act as nucleation templates for the formation of new hydroxyapatite crystals using Ca^{2+} and PO_4^{3-} in the extracellular fluid as substrate. The hydroxyapatite crystal is surrounded by a hydration shell that can incorporate other ions such as magnesium and fluoride.

Osteoblasts that differentiate into the thin and flattened bone-lining cells have not been well studied. They are known to cover the bone and act as a barrier to prevent osteoclast interaction with the matrix in areas not targeted for bone reabsorption. Alternatively, the osteoblast can undergo terminal differentiation, adopt stellate morphology, and become embedded in bone matrix as an osteocyte. In the past, osteocytes imprisoned in their tiny lacunae were considered relatively inert. It is now known that the osteocyte plays multiple regulatory roles in maintaining the bony skeleton. Although confined in their lacunae, osteocytes send out dendritic processes (40–100 per cell) through channels (canaliculi), forming a dense lacunar-canalicular network that allows communication with the extensive bone vasculature and neural networks, with surface osteoblasts and osteoclasts and with each

other. Canaliculi are filled with a fluid (poorly characterized as yet) that provides osteocytes with nutrients and oxygen and removes wastes and secretions. Osteocytes have mechanosensory abilities, not yet well understood, that detect muscle activity and mechanical loading and signal adaptive skeletal remodeling in response. Among their multiple functions, osteocytes sense effete bone, direct osteocyte homing and remodeling, stimulate osteoblast proliferation and action, and control mineralization of bone matrix (Manolagas and Parfitt, 2010) (Fig. 9.4).

Osteocytes participate in mineral homeostasis through the secretion of fibroblast growth factor 23 (FGF23) as discussed below and elsewhere (refer to Essentials III: Nutrients for Bone Structure and Calcification). In addition to osteoblasts, osteocytes also secrete receptor activator of nuclear factor kappa-B ligand (RANKL) that controls osteoclastic activity by binding to its osteoclast receptor, RANK. Sclerostin, a negative regulator of bone formation that antagonizes the Wnt/β-catenin signaling pathway is also secreted by osteocytes. In bone, Wnt signaling increases osteoblastic cell differentiation and bone formation, and also inhibits bone resorption by blocking the RANKL–RANK interaction that normally stimulates osteoclast activation (Baron and Kneissel, 2013). Although osteocytes are long-lived and can survive for decades within the bone matrix, their numbers have been reported to decline with aging, associated with an increase in the number of empty lacunae. The growing understanding of the disparate regulatory activities of the osteocytes has led many authors to suggest that decline in osteocyte numbers and activity may underlie age-associated bone quality deterioration (Manolagas and Parfitt, 2010; Delgado-Calle and

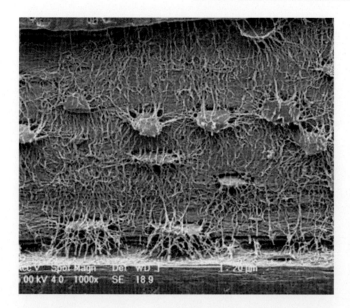

FIGURE 9.4 **Osteocytes and their lacunar-canalicular network.** Bone sections were acid-etched to remove the mineral and reveal the organic lacunar-canalicular network allowing osteocytes to communicate with each other and with the remodeling surface. *Source: Reproduced from Manolagas, S.C., Parfitt, A.M., 2010. What old means to bone. Trends Endocrinol. Metab. 21 (6), 369–374 with permission from Science Direct.*

Bellido, 2015). Thus osteocyte viability and survival appear to be key factors in repair and maintenance of bone structure throughout life (Dallas et al., 2013).

In contrast to the osteoblast and its derivatives, the osteoclast is derived from a hematopoietic stem cell (HSC) of the monocyte lineage. The HSC differentiates under the influence of macrophage-colony stimulating factor (M-CSF) into a multinuclear cell programmed for bone resorption as recently reviewed by Xu and Teitelbaum (2013). Human osteoclasts are large ($\sim 150-200\,\mu m$) and multinucleated with an average of five nuclei. The osteoclast is activated by RANKL, a member of the tumor necrosis factor (TNF) family. RANKL binds to its receptor (RANK) expressed on the cell surface of osteoclast precursors and triggers conversion of osteoclast precursors to functional osteoclasts that express a variety of osteoclastogenic factors. Osteoclast activity is regulated by a decoy protein, osteoprotegerin (OPG), also known as osteoclastogenesis inhibitory factor produced by osteoblasts as well as a large number of other fibroblast-like cells. OPG binds RANKL and prevents RANKL/RANK binding, thereby preventing osteoclastogenesis. RANKL/RANK binding stimulates the proinflammatory NFκB cascade that is central to osteoclast action. The importance of NFκB activation in triggering osteoclast activation is supported by the observations that other proinflammatory agents, including saturated fatty acids (lauric and palmitic) (Oh et al., 2010), supraphysiologic levels of vitamin D

hormone (1,25D; 1-25 (OH)$_2$ D$_3$) (Suda et al., 2003), lipopolysaccharide and bacterial products, parathyroid hormone (PTH), dexamethasone and inflammatory cytokines (Wada et al., 2006) also activate NFκB and contribute to osteoclast activation, survival, and function.

9.2.2 Bone Remodeling

Bone resorption occurs in two stages. First the osteoclasts home to the designated region of the bone, attach and dissolve crystalline hydroxyapatite. They then facilitate proteolytic cleavage of the organic component of bone matrix. Adjacent bone-lining cells are protected from destruction in the lytic resorption pit by two structural components of the osteoclast. The area of bone for osteoclast action is demarcated by matrix/osteocyte-derived signals. The osteoclast undergoes structural rearrangement of its actin skeleton that polarizes the osteoclast. An F-actin ring composed of microvilli is established in the area of membrane at the osteoclast—bone border; this ring will develop into the "ruffled border" at the lytic resorption pit. A clear "sealing" zone devoid of osteoclast organelles is formed by connective tissue proteins that encircle the central region and isolate it from the adjacent cells as shown in Fig. 9.5. Only when the osteoclast is attached to the extracellular mineralized matrix by integrin does the lytic resorption pit (Howship's lacuna) form.

To degrade the mineralized matrix, the osteoclast expresses H$^+$ATPase on the membrane of the ruffled border; this enzyme transports H$^+$ into the resorption pit. Protons are obtained by the action of carbonic anhydrase via the reaction ($CO_2 + H_2O \leftrightarrow H_2CO_3 \leftrightarrow H^+ + HCO_3^-$). Cl$^-$ ions are imported into the osteoclast from the extracellular fluid and cotransported across the ruffled border into the resorption pit coupled to the H$^+$ATPase. Bicarbonate leaves the cell as chloride enters by a passive chloride—bicarbonate exchanger in the basolateral border. The organic matrix is degraded by secretion of cathepsin K from the lysosomes and vacuoles. Cathepsin K is a protease that catabolizes elastin, collagen, and proteoglycans. Degradation products are transported from the bone surface and across the osteoclast by vesicles and released by exocytosis from the antiresorptive membrane by a transcytosis process (Xu and Teitelbaum, 2013; Fukunaga et al., 2014). Among these products are Bone morphogenic protein (BMP) isoforms that stimulate osteoblastogenesis by stimulating osteoblast proliferation to replace the recently lost bone. Osteocalcin (OCN) released from the pit enters the systemic circulation and regulates systemic glucose homeostasis as described below. At the same time, matrix insulin growth factor 1 stimulates osteoblastic differentiation of recruited MSCs by activation of mammalian

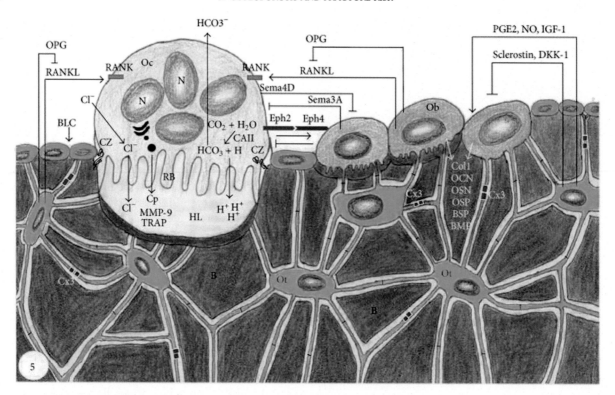

FIGURE 9.5 **Bone remodeling.** In this schematic summary, bone remodeling osteoblasts (Ob) secrete RANKL and OPG along with other signals that activate the osteoclast (Oc). The Oc cytoskeleton polarizes the cell creating the ruffled border (RB) and the surrounding clear zone (CZ) demarcating the specific area of the bone to be resorbed in the Howship lacuna (HL). The osteoclast pumps Cl^- and H^+ ions into the area adjacent to the ruffled border to dissolve the bone mineral. The passive chloride–bicarbonate exchanger is not pictured here. The osteoclast releases cathepsin (Cp), matrix metalloproteinase (MMP), and tartrate-resistant acid phosphatase (TRAP) to degrade the organic matrix. The extensive lacunar-canalicular network comprised of osteocytes (Ot) and their dendritic processes, as well as the flattened bone-lining cells (BLC) can be appreciated in this figure. See Florencio-Silva et al. (2015) for a description of the multiple proteins and signaling factors secreted by these cells. *Source: Reproduced from Florencio-Silva, R., et al., 2015. Biology of bone tissue: structure, function, and factors that influence bone cells. BioMed. Res. Inter. 2015, 421746 with permission from Biomed Research International.*

target of rapamycin (refer to Section 1.2), thus maintaining bone microarchitecture and mass (Xian et al., 2012).

9.2.3 ECM and Collagen Crosslinking

Bone matrix integrity is critical for strength and plastic deformation of bone. While extracellular matrix (ECM) is comprised of a complex mixture of diverse proteins that define structural integrity and function, collagen is the major component (~90%) of ECM proteins in bone. All members of the collagen family contain domains with proline-rich tripeptides (Gly-X-Y) that form trimeric collagen triple helices (Gelse et al., 2003). Within the osteoblast endoplasmic reticulum, procollagen and elastin molecules are hydroxylated at specific proline and lysine residues by ascorbate-dependent proline and lysyl hydroxylases (refer to Essentials I: Life in an Aerobic Environment. Ascorbate) and with the assistance of accessory proteins, three α-chains are assembled into trimeric procollagen monomers and packed into secretion vacuoles in the Golgi (Fig. 9.6).

Extracellular procollagen trimers are processed by Zn^{2+}-dependent proteinases that remove the C and N propeptides. In the extracellular milieu, copper-dependent lysyl oxidase (LOX) initiates covalent crosslinking of these fibers by oxidizing specific lysine residues to peptidyl α-aminoadipic-δ-semialdehyde. Subsequent spontaneous reactions form intermediate crosslinks between lysyl or hydroxylysyl residues on adjacent collagen molecules via aldol and Schiff-base type condensations. With time, more complex crosslinks form (Oxlund et al., 1995), thereby stabilizing polymeric collagen or elastin fibers in the ECM as reviewed by Kagan and Li (2003).

9.3 INTEGRATION OF BONE AND INTERMEDIARY METABOLISM

9.3.1 Osteocalcin and Energy Metabolism

Several points of interaction have been noted between bone and energy metabolism as recently

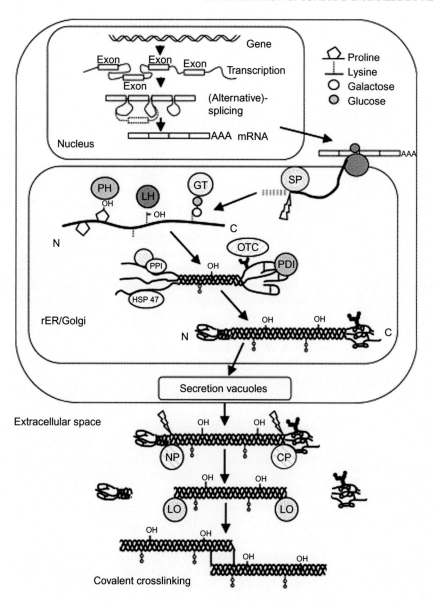

FIGURE 9.6 **Collagen biosynthesis and crosslinking.** Collagen biosynthesis is cell- and tissue-specific and highly regulated. In general, following nuclear transcription, mRNA processing and translation, collagen molecules are posttranslationally modified in the endoplasmic reticulum before secretion and final crosslinking and formation of the triple helical fibrils (Gelse et al., 2003). *Source: Reproduced from Gelse, K., et al., 2003. Collagens—structure, function, and biosynthesis. Adv. Drug Del. Rev. 55 (12), 1531–1546 with permission from Science Direct.*

reviewed by Brennan-Speranza and Conigrave (2015) and Wei and Karsenty (2015). Because bone remodeling requires substantial energy, not only during linear growth in childhood, but also in adult years, links between bone homeostasis and energy metabolism have been sought. Clinical observations have linked reduced energy intake with growth arrest and bone loss and early studies have demonstrated that mice deficient in uncarboxylated serum osteocalcin had increased visceral fat as well as decreased pancreatic β-cell proliferation, glucose intolerance, and insulin resistance (Lee et al., 2007). Osteocalcin, also referred to as bone Gla protein, is the most abundant noncollagenous protein found in the bone matrix. It is synthesized by osteocytes and osteoblasts and when carboxylated in three positions by a vitamin K-dependent, posttranslational modification, avidly binds free calcium in the extracellular fluid or in bone matrix. While the role of osteocalcin in the skeleton is unclear, recent work suggests that it "fine tunes" calcium deposition in the bone matrix and may play a role in bone remodeling.

The concept that bone is an endocrine organ is supported by an elegant series of studies describing the mechanisms as they are currently understood (Wei and Karsenty, 2015) as illustrated in Fig. 9.7. In brief, insulin signaling in the osteoblasts inhibits expression of OPG, the decoy that binds RANKL. When OPG secretion is low, RANKL/RANK binding occurs and the osteoclast is activated to resorb bone as described above. Carboxylated osteocalcin, deposited in the bone matrix by the osteoblast is released into the acidic resorption pit by osteoclast action and decarboxylated. The uncarboxylated osteocalcin

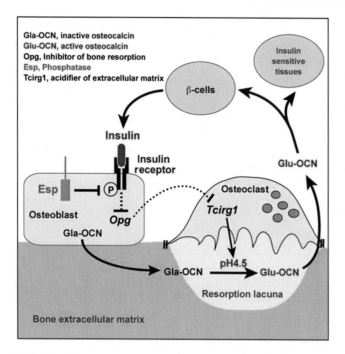

FIGURE 9.7 **Generation of active osteocalcin from bone.** During remodeling, the osteoblast secretes carboxylated osteocalcin (Gla-OCN) and other matrix proteins into the growing organic bone matrix. Insulin receptors on the osteoblast transmit signals that inhibit the decoy, osteoprotegerin (OPG), thereby triggering RANKL/RANK binding and osteoclast activation. To dissolve bone mineral, the osteoclast secretes acid into the acidic resorption lacuna. The acid medium decarboxylates OCN and releases uncarboxylated OCN (Glu-OCN) from the matrix for transport across the osteoclast. Glu-OCN enters the circulation where it enhances insulin secretion, insulin sensitivity, and energy expenditure. Osteoblasts also contain *Esp*, a gene coding for intracellular tyrosine phosphatase (OST-PTP). This gene is thought to act by increasing degradation of the insulin receptor, thereby reducing active Glu-OCN release. *Source: Reproduced from Ferron, M., et al., 2010. Insulin signaling in osteoblasts integrates bone remodeling and energy metabolism. Cell 142 (2), 296–308 (Ferron et al., 2010) with permission from Science Direct.*

product is transported across the osteoclast by vesicular translocation and enters the systemic circulation where it acts as an endocrine hormone. Not only does it stimulate insulin secretion from the pancreatic β cells, but it also increases adiponectin secretion from the adipocyte. Insulin signaling at the osteoblast appears to be a major, albeit indirect, factor in release of uncarboxylated osteocalcin to the systemic circulation. A recent series of experiments has revealed that mice made insulin resistant on a high-fat diet (58% fat plus sucrose) had a decrease in circulating uncarboxylated osteocalcin. Insulin resistance in the murine osteoblasts was shown to be secondary to high levels of free saturated fatty acids (especially stearic acid) due to upregulation of E3 ubiquitin-protein ligase and subsequent ubiquitination and proteosomic destruction of the osteoblast insulin receptor (Wei et al., 2014).

9.4 IMPACT OF HORMONES, OBESITY, AND AGING ON BONE

9.4.1 Estrogen Deficiency and Bone

Women, after the initial rapid bone loss in the perimenopausal period, lose bone gradually throughout life. The rapid bone loss in women following acute loss of sex steroids (from menopause or ovariectomy) has been attributed to an increase in the rate of bone remodeling with bone resorption outpacing formation. Mechanisms through which estrogens protect skeletal integrity have been reviewed (Manolagas, 2010). Estrogens maintain the balance between bone formation and resorption by slowing both osteoblastogenesis and osteoclastogenesis. Estrogens also exert a proapoptotic effect on osteoclasts and an antiapoptotic effect on osteoblasts and osteocytes. The authors suggest that sex steroids also protect bone by their ability to reduce reactive oxygen species (ROS) damage and prevent apoptosis; they cite studies in other organs such as the heart, in which estrogens increased the activity of thioredoxins and other enzymatic antioxidants (Manolagas, 2010).

Classical actions of sex steroids are mediated by receptor binding and regulation of gene transcription. Nongenotropic sex steroid actions include stimulation of mitogen-activated protein kinases (MAPK) that transduce chemical and physical signals from the cell surface and control proliferation, differentiation, and survival. The best-characterized MAPK subfamilies are the extracellular signal regulated kinases (ERKs), the c-Jun N-terminal kinases (JNKs), and the p38 kinases. In a process distinct from the classical genotrophic action, all sex steroids (estrogens or androgens) bind to their respective receptors (ERα, ERβ, and AR) and activate the ERK signaling pathway (Kousteni et al., 2001). ERK-dependent actions include upregulation of antioxidants such as thioredoxin reductase and glutathione. In osteoblastic cells, estrogen-dependent ERK signaling attenuates ROS effects by suppressing p66[shc] phosphorylation and NFκB activation. These changes minimize osteoblast and osteocyte apoptosis (Bhukhai et al., 2012).

9.4.2 Age-Associated Bone Loss

The "estrogen-centric" view of aging bone has complimented observations that bone is lost with aging in both sexes at similar rates. Histologic changes in bone with aging as recently reviewed by Almeida and O'Brien (2013) include reduction in cancellous (trabecular) bone and decreased cortical thickness, with an increase in cortical porosity. A consistent finding in bone biopsies of young compared with old persons is the decrease in mean wall thickness (MWT) in both cortical and trabecular bone (Justesen et al., 2001). MWT depends both on the number of osteoblasts recruited and the activity of each individual osteoblast; osteoblast recruitment depends on the availability of the mesenchymal stem cells (MSCs) and their response to growth and differentiation signals. Samples of cadaveric human bone revealed that porosity increased significantly with age, with little change in the mineral content and accounted for 76% of the age-associated reduction in bone strength (McCalden et al., 1993). Women rapidly lose trabeculae with age, while men develop trabecular thinning secondary to reduced bone formation. The relative preservation of trabeculae in men may help explain their lower lifetime risk of fractures. In men, trabecular bone loss starts early in life, in association with changes in the insulin-like growth factor 1 (IGF-1) regulating system, whereas cortical bone loss occurs later; 85% of cortical bone loss occurs after 50 years of age. Bone loss in aging men is associated with decreases in bioavailable testosterone and estrogen and increased bone remodeling (Ebeling, 2008) (Fig. 9.8).

In contrast to the short-lived osteoblasts and osteoclasts, the osteocyte can survive entombed in its lacuna for 50 years or more if not damaged (Manolagas, 2000; Rosen and Bouxsein, 2006; Almeida, 2010). Osteocytes die by apoptosis in both sexes. Osteocyte death has been reported as a function of age, but was not related to menopause in women (Qiu et al., 2002). Although osteocyte death does not appear to be related to changes in reduced ovarian estrogen in women, cell death is decreased by sex steroids and increased by glucocorticoid excess. Loss of mechanical strain, as induced by a sedentary lifestyle, also promotes osteocyte death by reducing bone hydration and communication. Since a number of nutrients play key roles in osteocyte maintenance and bone structure, the well-described reduction in nutrient status in an aging population may also increase osteocyte death as discussed below. Osteocyte ablation in a mouse model resulted in empty lacunae, thinned cortical bone, intracortical porosity, and compromised bone strength (Tatsumi et al., 2007), Increased empty lacunae have been related to decreased bone strength with aging, despite adequate BMD (McCalden et al., 1993) (Fig. 9.9).

The observation that bone undergoes age-associated changes similar to other organs and tissues has led to the proposal that age-associated actions that increase ROS exceed those that extinguish them (refer to Essentials I: Life in an Aerobic Environment). The net increase in ROS induces osteoblast/osteocyte apoptosis via activation of the p53/p66^{Shc} cascade with production of H_2O_2 and release of apoptotic cytochrome C through the mitochondrial permeability transition pore. The authors also propose that H_2O_2 stimulates proinflammatory cascades (NFκB; HIF-1) that inhibit

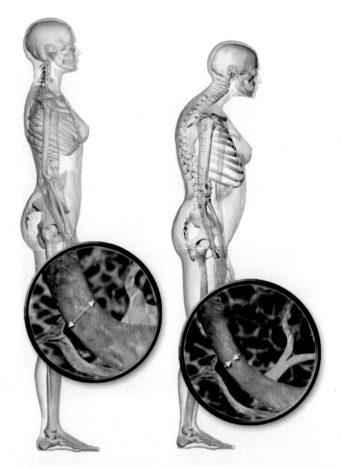

FIGURE 9.8 **Aging skeleton.** Note the reduced cortical thickness and increased porosity in aged, compared with young bone. *Source: From Blausen.com staff. Blausen gallery 2014. Wikiversity Journal of Medicine 1 (2). doi:10.15347/wjm/2014.010. ISSN 20018762 (own work) (CC BY 3.0 (http://creativecommons.org/licenses/by/3.0)), via Wikimedia Commons*

autophagy and oxidize other cellular components including the nucleoprotein, Nup93, that maintains the nuclear permeability barrier (Manolagas and Parfitt, 2010). These authors propose that oxidative stress is a key etiologic agent and that the osteocyte is the principal target. The age-associated increase in oxidative stress, lipid oxidation, reduction in sex steroids, together with increased levels and sensitivity to endogenous glucocorticoids increase osteocyte death and age-associated degradation of bone quality and strength.

9.4.3 Bone Hydration and Aging

Bone hydration, essential for bone structure and function, is reduced with aging (Timmins and Wall, 1977). At least 25% of the wet weight of bone is water, contained largely in the lacunar-canalicular system, bone vasculature, and bone marrow. About 10% of bone water is incorporated in collagen and exists as a

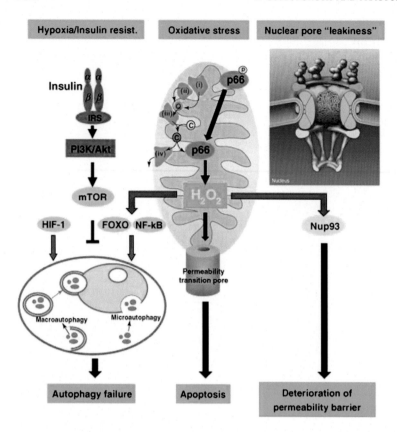

FIGURE 9.9 **Postulated changes in aging osteocytes.** Long-lived osteocytes are subject to several insults with age. Oxidative stress increases with advancing age, amplified by phosphorylation of p66 and release of oxidized cytochrome C from the permeability transition pore and apoptosis. Osteocytes attempt to adapt to increased ROS by increasing autophagy; these attempts are compromised with age and insulin resistance. Finally, osteocytes fail to maintain the nuclear pore diffusion barrier as a result of oxidation of nucleoproteins such as Nup93 (Manolagas and Parfitt, 2010). *Source: Reproduced from Manolagas, S.C., 2010. From estrogen-centric to aging and oxidative stress: a revised perspective of the pathogenesis of osteoporosis. Endocr. Rev. 31 (3), 266–300 with permission from Science Direct.*

hydration shell around hydroxyapatite crystals (Timmins and Wall, 1977; Manolagas and Parfitt, 2010). Submicroscopic channels allow water to percolate through what appears to be solid bone. New bone forms around capillaries and MSC progenitors are located near or on the surface of blood vessels. Bone fluid is also critical for providing a mineral reservoir for bone remodeling, for nutrient distribution, and for osteocyte sensing of mechanical stress. The gel-like matrix that surrounds osteocytes and their processes facilitates communication and metabolic activities. Decreased water content increases tensile and compression strength, elasticity, and bone hardness, but the bone becomes brittle and breaks under less deformation stress than a fully hydrated bone. The authors propose that reduced hydration may be a factor that predisposes aging bone to increased risk for fracture (Granke et al., 2015).

9.4.4 Bone Marrow Adipose Tissue and Aging

Bone marrow adipose tissue (MAT) is a unique fat depot that is not comparable to peripheral subcutaneous and abdominal fat depots (refer to Chapter 1: Obesity and Metabolic Syndrome: Adipose Tissue Form and Function). The metabolic profile of MAT cells resembles both white (WAT) and brown (BAT)

adipose tissue, reflecting its plasticity in responding to the microenvironment in the bone marrow. MAT has been called "yellow adipose tissue" and contains a moderate number of mitochondria. Whether MAT contains a mixed population of BAT and WAT adipocytes or whether MAT cells express activities common to both AT types is still unclear (Lecka-Czernik and Stechschulte, 2014).

MAT volume is increased in aging bone (Justesen et al., 2001). As illustrated in Fig. 9.10, the pluripotent mesenchymal stem (progenitor) cell (MSC; MPC) is capable of differentiating along both osteoblastic and adipogenic lines; mature osteoblasts have also been shown to transdifferentiate into adipocytes and "redifferentiate" along osteoblastic pathways under appropriate culture conditions as reviewed by Kassem et al. (2008). Key transcription factors have been identified for differentiation to osteoblasts (Runx2) and adipocytes (PPARγ); differentiation for each lineage appears to be mutually exclusive (Singh et al., 2016) but cellular plasticity has been suggested.

Preferential MSC differentiation along adipogenic lines has been demonstrated under conditions of oxidative stress. Activation of PPARγ both produces ROS and requires ROS to activate transcriptional machinery during adipogenic differentiation of MSC (Kanda et al., 2011; Wang et al., 2015). Thus, an age-associated increased oxidative environment could

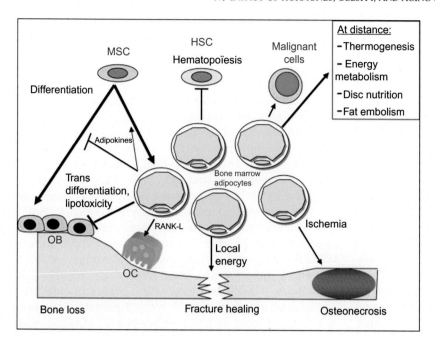

FIGURE 9.10 **Bone marrow adipose tissue.** Bone marrow adipose tissues (ATs) share similar functions with peripheral fat depots. They can store or release lipids; non-esterified fatty acids can induce lipotoxicity or induce transdifferentiation in osteoblast cultures. It should be noted that the concentrations used in vitro studies are much higher than those measured in vivo. MAT cells secrete cytokines and other signaling molecules and express RANKL, promoting osteoclast differentiation. Other, less well-described activities are reviewed in Hardouin et al. (2014). *Source: Reproduced from Hardouin, P., et al., 2014. Bone marrow fat. Joint Bone Spine 81 (4), 313–319 with permission from Science Direct.*

explain the increased MAT observed in aging bone (Bethel et al., 2013). MAT is also increased under metabolic stress conditions even in a young patient, as demonstrated in the increased MAT and osteoporotic bone seen in young patients with anorexia nervosa. MAT was highest in osteoporotic bone and intermediate in osteopenic bone compared with controls (Hardouin et al., 2014). The inverse relationship with MAT content and bone strength was demonstrated in a recent in vitro study that measured the "failure load" in cadaveric bone by destructive biomechanical testing (Karampinos et al., 2015). This study confirms earlier work and suggests that MAT accumulation explains, in part, the age-associated decrease in bone quality and strength.

In addition to its endocrine functions, MAT cells also store and release lipids for mitochondrial oxidation and ATP production. Lipid oxidation has been reported to account for 40–80% of osteoblast energy utilization (Adamek et al., 1987). Mitochondrial fat oxidation requires carnitine for long-chain fatty acid transport across the mitochondrial membranes as reviewed by Flanagan et al. (2010) and described elsewhere (refer to Chapter 5: Diabetic Cardiomyopathy and Congestive Heart Failure: Mitochondrial Production and Management of Reactive Oxygen Species; Carnitine). Carnitine depletion has been found in aging animal models as well as in aging humans (Costell et al., 1989). Carnitine analogs have been shown to accelerate bone recovery after experimental osteoporosis induction in pregnant mice (Patano et al., 2008). The authors suggested that improved fatty acid oxidation enhanced the energy available for protein synthesis. Since carnitine

was also shown to stimulate human osteoclast proliferation and differentiation in vitro (Colucci et al., 2005), the authors proposed that carnitine supplementation could benefit carnitine-deficient elderly patients and reduce age-related bone loss. This hypothesis was tested in a placebo-controlled double-blind study in 172 men (average age = 63) with primary osteoporosis (lumbar spine and femoral neck BMD T-score −3.0). Treatment with L-carnitine (4 g/day) for 2 years was associated with significant increases in BMD at the lumbar spine, femoral neck, total hip, and an increase in appendicular skeletal muscle mass compared with placebo (Lei and Chuan, 2015).

9.4.5 Bone Structure and Strength in Obesity

The positive association of body weight (usually quantitated as body mass index (BMI)) with improved BMD has been partially linked to the greater mechanical load carried by obese subjects as reviewed by Shapses and Sukumar (2012). However, obesity even with increased BMD did not protect against fracture in obese, postmenopausal women enrolled in the Global Longitudinal study of Osteoporosis in Women (GLOW) (Compston et al., 2011). Data obtained with newer assessment methods has revealed that bone structure is compromised in obesity. Specifically, while trabecular mass increased with BMI, cortical mass was reduced with obesity (Sukumar et al., 2011).

Obesity-associated metabolic dysregulation has been implicated in altered bone mass and quality in obese patients as reviewed by Shapses and Sukumar (2012).

Positive effects of obesity on bone integrity include increased secretion of insulin, amylin, resistin, and preptin from the pancreatic β cell, estrone production by adipocyte aromatization of adrenal androgens, secretion of leptin and adiponectin from the adipocyte (including from adipocytes in the bone marrow), and increased skeletal load-bearing stress in the obese patient. Leptin, produced primarily by adipocytes, regulates a spectrum of actions relating to energy and macronutrient homeostasis; leptin coordinates these effects through binding to its receptor in the central nervous system (Friedman and Halaas, 1998) (refer to Chapter 1: Obesity and Metabolic Syndrome: Nutrient Sensors and Downstream Targets). A lesser known homeostatic function for leptin is bone remodeling. Leptin controls bone remodeling by increasing bone sympathetic tone, signaling through the β2 adrenergic receptor expressed on osteoblasts (Elefteriou et al., 2005). The requirement of signals from intact ventromedial hypothalamic nuclei for bone remodeling was shown in animal studies using chemical lesioning.

Negative effects of obesity on bone quality have been examined. Inflammatory mediators, including secretions from proinflammatory visceral fat have been implicated. A recent study that examined bone quality in healthy, obese premenopausal women revealed that bone quality was compromised, possibly because of increased MSC differentiation along adipogenic rather than osteoblastic pathways (Rosen and Bouxsein, 2006; Cohen et al., 2013). Obesity can also modulate distribution of micronutrients required for bone integrity and crosslinking. For example, genetically obese mice had lower bone zinc, copper, and manganese concentrations (Kennedy et al., 1986) (Fig. 9.11).

Weight reduction has been associated with bone loss and increased fracture risk as reviewed by Shapses and Sukumar (2012). In peri- and menopausal women and older men, a 10% weight loss is associated with a 1−2% hip and total bone loss and a 3−4% loss in sites such as the trochanter and radius, rich in trabeculae. Bone loss with weight loss and/or disordered eating was also found in a longitudinal cohort study of healthy, physically active women in the US Military Academy ($n = 91$; average age = 18.4 years). BMD, measured by dual X-ray absorptiometry and calcaneal BMD, increased in the spine in 50% of these young adult subjects and was positively related to menstrual cycle frequency. BMD loss was related to subclinical eating disorders, weight loss and use of depot-medroxyprogesterone acetate (DMPA), suggesting that disordered eating behavior prevented these women from achieving maximal peak bone mass. (DMPA is a long-acting progestin-only hormonal contraceptive injected every 3 months. Bone loss has been reported as a side effect but not if taken after peak bone mass has been achieved.)

Several metabolic changes that occur with caloric restriction could lead to decreased bone mass (Shapses and Sukumar, 2012). These include decreased growth factors (IGF-1; GH), decreased weight bearing, and a change in hormonal signaling (reduced estrogen and increased cortisol. Reduced calcium intake and absorption with caloric restriction sufficient to lower ionized serum calcium and stimulate PTH activity could increase bone loss. Other changes for which the mechanism is less clear include changes in gut hormones (increased ghrelin; decreased GIP, GLP-1, GLP-2, Amylin), adipokines (increased adiponectin, resistin; decreased leptin, visfatin), and reduction in cytokines (IL-6, TNFα).

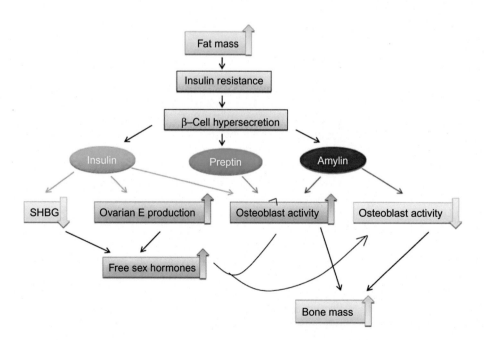

FIGURE 9.11 **Obesity and bone mass.** Increased fat mass with obesity increases insulin resistance and compensatory hypersecretion of insulin and other pancreatic peptides. Insulin stimulates ovarian estrogen production and reduces sex hormone-binding globulin, thereby increasing free sex hormones that increase bone accretion. Amylin also inhibits osteoclastogenesis (Reid, 2010). Insulin resistance and hyperinsulinemia are associated with elevated circulation of saturated fatty acids and reduced osteoclastogenesis (not shown) and In vitro studies revealed that fatty acids act directly on bone cells in ways predicted to increase bone mass; the effect was greatest for C14:0−C18:0 fatty acids. Double bonds reduced this effect, as did ω3 and ω6 fatty acids (Cornish et al., 2008). *Source: Reproduced from Reid, I.R., 2010. Fat and bone. Arch. Biochem. Biophys. 503 (1), 20−27 with permission from Science Direct.*

In contrast to the 1−2% bone loss associated with diet-induced weight loss, bariatric surgery, especially the malabsorptive procedures including the Roux-en-Y gastric bypass and duodenal switch induce a much greater bone loss, especially at the hip (Shapses and Sukumar, 2012; Stein and Silverberg, 2014; Yu, 2014). Measurement of bone mass in an obese patient is fraught with difficulties relating to the excess adipose tissue mass, the presence of arthritis, and the effects of rapid weight loss. Additionally, differences in assessment methodologies can produce conflicting results as reviewed by Yu (2014). Causes of bone loss following surgery include mechanical unloading, reduced calcium absorption, and low vitamin D status with compensatory rise in PTH, as well as altered production and action of gut and other hormones. Abnormalities in bone turnover indices have been shown to persist for years after bariatric surgery and clinical observations suggest bone loss is substantial. On the other hand, bariatric patients have higher BMD prior to surgery, benefits in terms of reduced risk for obesity comorbidities are well described, and evidence to suggest bone loss is clinically relevant is minimal (Shapses, 2009). For example, data from the few studies that examined fracture risk in bariatric surgery patients are inconclusive.

9.4.6 Osteoporosis in the Diabetic Patient

Increased fracture risk in the diabetic patient was confirmed by meta-analysis; relative risk was higher for type I diabetes (T1D) than for type II diabetes (T2D) (Vestergaard, 2007). The type of bone damage also differs between T1D and T2D. Patients with T1D show a moderately reduced BMD in both axial and appendicular skeleton, whereas patients with T2D often have higher BMD compared with control subjects; this is especially the case in overweight women (Carnevale et al., 2014). No clear relationship between bone mass measurements and biochemical parameters of mineral metabolism has been shown in the different types of diabetes.

The etiology of increased fracture risk among diabetics is poorly understood. Studies in humans and experimental animals have documented bone defects including reduced bone size and accrual for TID, and altered bone matrix and porous bone in T2D. Hyperglycemia and other hormonal dysregulations seen in both T1D and T2D inhibited the proliferation and differentiation of MSC to osteoblasts, and also suppressed osteoblast function and bone remodeling. Further, nonenzymatic glycation of collagen was also seen (Parajuli et al., 2015). It should be noted that treatment with thiazolidinediones (rosiglitazone and similar drugs), by providing a signal for adipogenesis, switches MSC differentiation from osteoblasts to adipocytes (Yaturu, 2009).

9.4.7 Etiology of Osteoporosis in Men

Osteoporosis in men continues to be under-recognized, as reviewed by Ebeling (2008). One-third of all hip fractures worldwide occur in men; more men than women die in the year after a hip fracture, with a mortality rate in men of up to 37.5%. Almost half of hip fractures in men occur before age 80, and about 40% of these fractures occur in long-term care facilities. Twenty percent of male patients with one fracture will sustain a second hip fracture. Vertebral fractures are also common among elderly men; most vertebral fractures (70−85%) are painless but are associated with height loss, reduced quality of life, respiratory dysfunction, increased risk of death, and subsequent hip and other fractures.

The majority of fractures occur in men whose BMD measurements are not in the osteoporotic range. This observation underscores the importance of factors other than BMD in determining the risk of fracture. Osteoporosis in men often has secondary causes; the most frequent secondary causes are endogenous production of corticosteroids and excessive alcohol use. Hypogonadism, as a result of age-impaired Leydig cell production of testosterone afflicts 5 million American men; this condition has been linked to increased visceral fat and metabolic syndrome (Beattie et al., 2015). Other secondary causes include vitamin D deficiency; serum levels of 25-hydroxyvitamin D below 25 ng/mL (62.5 nM/L) are associated with an increased risk of hip fracture in men and women older than 65 years. Estrogen levels are also important for the male skeleton; testosterone exerts indirect effects on bone through its aromatization to estrogen. In rare cases, mutations of the estrogen receptor or the aromatase enzyme have been associated with severe osteoporosis in men. In up to 40% of cases in men, no secondary cause is identified, and the osteoporosis is considered to be primary or idiopathic.

9.5 CLINICAL DIAGNOSIS AND MANAGEMENT OF THE FRACTURE-RISK PATIENT

Several diseases, drugs, and altered metabolic states have been associated with risk for osteoporosis as listed in Table 9.1. Conditions with primary dietary implications are marked with italics in Table 9.1 and discussed below (Fig. 9.12).

TABLE 9.1   Diseases and Other Secondary Causes of Osteoporosis[a]

Endocrine	Gastrointestinal	Bone marrow	Connective tissue	Drugs	Miscellaneous
Hypogonadism—male/female	Subtotal gastrectomy	Multiple myeloma	Osteogenesis imperfecta	*Alcohol*	Immobilization
Hyperthyroidism	*Malabsorption syndromes*	Lymphoma	Ehlers–Danlos syndrome	Heparin Inappropriate thyroid therapy	Rheumatoid arthritis
Hyperparathyroidism	Chronic obstructive Jaundice	Leukemia	Marfan's syndrome	Glucocorticoids	Renal tumor acidosis
Hypercortisolism	Primary biliary cirrhosis	Hemolytic anemias	*Homocysteinuria*	*Thyroxine*	
Growth hormone deficiency	Other cirrhosis	Systemic mastocytosis		*Anticonvulsants*	
Vitamin D deficiency		Disseminated carcinoma		Gonadotropin RH agonists	
Idiopathic hypercalciuria				Cyclosporine	
Diabetes mellitus				Tacrolimus thiazolidinediones	
Cancer				Chemotherapy	

[a]*Table abstracted from Cecil: Table 264-2, p. 1882.*

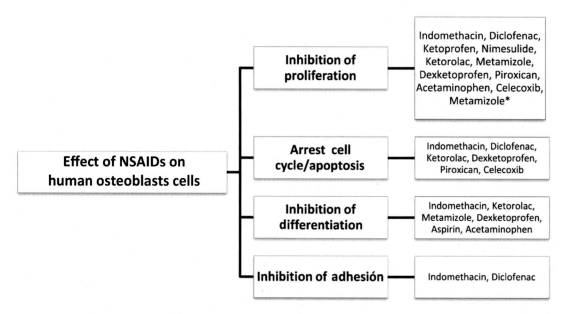

FIGURE 9.12   **Effects of NSAIDs on osteoblasts.** Nonsteroidal antiinflammatory drugs (NSAIDs) can act by affect bone by modulating osteoblast proliferation, differentiation, adhesion, and migration. Some, but not all, NSAIDs have these effects as indicated (García-Martínez et al., 2015). *No consensus is available in scientific publications about the effect of metamizole on osteoblasts. *Source: Reproduced from García-Martínez, O., et al., 2015. Repercussions of NSAIDs drugs on bone tissue: the osteoblast. Life Sci. 123, 72–77 with permission from Science Direct.*

Specific drugs known to increase bone loss include (National Osteoporosis Foundation: http://www.nof.org/advocacy/resources/prevalencereport):

- Aluminum-containing antacids;
- Antiseizure medicines (only some) such as Dilantin or Phenobarbital;
- Aromatase inhibitors such as Arimidex, Aromasin, and Femara;
- Cancer chemotherapeutic drugs;
- Cyclosporine A and FK506 (tacrolimus);
- Gonadotropin-releasing hormone (GnRH) such as Lupron and Zoladex;

- Heparin;
- Lithium;
- Medroxyprogesterone acetate for contraception (Depo-Provera);
- Methotrexate;
- Proton pump inhibitors (PPIs) such as Nexium, Prevacid, and Prilosec;
- Selective serotonin reuptake inhibitors (SSRIs) such as Lexapro, Prozac, and Zoloft;
- Steroids (glucocorticoids) such as cortisone and prednisone:
  - Common steroid medicines are cortisone, dexamethasone (Decadron), methylprednisolone (Medrol) and prednisone. Intravenous forms include methylprednisolone sodium succinate (Solu-Medrol);
- Tamoxifen (premenopausal use);
- Thiazolidinediones such as Actos and Avandia;
- Thyroid hormones in excess.

9.5.1 Assessment of Bone Density

A history of a minimal trauma fracture after the age of 50 years is the strongest clinical risk factor for a subsequent fracture. In patients without a history of fracture, BMD, measured by dual-energy X-ray absorptiometry (DEXA), is considered a robust predictor of fractures. Each decrease of 1 standard deviation (SD) in hip BMD is associated with an increased relative risk of hip fracture of 2.6. Epidemiologic data suggest that for any given absolute BMD value at the spine or hip, the risk of fracture is similar among men and women of the same age. Using male-specific cutoffs for hip BMD, the National Health and Nutrition Examination Survey III study showed that 6% of US men who were 50 years of age or older had osteoporosis and 47% had osteopenia; corresponding prevalence in women is 18% and 50%, respectively. Bone densitometry is recommended in men 70 years of age or older; assessment is recommended at an earlier age in men with major risk factors for osteoporosis (Table 9.1).

Femoral neck BMD measures are preferable to spinal measures. The National Osteoporosis Risk Assessment Cohort studied 200,150 postmenopausal women without known osteoporosis and identified over 50% with low BMD, including ~7% with osteoporosis, as defined by WHO criteria (Siris et al., 2001). Given the difficulty of assessing fracture risk from altered bone quantity (eg, BMD) as described above, patients should be assessed routinely for risk factors suggesting altered bone quality including age, previous fractures, and family history, and for clinical signs of secondary causes. The WHO has established country- and ethnic group-specific risk-assessment tools (FRAX; FRAX can be accessed at http://www.sheffield.ac.uk/FRAX/tool.jsp)

incorporating clinical risk factors with age and BMD measurements to calculate the risk of fracture in patients 50 years of age or older. Clinical nomograms have been developed for predicting 5- and 10-year fracture risks for older men and women. Risk factors included age, femoral neck BMD, fracture and fall history, and body weight (Nguyen et al., 2008). Results in brief suggest that virtually all 80-year-old men with BMD T-scores <-1.0 or 80-year-old women with T-scores <-2.0 were predicted to be in the high-risk group (5-year fracture risk 10% or greater). A 60-year-old woman's risk was considered high risk only if her BMD T-scores were ≤ -2.5 and she had a history of a prior fracture. Using this nomogram, 60-year-old men would not be at high risk regardless of their BMD and risk profile. Biomarkers such as low testosterone levels and low levels of 25-hydroxyvitamin D are also useful in predicting an increased risk of fracture. Other risk factors such as strong family history, recent significant weight loss, and secondary causes as listed in Table 9.1 should also enter the clinical decision making.

9.5.2 Laboratory Tests

Further testing is indicated to rule out secondary causes when the DEXA z score is below -2.0; this value is 2 SD below the age-specific mean. Markers of osteoclast-mediated bone resorption include urinary pyridinoline and deoxypyridinoline, as well as urine crosslinks of the C- and N-terminal peptides of type 1 collagen (C-TX and N-Tx). Measures of osteoblast function include bone-specific alkaline phosphatase (ALPL), osteocalcin, type 1 procollagen amino terminal propeptide, and type 1 procollagen carboxy terminal propeptide (Seibel, 2005). It should be noted that many of these biomarkers are also produced by non-bony tissues. Thus changes in their circulating levels are not bone-specific but can reflect alterations in other systems. Results of bone marker measurements should always be interpreted against the background of their basic science and the clinical picture. For example, both types of bone turnover markers increase dramatically after estrogen cessation for whatever reason, and can be especially elevated during states of secondary hyperparathyroidism. Estrogen and bisphosphonate therapy significantly lower both urinary and serum concentrations of turnover markers. Unfortunately, these measures are associated with a high degree of variability that has dampened enthusiasm for their use either to diagnose bone loss or to monitor treatment.

To identify possible secondary risks, the case discussant (Ebeling, 2008) recommends routine laboratory tests including serum calcium and creatinine levels, liver-function tests, measurement of the thyrotropin level, and a complete blood count. (It should be noted that all serum chemistry values can be normal in osteoporosis patients.) If clinically indicated, serum protein electrophoresis and tests for urinary Bence–Jones protein (to check for monoclonal gammopathy), antitissue transglutaminase antibodies (to check for celiac sprue), 24-hour urinary cortisol or calcium, and human immunodeficiency virus antibodies should be performed. Measurement of the total testosterone level is recommended in all men with osteoporosis. Sex hormone-binding globulin levels may provide additional information, especially in men with insulin resistance or obesity, in whom low levels of sex hormone-binding globulin may complicate interpretation of total testosterone levels. Serum levels of 25-hydroxyvitamin D (25D) should also be measured. Levels below 30 ng/mL (75 nM/L) should be treated.

9.5.3 Current Treatment

To treat the current patient, the case discussant (Ebeling, 2008) suggests that the patient be instructed to take in at least 1200 mg/day of calcium from diet and/or supplements, as well as vitamin D supplementation of at least 800 IU/day; more vitamin D may be needed to increase 25D levels to an acceptable range. Regular, weight-bearing exercise would be highly recommended. A T score of −2.5 or less indicates bone loss; evidence of vertebral fracture would be an indication for pharmacologic therapy. An oral bisphosphonate, currently considered first-line treatment for osteoporosis in men, should be recommended, with patient education regarding potential side effects.

Aggressive intervention has been shown to reduce fracture risk and improve quality of life for patients with low BMD and/or history of vertebral or minimal trauma fractures. Two types of drugs have been shown to reduce risk.

- *Antiresorptives*—block bone resorption by inhibiting osteoclasts;
- *Anabolics*—stimulate bone formation by acting on osteoblasts.

9.5.4 Antiresorptive Agents

These drugs inhibit bone resorption by suppressing osteoclast activity, thereby slowing the remodeling cycle and allowing bone formation to catch up with resorption. They also stabilize trabecular microarchitecture, increase BMD, and reduce fracture risk.

- *Estrogen*
  Sex steroids inhibit cytokine signaling from osteoblast to osteoclast in both cortical and trabecular bone allowing BMD increase of up to 3–5% within 3 years. While addition of progesterone does not appear additive with respect to bone, it prevents endometrial hyperplasia and carcinoma in the woman with a uterus. Significant risk for breast cancer, myocardial infarction, and thromboembolic disease has reduced use of estrogen or combination preparations as antiresorptive agents.

- *Selective estrogen receptor modulators*
  Tamoxifen and raloxifene bind estrogen receptors and inhibit bone resorption by blocking cytokine release from the osteoblast. Only raloxifene has been approved by the US Food and Drug Administration for prevention and treatment of osteoporosis. Both agents block the action of estrogen at the breast, but act as agonists in bone. Tamoxifen, but not raloxifene, has estrogen agonist activities at the uterus and is associated with risk for endometrial carcinoma. Raloxifene increases spine BMD slightly and lowers risk for vertebral fractures, but has no effect on nonvertebral fracture risk.

- *Bisphosphonates*
  These drugs are carbon-substituted analogs of pyrophosphate and bind strongly to hydroxyapatite crystals. They are thought to inhibit resorption by inhibiting osteoclast structural changes and stimulating osteoclast apoptosis (Drake et al., 2008). Second-generation nitrogen-containing bisphosphonates, alendronate (fosomax) and risedronate, have been shown to reduce vertebral, hip, and nonvertebral fractures by nearly 50% in postmenopausal woman. Newer drugs of this type (zoledronic acid) are under investigation to address an erosive esophagitis, a serious complication of these drugs. It should be noted that bisphosphonate use requires adequate calcium and vitamin D intake (Kennel and Drake, 2009)

- *Calcitonin*
  This peptide normally produced by the thyroid C cells binds receptors on the osteoclast and inhibits bone resorption. Although nasal and subcutaneous calcitonin are approved for treatment of postmenopausal osteoporosis, evidence for a strong effect on reducing bone loss or fracture risk is lacking.

9.5.5 Anabolic Agents

Anabolic agents are targeted to the osteoblast as reviewed by Almeida et al. (2009). Age-associated bone loss is characterized by a decline in osteoblast number and function and preferential MSC differentiation to adipocytes. Anabolic agents promote osteoblastogenesis and reduce adipogenic differentiation, enhance preosteoblast replication or function, or increase osteoblast survival.

- *Drugs that target Wnt/β-catenin signaling*
  Once MSCs have reached the bone surface, they differentiate into osteoblasts and produce mineralized

bone matrix. The Wnt/β-catenin signaling pathway switches MSC differentiation away from adipocytes and promotes osteoblast differentiation and activation (Milat and Ng, 2009). Wnt signaling also promotes osteoblast survival and interacts with BMP2 and PTH signaling to increase osteoblastogenesis. Mechanical bone loading upregulates Wnt signaling in MSCs, suggesting that age-associated bone loss may be due, in part, to the combination of reduced β-catenin signaling and decreased physical exercise. While modification of the Wnt signaling pathway is a potential target for treating osteoporosis, its ubiquitous actions and the association of Wnt upregulation with cancer suggests that antagonists of Wnt inhibitors may be safer therapeutic agents. At this time, no pharmacologic agents to modify this system are currently available.

• *Parathyroid hormone*

PTH is the only anabolic agent approved for osteoporosis. Intermittent, but not sustained, low-dose PTH has been shown to increase bone formation more than bone resorption, leading to increased bone mass. PTH 1-34 (teriparatide) has been shown to increase BMD and reduce fracture risk. Its duration of use has been limited, because rats treated with high doses of the drug developed osteosarcoma. No associations between osteosarcoma and primary or secondary hyperparathyroidism have been found in patients treated with PTH 1-34. Recombinant human PTH (PTH_{1-84}) has also been found to be efficacious, but its use is limited to 2 years in patients with moderate to severe osteoporosis. Use is also limited by cost and compliance. Unlike bisphosphonates, discontinuation of PTH can result in 3–4% bone loss in the first year after cessation.

• *Testosterone*

Studies of testosterone in men with osteoporosis are limited, and none to date have used fractures as a primary end point. The risks of testosterone therapy, including polycythemia, sleep apnea, benign prostate enlargement, and possibly prostate cancer, argue against its use in eugonadal men with osteoporosis at this time.

9.6 TARGETS FOR DIETARY INTERVENTION IN OSTEOPOROSIS AND FRACTURE RISK

9.6.1 Nutrients for Bone Mineral Integrity

Bone composition varies with age, maturation, and turnover, but it is generally 50–70% mineral and 20–40% matrix. Bone crystal contains large amounts of calcium and phosphorus, with lesser amounts of cations

and anions including magnesium, carbonates, and trace elements. Therapy to foster bone deposition requires that nutrients for bone formation be available in adequate quantity. Mineral homeostasis is maintained through a dynamic state of mineral ion balance maintained by the calciotropic hormones PTH, FGF23, and $1,25 (OH)_2D_3$ (1,25D) as described elsewhere (refer to Essentials III: Nutrients for Bone Structure and Calcification). A recent Cochrane Review of the efficacy of supplements of vitamin D with or without calcium for prevention of fractures in older men and women revealed high-quality evidence that vitamin D plus calcium resulted in a small reduction in fracture risk (Avenell et al., 2014). Risk reduction was greater in high-risk institutionalized populations (nine fractures/1000 older adults) than in low-risk community populations (one fracture/1000 older adults). Mortality was not adversely affected by either vitamin D or vitamin D with calcium.

Dietary intake data from the Framingham Osteoporosis Study was examined in terms of fracture risk, BMD, and other laboratory data to determine the influence of diet components on bone. Modest improvements in fracture risk were reported with several nutrients (vitamin C, carotenoids, folate, vitamin B12, vitamin K, potassium, magnesium, alkaline diets, omega 3 fatty acids, dairy products, seafood, beer, wine). However, since food contains a mixture of nutrients that can amplify or inhibit biological effects, the association of specific foods and nutrients on fracture risk is difficult to interpret. The authors therefore assessed risk associated with diet patterns. In general, the diet pattern most likely to reduce fracture risk included fruits, vegetables, and fish, while the pattern most associated with bone loss and fracture risk was high in fat and processed foods (Sahni et al., 2015).

9.6.2 Vitamin D

In addition to its effects on serum calcium homeostasis, vitamin D plays critical roles in osteoblast differentiation and function as reviewed by van de Peppel and Van Leeuwen (2014). Active 1,25D binds to the vitamin D receptor (VDR) expressed in osteoblasts and modifies expression of genes related to osteoblast differentiation from MSC, activation, and mineralization. VDR expression is regulated by 1,25D, as well as PTH and related proteins, glucocorticoids, TGF-β, and epidermal growth factor. Genes expressed by 1,25D/VDR include alkaline phosphatase liver/bone/kidney (ALPL), osteocalcin (OCN), and osteopontin (OSP). It should be noted that while human and rat osteoblasts were activated in vitro, murine osteoblasts were inhibited by 1,25D. The methodologic or species variables that explain these differences are not yet understood. 1,25D/VDR may also modulate osteoclast formation. For example, murine macrophage-like RAW264.7 cells were induced to form osteoclasts when cultured with RANKL and M-CSF. Addition of 1,25D enhanced the numbers of osteoclasts formed and bone resorption ability, as well as expression of osteoclast-specific proteins required for bone resorption (Gu et al., 2015).

Human osteoblasts express 1α-hydroxlase, as well as the vitamin D-binding protein receptors megalin and cubilin and are able to use 25D as substrate. The classical renal regulators of 1α-hydroxylase, PTH, and calcium levels do not regulate 1α-hydroxylase in osteoblasts. However interleukin 1 (IL-1β) strongly induced 1α-hydroxylase in osteoblasts and in osteoclasts. Thus, bone responds to vitamin D in an autocrine as well as a paracrine manner (van Driel et al., 2006). At the same time, 1,25D stimulates biosynthesis of activin A, a protein that inhibits osteoblast differentiation and mineralization in human osteoblasts (Woeckel et al., 2013). These authors propose a feedback loop, in which 1,25D stimulates human osteoblast differentiation and mineralization and also stimulates the production of a mineralization inhibitor, thereby preventing overmineralization of bone.

9.6.3 Calcium

Evidence that calcium supplementation alone reduces fracture risk is uneven. A large meta-analysis (Shea et al., 2002) showed a small positive effect on bone density and a trend toward reduction in vertebral fractures, but did not address the possible effect of calcium on reducing the incidence of nonvertebral fractures. On the other hand, a meta-analysis of randomized, placebo-controlled trials with >500 mg supplemental calcium without vitamin D revealed an increased risk for myocardial infarction and cardiovascular events (Bolland et al., 2010). More recently, a large prospective study (Warensjö et al., 2011) investigated the relationship between calcium intake and risk for fractures and overall bone health in 61,000 women followed for 19 years. Primary outcome measures were incident fractures of any type, identified from registry data. A subcohort of 5022 women was studied for osteoporosis diagnosed by DEXA and repeated food frequency questionnaires (FFQs). Investigators concluded that only women with calcium intakes <750 mg/day were at increased risk for fracture, and that women who consumed >1100 mg/day actually had a slightly increased risk for hip fracture. This well-controlled study suggests that most of the calcium should be obtained through diet (at least ~700 mg/day), and that total calcium intake should be <1100 mg/day.

9.6.4 Phosphorus (P)

Phosphate is an integral part of the hydroxyapatite molecule and is required for optimal bone mineralization. The tight regulation of serum calcium and phosphorus concentrations has been discussed elsewhere (refer to Essentials III: Nutrients in Bone Structure and Calcification: Phorphorus). It should be noted that phosphorus intake may be several-fold higher than calculated in individuals who consume foods processed with phosphate additives. High intakes of phosphorus have potential to alter the PTH−vitamin D−FGF23 axis with deleterious effects. High phosphorus concentrations in the intestinal lumen form insoluble Ca-P complexes with calcium, reducing calcium absorption. If dietary phosphorous is low, calcium supplementation can result in phosphorus malabsorption. (The ability of dietary calcium to bind phosphorus has been used therapeutically to reduce phosphorus absorption in dialysis patients.)

A recent National Health and Nutrition Examination Survey revealed that substantial fractions of the older population have phosphorus intakes less than 70% of the adult Dietary Reference Intake. These observations prompted a clinical trial (Heaney et al., 2010) in older osteoporotic women, comparing 1800 mg Ca/day as tricalcium phosphate or calcium carbonate as adjuvant therapy with teriparatide (synthetic PTH) and 1000 IU cholecalciferol. Lumbar spine and hip BMD increased in both groups, suggesting that calcium−phosphorus salt might be effective for patients with low dietary phosphorus intake.

9.6.5 Magnesium (Mg)

Data from NHANES and other dietary surveys have repeatedly shown that while dietary Mg intake is low in all age groups, elderly individuals (>71 years) have especially inadequate intakes (Food and Nutrition Board, 1997; Ford and Mokdad, 2003; King et al., 2005). Furthermore, dietary Mg intake, as well as dietary Ca and P, were significantly reduced in osteoporotic women and correlated with BMD (Tranquilli et al., 1994; New et al., 2000). Mg food sources and metabolism are discussed elsewhere (refer to Essentials III: Nutrients in Bone Structure and Calcification: Magnesium).

Bone contains ~50−60% of all Mg^{2+} ions in the body. Approximately 30% of bone Mg^{2+} is bound at the surface of the hydroxyapatite crystals and modulates the rate and stability of bone crystals. Mg^{2+} bound to the bone surface fell rapidly with dietary depletion, suggesting that it forms an "exchangeable pool" that equilibrates with serum Mg^{2+} to support serum levels (Alfrey et al., 1974). Mg^{2+} stimulates osteoblast proliferation and also inhibits secretion of IL-1β, TNF-α, and other proinflammatory mediators which stimulate excessive osteoclast activity. In the developing bone, Mg^{2+} stimulates the formation of the chondrocyte column (de Baaij et al., 2015). Mg^{2+} is also required for the release of calcium and phosphate from the bone in response to PTH.

Mg deficiency impairs PTH secretion and, presumably because it is required for cAMP activity, also renders target organs refractory to PTH. Thus, in the Mg-deficient kidney, PTH signaling and activation of tubule 1-α hydroxylase activity is compromised, reducing 1,25D formation. In animal models, Mg deficiency impairs bone growth and structure. Human magnesium deficiency has been associated with hypoparathyroidism, low vitamin D conversion to 1,25D, end-organ resistance to PTH and 1,25D, and increased risk for fractures (New et al., 2000; Castiglioni et al., 2013; Hayhoe et al., 2015). These effects can lead to hypocalcemia without a compensatory increase in PTH (Fig. 9.13).

Mg deficiency is frequently seen in obesity and its comorbid conditions. In addition to dietary inadequacy, Mg can be lost with disorders of the gastrointestinal tract; diuretic therapy also potentiates Mg loss in the urine (Kareeann Sok Fun et al., 2014). Mg malabsorption has been reported secondary to proton pump inhibitors (Toh et al., 2015), alcohol intake, and use of other drugs as described elsewhere (refer to Essentials III: Nutrients in Bone Structure and Calcification: Magnesium). Diabetes, endocrine and renal disorders cause magnesium wasting. Several studies have reported that oral Mg supplements are associated with significant increases in bone density, suggesting that increased Mg intake may have potential benefits in the prevention and/or treatment of osteopenia and/or osteoporosis (Castiglioni et al., 2013).

9.7 NUTRIENTS FOR BONE MATRIX AND COLLAGEN CROSSLINKING

As discussed above, collagen fibers of compact bone lie adjacent to and are firmly bound to hydroxyapatite crystals. This intimate relationship prevents "shear" in the bone; by inhibiting crystals and fibers from slipping out of place, collagen binding maintains bone strength. The abundant crosslinks in the collagen fibers in bone give great tensile strength, whereas the mineral crystals provide compressional strength. Both properties are required to reduce fracture risk.

9.7.1 Copper

Copper status is rarely assessed because an appropriate biomarker has not yet been identified, as discussed elsewhere (refer to Essentials II: Heavy Metals, Retinoids and Precursors: Copper). Copper is required for the enzyme, lysyl oxidase (LOX), that initiates collagen crosslinks in the extracellular space. The extent of covalent crosslinks in mature collagen is a major determinant of bone strength. Animals with copper deficiency have deformed bones, hypoplasia, brittle

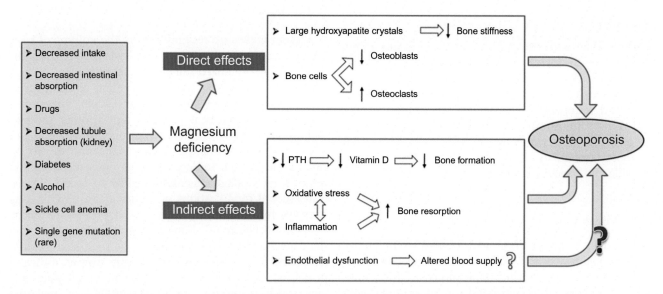

FIGURE 9.13 **Causes and consequences of Mg depletion.** Reduced Mg intake, combined with drugs and obesity comorbidities such as diabetes have potential to compromise bone structure and lead to osteoporosis (Castiglioni et al., 2013). *Source: From Castiglioni, S., et al., 2013. Magnesium and osteoporosis: current state of knowledge and future research directions. Nutrients 5 (8), 3022. Distributed under Creative Commons Attribution License (CC BY).*

bones, and frequent fractures. Bones from rats injected with an inhibitor of LOX, β-amino-propionitrile, had 45% fewer crosslinks, 31% decrease in stability of bones exposed to acetic acid and pepsin, and reduced resistance to deflection strain. No changes in mineral ash or collagen concentrations were noted (Oxlund et al., 1995). LOX activity requires both copper for catalytic activity (Gacheru et al., 1990) and vitamin B6 cofactor (Bird and Levene, 1982) as evidenced by deformed bones reported in vitamin B6-deficient chicks (Massé et al., 1996) and rats (Fujii et al., 1979).

Few human studies have evaluated the influence of copper status on bone strength. Symptoms of human copper deficiency reported in the literature include decreased bone strength, impairment of bone formation and growth, reduced bone mineralization, reduced ossification of growth centers, and compromised cartilage integrity. Copper absorption in the enterocyte is impaired by binding to zinc-induced metallothionein; thus zinc supplementation and a high-zinc/low-copper diet (meat and potatoes versus vegetables and legumes) can impair Cu availability as discussed elsewhere (refer to Essentials II: Heavy Metals, Retinoids and Precursors: Copper). Supplementation for 2 years in postmenopausal women reduced the rate of vertebral trabecular BMD decline; women receiving placebo showed significant decline over the experimental period (Eaton-Evans et al., 1996). It should be noted that oxidative stress is associated with intracellular Cu deficiency as described elsewhere (refer to Chapter 5: Diabetic Cardiomyopathy and Congestive Heart Failure: Targets for Myocardial Protection: Cardiac Remodeling).

9.7.2 Zinc

Elderly men with osteoporosis at the hip and spine had significantly lower dietary intakes of zinc (dietary and supplemental) than men without osteoporosis. BMD was lower in men at the lowest zinc quartile of intake and independent of BMI or age (Hyun et al., 2004). Zinc sources, absorption, utilization, and interference with copper absorption have been discussed elsewhere (refer to Essentials II: Heavy Metals, Retinoids and Precursors: Zinc).

9.7.3 Vitamin C

The relationship of fruit and vegetable intake to BMD was suggested in a study of middle-aged women (45−55 years) by measuring bone indices, DEXA and peripheral quantitative computed tomography as well as metabolites such as pyridinium crosslinks as markers of bone resorption and osteocalcin for bone formation. Total energy intake and individual nutrients were assessed with a validated FFQ; lifestyle activity was also determined. Results indicated that low intakes of potassium, β-carotene, magnesium, and vitamin C, all richly supplied especially by vegetables, were associated with increased bone resorption (New et al., 2000).

Ascorbate (AA) has many actions on bone metabolism. In osteoblasts, AA is required for proline and lysine hydroxylation to form crosslinks in normal procollagen residues (Peterkofsky, 1991). As described elsewhere (refer to Essentials I: Life in an Aerobic World: Single Unit Antioxidant Nutrients: Ascorbate), AA maintains iron in the reduced form required as a cofactor for prolyl and lysyl hydroxylases. AA also modulates nuclear translocation of several transcription factors important for bone cell differentiation; AA partly regulates expression of the osteoblast transcription factor, osterix, via nuclear factor (erythroid-derived 2)-related factor 2 (Nrf2) upregulation of genes critical for osteoblast differentiation (Gabbay et al., 2010). AA modulates hypoxia-inducible factor 1α (HIF-1α) required for angiogenesis and osteogenesis; more work is necessary to determine whether this action of AA is important in human bone integrity. Additional in vivo studies by these authors demonstrated that bone loss caused by low sex steroid levels worsened when AA levels were limited, especially when oxidative stress is present. Thus, AA may be important in states associated with increased inflammation and oxidative stress (aging, obesity, and diabetes) especially in individuals with low sex steroid activity.

9.7.4 Protein

Protein makes up roughly 50% of the volume of bone and about one-third its mass. The bone matrix is unique for its highly organized mineralization phase and lifelong process of regeneration and replacement or turnover as reviewed by Heaney and Layman (2008). Ninety-five percent of the matrix is collagen type I; other proteins include osteocalcin with its high affinity for calcium (see vitamin K below), anionic glycoproteins, and proteoglycans. Bone protein undergoes continuous turnover and remodeling. Collagen is crosslinked by a process of post-translational modifications of amino acids. Crosslinking involves hydroxylation of lysine and proline, requiring copper and vitamin C. Bone resorption releases many of the collagen fragments that cannot be reutilized to build new bone matrix. Accordingly, a daily supply of dietary protein is required for bone maintenance. Adequate dietary protein in children and young adults promotes peak bone mass and appears to favorably maintain bone mass, BMD, and bone strength throughout life and into old age as reviewed by Conigrave et al. (2008).

Early suggestions that high protein intake was detrimental to optimal bone structure were based on the fact that protein, whether from plant or animal, is a source of metabolic acid. Protein intake lowers urine pH and consumption of sulfur-containing amino acids (methionine and cysteine) provides substrate for sulfuric acid. Furthermore, dietary protein was reported to increase renal calcium excretion (Itoh et al., 1998), but this and similar reports suffered methodological bias (Lausen, 1999). On the other hand, several early animal studies reported a positive association between dietary protein intake and increased bone mineral content or decreased risk of fracture. Similarly, despite observations that animal protein increased calcium excretion, more recent studies have not shown a deleterious effect on mineral homeostasis from meat or vegetable protein.

The influence of protein intake on bone integrity has been reviewed (Conigrave et al., 2008). Overall, there is general agreement that diets moderate to high in protein ($\approx 1.0-1.5$ g/kg/day) are associated with improved calcium absorption, however, low protein intakes (<0.8 g/kg/day) reduce intestinal calcium absorption, increase PTH secretion and release calcium from the bone. Higher protein intake increased bone growth and peak bone mass in children, increased BMD and reduced the rate of bone loss in adults. Evidence that dietary protein modifies fracture risk is less clear, although studies cited have concluded that dietary protein reduces the risk of hip fractures and that protein supplements reduce complications following a hip fracture. The authors discussed several mechanisms that link circulating amino acids with control of calcium homeostasis and maintenance of bone remodeling.

9.7.5 Vitamin K

Vitamin K acts by a tissue- and cell-specific γ-carboxylation/decarboxylation process of amino-γ-carboxy glutamic acid (Gla) proteins as reviewed by Shearer and Newman (2008). These proteins have structural and regulatory functions in bone as well as in nonmineralized tissues. The matrix Gla protein (MGP) that inhibits arterial calcification has been described elsewhere (refer to Essentials III: Nutrients for Bone Structure and Calcification: Vitamin K and Chapter 6: Atherosclerosis and Arterial Calcification: Targets for Atherosclerotic Prevention and Control).

Vitamin K action in bone is associated with promotion of collagen accumulation, cell–matrix interactions, matrix mineralization, mineral maturation, osteocyte differentiation, and downregulated osteoclastogenesis, thereby maintaining optimal bone structure and function as reviewed. Bone has long been known to contain

at least two Gla-containing proteins, osteocalcin (OCN—bone Gla protein), MGP; several other Gla proteins have recently been identified (Rubinacci, 2009). Gla proteins (OCN, MGP, periostin, and others) have γ-carboxylation-dependent calcium-binding properties. As described above, carboxylated OCN is deposited in the bone matrix by osteoblasts and is associated with bone structure. Osteoclast-mediated bone resorption decarboxylates OCN and the uncarboxylated protein enters the circulation where it is reported to target adipocytes and pancreatic β-cells and participate in glucose homeostasis (Wei and Karsenty, 2015). Carboxylated MGP is implicated in the regulation of mineralization in the developing skeleton.

Vitamin K also promotes bone matrix deposition and integrity by fostering the osteoblast-to-osteocyte transition and by decreasing the osteoclastogenic actions (Atkins et al., 2009). These studies demonstrate that vitamin K promotes matrix mineralization and osteocyte differentiation, thereby modulating factors that contribute to cell strength, mechanical competence, and reduced fracture risk. It is possible that mechanisms other than carboxylation are involved. Preliminary studies have determined that vitamin K is a ligand to the steroid and xenobiotic receptor (Azuma and Inoue, 2009) and can promote transcription of several proteins important for bone integrity. This alternate mechanism may help explain the diverse functions of vitamin K in bone.

Inadequate availability of vitamin K for these functions can occur in multiple ways. Although the estimated dietary vitamin K adequate intakes are 90–120 μg/day for female and male adults respectively, these quantities may not be achieved by all persons, especially older adults. Elderly patients may dislike dark green leafy vegetables or may not be inclined to spend the time required for preparation, indeed low intake of fruits and vegetables in this age group have been well described. Furthermore, since vitamin K is fat-soluble, any condition that limits fat intake or absorption increases risk for deficiency. Finally, Dicoumarol (Warfarin) acts by limiting vitamin K carboxylation of coagulation factors; patients are also asked to limit dietary vitamin K intake that could interfere with drug action.

Vitamin K appears to exert its effects on bone matrix and bone strength rather than increasing BMD (Booth et al., 2000). This observation was supported by the Framingham Osteoporosis study, that reported a 65% lower hip fracture risk in subjects in the highest quartile of dietary vitamin K intake (254 μg/day); no relationship between vitamin K intake and BMD was seen at baseline or in the extent of BMD loss (Sahni et al., 2015). It should also be noted that uncarboxylated osteocalcin taken as an index of vitamin

K-dependent activity in this report may not actually address the function of vitamin K in bone matrix. Since osteocalcin is decarboxylated in the process of bone remodeling and uncarboxylated osteocalcin circulates as an endocrine hormone and regulates glucose homeostasis (Wei and Karsenty, 2015), the interpretation of these data may be difficult.

9.7.6 Resistance Exercise

Whether the treatment goal is primary fracture prevention or treatment of established disease, all patients benefit from progressive resistance training and weight-bearing exercise. Bed rest or immobility, especially in an older population, can result in rapid bone loss. Observational data suggest a lower risk among older men who maintain an active lifestyle and meta-analyses of trials show that balance and strengthening exercises reduce the risk of falls among older adults.

9.8 ANTIRESORPTIVE EFFECTS OF DIETARY FACTORS

In recent years, the antiosteoclastogenic effects of dietary components have been under intense investigation. Since oxidative stress has been linked to osteoclastic proliferation and activation, dietary antioxidants and nonnutrient components that modify key oxidative and inflammatory pathways have potential to modulate this process. While cell and animal models have shown potential, conclusive data from clinical trials is not yet available. Several dietary components such as catechins in green tea and chocolate as well as curcumin, from the yellow spice turmeric have been shown to inhibit inflammatory pathways and downregulate the OPG/RANK/RANKL pathway as reviewed by Bharti et al. (2004) and Lee et al. (2010). Resveratrol from red wine promotes osteoblastic differentiation from MSCs via a Wnt signaling pathway (Zhou et al., 2009). On the other hand, oxidized lipids and high sucrose in processed foods increase ROS and bone resorption (London et al., 2007; Goicoechea et al., 2008).

9.8.1 Dietary Phytoestrogens

The soy isoflavone, genistein, increased OPG mRNA levels and protein secretion by human osteoblastic cells by up to two- to sixfold in a dose- ($P < 0.0001$) and time-dependent ($P < 0.0001$) fashion as reviewed by Sacks et al. (2006). The stimulatory effect appeared to be mediated by the estrogen receptor (ER) and was abrogated by treatment with an ER agonist. Genestein upregulated production of OPG by human osteoblasts, reducing RANK activation (Viereck et al., 2002). Thus, dietary sources of phytoestrogens may help to prevent bone resorption and bone loss by enhanced osteoblastic production of OPG.

The authors review population studies and animal research on osteoporosis models. Soy isoflavones have been shown to reduce bone loss after menopause, however, to date, clinical trials have been too small and/or too short to be conclusive, and results have varied. To test the effectiveness of soy isoflavones ranging from 54 to 300 mg/day, direct measurements of bone mineral content and density in the spine and hip and estimated bone status from biochemical indices of bone resorption or formation have been studied. Interventions lessened bone loss over 6–24 months in some studies, whereas other trials did not show a benefit. Some studies showed favorable effects in the spine but not the hip, and vice versa. In some cases, bone mineral content, but not BMD, improved. These preliminary studies show promise, but more work is needed to clarify the impact of supplemental phytoestrogens on other estrogen-dependent tissues including breast and uterus, especially in postmenopausal women.

9.8.2 Omega 3 Fatty Acids

Omega 3 and omega 6 polyunsaturated fatty acids (PUFAs) may modify bone status by several mechanisms, including opposing effects on inflammatory cytokines, modulation of prostaglandin E_2 production, enhancement of calcium transport, and reducing urinary calcium excretion as reviewed by Farina et al. (2011). It is known that PUFAs and their derivatives can serve as ligands for PPAR α and γ, known to inhibit proinflammatory nuclear transcription factor κB (NFκB) and modulate differentiation of MSCs to adipocytes or osteoblasts.

Long-chain PUFAs also serve as precursors in the production of proresolving lipid mediators, including lipoxins synthesized from arachidonic acid, and E-series and D-series resolvins synthesized from eicosapentaenoic acid (EPA) and docosahexaenoic acid (DHA), respectively. Lipoxins and resolvins appear to have a myriad of effects that promote the resolution of inflammation, and both classes of lipid mediators have been found to reduce bone loss induced by periodontitis in animal models. DHA was a more potent inhibitor of osteoclast differentiation in RAW 264.7 cells than EPA (Rahman et al., 2008). At this time, although EPA and DHA have potential to reduce age-associated bone loss, evidence from clinical studies is fragmentary and conflicting.

The Framingham Osteoporosis Study measured BMD at the femoral neck (FN-BMD) at baseline (1988–89; $n = 854$) and 4 years later in adults ($n = 623$) with a mean age of 75 years (Sahni et al., 2015). Despite limitations that included the questionable accuracy of diet analysis done using a food-frequency measure, results support the hypothesis that fish intake, particularly intake of oily fish that contain omega 3 FA is protective against loss of FN-BMD in the elderly. A protective effect was also seen with arachidonic acid, but only if EPA + DHA intake was high.

9.9 FUTURE RESEARCH NEEDS

Age-associated bone loss and reduction in strength is a major source of morbidity and economic hardship, especially in an increasingly "graying" population. It is well known that optimal peak bone mass in early adult life extends the fracture-free period in old age. It is also known that optimal bone mass depends not only on calcium and vitamin D, but as described in this chapter, nutrients not usually considered such as ascorbate, vitamin K, phosphorus, magnesium, copper, and protein. Further, the recent Framingham Osteoporosis Study revealed that diet patterns that include processed food result in bone loss. The repeated observations that subclinical eating disorders can prevent fit young adults from achieving optimal bone mass is of concern in a social environment that glorifies the extremely thin physique. At the same time, the observation that weight loss and bariatric surgery results in an extended period of bone loss should be more closely examined, especially to determine whether bone loss with weight loss is associated with increased risk for fractures. Finally, while some investigators are assessing the quality of bone (bone strength) as a characteristic separate from BMD, this distinction needs to be standardized. As illustrated by the vitamin K data, bone strength was improved by increased dietary intake, although bone mineral density was not changed. It is likely that the influence of other nutrients will also be reflected more in bone quality than mineral quantity.

References

Adamek, G., Felix, R., et al., 1987. Fatty acid oxidation in bone tissue and bone cells in culture. Characterization and hormonal influences. Biochem. J. 248 (1), 129–137.

Alfrey, A.C., Miller, N.L., et al., 1974. Effect of age and magnesium depletion on bone magnesium pools in rats. J. Clin. Invest. 54 (5), 1074–1081.

Almeida, M., 2010. Aging and oxidative stress: a new look at old bone. IBMS BoneKEy. 7 (10), 340–352.

Almeida, M., Ambrogini, E., et al., 2009. Increased lipid oxidation causes oxidative stress, increased peroxisome proliferator-activated receptor-γ expression, and diminished pro-osteogenic Wnt signaling in the skeleton. J. Biol. Chem. 284 (40), 27438–27448.

Almeida, M., O'Brien, C.A., 2013. Basic biology of skeletal aging: role of stress response pathways. J. Gerontol. A Biol. Sci. Med. Sci. 68 (10), 1197–1208.

Anderson, H.C., 2003. Matrix vesicles and calcification. Curr. Rheumatol. Rep. 5 (3), 222–226.

Atkins, G.J., Welldon, K.J., et al., 2009. Vitamin K promotes mineralization, osteoblast-to-osteocyte transition, and an anticatabolic phenotype by γ-carboxylation-dependent and -independent mechanisms. Am. J. Physiol. Cell Physiol. 297 (6), C1358–C1367.

Avenell, A., Mak Jenson, C.S., et al., 2014. Vitamin D and vitamin D analogues for preventing fractures in post-menopausal women and older men. Cochrane Database Sys. Rev. Available from: http://dx.doi.org/10.1002/14651858.CD000227.pub4.

Azuma, K., Inoue, S., 2009. Vitamin K function mediated by activation of steroid and xenobiotic receptor. Clin. Calcium. 19 (12), 1770–1778.

Baron, R., Kneissel, M., 2013. WNT signaling in bone homeostasis and disease: from human mutations to treatments. Nat. Med. 19 (2), 179–192.

Beattie, M.C., Adekola, L., et al., 2015. Leydig cell aging and hypogonadism. Exp. Gerontol. 68, 87–91.

Bethel, M., Chitteti, B.R., et al., 2013. The changing balance between osteoblastogenesis and adipogenesis in aging and its impact on hematopoiesis. Curr. Osteoporos. Rep. 11 (2), 99–106.

Bharti, A.C., Takada, Y., et al., 2004. Curcumin (diferuloylmethane) inhibits receptor activator of NF-κB ligand-induced NF-κB activation in osteoclast precursors and suppresses osteoclastogenesis. J. Immunol. 172 (10), 5940–5947.

Bhukhai, K., Suksen, K., et al., 2012. A phytoestrogen diarylheptanoid mediates estrogen receptor/Akt/glycogen synthase kinase 3β protein-dependent activation of the Wnt/β-catenin signaling pathway. J. Biol. Chem. 287 (43), 36168–36178.

Bird, T.A., Levene, C.I., 1982. Lysyl oxidase: evidence that pyridoxal phosphate is a cofactor. Biochem. Biophys. Res. Commun. 108 (3), 1172–1180.

Bolland, M.J., Avenell, A., et al., 2010. Effect of calcium supplements on risk of myocardial infarction and cardiovascular events: meta-analysis. BMJ. 341, c3691.

Booth, S.L., Tucker, K.L., et al., 2000. Dietary vitamin K intakes are associated with hip fracture but not with bone mineral density in elderly men and women. Am. J. Clin. Nutr. 71 (5), 1201–1208.

Brennan-Speranza, T., Conigrave, A., 2015. Osteocalcin: an osteoblast-derived polypeptide hormone that modulates whole body energy metabolism. Calcif. Tissue Int. 96 (1), 1–10.

Carnevale, V., Romagnoli, E., et al., 2014. Bone damage in type 2 diabetes mellitus. Nutr. Metab. Cardiovas. Dis. 24 (11), 1151–1157.

Castiglioni, S., Cazzaniga, A., et al., 2013. Magnesium and osteoporosis: current state of knowledge and future research directions. Nutrients. 5 (8), 3022.

Cohen, A., Dempster, D.W., et al., 2013. Abdominal fat is associated with lower bone formation and inferior bone quality in healthy premenopausal women: a transiliac bone biopsy study. J. Clin. Endocrinol. Metab. 98 (6), 2562–2572.

Colucci, S., Mori, G., et al., 2005. L-Carnitine and isovaleryl L-carnitine fumarate positively affect human osteoblast proliferation and differentiation in vitro. Calcif. Tissue Int. 76 (6), 458–465.

Compston, J.E., Watts, N.B., et al., 2011. Obesity is not protective against fracture in postmenopausal women: glow. Am. J. Med. 124 (11), 1043–1050.

Conigrave, A.D., Brown, E.M., et al., 2008. Dietary protein and bone health: roles of amino acid–sensing receptors in the control of calcium metabolism and bone homeostasis. Ann. Rev. Nutr. 28 (1), 131–155.

Consensus, 1993. Consensus development conference: diagnosis, prophylaxis, and treatment of osteoporosis. Am. J. Med. 94 (6), 646–650.

Cornish, J., MacGibbon, A., et al., 2008. Modulation of osteoclastogenesis by fatty acids. Endocrinology. 149 (11), 5688–5695.

Costell, M., O'Connor, J., et al., 1989. Age-dependent decrease of carnitine content in muscle of mice and humans. Biochem. Biophys. Res. Commun. 161, 1135–1143.

Dallas, S.L., Prideaux, M., et al., 2013. The osteocyte: an endocrine cell … and more. Endocr. Rev. 34 (5), 658–690.

de Baaij, J.H.F., Hoenderop, J.G.J., et al., 2015. Magnesium in man: implications for health and disease. Physiol. Rev. 95 (1), 1–46.

Delgado-Calle, J., Bellido, T., 2015. Osteocytes and skeletal pathophysiology. Curr. Mol. Biol. Rep. 1 (4), 157–167.

Drake, M.T., Clarke, B.L., et al., 2008. Bisphosphonates: mechanism of action and role in clinical practice. Mayo Clin. Proc. 83 (9), 1032–1045.

Eaton-Evans, J., McLlrath, E.M., et al., 1996. Copper supplementation and the maintenance of bone mineral density in middle-aged women. J. Trace Elements Exp. Med. 9 (3), 87–94.

Ebeling, P.R., 2008. Osteoporosis in men. N. Engl. J. Med. 358 (14), 1474–1482.

Elefteriou, F., Ahn, J.D., et al., 2005. Leptin regulation of bone resorption by the sympathetic nervous system and CART. Nature. 434 (7032), 514–520.

Farina, E.K., Kiel, D.P., et al., 2011. Protective effects of fish intake and interactive effects of long-chain polyunsaturated fatty acid intakes on hip bone mineral density in older adults: the Framingham Osteoporosis Study. Am. J. Clin. Nutr. 93 (5), 1142–1151.

Ferron, M., Wei, J., et al., 2010. Insulin signaling in osteoblasts integrates bone remodeling and energy metabolism. Cell. 142 (2), 296–308.

Flanagan, J., Simmons, P., et al., 2010. Role of carnitine in disease. Nutr. Metab. (Lond.) 7 (1), 30.

Florence, R., Allen, S., et al., 2013. Diagnosis and Treatment of Osteoporosis. I. f. C. S. I. (ICSI). Agency for Healthcare Research and Quality (AHRQ): Institute of Medicine (IOM) National Healthcare Quality Report Categories, Rockville, MD.

Florencio-Silva, R., Sasso, G., et al., 2015. Biology of bone tissue: structure, function, and factors that influence bone cells. BioMed Res. Int. 2015, 421746.

Food and Nutrition Board, 1997. Dietary Reference Intakes: Calcium, Phosphorus, Magnesium, Vitamin D and Fluoride. National Academy Press, Washington, DC.

Ford, E.S., Mokdad, A.H., 2003. Dietary magnesium intake in a national sample of U.S. adults. J. Nutr. 133 (9), 2879–2882.

Friedman, J.M., Halaas, J.L., 1998. Leptin and the regulation of body weight in mammals. Nature. 395 (6704), 763–770.

Fujii, K., Kajiwara, T., et al., 1979. Effect of vitamin B6 deficiency on the crosslink formation of collagen. FEBS Lett. 97 (1), 193–195.

Fukunaga, T., Zou, W., et al., 2014. Vinculin regulates osteoclast function. J. Biol. Chem. 289 (19), 13554–13564.

Gabbay, K.H., Bohren, K.M., et al., 2010. Ascorbate synthesis pathway: dual role of ascorbate in bone homeostasis. J. Biol. Chem. 285 (25), 19510–19520.

Gacheru, S.N., Trackman, P.C., et al., 1990. Structural and catalytic properties of copper in lysyl oxidase. J. Biol. Chem. 265 (31), 19022–19027.

García-Martínez, O., De Luna-Bertos, E., et al., 2015. Repercussions of NSAIDs drugs on bone tissue: the osteoblast. Life Sci. 123, 72–77.

Gelse, K., Pöschl, E., et al., 2003. Collagens—structure, function, and biosynthesis. Adv. Drug Del. Rev. 55 (12), 1531–1546.

Goicoechea, E.N., Van Twillert, K., et al., 2008. Use of an in vitro digestion model to study the bioaccessibility of 4-hydroxy-2-nonenal and related aldehydes present in oxidized oils rich in omega-6 acyl groups. J. Agric. Food. Chem. 56 (18), 8475–8483.

Granke, M., Does, M.D., et al., 2015. The Role of Water Compartments in the Material Properties of Cortical Bone. Calcif. tissue int. 97 (3), 292–307.

Gu, J., Tong, X.-S., et al., 2015. Effects of $1\alpha,25$-(OH)2D3 on the formation and activity of osteoclasts in RAW264.7 cells. J. Steroid Biochem. Mol. Biol. 152, 25–33.

Hardouin, P., Pansini, V., et al., 2014. Bone marrow fat. Joint Bone Spine. 81 (4), 313–319.

Hayhoe, R.P., Lentjes, M.A., et al., 2015. Dietary magnesium and potassium intakes and circulating magnesium are associated with heel bone ultrasound attenuation and osteoporotic fracture risk in the EPIC-Norfolk cohort study. Am. J. Clin. Nutr. 102 (2), 376–384.

Heaney, R.P., Layman, D.K., 2008. Amount and type of protein influences bone health. Am. J. Clin. Nutr. 87 (5), 1567S–1570S.

Heaney, R.P., Recker, R.R., et al., 2010. Phosphate and carbonate salts of calcium support robust bone building in osteoporosis. Am. J. Clin. Nutr. 92 (1), 101–105.

Hyun, T.H., Barrett-Connor, E., et al., 2004. Zinc intakes and plasma concentrations in men with osteoporosis: the Rancho Bernardo Study. Am. J. Clin. Nutr. 80 (3), 715–721.

Itoh, R., Nishiyama, N., et al., 1998. Dietary protein intake and urinary excretion of calcium: a cross-sectional study in a healthy Japanese population. Am. J. Clin. Nutr. 67 (3), 438–444.

Justesen, J., Stenderup, K., et al., 2001. Adipocyte tissue volume in bone marrow is increased with aging and in patients with osteoporosis. Biogerontology. 2 (3), 165–171.

Kagan, H.M., Li, W., 2003. Lysyl oxidase: properties, specificity, and biological roles inside and outside of the cell. J. Cell Biochem. 88 (4), 660–672.

Kanda, Y., Hinata, T., et al., 2011. Reactive oxygen species mediate adipocyte differentiation in mesenchymal stem cells. Life Sci. 89 (7–8), 250–258.

Kanis, J.A., McCloskey, E.V., et al., 2008. A reference standard for the description of osteoporosis. Bone. 42 (3), 467–475.

Karampinos, D.C., Ruschke, S., et al., 2015. Association of MRS-based vertebral bone marrow fat fraction with bone strength in a human in vitro model. J. Osteoporos. 2015, 152349.

Kareeann Sok Fun, K., Su Yin, L., et al., 2014. Diuretic-associated electrolyte disorders in the elderly: risk factors, impact, management and prevention. Curr. Drug Saf. 9 (1), 2–15.

Karsenty, G., 2008. Transcriptional control of skeletogenesis. Ann. Rev. Genomics Hum. Genet. 9 (1), 183–196.

Kassem, M., Abdallah, B.M., et al., 2008. Osteoblastic cells: differentiation and trans-differentiation. Arch. Biochem. Biophys. 473 (2), 183–187.

Kearns, A.E., Khosla, S., et al., 2008. Receptor activator of nuclear factor {kappa}B ligand and osteoprotegerin regulation of bone remodeling in health and disease. Endocr. Rev. 29 (2), 155–192.

Kennedy, M.L., Failla, M.L., et al., 1986. Influence of genetic obesity on tissue concentrations of zinc, copper, manganese and iron in mice. J. Nutr. 116 (8), 1432–1441.

Kennel, K.A., Drake, M.T., 2009. Adverse effects of bisphosphonates: implications for osteoporosis management. Mayo Clin. Proc. 84 (7), 632–638.

King, D.E., Mainous, A.G., et al., 2005. Dietary magnesium and C-reactive protein levels. J. Am. Coll. Nutr. 24 (3), 166–171.

Kousteni, S., Bellido, T., et al., 2001. Nongenotropic, sex-nonspecific signaling through the estrogen or androgen receptors. Cell. 104 (5), 719−730.

Lausen, B., 1999. No evidence for dietary protein and dietary salt as main factors of calcium excretion in healthy children and adolescents. Am. J. Clin. Nutr. 69 (4), 742−743.

Lecka-Czernik, B., Stechschulte, L.A., 2014. Bone and fat: a relationship of different shades. Arch. Biochem. Biophys. 561, 124−129.

Lee, J.-H., Jin, H., et al., 2010. Epigallocatechin-3-gallate inhibits osteoclastogenesis by down-regulating c-fos expression and suppressing the nuclear factor-κB signal. Mol. Pharmacol. 77 (1), 17−25.

Lee, N.K., Sowa, H., et al., 2007. Endocrine regulation of energy metabolism by the skeleton. Cell. 130 (3), 456−469.

Lei, W., Chuan, W., 2015. Efficacy of l-carnitine in the treatment of osteoporosis in men. Int. J. Pharmacol. 11 (2), 148−151.

London, E., Lala, G., et al., 2007. Sucrose access differentially modifies 11ß-hydroxysteroid dehydrogenase-1 and hexose-6-phosphate dehydrogenase message in liver and adipose tissue in rats. J. Nutr. 137 (12), 2616−2621.

Manolagas, S.C., 2000. Birth and death of bone cells: basic regulatory mechanisms and implications for the pathogenesis and treatment of osteoporosis. Endocr. Rev. 21 (2), 115−137.

Manolagas, S.C., 2010. From estrogen-centric to aging and oxidative stress: a revised perspective of the pathogenesis of osteoporosis. Endocr. Rev. 31 (3), 266−300.

Manolagas, S.C., Parfitt, A.M., 2010. What old means to bone. Trends Endocrinol. Metab. 21 (6), 369−374.

Massé, P.G., Rimnac, C.M., et al., 1996. Pyridoxine deficiency affects biomechanical properties of chick tibial bone. Bone. 18 (6), 567−574.

McCalden, R.W., McGeough, J.A., et al., 1993. Age-related changes in the tensile properties of cortical bone. The relative importance of changes in porosity, mineralization, and microstructure. J. Bone Joint Surg. 75 (8), 1193−1205.

Milat, F., Ng, K.W., 2009. Is Wnt signalling the final common pathway leading to bone formation? Mol. Cell. Endocrinol. 310 (1−2), 52−62.

Modica, S., Wolfrum, C., 2013. Bone morphogenic proteins signaling in adipogenesis and energy homeostasis. BBA Mol. Cell Biol. Lipid. 1831 (5), 915−923.

New, S.A., Robins, S.P., et al., 2000. Dietary influences on bone mass and bone metabolism: further evidence of a positive link between fruit and vegetable consumption and bone health? Am. J. Clin. Nutr. 71 (1), 142−151.

Nguyen, N.D., Frost, S.A., et al., 2008. Development of prognostic nomograms for individualizing 5-year and 10-year fracture risks. Osteoporos. Int. 19 (10), 1431−1444.

NIH Consensus Development Panel on Osteoporosis Prevention, Diagnosis, and Therapy, 2001. Osteoporosis prevention, diagnosis, and therapy. JAMA. 285 (6), 785−795.

Oh, S.-R., Sul, O.-J., et al., 2010. Saturated fatty acids enhance osteoclast survival. J. Lipid. Res. 51 (5), 892−899.

Oxlund, H., Barckman, M., et al., 1995. Reduced concentrations of collagen cross-links are associated with reduced strength of bone. Bone. 17 (Suppl. 4), S365−S371.

Parajuli, A., Liu, C., et al., 2015. Bone's responses to mechanical loading are impaired in type 1 diabetes. Bone. 81, 152−160.

Patano, N., Mancini, L., et al., 2008. L: -carnitine fumarate and isovaleryl-L: -carnitine fumarate accelerate the recovery of bone volume/total volume ratio after experimetally induced osteoporosis in pregnant mice. Calcif. Tissue Int. 82, 221−228.

Peterkofsky, B., 1991. Ascorbate requirement for hydroxylation and secretion of procollagen: relationship to inhibition of collagen synthesis in scurvy. Am. J. Clin. Nutr. 54 (6), 1135S−1140S.

Qiu, S., Rao, D.S., et al., 2002. Age and distance from the surface but not menopause reduce osteocyte density in human cancellous bone. Bone. 31 (2), 313−318.

Rahman, M., Bhattacharya, A., et al., 2008. Docosahexaenoic acid is more potent inhibitor of osteoclast differentiation in RAW 264.7 cells than eicosapentaenoic acid. J. Cell. Physiol. 214 (1), 201−209.

Reid, I.R., 2010. Fat and bone. Arch. Biochem. Biophys. 503 (1), 20−27.

Rosen, C.J., Bouxsein, M.L., 2006. Mechanisms of disease: is osteoporosis the obesity of bone? Nat. Clin. Pract. Rheumatol. 2 (1), 35−43.

Rubinacci, A., 2009. Expanding the functional spectrum of vitamin K in bone. Focus on: "Vitamin K promotes mineralization, osteoblast to osteocyte transition, and an anti-catabolic phenotype by {gamma}-carboxylation-dependent and -independent mechanisms". Am. J. Physiol. Cell. Physiol. 297 (6), C1336−C1338.

Sacks, F.M., Lichtenstein, A., et al., 2006. Soy protein, isoflavones, and cardiovascular health: an American heart association science advisory for professionals from the nutrition committee. Circulation. 113 (7), 1034−1044.

Sahni, S., Mangano, K.M., et al., 2015. Dietary approaches for bone health: lessons from the Framingham Osteoporosis Study. Curr. Osteoporos. Rep. 13 (4), 245−255.

Seibel, M.J., 2005. Biochemical markers of bone turnover part I: biochemistry and variability. Clin. Biochem. Rev. 26 (4), 97−122.

Shapses, S.A., 2009. Is bone loss after gastric bypass surgery associated with the extent of weight loss? Nat. Clin. Pract. Endocrinol. Metab. 5 (2), 80−81.

Shapses, S.A., Sukumar, D., 2012. Bone metabolism in obesity and weight loss. Ann. Rev. Nutr. 32 (1), 287−309.

Shea, B., Wells, G., et al., 2002. VII. meta-analysis of calcium supplementation for the prevention of postmenopausal osteoporosis. Endocr. Rev. 23 (4), 552−559.

Shearer, M.J., Newman, P., 2008. Metabolism and cell biology of vitamin K. Thromb. Haemost. 100, 530−547.

Singh, L., Brennan, T.A., et al., 2016. Aging alters bone-fat reciprocity by shifting in vivo mesenchymal precursor cell fate towards an adipogenic lineage. Bone. 85, 29−36.

Siris, E.S., Miller, P.D., et al., 2001. Identification and fracture outcomes of undiagnosed low bone mineral density in postmenopausal women: results from the national osteoporosis risk assessment. JAMA. 286 (22), 2815−2822.

Stein, E.M., Silverberg, S.J., 2014. Bone loss after bariatric surgery: causes, consequences and management. Lancet Diabetes Endocrinol. 2 (2), 165−174.

Suda, T., Ueno, Y., et al., 2003. Vitamin D and bone. J. Cell. Biochem. 88 (2), 259−266.

Sukumar, D., Schlussel, Y., et al., 2011. Obesity alters cortical and trabecular bone density and geometry in women. Osteoporos. Int. 22 (2), 635−645.

Tatsumi, S., Ishii, K., et al., 2007. Targeted ablation of osteocytes induces osteoporosis with defective mechanotransduction. Cell. Metab. 5 (6), 464−475.

Timmins, P.A., Wall, J.C., 1977. Bone water. Calcif. Tissue. Res. 23 (1), 1−5.

Toh, J.W.T., Ong, E., et al., 2015. Hypomagnesaemia associated with long-term use of proton pump inhibitors. Gastroenterol. Rep. 3 (3), 243−253.

Tranquilli, A.L., Lucino, E., et al., 1994. Calcium, phosphorus and magnesium intakes correlate with bone mineral content in postmenopausal women. Gynecol. Endocrinol. 8 (1), 55−58.

van de Peppel, J., Van Leeuwen, J., 2014. Vitamin D and gene networks in human osteoblasts. Front. Physiol. 5, 137.

van Driel, M., Koedam, M., et al., 2006. Evidence for auto/paracrine actions of vitamin D in bone: 1α-hydroxylase expression and activity in human bone cells. FASEB J. 20 (13), 2417–2419.

Vestergaard, P., 2007. Discrepancies in bone mineral density and fracture risk in patients with type 1 and type 2 diabetes—a meta-analysis. Osteoporos. Int. 18 (4), 427–444.

Viereck, V., Gründker, C., et al., 2002. Phytoestrogen genistein stimulates the production of osteoprotegerin by human trabecular osteoblasts. J. Cell. Biochem. 84 (4), 725–735.

Wada, T., Nakashima, T., et al., 2006. RANKL–RANK signaling in osteoclastogenesis and bone disease. Trends Mol. Med. 12 (1), 17–25.

Wang, W., Zhang, Y., et al., 2015. Mitochondrial reactive oxygen species regulate adipocyte differentiation of mesenchymal stem cells in hematopoietic stress induced by arabinosylcytosine. PLoS One. 10 (3), e0120629.

Warensjö, E., Byberg, L., et al., 2011. Dietary calcium intake and risk of fracture and osteoporosis: prospective longitudinal cohort study. BMJ. 342, d1473.

Wei, J., Ferron, M., et al., 2014. Bone-specific insulin resistance disrupts whole-body glucose homeostasis via decreased osteocalcin activation. J. Clin. Invest. 124 (4), 1781–1793.

Wei, J., Karsenty, G., 2015. An overview of the metabolic functions of osteocalcin. Rev. Endocr. Metab. Disord. 16 (2), 93–98.

Woeckel, V.J., van der Eerden, B.C.J., et al., 2013. 1α, 25-dihydroxyvitamin D3 stimulates activin A production to fine-tune osteoblast-induced mineralization. J. Cell. Physiol. 228 (11), 2167–2174.

Xian, L., Wu, X., et al., 2012. Matrix IGF-1 maintains bone mass by activation of mTOR in mesenchymal stem cells. Nat. Med. 18 (7), 1095–1101.

Xu, F., Teitelbaum, S.L., 2013. Osteoclasts: new insights. Bone Res. 1 (1), 11–26.

Yaturu, S., 2009. Diabetes and skeletal health. J. Diabetes. 1 (4), 246–254.

Yu, E.W., 2014. Bone metabolism after bariatric surgery. J. Bone Mineral Res. 29 (7), 1507–1518.

Zhou, H., Shang, L., et al., 2009. Resveratrol augments the canonical Wnt signaling pathway in promoting osteoblastic differentiation of multipotent mesenchymal cells. Exp. Cell. Res. 315 (17), 2953–2962.

Further Reading

Ebeling, P.R., 2008. Osteoporosis in men. NEJM. 358, 1474–1482.

Kumar, et al., 2015. Robbins and Cotran Pathologic Basis of Disease, ninth ed. Saunders Elsevier, Philadelphia, PA, pp. 1180–1194. Chapter 26. Bones, Joints and Soft Tissue Tumors.

Essentials I: Life in an Aerobic World

All substances are poisons; there is none that is not a poison. The right dose differentiates a poison from a remedy.
Paracelsus (1493–1541)

On or about 3.5 billion years ago, bacteria living on our hot and barren planet developed the ability to use energy from the sun to create food. Waste products from these prechloroplast organisms included the toxic gas, oxygen. As the oxygen content in the atmosphere increased, aerobic bacteria evolved with the ability to utilize oxygen to break down food and to pack the energy generated into high-energy phosphate bonds, creating abundant ATP. In a process termed endosymbiosis, other amoeba-like, anaerobic bacteria engulfed the aerobic bacteria and permanently lodged them into their cytoplasm as mitochondrial power sources.

The integration of symbiogenesis into evolutionary theory has been ascribed to a Russian scientist, Boris Mikhaylovich Kozo-Polyansky (1890–1957). This concept was been further explored and characterized (Sagan, 1967; Margulis, 2010) and is now well accepted. In addition to mitochondria, endosymbiotic theory suggests that chloroplasts, flagella, and other organelles also resulted from cooperative ventures between organisms (Fig. EI.1).

Oxygen gas (O_2) can exert toxic effects on biological organisms, especially when it is partially reduced to highly reactive forms such as superoxide (O_2^-), hydrogen peroxide (H_2O_2), and the hydroxyl radical ($\bullet OH$). The concept that these reactive oxygen species (ROS) and their metabolites could damage cellular components and play a major role in the cumulative process of aging and chronic disease was proposed over 60 years ago (Harman, 1956) and sparked a search for both endogenous and exogenous "antioxidants" that could reduce or quench ROS and minimize tissue damage.

Since then, several types of antioxidants have been characterized. Isoprenoids (terpenoids) such as beta carotene assimilate the unpaired electron in ROS into their conjugated double-bond structure and thus "quench" the reactive species (refer to Essentials II: Heavy Metals, Retinoids, and Precursors: Carotenoids). Reducing agents such as ascorbate and glutathione donate a proton to the ROS radical, reducing it and, in the process, becoming oxidized. Additionally, since divalent cations such as iron and copper can participate in the Fenton reaction (Fenton, 1894), biological organisms have developed systems that bind and store these metals safely to avoid forming ROS. In the Fenton reaction, divalent cations, especially iron and also copper react with hydrogen peroxide to form the dangerous hydroxyl radical. (Refer to Essentials II: Heavy Metals, Retinoids, and Precursors: Iron.) Because they are readily available as dietary supplements, vitamins E, C, beta carotene, and other dietary "antioxidants" have been used as therapeutic interventions to slow oxidative damage observed with obesity, aging, and chronic disease (Balaban et al., 2005). In contrast to expectations, large, multi-institutional clinical trials have repeatedly demonstrated that supplementation with antioxidant nutrients is not effective in reducing disease risk, do not effectively reduce premature mortality, and may produce deleterious consequences (Albanes et al., 1996; Leo and Lieber, 1999; Bjelakovic et al., 2013).

These negative clinical observations support the concept of hormesis, the adaptive response to stress (Calabrese and Baldwin, 2002), and more specifically, mitohormesis, a term that describes mitochondrial production of ROS in response to stress as reviewed by Tapia (2006), Ristow et al. (2009), Ristow (2014) and Yun and Finkel (2014). This hypothesis holds that sublethal mitochondrial stress and superoxide production rapidly induce adaptive mitochondrial antioxidant production. H_2O_2 is continuously generated in the cell and maintained at low concentration by a variety of peroxidases, including glutathione peroxidase. At these low levels, H_2O_2 activates signaling pathways by modifying redox-dependent posttranslational proteins, including protein phosphatases and transcription factors as comprehensively discussed by Olsen et al. (2015). For example, modification of cysteine SH groups on kelch-like ECH-associated protein 1 (KEAP1) allows nuclear factor (erythroid-derived 2)-related factor 2 (Nrf2) to enter the nucleus and upregulate production of cytoprotective enzymes as described below (Dodson et al., 2015). H_2O_2 may also shift the glutathione ratio in favor of oxidized glutathione and promote S-glutathionylation. It has been proposed that reversible oxidation of

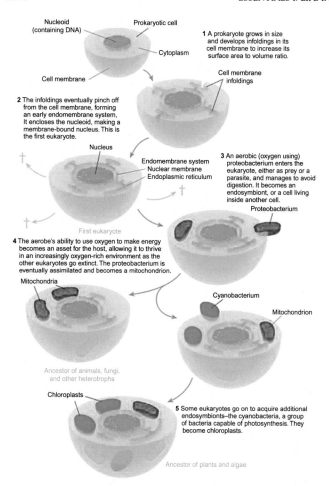

1 A prokaryote grows in size and develops infoldings in its cell membrane to increase its surface area to volume ratio.

2 The infoldings eventually pinch off from the cell membrane, forming an early endomembrane system, It encloses the nucleoid, making a membrane-bound nucleus. This is the first eukaryote.

3 An aerobic (oxygen using) proteobacterium enters the eukaryote, either as prey or a parasite, and manages to avoid digestion. It becomes an endosymbiont, or a cell living inside another cell.

4 The aerobe's ability to use oxygen to make energy becomes an asset for the host, allowing it to thrive in an increasingly oxygen-rich environment as the other eukaryotes go extinct. The proteobacterium is eventually assimilated and becomes a mitochondrion.

5 Some eukaryotes go on to acquire additional endosymbionts–the cyanobacteria, a group of bacteria capable of photosynthesis. They become chloroplasts.

FIGURE EI.1 **Endosymbiotic theory.** One model suggests that the initial prokaryotic cell developed a membrane system that encased the nucleus and became the prototype for organelle formation. When the aerobic microbe entered this cell, either as prey or parasite, it was trapped in the membrane system and retained because its ability to use oxygen to generate energy was an asset to the host. For information on other models, see https://en.wikipedia.org/wiki/Symbiogenesis#cite_note-Kimball-30. *Source: Figure by Kelvinsong (own work) (CC BY-SA 3.0 (http://creativecommons.org/licenses/by-sa/3.0)), via Wikimedia Commons.*

cysteine SH groups acts as an "on–off" switch, changing protein structure, localization, and half-life (Jones, 2008; Olsen et al., 2015). This hypothesis also predicts that excess H_2O_2 production by peroxisomal proliferators, ER stress, and a variety of cellular oxidases, as well as other species of nonfree-radical oxidants including lipid hydroperoxides and epoxides, aldehydes and quinones can disrupt thiol redox circuits that normally function in physiological regulation and lead to disease (Fig. EI.2).

Abundant evidence is accumulating that exogenous antioxidant supplements reduce the availability of oxygen radicals such as superoxide, normally dismutated to H_2O_2, and thereby inhibit H_2O_2 signaling to cytoplasmic and nuclear targets. For example,

supplementation with vitamins E and C prevented the beneficial effects of exercise on insulin sensitivity in both untrained and trained human subjects (Ristow et al., 2009). While chronic oxidative stress that damages cell structures is known to play a major role in the pathology of most chronic disease states, effective therapeutic modalities need to target the specific dysregulation that causes the oxidative stress, rather than simply administering supraphysiologic does of "antioxidant" nutrients that may, in the pharmacological doses provided, have diverse and sometimes adverse effects on the organism.

In healthy organisms, a balance of oxidants and antioxidants is maintained by controlled production of oxidants and reducing networks that include single-unit antioxidants such as vitamin E, antioxidant enzymes such as selenium-requiring glutathione peroxidase, and transcription factor cascades such as Nrf2-Keap1 described below. A complex picture has emerged that supports the concept that the oxidation/reduction state may be evolutionarily utilized to regulate biological processes and play an active role in metabolic processes. The "Redox chain" model depicted in Fig. EI.3 is an attempt to place the cellular oxidants and antioxidants with their respective enzymes and transcription factors into a logical framework to characterize the dynamic relationships that determine the redox status of an organism (Wang and Hai, 2015).

It is clear that evolution of complex antioxidant networks was essential for biological organisms to adapt to an oxygen-rich environment. In hindsight, the concept that supplemental dietary antioxidants could ameliorate the oxidative damage associated with obesity and its comorbid, chronic diseases states was somewhat naïve. Using newer technology, it should be possible to better characterize the molecular interactions and mechanisms through which an antioxidant species functions and determine the risk/benefit potential for each (Vrolijk et al., 2015). This knowledge should allow more effective targeted treatment for the individual patient based on genetic profile, disease state, dietary patterns, and environmental exposures. The nutrients known to play critical roles in maintenance of normal redox status and dysregulation in obesity and its comorbidities are discussed below.

EI.1 NUTRIENTS, MITOCHONDRIAL ANTIOXIDANTS, AND ANTIOXIDANT PROTECTION

Under normal conditions, mitochondria produce oxidants in the course of ATP biosynthesis. At the same time, an efficient antioxidant system assures that

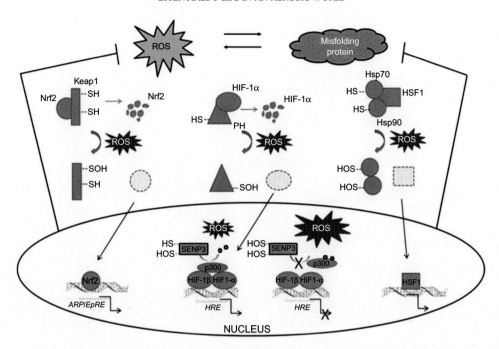

FIGURE EI.2 **Systems of cell adaptation.** In addition to the Nrf2 KEAP1 system described above, the cell maintains defense mechanisms such as heat shock proteins (Hsp70, 90) that bind misfolded and damaged proteins and release a heat shock transcription factor (HSF) that amplifies the response. Hypoxia-inducible factor 1α (HIF1α) initiates a switch to glycolysis in response to hypoxia. HIF1α may also stimulate a Warburg-like glycolysis without hypoxia, thereby reducing oxidant production from a damaged electron transport chain (ETC) (Olsen et al., 2013, 2015). *Source: Reproduced from Olsen, R.K.J., Cornelius, N., et al., 2013. Genetic and cellular modifiers of oxidative stress: what can we learn from fatty acid oxidation defects? Mol. Genet. Metab. 110 (Supplement), S31−S39 with permission from Science Direct.*

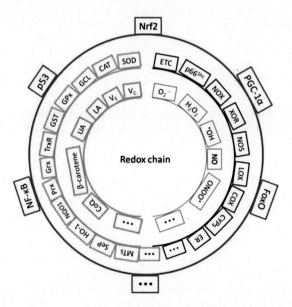

FIGURE EI.3 **A model of the complex redox chain.** In the first tier, oxidant species (black line) face nonenzymatic antioxidants (green line); the second tier includes the array of antioxidant enzymes (blue line) versus ROS-producing enzymes (black line). Finally, the superdirectors (red line) transcriptionally regulate both ROS-generating and antioxidant enzymes. These enzyme cascades can be activated in temporal and spatial sequence in response to physiological and/or pathological insult and modify gene expression (Wang and Hai, 2015). *Source: Reproduced from Wang, X., Hai, C., 2015. Redox modulation of adipocyte differentiation: hypothesis of "Redox Chain" and novel insights into intervention of adipogenesis and obesity. Free Radic. Biol. Med. 89, 99−125 with permission from Science Direct.*

few, if any, oxidants escape to damage surrounding organelles and membranes as described in Fig. EI.4 (refer to Chapter 5: Diabetic Cardiomyopathy and Congestive Heart Failure: Cardiomyocyte Excitation−Contraction). Unfortunately, many of these studies were conducted in vitro using isolated mitochondria and results may be different in vivo (Andreyev et al., 2015). At least two insults to the electron transport chain (ETC) can lead to increased production of superoxide radicals (O_2^-). Nutrient overload can create high NADH production that over-reduces the ETC. For example, in the obese and/or sedentary individual, too much fuel is provided and too little ATP is used. In consequence, the excessive NADH produced swamps the ETC, allowing unpaired electrons to "escape" and directly reduce molecular O_2, forming $O_2^{\bullet-}$. The second condition occurs when ATP cannot be produced. For example, the final ETC complex, cytochrome oxidase, contains copper and iron that tightly bind O_2 until it is safely reduced to H_2O. In copper deficiency, cytochrome oxidase is dysfunctional (Sena and Chandel, 2012; Ristow, 2014) and partially reduced oxygen radicals can escape (refer to Chapter 5: Diabetic Cardiomyopathy and Congestive Heart Failure: Targets for Myocardial Protection: Cardiac Remodeling).

H_2O_2 produced by the mitochondria can reversibly oxidize critical protein thiol groups, as described below. These modifications alter the activity of kinases, phosphatases, and especially transcription factors. H_2O_2 is paradoxically produced in the mitochondria

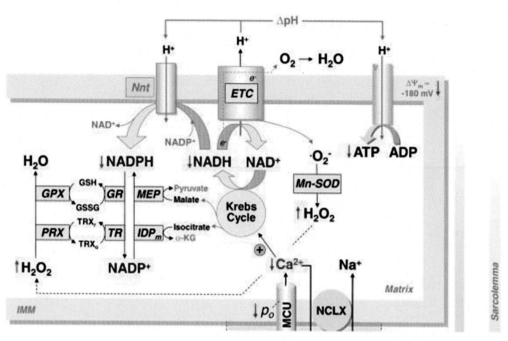

FIGURE EI.4 Mitochondrial redox balance. Several overlapping antioxidant and radical quenching systems minimize ROS and maintain redox balance. Any O_2^- produced is efficiently dismutated to H_2O_2 by mitochondrial Mn^{2+}-superoxide dismutase (Mn-SOD), and by zinc–copper SOD (Zn-CU SOD) in the intermembranous space. Several systems reduce H_2O_2 to water; mitochondrial glutathione peroxidases that requires selenium and peroxisomal catalase (requiring iron) are prominent. To maintain glutathione in its reduced state (GSH), some of the high-energy electrons carried on NADH are transferred to $NADP^+$ by nicotinamide nucleotide transhydrogenase (NNT) (Bay et al., 2013). Peroxiredoxins (PRX) are a ubiquitous family of antioxidant enzymes containing a reactive cysteine that also reduce H_2O_2 to water. PRX enzymes control cytokine-induced peroxide levels and mediate signal transduction in mammalian cells (Rhee et al., 2005). Thioredoxin (TRX) is also reduced by NADPH (Nickel et al., 2014). It should be noted that the riboflavin cofactor, flavin adenine dinucleotide (FAD), is required by glutathione reductase to transport H^+ from NADPH to oxidized glutathione (GSSG), thereby reducing it to its active, sulfhydryl form, GSH (Ashoori and Saedisomeolia, 2014). *Source: Reproduced from Bay, J., Kohlhaas, M., et al., 2013. Intracellular Na^+ and cardiac metabolism. J. Mol. Cell. Cardiol. 61, 20–27 with permission from Science Direct.*

under low-oxygen conditions. Hypoxia inducible factor 1 alpha (HIF1α) is normally held in the cytoplasm and degraded by the action of prolyl hydroxylase (PDH) and O_2 in a reaction requiring ascorbate (see Ascorbate section below). H_2O_2 inactivates PDH, allowing HIF1α to dimerize with HIF-1β and translocate to the nucleus where it induces transcription of hypoxia-sensitive genes (Murphy, 2009).

The mitochondrial defense system inactivates ROS and its metabolites (eg, peroxyradicals, lipid peroxides) that escape the ETC or are produced by membrane-bound enzymes including the NADPH oxidase family (NOX). In mammals, $O_2^{•-}$ is dismutated by three distinct superoxide dismutase (SOD) isoforms (McCord and Fridovich, 1988; Zelko et al., 2002). SOD2 (MnSOD) contains manganese (Mn) and is found in prokaryotic cells and in the mitochondrial matrix of eukaryotic cells (St. Clair et al., 1994). SOD1 and SOD3 have Cu and Zn atoms at their catalytic centers. Cu/Zn-SOD (SOD1) is located in the mitochondrial intermembranous space, cytoplasm, nuclear compartments, and lysosomes of mammalian cells. The most recently discovered isoform, SOD3, is secreted and circulates in the plasma (Culotta

et al., 2006); it can remain tethered to the extracellular matrix and has also been detected in lymph, ascites, and cerebrospinal fluids. SOD enzymes are heavily dependent on the availability of their respective metals. SOD3 and SOD1 activities are both compromised by depletion of both copper and zinc (Itoh et al., 2009).

Hydrogen peroxide (H_2O_2) is decomposed to water by a number of mitochondrial enzymes including selenium-dependent glutathione peroxidase (GPX) and peroxiredoxins (PRXs) using a redox-active cysteine in the active site. GPX isoforms are oxidized by H_2O_2 and reduced by glutathione (GSH). PRXs function by undergoing oxidation by H_2O_2 at an active site cysteine with subsequent reduction by thioredoxin, thioredoxin reductase, and NADPH. Catalase is found in peroxisomes. These antioxidant enzymes are characterized by the rate in which they react with H_2O_2 (rate constant) and the concentration of H_2O_2 and enzyme in vivo. PRXs, with a high rate constant and high abundance, are thought to scavenge nanomolar levels of H_2O_2 associated with signaling. GPXs have similar rate constants, but are less abundant and therefore are likely only important at higher intracellular concentrations of H_2O_2 when GPXs can begin to compete with PRXs for substrate (Winterbourn and Hampton, 2008). These observations suggest that PRXs turn off signaling when $[H_2O_2]$ is low, while GPXs buffer high levels of H_2O_2 to bring them to a safe level at which cell signaling can occur. The affinity of catalase for H_2O_2 is much lower than either GPX or RPX and may be restricted to peroxisomes (Sena and Chandel, 2012).

EI.2 NUTRIENTS AND TRANSCRIPTIONAL CONTROL OF ANTIOXIDANT PROTECTION

EI.2.1 Nrf2-KEAP1-Dependent Transcription

Nuclear factor (erythroid-derived 2)-related factor 2 (Nrf2) is a transcription factor that translocates to the nucleus, heterodimerizes with other small transcription factors, binds to specific response elements such as electrophile-response element (ERE); antioxidant response element (ARE) and others to facilitate expression of ARE-regulated genes as recently reviewed by Hayes and Dinkova-Kostova (2014). These authors explore factors that control the transcription of the ~200 genes upregulated by Nrf2, including NADPH quinone oxidoreductase (NQO1), heme oxygenase-1 (HO-1), glutathione-S-transferase, glutamate cysteine ligase (GCL), SOD, catalase (CAT), as well as phase II xenobiotic detoxification enzymes. Many of the genes upregulated include those that control intermediary metabolism, thus activation of Nrf2 genes is involved in regulation of adipocyte differentiation and lipid storage as well as reduction of obesity-associated inflammation. It should be noted that Nrf2 also increases transcription of multidrug-resistance proteins (mrps), often upregulated in cancer cells (Kensler and Wakabayashi, 2010).

In the unstressed organism, Nrf2 is retained in the cytoplasm bound to Kelch-like ECH associated protein 1 (KEAP1) and the Cul3-based E2-ubiquitin ligase complex (Cul3). Together, these complexes label Nrf2 with ubiquitin and target it for proteasomal destruction. KEAP1 is maintained at physiological level by autophagic degradation (Taguchi et al., 2012). A variety of endogenous activators, including ROS, reactive nitrogen species, and lipid aldehydes, stimulate the release of Nrf2 from KEAP1, allowing it to enter the nucleus and upregulate gene transcription. KEAP1 contains at least 25 cysteines, some of which are modified by activators and undergo a conformational change that prevents Nrf2 ubiquitination, allowing its release from KEAP1 and translocation to the nucleus (Dodson et al., 2015) (Fig. EI.5).

Several bioactive components in the human diet activate the Nrf2 signaling pathway. Members of the vast flavonoid family in fruits and vegetables, including isoflavones in soy beans and phenylpropenoids in green tea, coffee, and chocolate (Boettler et al., 2011) are metabolized to electrophilic quinones that bind cysteine sensors and activate Nrf2 signaling (Stefanson and Bakovic, 2014). Curcumin is a diarylheptanoid found in the yellow spice, turmeric. It is also known to activate Nrf2 signaling (Tapia et al., 2012). Vitamin D bound to its receptor (VDR) upregulates Nrf2 gene expression and Nrf2 protein biosynthesis, thereby maintaining adequate protein to stabilize the signaling pathway (Nakai et al., 2014). Several authors cited evidence that the protective effects of dietary Nrf2 activators required chronic dosage (Aggarwal and Sung, 2009; Fattori et al., 2015). The authors postulated that chronic transient Nrf2 activation due to dietary activator consumption can upregulate cytoprotective gene transcription in the absence of actual oxidative or xenobiotic stress that would also activate the more intensive inflammatory NFκB system described below.

EI.2.2 Hydrogen Sulfide (H_2S) in Thiol Redox Signaling

…these fetid streams of subterranean slime which the pavement hides from you, do you know what all this is? It is the flowering meadow … it is perfumed hay, it is golden corn, it is bread on your table, it is warm blood in your veins, it is health, it is joy, it is life. *Victor Hugo, 1862, "Les Miserables" Second Book; from The Intestine of Leviathan. Boston: Little, Brown, 1887.*

Another activator of Nrf2 translocation is the toxic gas hydrogen sulfide (H_2S). While the toxic properties of H_2S gas emitted during volcanic eruptions and concentrated in Paris sewers have long been recognized, its multiple biological activities are only now being appreciated. H_2S meets the criterion of an essential "gasotransmetter," together with CO and NO (Mustafa et al., 2009) and acts as a physiologic vasorelaxant. The gas is produced in endothelial cells, vascular smooth muscle cells, and perivascular adipose tissue and dilates blood vessels by activating ATP-sensitive potassium channels. It also modifies cysteines in a large number of proteins by S-sulfhydration. Proteins modified by S-sulfhydration include actin, tubulin, and glyceraldehyde-3-phosphate dehydrogenase. Recently it has been shown that H_2S also creates a disulfide on selected cysteine residues on KEAP1 (Yang et al., 2012), releasing Nrf2 and providing cytoprotection as described above.

H_2S is produced endogenously in animals from cysteine, a nonessential amino acid formed de novo from methionine by reverse transsulfuration as illustrated in Fig. EI.6. Its substrate, methionine can be obtained

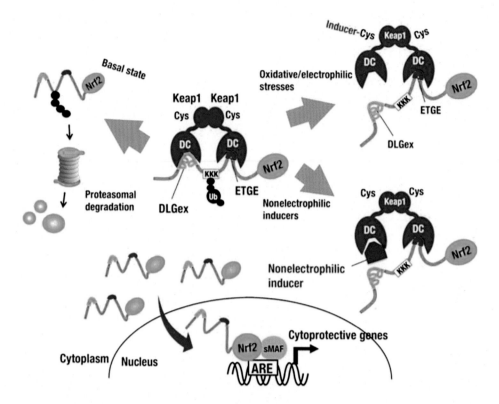

FIGURE EI.5 **A model for Nrf2 binding and activation.** KEAP1 is a homodimer that binds to a single Nrf2 molecule through two distinct binding sites (DLG and ETGE motifs). Binding of the high-affinity ETGE motif and low-affinity DLG motif provides a hinge and latch, which facilitate optimal positioning of the lysine (K) residues between the two motifs for ubiquitin conjugation. The KEAP1-Nrf2-ARE signaling pathway is activated by electrophilic small molecules that disrupt the association of Nrf2 with Keap1 in at least two ways. Activators can bind to specific reactive cysteine sensors, modify KEAP1 structure, and release Nrf2. Nonelectrophilic activators bind at alternative sites and also disrupt KEAP1 binding. Released Nrf2 is stabilized and translocates to the nucleus where it heterodimerizes and binds to the ARE in the upstream promoter regions of cytoprotective genes. (Suzuki and Yamamoto, 2015). *Source: Reproduced from Suzuki, T., Yamamoto, M., 2015. Molecular basis of the Keap1—Nrf2 system. Free Radic. Biol. Med. 88 (Part B), 93—100 with permission from Science Direct.*

from dietary protein, or by transmethylation, in which methyl is transferred from 5-methyltetrahydrofolate (5-meTHF) to homocysteine by methionine synthase (MS) in a process that requires cobalamin (vitamin B12). Homocysteine is converted to cystathionine by cystathionine beta synthase (CBS) and ultimately to cysteine and finally to H_2S by cystathionine-gamma-lyase (CSE); CSE is used for both reactions and requires pyridoxine (vitamin B6). Reverse transsulfuration is believed to be the sole route for cysteine biosynthesis in animals. In order for it to be utilized for glutathione and protein biosynthesis, the extracellular oxidized form (cysteine) must be transported into the cell where it is reduced to cysteine. A comprehensive review of cysteine metabolism is available (Yin et al., 2016). Cysteine is rate-limiting for glutathione biosynthesis via GCL and for incorporation into newly synthesized proteins. It is also important for cell signaling and has been characterized as a functionally essential amino acid.

In normal humans, homocysteine concentration is tightly regulated by MS and reverse transsulfuration

to maintain levels below a toxic threshold, but adequate to provide substrate for cysteine formation (refer to Chapter 6: Atherosclerosis and Arterial Calcification; Metabolic Targets for Diet Modification). Homocysteine in excess of this threshold enters the plasma where it can form toxic intermediates as comprehensively reviewed by Sibrian-Vazquez et al. (2010) and Wang (2012). Although a high plasma homocysteine concentration has been proposed as an independent risk factor for cardiovascular and other diseases (Clarke et al., 1991; McCully, 1993), recent evidence suggests that the pathology is more complex and that dysfunctional homocysteine homeostasis increases oxidative stress (Barroso et al., 2014), potentiates tissue copper depletion (Zuo et al., 2013) (refer to Chapter 5: Diabetic Cardiomyopathy and Congestive Heart Failure: Targets for Myocardial Protection: Cardiac Remodeling), and leads to subsequent disease. It is of interest that diabetic patients without evidence of diabetic nephropathy maintain a low serum homocysteine, possibly because of relative hyperfiltration. As the kidney fails, serum homocysteine concentrations rise

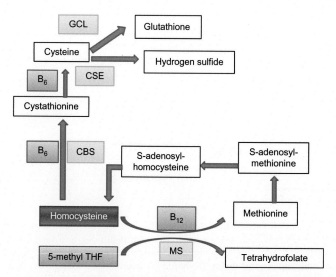

FIGURE EI.6 **Homocysteine homeostasis.** Cysteine provides the primary substrate for hydrogen sulfide (H_2S) in mammals. Homocysteine concentration is tightly regulated via formation of methionine and S-adenosylmethionine (SAM) by methionine synthase (MS) activity that requires both folate and vitamin B12. Homocysteine is converted to cystathione by cystathionine beta synthase (CBS) activity that requires vitamin B6 and serine (not shown). Cystathionine is converted to cysteine by a second B6-dependent enzyme, cystathionine gamma lyase (CSE). Cysteine is also substrate for CSE in the formation of hydrogen sulfide (Sun et al., 2009). CSE is upregulated by Nrf2 during oxidative stress. A third enzyme (not shown), 3-mercaptopyruvate sulphurtransferase (3-MST), is also expressed in the majority of human cell types and produces hydrogen sulfide. The enzyme gamma glutamyl cysteine ligase (GCL) binds cysteine to glutamate and is the limiting enzyme for glutathione biosynthesis. This enzyme is also upregulated by Nrf2. *Source: Figure by Susan Ettinger.*

together with the risk for comorbidities (Wollesen et al., 1999).

Sulfur-containing isothiocyanates in cruciferous vegetables such as cabbage and kale, as well as organosulfur compounds in members of the allium family (garlic and onions) modify cysteine residues on KEAP1. These foods contain glucosinolates (β-thioglucose N-hydroxysulfates) that must be hydrolyzed by myrosinase (β-thioglucosidase) prior to food ingestion. (Myrosinase is activated by cutting, crushing, or bruising the vegetables 5—10 minutes before cooking.) Released isothionates react with specific cysteine residues on KEAP1 to yield thionoacyl adducts that release Nrf2 (Kensler et al., 2013). Cruciferous and members of the allium family, as well as other foods such as egg yolk, fermented fish, and durian fruit are sulfur donors (Wang, 2012) and sources of exogenous H_2S.

EI.2.3 NFκB-Dependent Transcription

Inflammatory cytokines, toll-like receptors, pattern recognition receptors, and other signals of cell injury stimulate the ancient proinflammatory signaling pathway, nuclear factor kappa-light-chain-enhancer of activated B cells (NFκB). This complex is comprised of an evolutionarily conserved family of proteins that can form homodimers or heterodimers and control DNA transcription of a distinct, but overlapping set of genes involved in inflammation, oncogenesis, and tumor progression as reviewed by Hayden and Ghosh (2004). Diversity in combinational arrangement and of DNA binding allows the complex to respond specifically to individual target promoters to be modified under physiological conditions. In an unstressed state, individual NFκB transcription factors are bound to inhibitory IκB proteins that block the nuclear localization signal of their respective targets, and sequester the NFκB complex in the cytoplasm in a latent, inactive form.

NFκB activation occurs in at least two ways. The classical (canonical) pathway is activated by inflammatory signals such as IL-1β and TNFα that phosphorylate IκB protein, freeing NFκB to translocate to the nucleus and regulate expression of inflammatory and pro-oncogenic genes. Some of these genes code for inflammatory cytokines, thereby creating a feedforward loop that amplifies the signal. The alternative activation pathway allows different NFκB proteins to translocate into the nucleus and regulate expression of genes required for development and maintenance of secondary lymphoid organs. NFκB activation has also been reported in response to posttranslational epigenetic modification including methylation, phosphorylation, acetylation, and ubiquitination (Lu and Stark, 2015).

Several bioactive components of food have been shown to inhibit NFκB signaling and ameliorate some of the damage associated with high-fat diets (Zeng et al., 2015). Some components reduce production of inflammatory mediators, thereby preventing the signal, while others inhibit NFκB activation, release, or DNA binding. Examination of 36 naturally occurring flavonoids for their ability to inhibit NFκB-induced NO production by macrophages exposed to lipopolysaccharide (LPS) revealed that isoflavones (daidzein and genistein), flavonols (isorhamnetin, kaempferol and quercetin), flavanone (naringenin), and an anthocyanin (pelargonidin) inhibited NFκB-induced iNOS protein and mRNA expression in a dose-dependent manner. Genistein, kaempferol, quercetin, and daidzein also inhibited the activation of the signal transducer and activator of transcription 1 (STAT-1), another important transcription regulator of iNOS activity (Hämäläinen et al., 2007).

Curcumin from turmeric inhibited the pathway leading from activation of IκBα kinase and IκBα phosphorylation to IκBα degradation (Bharti et al., 2004); curcumin also inhibited NFκB binding to DNA in dendritic cells from ob/ob mice (Yekollu et al., 2011). Genistein, an isoflavone in soybean exhibited

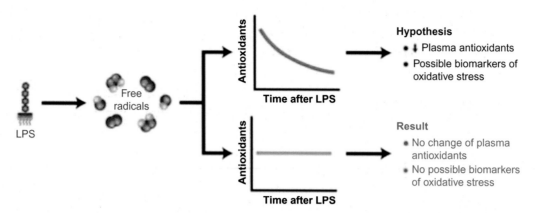

FIGURE EI.7   Serum antioxidants post LPS. Lipopolysaccharide (LPS) was administered to Göttingen mini pigs. In contrast to expectations, total serum antioxidants (ascorbic acid, tocopherols (α, δ, γ), ratios of GSH/GSSG and cysteine/cystine, mixed disulfides, and total antioxidant capacity) were not changed with time. However, uric acid, total GSH, and total Cys were elevated, possibly as a result of cell injury and liver damage as quantitated histologically as oxidative degeneration and hepatic cell death. It should be noted that although the pig model closely resembles human physiology, the pig, in contrast to the human, retains the capacity to synthesize ascorbate (Kadiiska et al., 2015). *Source: Reproduced from Kadiiska, M.B., Peddada, S., et al., 2015. Biomarkers of oxidative stress study VI. Endogenous plasma antioxidants fail as useful biomarkers of endotoxin-induced oxidative stress. Free Radic. Biol. Med. 81, 100–106 with permission from Science Direct.*

NFκB-dependent and NFκB-independent apoptotic control via ROS generation; genestein activities depended on genetic cell types (Lee and Park, 2013). Resveratrol, a polyphenolic phytoalexin found in grapes and red wine, as well as polyphenols and other bioactive components in green tea also inhibit NFκB signaling as reviewed by Gupta et al. (2011).

Terpenoids (also called isoprenoids) are derived from five-carbon isoprene units (C_5H_8); over 20,000 different terpenoids are found nature. Several isoprenoids have been shown to reduce NFκB signaling as comprehensively reviewed by Salminen et al. (2008). Many isoprenoids are found in the human food supply and many more are being investigated for their pharmaceutical activities. Examples of NFκB inhibition by isoprenoids includes inhibition of DNA binding by monterpenes such as limonene in citrus zest and by perillyl alcohol, its metabolite found in essential oils of lemon grass and sage. Zerumbone, a sequiterpene from ginger, reduced phosphorylation and degradation of IκB, leading to decreased nuclear translocation. Beta carotene, the orange pigment in squash and carrots, as well as many of the 600 or more carotenoids in the food supply such as lycopene, the red pigment in tomato and lutein, the yellow pigment in kale and other green leafy vegetables decreased NFκB translocation and DNA binding. It should be noted that terpenoids are fat-soluble and require the presence of fat in the diet for optimal absorption.

EI.2.4 Assessment of Total Antioxidant Capacity

Antioxidant balance is maintained in health, but excess production of oxidants can overwhelm the defense system and result in cell injury, inflammation, and death. Thus, efforts have been made to quantitate total antioxidant capacity (TAC) in serum, urine, and other bodily fluids as a predictor of disease risk and survival in critically ill patients (Ghiselli et al., 2000). The TAC biomarker is calculated on the ability of plasma to trap a flow of water-soluble peroxyl radicals produced at a constant rate (Ghiselli et al., 1995). Unfortunately, TAC results can be difficult to interpret. Measurement of TAC in septic patients revealed that nonsurvivors actually had higher TAC levels and that TAC was a positive predictor of mortality (Lorente et al., 2015). The authors discussed possible methodologic reasons for this result (eg, measurement in serum vs plasma). Since TAC represents a mixture of antioxidants, it was difficult to separate the influence of each species. For example, the authors noted that urate, a significant component of TAC, might actually rise as a result of oxidative damage. Serum urate was a strong predictor of mortality in septic patients. This result is also seen in the unexpectedly unchanged levels of "antioxidant species" following administration of LPS to mini pigs as described in Fig. EI.7.

EI.3 SINGLE-UNIT ANTIOXIDANT NUTRIENTS

In the 1970s, work on antioxidant nutrients assumed that scavenging or quenching a free radical was sufficient to absorb their reactivity. The intervening years have deepened our understanding of antioxidant protection, as reviewed by Vrolijk et al. (2015).

At the molecular level, scavenging is only the first step. The second is the sequestration of the radical within the antioxidant, creating an antioxidant radical. Finally, the radical is transferred to the antioxidant network. For example, vitamin E scavenges lipid peroxyl radicals (LOO•) and delocalizes the radical over the antioxidant, creating a relatively safe LOOH. The vitamin E radical integrated in the lipoprotein biomembrane is reduced by vitamin C in the plasma (Packer et al., 1979) to form dehydroascorbate which can react a reducing agent (NADH; glutathione) to regenerate ascorbate. In this way radical reactivity is totally absorbed (Fig. EI.8).

Antioxidant nutrients are not entirely benign, and can exert pro-oxidant or other deleterious effects, as recently reviewed by Vrolijk et al. (2015). Glutathione-S-transferase (GST) isoforms found in the erythrocytes, lungs, and skin can protect these tissues by conjugating reduced glutathione to xenobiotic substrates, thereby facilitating their detoxification. Both vitamin E and β carotene have been shown to inhibit GST resulting in the more frequent DNA adducts (van Haaften et al., 2003). Pro-oxidant activities have been reported with vitamin C (Rietjens et al., 2002) as well as aromatic components found in foods including well-known "antioxidants" such as curcumin, resveratrol, and quercetin (León-González et al., 2015).

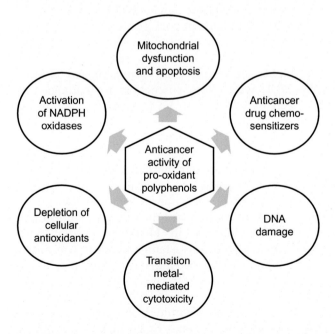

FIGURE EI.8 **Pro-oxidant activities of polyphenols may contribute to their observed anticancer actions.** *Source: Reproduced from León-González, A.J., Auger, C., et al., 2015. Pro-oxidant activity of polyphenols and its implication on cancer chemoprevention and chemotherapy. Biochem. Pharmacol. 98 (3), 371–380 with permission from Science Direct.*

EI.4 TOCOPHEROLS; TOCOTRIENOLS (VITAMIN E)

Vitamin E was first characterized as a lipid-soluble vitamin required for reproduction almost a century ago (Evans and Bishop, 1922). The name tocopherol comes from the Greek *tŏkos*, birth, and *phérein*, to bear or carry (eg, to carry a pregnancy). Naturally occurring vitamin E exists in eight chemical forms (alpha-, beta-, gamma-, and delta-tocopherol and alpha-, beta-, gamma-, and delta-tocotrienol) that have varying levels of biological activity. As discussed below, α tocopherol is the predominant form in the blood, however other forms (β, γ, δ tocopherols) are found more frequently in the food supply. The most potent sources of all tocopherols are nuts and oils, as well as green leafy vegetables. Vitamin E isomers are fat-soluble and are efficiently absorbed when consumed with fat (Fig. EI.9).

Incorporation of vitamin E into micelles containing taurocholate and oleic acid appears to be essential for absorption. Furthermore, like other fat-soluble esters, vitamin E must be hydrolyzed in the intestinal lumen prior to absorption as reviewed by Reboul and Borel (2011) and Reboul et al. (2012). Although evidence is fragmentary, vitamin E forms appear to be transported into the enterocyte by the salvage receptor b1 (SR-B1) and cluster of differentiation 36 (CD36) in the brush border of the enterocyte. SR-BI is a single-chain transmembrane glycoprotein found in diverse cell types such as liver, testes, ovaries, and macrophages and facilitates transport of lipid-soluble species. SR-B1 has been shown to transport a variety of carotenoids (lutein, lycopene) as well as vitamins E, D, and K. CD 36 is a multiligand transmembrane glycoprotein, expressed on multiple cell types and involved in cellular adhesion, angiogenesis, lipoprotein binding, endocytosis, and lipid transport (Goncalves et al., 2014). Several other proteins appear to modulate fat-soluble vitamins and lipid transport including the efflux transporter, ATP-binding cassette (ABC), and Niemann−Pick C1-like 1 (NPC1L1). It is likely that one or more of these proteins facilitate vitamin E integration into chylomicrons and transport in the lymphatics. Vitamin E absorption likely takes place in the distal intestine but significant interactions exist between

FIGURE EI.9 **Structure of alpha tocopherol (αToc).** *Source: Figure by Calvero (selfmade with ChemDraw) (Public domain), via Wikimedia Commons.*

fat-soluble transport proteins (Goncalves et al., 2015). Since these proteins have the capacity to bind fat-soluble vitamins, lipids, and carotenoids, their actions can explain, in part, some of the well-described competitive interactions (eg, between carotenoids and vitamin E; between α and γ tocopherol and cholesterol; between vitamin D and vitamin E and vitamin K).

Transport and distribution of vitamin E has been comprehensively reviewed (Jiang, 2014). All forms of vitamin E are transported by chylomicrons to peripheral tissues such as muscle, bone marrow, adipose tissue, skin, and brain, where they are apparently released by lipoprotein lipase; peripheral transport into tissues such as skin and fat prior to liver uptake may account for the presence of non-αToc forms in these tissues where they may have alternative functions (Jiang et al., 2008). In the liver, αToc is preferentially bound to, and protected by, αToc-binding protein (TTP); together with ABC-A1, TTP incorporates αToc into lipoproteins, primarily VLDL, for transport. Other vitamin E forms poorly bound by TTP are catabolized by hepatic cytochrome P450 enzyme, CYP4F2. Supplements of αToc may result in increased utilization of TTP sites, leaving fewer sites available for non-αToc forms. Sesamin, a lignin in sesame seeds inhibits CYP4F2 and increases retention of non-αToc forms. In contrast to previous assumptions, ~80% of unmetabolized tocopherols and tocotrienols are discarded via biliary excretion.

Vitamin E transport in lipoproteins has been reviewed (Traber, 2013). All lipoproteins carry αToc; VLDL appears to be preferentially enriched, although the enrichment does not occur in the hepatocyte. VLDL conversion to LDL enriches the αToc concentration in the latter lipoprotein. VLDL is dependent on adequate αToc to prevent lipid peroxidation that can increase apoB degradation. HDL may be enriched with αToc by specific mechanisms as described by the author. HDL contains ~40% of circulating αToc and plays a major role in reverse cholesterol transport and cholesterol excretion. It is not known whether HDL targets its αToc to the liver for excretion in the bile or for recycling into VLDL forms.

EI.4.1 Vitamin E Function

All vitamin E forms scavenge lipid peroxyl radicals by donating hydrogen from the phenolic group. However, tocotrienol forms are more evenly distributed in the phospholipid bilayer and their interactions with lipid peroxyl radicals are more effective than the tocopherols. Of the tocopherols, the structure of γToc is more effective than αToc in trapping electrophiles such as reactive nitrogen species. Vitamin E forms also have antiinflammatory properties (Reiter et al., 2007). These effects are likely to involve cell signaling and include inhibition of proinflammatory cyclooxygenase and lipoxygenase activities (Jiang et al., 2000, 2008). However, non-αToc forms appear to have greater anti inflammatory actions and elevated levels of αToc availability (eg, through supplementation) depress plasma and ultimately tissue levels of γToc. Thus, it is possible that providing large-dose vitamin E supplements (αToc) without first assessing vitamin E status (by measuring serum αToc levels) may potentiate an increase in catabolism of the more effective anti-inflammatory vitamin E forms.

EI.4.2 Adverse Effects of Vitamin E

Although the metabolism and actions of natural and synthetic forms of vitamin E are still largely uncharacterized, meta-analyses of clinical trials of αToc supplementation at a 400 mg/day dose have revealed a small, but significant increase in all-cause mortality (Miller et al., 2005). Several causes for proatherogenic effects of αToc supplementation have been proposed. Supplementation may reduce the antioxidant impact of HDL2 and HDL3 (Wade et al., 2013). Although in vitro investigations showed that increasing concentrations of αToc protected HDL from oxidation, if the HDL were isolated from serum pre-incubated with αToc, HDL concentrations of αToc were increased, but the HDL exhibited increased susceptibility to oxidation. Ex vivo studies conducted in young volunteers supplemented with 400 mg/day of αToc for 6 months supported these observations. HDL2 and HDL3 concentrations were higher in serum from supplemented subjects, while HDL were more susceptible to oxidation and had decreased levels of HDL-associated antioxidant enzymes, paroxonase-1 (PON-1), cholesteryl ester transfer protein (CETP), and lecithin cholesteryl acyltransferase (LCAT). It is also possible that the observed increase in cardiovascular mortality with αToc supplementation was related to modification of xenobiotic metabolizing systems and transporters, thereby counteracting actions of concomitant medications. For example, the beneficial increase in HDL2 in cardiovascular patients treated with simvastatin and niacin was attenuated by vitamins E (1600 mg/day αToc) and C plus selenium (Brown et al., 2001) (Fig. EI.10).

Additional adverse effects of supplemental αToc have been reported. Vitamin E supplementation has been associated with hemorrhage in patients on oral coagulation agents, especially vitamin K antagonists (Pastori et al., 2013). The mechanisms through which vitamin E increases bleeding are unclear, although it has been proposed that vitamin E interferes with

FIGURE EI.10 **Structure of alpha tocotrienol.** *Source: Figure by Calvero (selfmade with ChemDraw) (Public domain), via Wikimedia Commons.*

FIGURE EI.11 **Sodium ascorbate.** Ascorbic acid (AA) is synthesized from glucose by all species except primates, humans, guinea pigs, and some fruit bats, fish, and birds. *Source: Figure by Yikrazuul (own work) (Public domain), via Wikimedia Commons.*

vitamin K activation and/or degradation (Traber, 2013). Since lipid-soluble vitamins and carotenoids compete for absorption and transport (Goncalves et al., 2015), competition between vitamins E and K for absorption could explain the bleeding complications from αToc supplementation. It should be noted that no adverse effects have been reported from subjects consuming dietary sources rich in vitamin E.

In summary, while the antioxidant behavior of vitamin E forms has been well established, preliminary studies strongly implicate that vitamin E has other biological actions including cell signaling and modulation of gene expression. Current knowledge of the forms, distribution, structural role, and interactions of vitamin E forms with phospholipids, cholesterol, membrane-bound proteins, and other components such as lysolipids is fragmentary (Yoshida et al., 2007). At this time, the recommended dietary intake for vitamin E is based on serum levels of αToc required to prevent erythrocyte lysis ex vivo when exposed to hydrogen peroxide (Panel on Dietary Antioxidants and Related Compounds, 2000). The RDA for adults is 15 mg/day or 22.4 international units (IU) and a tolerable upper limit of intake (UL) of 1000 mg/day has been set for all adults. Additional information on vitamin E can be found at the Vitamin E-Health Professional Factsheet (https://ods.od.nih.gov/factsheets/VitaminE-HealthProfessional/).

EI.5 ASCORBATE (VITAMIN C)

The medical profession itself took a very narrow and very wrong view. Lack of ascorbic acid caused scurvy, so if there was no scurvy there was no lack of ascorbic acid. Nothing could be clearer than this. The only trouble was that scurvy is not a first symptom of a lack but a final collapse, a premortal syndrome and there is a very wide gap between scurvy and full health. *Albert Szent-Gyorgyi*

The ascorbic acid (AA) deficiency disease, scurvy, was recognized by the ancient Egyptians, described by Hippocrates, and decried as the scourge of sailors and explorers long before the disease was characterized as deficiency of a six-carbon reducing substance by Szent-Gyorgi in 1928. The literal meaning of

"ascorbate" is "against scurvy" De Tullio and Arrigoni (2004). Since its characterization, the function of ascorbate has been intensively studied but remains a mystery (De Tullio, 2012). Although ascorbate is usually regarded as an antioxidant due to its reducing capacity, it is also able to reduce transition metals such as copper and iron, thereby increasing their pro-oxidant character. At low doses, ascorbate tends to be a pro-oxidant, whereas at high concentrations, it functions as an antioxidant. The concentration of free metal appears to determine the point at which ascorbate changes from pro- to antioxidant (Buettner and Jurkiewicz, 1996) (Fig. EI.11).

An estimated 40 million years ago, the gene for L-gulonolactone oxidase became nonfunctional in humans and selected species due to accumulation of mutations. Thus, for these species, ascorbate became an essential nutrient. For primative peoples, ascorbate was easily obtained from readily available plant sources, thus deficiency was uncommon. Unfortunately, with the discovery of agriculture and increased reliance on grains and processed food, ascorbate deficiency has become more common. Statistics published in the Morbidity and Mortality Report from the Center for Disease Control (http://www.cdc.gov/mmwr/preview/mmwrhtml/mm5935a1.htm) indicate that only ~30% of the US population actually consume the recommended five portions of fruits and vegetables per day. Furthermore, for those who eat the recommended allotment of fruits and vegetables, foods rich in ascorbate (broccoli, deep green vegetables) are seldom consumed (<2%) while poor-ascorbate sources (iceberg lettuce, French fried potatoes) are more frequently consumed (>30%) (Johnston et al., 2000). Ascorbate depletion is compounded by an increase in oxidant load associated with smoking, obesity, disease, and aging. Food preparation methods, such as exposure to light and heat, long storage, and cooking in iron pots are all known to lower ascorbate concentration in food. It must be noted that the small residue of vitamin C that remains in the food after cooking is as bioavailable as supplemental ascorbate.

AA absorption, distribution, and excretion are tightly controlled by sodium-dependent transporters, SVCT1 and SVCT2. In the process of donating an electron, AA is oxidized to dehydroascorbate. Dehydroascorbate is taken into cells by GLUT transporters and efficiently reduced to ascorbate by intracellular glutathione or the selenoprotein, thioredoxin reductase (Rumsey and Levine, 1998). AA filtered in the kidney is actively reabsorbed in the proximal tubule by SVCT2. In the healthy individual, AA is maintained in steady state by AA absorption, distribution, and renal reabsorption. The recommended dietary allowance (RDA) was based on intake trials to determine the amount of vitamin C required to achieve near-saturation of plasma and leukocytes with minimal urinary excretion. The RDA of 75 mg for women and 90 mg for men was established in 2000 by the US Institute of Medicine. More recent trials have suggested that an optimal intake to prevent scurvy and minimize chronic disease was higher, at approximately 200 mg/day. This reflects the approximate vitamin C content in the "five-a-day" fruit and vegetable recommendation indicated above (Frei et al., 2012).

The argument for increasing the recommended ascorbate intake was based on a trial in healthy subjects that revealed intake in excess of ~400 mg/day resulted in decreased absorption and increased urinary excretion to reach a homeostatic state with a maximal plasma concentration of ~60−90 μM/L. Homeostasis was associated with an intracellular concentration approximately 14-fold higher than in plasma, ranging from 0.5 to 10 mM/L depending on the cell type (Levine et al., 1996). Subjects did not excrete oxalate (degradation product of ascorbate) until ascorbate intake was 1000 mg/day.

EI.5.1 Ascorbate Function

Many AA-dependent enzymes belong to a large class of 2-oxoglutarate-dependent dioxygenases (2-ODD) that catalyze the incorporation of O_2 into an organic substrate. Different 2-ODDs catalyze multiple reactions, including hydroxylation, desaturation, oxidative ring-closure, and expansion. The first identified cause of scurvy was the inactivation of the 2-ODD, proline hydroxylase by deficiency of AA (Meister and Bukenberger, 1962). Further studies revealed that the reaction was extremely complex and that AA is required to reduce the Fe^{3+} in proline hydroxylase to Fe^{2+}, facilitating posttranslational hydroxylation of proline (and lysine) essential to crosslink collagen molecules into fibrils with enormous tensile strength and thermal

stability. Since scurvy is also accompanied by reduction of procollagen biosynthesis, it has been suggested that AA deficiency in scurvy also produced anorexia and inhibition of insulin-like growth factor (a starvation-like effect). This change could explain the reduction in protein synthesis in scurvy (Peterkofsky, 1991).

Ascorbate is also reported to be required for the formation of sulfated proteoglycans in connective tissue. Early work with scorbutic cells revealed that AA catalyzes the oxidation of the sulfur-containing amino acid, homocysteine, in vivo, and that homocysteic acid is the precursor of the esterified sulfate of connective tissue proteoglycans (McCully, 1971). Children with inborn errors in sulfur-containing amino acid metabolism have high levels of homocysteine due to altered cystathionine synthase activity, as illustrated in Fig. EI.6. These children also develop premature atherosclerosis resulting from elevated sulfated proteoglycans and increased homocysteic acid in their vessel walls (refer to Chapter 6: Atherosclerosis and Arterial Calcification: Metabolic Targets for Diet Modification).

AA is essential for other functions, as recently reviewed by De Tullio (2012). Protein folding in the ER requires AA, as evidenced by ER stress in guinea pigs chronically AA-depleted (Margittai et al., 2005; Mandl et al., 2009). Two AA-dependent 2-ODDs are required for the biosynthesis of carnitine, required to chaperone long-chain fatty acids across the mitochondrial membranes. Additionally, AA-dependent enzymes are essential in the biosynthesis of neurotransmitters, serotonin, and norepinephrine and for activation of diverse peptide hormones and neuropeptides that require amidation of their C terminus to become biologically active. In thyroxine metabolism, ascorbate appears to protect an intermediary enzyme from oxidation by its substrate (La Du and Zannoni, 1961). AA-dependent ODDs have been reported that catalyze repair of methylated DNA bases (eg, 3-methylcytosine), oxidizing and releasing them as formaldehyde (Koivisto et al., 2004). These observations have led to the investigation of AA-dependent ODD function in reprogramming, stem cell differentiation, and epigenetic regulation (Monfort and Wutz, 2013).

Since ODDs cannot function without oxygen, animal cells evolved an elegant mechanism coupling the capacity of the ODD to sense oxygen with induction of cellular defenses against hypoxia (oxygen deprivation). When oxygen is available, proline residues on HIF1α are hydroxylated by four ODDs (Schofield and Ratcliffe, 2004). Hydroxylated HIF1α binds in a multiprotein complex with the von Hippel−Lindau tumor suppression protein and the complex undergoes proteasome-mediated degradation. Limiting oxygen availability causes the hydroxylase activity to fall.

HIFα levels rise and dimerize with HIFβ. The dimer enters the nucleus and binds to the hypoxia response element (HRE) in the promoter region of genes that support metabolic response during stress, including erythropoietin, transferrin, ceruloplasmin, VEGF, NOS, carbonic anhydrase, insulin-like growth factor, and others. While the exact role AA plays in HIFα hydroxylation and other ODD activity is still unclear, there is experimental evidence that AA suppresses the HIF response (Kuiper et al., 2014) and that further research will uncover other roles in regulating gene expression (Mandl et al., 2009; Lykkesfeldt et al., 2014) in addition to its known function as a reducing agent (De Tullio, 2012).

EI.5.2 Assessment of Vitamin C Status and Caveats for Experimental Design

Ascorbate is accumulated against a concentration gradient by most tissues with the exception of red blood cells (Levine et al., 1996). Since ascorbate distribution is tightly regulated and is interdependent on other antioxidant moieties (see above), it is likely that the antioxidant activity in the serum will be relatively stable in the healthy individual, but low AA plasma levels have been seen in smokers, aging individuals, and subjects with chronic disease. Plasma contains other antioxidants including urate, tocopherols, and ubiquinone, however ascorbate has been shown to have an important role as the first line of defense against oxidative stress (Frei et al., 1990). Vitamin C is measured in the fasting state; the normal range of vitamin C in heparinized plasma is 0.3−0.6 mg/dL.

It is important to recognize that vitamin C research is subject to serious methodologic flaws. Culture of human cells in a high-oxygen environment and in the presence of transition metal ions rapidly depletes ascorbate in the medium. This is also true with the use of cell lines grown in the absence of supplemental vitamin C. The extent of pro-oxidative cell death in culture with ascorbate supplementation has been linked to the differing composition of the medium (Clément et al., 2001). Several studies use rodents as experimental models, despite the fact that ascorbate is not an essential vitamin for these animals. Mutant rodents are available that lack the enzyme required for ascorbate biosynthesis but these animals do not precisely model the effects of ascorbate depletion/supplementation in the human. Study design flaws, especially in population studies as well as controlled intervention trials prevent accurate interpretation of the data (Michels and Frei, 2013).

FIGURE EI.12 **Selenomethionine structure.** *Source: By Fuse809 (own work) (Public domain), GFDL (http://www.gnu.org/copyleft/fdl.html) or CC BY-SA 4.0-3.0-2.5-2.0-1.0 (http://creativecommons.org/licenses/by-sa/4.0-3.0-2.5-2.0-1.0), via Wikimedia Commons.*

EI.6 SELENIUM (Se)

Selenium transport and metabolism have recently been reviewed (Burk and Hill, 2015). In brief, organic Se forms (selenomethionine and selenocysteine) and inorganic forms (selenite and selenate) are all well absorbed from the human diet. Selenomethionine is a naturally occurring amino acid synthesized by plants such as Brazil nuts, cereal grains, soybeans, and legumes and incorporated by the plant into proteins as a methionine. Almost >90% of selenium in plants is selenomethioneine (Fig. EI.12).

Most plants do not require selenium and can grow well in soil depleted of selenium. Although the plants are healthy, they provide little selenium to animals and humans that consume them. On the other hand, plants that grow in soil with high selenium levels efficiently absorb the highly bioavailable metal, sometimes to levels that can produce toxic symptoms in animals that consume them. Plants that concentrate selenium are called "selenium accumulators." They detoxify selenocysteine by adding a methyl group, producing Se-methylselenocysteine which not only is not incorporated into protein and but also reduces selenomethionine production in the plant. These plants, predominately grassland legumes (weeds), are not usually consumed by humans, but grow wild on grazing land. Several instances of chronic selenium toxicity have been reported in horses and other grazing animals which consume plants containing high levels of Se-methylselenocysteine (Olson, 1986). After reviewing reports of toxicity in grazing animals, experimental animals, and man, the author suggested that the maximum tolerable oral intake for adults should be 0.05 mg Se/kg.

EI.6.1 Human Selenium Metabolism

Following Se ingestion and transport via intestinal methionine transporters, diet-derived Se, primarily selenomethionine, enters the methionine pool and is randomly inserted into human proteins as a function of its proportion of the methionine pool. Healthy human protein selenium is in a ratio of one selenium atom to 8000 methionine residues or one selenomethionine per 1100 albumin molecules (Burk et al., 2001). Selenium supplementation enriches the selenomethionine pool. The only known biological function of selenomethionine is to serve as a reserve pool of selenium. The liver is the primary organ for Se distribution. In the liver, selenomethionine is metabolized via the methionine synthase (refer to Chapter 6: Atherosclerosis and Arterial Calcification: Metabolic Targets for Diet Modification) and reverse transsulfuration pathways (described above) to yield selenocysteine. Selenocysteine is an analog of the amino acid, cysteine, but contains a senol group in place of the thiol (sulfur) group. This selenium form is highly reactive and is the major selenium form in animal proteins, especially fatty fish and animal flesh. Free selenocysteine is maintained at a low concentration by selenocysteine lyase that catabolizes it into selenide and alanine.

Se availability is regulated at two levels. At the whole-body level, selenium is transported from the liver to the other tissues (brain, testes) by selenoprotein P (Sepp1). Sepp1 contains 10 Se per protein molecule and is the major circulating selenoprotein. Sepp1 is taken up by apolipoprotein E receptor 2 (ApoER2), especially abundant on plasma membranes in the brain and testes. The extent of expression of apoER2 on the cell membrane is thought to determine the ability of most tissues to take up selenium. Sepp1 is filtered by the kidney and taken up by megalin, a lipoprotein receptor in the proximal tubule epithelium (Olson et al., 2008). The kidney synthesizes glutathione peroxidase 3 (GPx3) and secretes some of this selenoprotein into the extracellular fluid. Low renal Se concentrations lead to reduced kidney-derived circulating GPx3 (Schweizer et al., 2005). Dietary selenium in excess of need is incorporated into excretory metabolites and released into the urine; some dietary Se is excreted in the feces. With the exception of selenomethionine randomly bound to protein, there does not appear to be a regulated reserve pool of selenium as, for example, iron storage as ferritin.

EI.6.2 Selenium Function

Most selenoproteins are redox enzymes. Thus far, eight glutathione peroxidases (Gpx) have been identified in the human. They are produced by different genes and function in different cells and locations. The general function of a GPx is to reduce lipid hydroperoxides to alcohols and to reduce hydrogen peroxide to water. Three thioredoxin reductases (TrxR) are also selenoproteins and regulate thiol reductase status. TrxRs reduce cellular disulfides and also serve as electron donors. Glutaredoxins catalyze reduction of oxidized glutathione using electrons from NADPH. TrxR isoforms play a major role in maintaining the redox status of the mitochondria (Arnér and Holmgren, 2000). The three iodothyronine deiodinases are also selenoproteins. These enzymes regulate thyroid hormone activity in tissues. Additional selenoproteins are directly involved in the metabolism of selenium itself. Sepp1 transports selenium from the liver to other tissues. Several other selenoproteins exist, but their functions are not yet fully understood, as recently reviewed by Labunskyy et al. (2014).

EI.6.3 Selenium Hierarchy and Distribution

Triage theory holds that when dietary availability of a nutrient is inadequate, those functions essential for short-term survival and reproduction are protected at the expense of those less essential (McCann and Ames, 2011). Selenium distribution provides a clear example of evolutionary triage. Two layers of triage have been identified: the first is the preferential distribution of Se to some tissues (brain, reproductive organs), while the second layer involves the tissue hierarchy among similar enzymes. For example, the glutathione peroxidases all use glutathione to reduce peroxides. However, when Se was limiting, rat intracellular glutathione peroxidase 1 (Gpx1) activity fell by 90% while phospholipid hydroperoxide glutathione peroxidase (Gpx4) was maintained (at 25–50% of normal) in liver, kidney, testes, and heart (Lei et al., 1995). Gpx4 reduces bulky phospholipid hydroperoxides that are not substrate for Gpx1. The evolutionary significance of preserved Gpx4 in the face of Se deficiency is that cytosolic Gpx4 is essential for embryonic development and mitochondrial Gpx4 plays a dual role in spermatozoa maturation and implantation (Ursini et al., 1999).

EI.6.4 Assessment of Selenium in Humans and Recommended Intakes

Whole-body selenium estimates range widely from <5 to >20 mg, largely attributed to the unregulated selenomethionine pool. Results from supplementing selenium-deficient human subjects indicated that a

70-kg man maintains ~10–15 mg in the regulated pool. Note that total whole-body selenium concentrations would be greater because they include the unregulated pool of selenomethionine (Burk and Hill, 2015).

Selenium status in human subjects has been estimated by measuring elemental selenium in hair, toenails, and urine. Plasma (or serum) has an advantage in that it contains two selenoproteins, GPx3 (of kidney origin) and Sepp1, synthesized in the liver. Since the liver provides selenium in response to tissue need, Sepp1 is taken to mirror the selenium content of the liver and, by extension, the selenium availability to extrahepatic tissues. In health, Sepp1 contains ~65 μg Se/L plasma and GPx3 ~17 μg Se/L plasma. Thus, the total regulated pool is ~85 μg Se/L plasma. Plasma selenium in excess of this value represents contributions from the unregulated selenomethionine pool.

Feeding studies have demonstrated that GPx3 biomarkers reached a plateau before Sepp1 maximum was reached. This finding suggests that the peripheral tissues (GPx3 from kidney) become replete before the liver reaches its maximal production of Sepp1. Sepp1 concentration has therefore been proposed as a better indicator of selenium repletion than GPx3. The US dietary allowance (RDA; https://ods.od.nih.gov/factsheets/Selenium-HealthProfessional/) for adults of 55 mg Se/day was calculated based on trials of selenium supplementation necessary to fill the regulated pool. When supplemental selenium fills the selenomethionine pool and a plateau is reached, additional selenium is converted into small molecular forms and excreted in the urine. The feeding studies also compared the bioavailability of selenomethionine with selenite. In each study, the supplements were administered as tablets. Selenite was ~66% as bioavailable as selenomethionine. Thus, to achieve equivalent bioavailability, the selenite supplement should be ~150% that of selenomethionine.

EI.6.5 Importance of Selenium Regulation to Health

Mechanistic studies have largely been carried out with experimental animals. Data reviewed by Burk and Hill (2015) indicate that severe selenium deficiency in the mouse causes male infertility, with no other observable effects. If selenium transport is impaired by Sepp1 or apoER2 knock-out, mild selenium deficiency causes lethal brain injury. Human studies cannot ethically elicit a severe selenium deficiency and overt health effects from mild deficiency have been difficult to identify. However, patients with added stress (eg, sepsis) or genetic mutation (regulatory protein SBP2) develop more severe sequelae. The authors conclude that whole-body selenium hierarchy efficiently adapts to maintain selenoprotein status, even with mild deficiency or excess.

EI.7 RIBOFLAVIN (VITAMIN B2)

Riboflavin (vitamin B2) was initially isolated as a yellow pigment in the milk protein, whey, in 1879, but is also found in eggs, lean meat, and green leafy vegetables. Its characteristic yellow color is noted in the urine following riboflavin supplementation. The RDA for riboflavin is 1.3 mg in men and 1.1 mg in women (Food and Nutrition Board, 1998). Riboflavin is absorbed in the jejunum by a transport system that is saturated with a single dose of ~30 mg. Once absorbed it is weakly bound to albumin with minimal binding to immunoglobulins. It enters the cell by carrier-mediated transport (Fig. EI.13).

The actions of riboflavin as reviewed by Powers (2003) are primarily through its major coenzymes, flavin mononucleotide (FMN) and flavin adenine dinucleotide (FAD). Specifically, FMN and FAD act as cofactors for flavoproteins in the electron transport chain (ETC) and in metabolic conversions of other nutrients (folate, pyridoxine, niacin). Riboflavin cofactors are also required for detoxifying systems and for glutathione reduction. In addition, riboflavin, through its conjugated ring structure, can act as a free radical scavenger (Ashoori and Saedisomeolia, 2014).

FIGURE EI.13 Chemical structure of riboflavin. *Source: Figure by Calvero (selfmade with ChemDraw) (Public domain), via Wikimedia Commons.*

Riboflavin is excreted in the urine and deficiency can occur rapidly if dietary intake is insufficient. A poor diet is also likely to be deficient in other "B" vitamins (such as thiamin, folate, niacin). Since riboflavin is also required for metabolism of these vitamins, riboflavin deficiency is usually complicated by deficiency of other nutrients. Subclinical deficiency has been observed with estrogen supplementation as in high-dose oral contraceptive agents or hormone replacement therapy. Riboflavin deficiency and suboptimal levels occur in persons with alcoholism and in patients with inflammatory bowel syndrome and diabetes, as well as in elderly persons. Depending on the severity and duration, deficiency can lead to degenerative changes in the nervous system, endocrine dysfunction, anemia, and skin disorders, as well as inflammation of the lining of the mouth, tongue, and throat; cracks at the corners of the mouth (angular cheilitis); and red and itchy eyes (from vascularization of the cornea).

EI.8 COENZYME Q-10 (CoQ10)

Coenzyme Q-10 was first isolated over 50 years ago (Festenstein et al., 1955). It is synthesized from acetyl CoA in all tissues and is present in all membranes of biological organisms (Turunen et al., 2004). CoQ10 is a high-efficiency lipid-soluble antioxidant because of its localization in the membrane and relatively high concentration, as reviewed by Bentinger et al. (2007). It has the capacity to inhibit both initiation and propagation of lipid and protein oxidation; this property is not found with other lipid-soluble antioxidants, for example, vitamin E. Since it is formed in all tissues, redistribution from the serum is not needed, thus serum CoQ10 concentration may not reflect the tissue CoQ10 concentration (Fig. EI.14).

CoQ10 is an obligatory component of the mitochondrial ETC. It is a membrane-bound carrier for high-energy electrons as they transverse the inner mitochondrial membrane through complexes I, II, and III; it also stabilizes respiratory complexes, controls electron flow, and regulates the flow of reducing equivalents as discussed elsewhere (refer to Chapter 5: Diabetic Cardiomyopathy and Congestive Heart Failure: Mitochondrial Production and Management of Reactive Oxygen Species). Other known functions of CoQ10, in addition to its function as a lipid-soluble

FIGURE EI.14 **The chemical structure of ubiquinone (coenzyme Q).** CoQ10 is a 1,4-benzoquinone. Q refers to the quinone group. Human ubiquinone has 10 isoprene units in its tail. *Source: Figure by Krishnavedala (own work) (Public domain), via Wikimedia Commons.*

antioxidant, include the following (Turunen et al., 2004; Bentinger et al., 2007):

- Electron carrier in the mitochondrial respiratory chain;
- Extramitochondrial electron transport
  - as in the plasma membrane and lysosomes;
- Regulator of the mitochondrial permeability transition pores
  - prevents oxidative stress-induced activation of apoptotic cascades;
- Regulator of mitochondrial uncoupling protein activation;
- Regulator of physiochemical properties of biomembranes;
- Modulator of B2-integrin expression on blood monocyte surface;
- Possible role in endothelial function
  - by increasing nitric oxide;
- Reported to participate in sulfide oxidation (in yeast) and introduction of disulfide bonds (in bacteria) (Fig. EI.15).

Dietary sources of CoQ10 include organ meats such as liver, kidney, and heart, as well as beef, sardines, and mackerel. Vegetable sources include spinach, broccoli, and cauliflower; legumes such as peanuts and soybeans are the best vegetarian protein sources. As indicated in Fig. EI.15, most CoQ10 is endogenously synthesized through the mevalonate pathway. A recent meta-analysis revealed that statin drugs significantly inhibit CoQ10 formation (Banach et al., 2015), but the clinical significance of this finding is not clear. Controversy remains as to supplemental recommendations for patients with diseases involving oxidative stress such as for patients with metabolic syndrome, cardiovascular disease, and diabetes (Stocker and Macdonald, 2013; Mortensen et al., 2014).

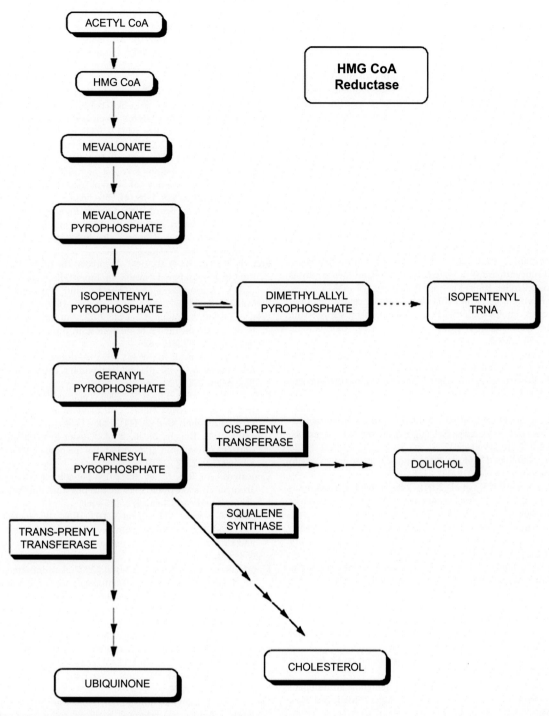

FIGURE EI.15 **Biosynthesis of ubiquinone (CoQ10).** CoQ10 is synthesized from acetyl CoA as illustrated. The rate-limiting enzyme for mevalonate synthesis is HMGCoA reductase; this intermediate is a precursor for synthesis, not only of cholesterol, but of ubiquinone and other essential isoprenoids. HMGCoA reductase is the target of statin drugs used widely to reduce serum cholesterol levels. *Source: Figure by Rnair (own work) (Public domain), via Wikimedia Commons.*

References

Aggarwal, B.B., Sung, B., 2009. Pharmacological basis for the role of curcumin in chronic diseases: an age-old spice with modern targets. Trends Pharmacol. Sci. 30 (2), 85–94.

Albanes, D., Heinonen, O.P., et al., 1996. α-tocopherol and β-carotene supplements and lung cancer incidence in the alpha-tocopherol, beta-carotene cancer prevention study: effects of base-line characteristics and study compliance. J. Natl. Cancer Inst. 88 (21), 1560–1570.

Andreyev, A.Y., Kushnareva, Y.E., et al., 2015. Mitochondrial ROS metabolism: 10 years later. Biochemistry (Mosc). 80 (5), 517–531.

Arnér, E.S.J., Holmgren, A., 2000. Physiological functions of thioredoxin and thioredoxin reductase. Eur. J. Biochem. 267 (20), 6102–6109.

Ashoori, M., Saedisomeolia, A., 2014. Riboflavin (vitamin B2) and oxidative stress: a review. Br. J. Nutr. 111 (11), 1985–1991.

Balaban, R.S., Nemoto, S., et al., 2005. Mitochondria, oxidants, and aging. Cell 120 (4), 483–495.

Banach, M., Serban, C., et al., 2015. Statin therapy and plasma coenzyme Q10 concentrations—a systematic review and meta-analysis of placebo-controlled trials. Pharmacol. Res. 99, 329–336.

Barroso, M., Florindo, C., et al., 2014. Inhibition of cellular methyltransferases promotes endothelial cell activation by suppressing glutathione peroxidase 1 protein expression. J. Biol. Chem. 289 (22), 15350–15362.

Bay, J., Kohlhaas, M., et al., 2013. Intracellular Na^+ and cardiac metabolism. J. Mol. Cell. Cardiol. 61, 20–27.

Bentinger, M., Brismar, K., et al., 2007. The antioxidant role of coenzyme Q. Mitochondrion 7 (Supplement), S41–S50.

Bharti, A.C., Takada, Y., et al., 2004. Curcumin (diferuloylmethane) inhibits receptor activator of NF-κB ligand-induced NF-κB activation in osteoclast precursors and suppresses osteoclastogenesis. J. Immunol. 172 (10), 5940–5947.

Bjelakovic, G., Nikolova, D., et al., 2013. Antioxidant supplements to prevent mortality. JAMA 310 (11), 1178–1179.

Boettler, U., Sommerfeld, K., et al., 2011. Coffee constituents as modulators of Nrf2 nuclear translocation and ARE (EpRE)-dependent gene expression. J. Nutr. Biochem. 22 (5), 426–440.

Brown, B.G., Zhao, X.-Q., et al., 2001. Simvastatin and niacin, antioxidant vitamins, or the combination for the prevention of coronary disease. N Engl. J. Med. 345 (22), 1583–1592.

Buettner, G.R., Jurkiewicz, B.A., 1996. Catalytic metals, ascorbate and free radicals: combinations to avoid. Radiat. Res. 145 (5), 532–541.

Burk, R.F., Hill, K.E., et al., 2001. Plasma selenium in specific and non-specific forms. Biofactors 14 (1–4), 107–114.

Burk, R.F., Hill, K.E., 2015. Regulation of selenium metabolism and transport. Annu. Rev. Nutr. 35 (1), 109–134.

Calabrese, E.J., Baldwin, L.A., 2002. Defining hormesis. Hum. Exp. Toxicol. 21 (2), 91–97.

Clarke, R., Daly, L., et al., 1991. Hyperhomocysteinemia: an independent risk factor for vascular disease. N Engl. J. Med. 324 (17), 1149–1155.

Clément, M.-V., Ramalingam, J., et al., 2001. The in vitro cytotoxicity of ascorbate depends on the culture medium used to perform the assay and involves hydrogen peroxide. Antioxid. Redox. Signal. 3 (1), 157–163.

Culotta, V.C., Yang, M., et al., 2006. Activation of superoxide dismutases: putting the metal to the pedal. Biochim. Biophys. Acta (BBA)-Mol. Cell Res. 1763 (7), 747–758.

De Tullio, M.C., 2012. Beyond the antioxidant: the double life of vitamin C. Subcell. Biochem. 56, 49–65.

De Tullio, M.C., Arrigoni, O., 2004. Hopes, disillusions and more hopes from vitamin C. Cell. Mol. Life Sci. 61 (2), 209–219.

Dodson, M., Redmann, M., et al., 2015. KEAP1–NRF2 signalling and autophagy in protection against oxidative and reductive proteotoxicity. Biochem. J. 469 (3), 347–355.

Evans, H.M., Bishop, K.S., 1922. On the existence of a hitherto unrecognized dietary factor essential for reproduction. Science. 56 (1458), 650–651.

Fattori, V., Pinho-Ribeiro, F.A., et al., 2015. Curcumin inhibits superoxide anion-induced pain-like behavior and leukocyte recruitment by increasing Nrf2 expression and reducing NF-κB activation. Inflamm. Res. 64 (12), 993–1003.

Fenton, H.J.H., 1894. LXXIII.-oxidation of tartaric acid in presence of iron. J. Chem. Soc. Trans. 65 (0), 899–910.

Festenstein, G.N., Heaton, F.W., et al., 1955. A constituent of the unsaponifiable portion of animal tissue lipids (lambda max. 272 m mu). Biochem. J. 59 (4), 558–566.

Food and Nutrition Board, 1998. Dietary Reference Intakes: Thiamin, Riboflavin, Niacin, Vitamin B6, Folate, Vitamin B12, Pantothenic Acid, Biotin, and Choline. Institute of Medicine, National Academy Press, Washington, DC.

Frei, B., Stocker, R., et al., 1990. Ascorbate: the most effective antioxidant in human blood plasma. In: Emerit, I., Packer, L., Auclair, C. (Eds.), Antioxidants in Therapy and Preventive Medicine, vol. 264. Springer, Boston, MA, pp. 155–163.

Frei, B., Birlouez-Aragon, I., et al., 2012. Authors' perspective: what is the optimum intake of vitamin C in humans? Crit. Rev. Food Sci. Nutr. 52 (9), 815–829.

Ghiselli, A., Serafini, M., et al., 1995. A fluorescence-based method for measuring total plasma antioxidant capability. Free Radic. Biol. Med. 18 (1), 29–36.

Ghiselli, A., Serafini, M., et al., 2000. Total antioxidant capacity as a tool to assess redox status: critical view and experimental data. Free Radic. Biol. Med. 29 (11), 1106–1114.

Goncalves, A., Roi, S., et al., 2014. Cluster-determinant 36 (CD36) impacts on vitamin E postprandial response. Mol. Nutr. Food Res. 58 (12), 2297–2306.

Goncalves, A., Roi, S., et al., 2015. Fat-soluble vitamin intestinal absorption: absorption sites in the intestine and interactions for absorption. Food Chem. 172, 155–160.

Gupta, S.C., Kim, J.H., et al., 2011. Role of nuclear factor-κB-mediated inflammatory pathways in cancer-related symptoms and their regulation by nutritional agents. Exp. Biol. Med. 236 (6), 658–671.

Hämäläinen, M., Nieminen, R., et al., 2007. Anti-inflammatory effects of flavonoids: genistein, kaempferol, quercetin, and daidzein inhibit STAT-1 and NF-κB activations, whereas flavone, isorhamnetin, naringenin, and pelargonidin inhibit only NF-κB activation along with their inhibitory effect on iNOS expression and NO production in activated macrophages. Mediat. Inflamm. 2007, 10.

Harman, D., 1956. Aging: a theory based on free radical and radiation chemistry. J. Gerontol. 11 (3), 298–300.

Hayden, M.S., Ghosh, S., 2004. Signaling to NF-κB. Genes Dev. 18 (18), 2195–2224.

Hayes, J.D., Dinkova-Kostova, A.T., 2014. The Nrf2 regulatory network provides an interface between redox and intermediary metabolism. Trends Biochem. Sci. 39 (4), 199–218.

Itoh, S., Ozumi, K., et al., 2009. Novel mechanism for regulation of extracellular SOD transcription and activity by copper: role of antioxidant-1. Free Radic. Biol. Med. 46 (1), 95–104.

Jiang, Q., 2014. Natural forms of vitamin E: metabolism, antioxidant and anti-inflammatory activities and the role in disease prevention and therapy. Free Radic. Biol. Med. 72, 76–90.

Jiang, Q., Elson-Schwab, I., et al., 2000. γ-tocopherol and its major metabolite, in contrast to α-tocopherol, inhibit cyclooxygenase activity in macrophages and epithelial cells. Proc. Natl. Acad. Sci. 97 (21), 11494–11499.

Jiang, Q., Yin, X., et al., 2008. Long-chain carboxychromanols, metabolites of vitamin E, are potent inhibitors of cyclooxygenases. Proc. Natl. Acad. Sci. 105 (51), 20464–20469.

Johnston, C.S., Taylor, C.A., et al., 2000. More Americans are eating "5 a day" but intakes of dark green and cruciferous vegetables remain low. J. Nutr. 130 (12), 3063–3067.

Jones, D.P., 2008. Radical-free biology of oxidative stress. Am. J. Physiol. Cell Physiol. 295 (4), C849–C868.

Kadiiska, M.B., Peddada, S., et al., 2015. Biomarkers of oxidative stress study VI. Endogenous plasma antioxidants fail as useful biomarkers of endotoxin-induced oxidative stress. Free Radic. Biol. Med. 81, 100–106.

Kensler, T.W., Wakabayashi, N., 2010. Nrf2: friend or foe for chemoprevention? Carcinogenesis 31 (1), 90–99.

Kensler, T.W., Egner, P.A., et al., 2013. Keap1-Nrf2 signaling: a target for cancer prevention by sulforaphane. Top. Curr. Chem. 329, 163–177.

Koivisto, P., Robins, P., et al., 2004. Demethylation of 3-methylthymine in DNA by bacterial and human DNA dioxygenases. J. Biol. Chem. 279 (39), 40470–40474.

Kuiper, C., Dachs, G.U., et al., 2014. Intracellular ascorbate enhances hypoxia-inducible factor (HIF)-hydroxylase activity and preferentially suppresses the HIF-1 transcriptional response. Free Radic. Biol. Med. 69, 308–317.

Labunskyy, V.M., Hatfield, D.L., et al., 2014. Selenoproteins: molecular pathways and physiological roles. Physiol. Rev. 94 (3), 739–777.

La Du, B.N., Zannoni, V.G., 1961. The role of ascorbic acid in tyrosine metabolism. Ann. N. Y. Acad. Sci. 92 (1), 175–191.

Lee, Y.-K., Park, O.J., 2013. Soybean isoflavone genistein regulates apoptosis through NF-κB dependent and independent pathways. Exp. Toxicol. Pathol. 65 (1–2), 1–6.

Lei, X.G., Evenson, J.K., et al., 1995. Glutathione peroxidase and phospholipid hydroperoxide glutathione peroxidase are differentially regulated in rats by dietary selenium. J. Nutr. 125 (6), 1438–1446.

Leo, M.A., Lieber, C.S., 1999. Alcohol, vitamin A, and ß-carotene: adverse interactions, including hepatotoxicity and carcinogenicity. Am. J. Clin. Nutr. 69 (6), 1071–1085.

León-González, A.J., Auger, C., et al., 2015. Pro-oxidant activity of polyphenols and its implication on cancer chemoprevention and chemotherapy. Biochem. Pharmacol. 98 (3), 371–380.

Levine, M., Conry-Cantilena, C., et al., 1996. Vitamin C pharmacokinetics in healthy volunteers: evidence for a recommended dietary allowance. Proc. Natl. Acad. Sci. USA 93 (8), 3704–3709.

Lorente, L., Martín, M.M., et al., 2015. Association between serum total antioxidant capacity and mortality in severe septic patients. J. Crit. Care 30 (1), 217.e212–217.e217.

Lu, T., Stark, G.R., 2015. NF-κB: regulation by methylation. Cancer Res. 75 (18), 3692–3695.

Lykkesfeldt, J., Michels, A.J., et al., 2014. Vitamin C. Adv. Nutr. Int. Rev. J. 5 (1), 16–18.

Mandl, J., Szarka, A., et al., 2009. Vitamin C: update on physiology and pharmacology. Br. J. Pharmacol. 157 (7), 1097–1110.

Margittai, É., Bánhegyi, G., et al., 2005. Scurvy leads to endoplasmic reticulum stress and apoptosis in the liver of Guinea pigs. J. Nutr. 135 (11), 2530–2534.

Margulis, L., 2010. Symbiogenesis. A new principle of evolution rediscovery of Boris Mikhaylovich Kozo-Polyansky (1890–1957). Paleontol. J. 44 (12), 1525–1539.

McCann, J.C., Ames, B.N., 2011. Adaptive dysfunction of selenoproteins from the perspective of the triage theory: why modest selenium deficiency may increase risk of diseases of aging. FASEB J. 25 (6), 1793–1814.

McCord, J.M., Fridovich, I., 1988. Superoxide dismutase: the first twenty years (1968–1988). Free Radic. Biol. Med. 5 (5), 363–369.

McCully, K.S., 1971. Homocysteine metabolism in scurvy, growth and arteriosclerosis. Nature 231 (5302), 391–392.

McCully, K., 1993. Chemical pathology of homocysteine. I. Atherogenesis. Ann. Clin. Lab. Sci. 23 (6), 477–493.

Meister, A., Bukenberger, M.W., 1962. Enzymatic conversion of D-glutamic acid to D-pyrrolidone carboxylic acid by mammalian tissues. Nature 194 (4828), 557–559.

Michels, A.J., Frei, B., 2013. Myths, artifacts, and fatal flaws: identifying limitations and opportunities in vitamin C research. Nutrients 5 (12), 5161–5192.

Miller III, E.R., Pastor-Barriuso, R., et al., 2005. Meta-analysis: high-dosage vitamin e supplementation may increase all-cause mortality. Ann. Intern. Med. 142 (1), 37–46.

Monfort, A., Wutz, A., 2013. Breathing-in epigenetic change with vitamin C. EMBO Rep. 14 (4), 337–346.

Mortensen, S.A., Rosenfeldt, F., et al., 2014. The effect of coenzyme Q10 on morbidity and mortality in chronic heart failure: results from Q-SYMBIO: a randomized double-blind trial. JACC Heart Fail. 2 (6), 641–649.

Murphy, M.P., 2009. How mitochondria produce reactive oxygen species. Biochem. J. 417 (1), 1–13.

Mustafa, A.K., Gadalla, M.M., et al., 2009. H2S signals through protein S-sulfhydration. Sci. Signal. 2 (96), ra72-ra72.

Nakai, K., Fujii, H., et al., 2014. Vitamin D activates the Nrf2-Keap1 antioxidant pathway and ameliorates nephropathy in diabetic rats. Am. J. Hypertens. 27 (4), 586–595.

Nickel, A., Kohlhaas, M., et al., 2014. Mitochondrial reactive oxygen species production and elimination. J. Mol. Cell. Cardiol. 73, 26–33.

Olsen, R.K.J., Cornelius, N., et al., 2013. Genetic and cellular modifiers of oxidative stress: what can we learn from fatty acid oxidation defects? Mol. Genet. Metab. 110 (Supplement), S31–S39.

Olsen, R.J., Cornelius, N., et al., 2015. Redox signalling and mitochondrial stress responses; lessons from inborn errors of metabolism. J. Inherit. Metab. Dis. 38 (4), 703–719.

Olson, O.E., 1986. Selenium toxicity in animals with emphasis on man. Int. J. Toxicol. 5 (1), 45–70.

Olson, G.E., Winfrey, V.P., et al., 2008. Megalin mediates selenoprotein P uptake by kidney proximal tubule epithelial cells. J. Biol. Chem. 283 (11), 6854–6860.

Packer, J.E., Slater, T.F., et al., 1979. Direct observation of a free radical interaction between vitamin E and vitamin C. Nature 278 (5706), 737–738.

Panel on Dietary Antioxidants and Related Compounds, Standing Committee on the Scientific Evaluation of Dietary Reference Intakes, Standing Committee on the Scientific Evaluation of Dietary Reference Intakes, Food and Nutrition Board, Institute of Medicine, 2000. Dietary Reference Intakes: Vitamin C, Vitamin E, Selenium, and Carotenoids. National Academies Press, Washington, DC.

Pastori, D., Carnevale, R., et al., 2013. Vitamin E serum levels and bleeding risk in patients receiving oral anticoagulant therapy: a retrospective cohort study. J. Am. Heart Assoc. 2 (6), e000364.

Peterkofsky, B., 1991. Ascorbate requirement for hydroxylation and secretion of procollagen: relationship to inhibition of collagen synthesis in scurvy. Am. J. Clin. Nutr. 54 (6), 1135S–1140S.

Powers, H.J., 2003. Riboflavin (vitamin B-2) and health. Am. J. Clin. Nutr. 77 (6), 1352–1360.

Reboul, E., Borel, P., 2011. Proteins involved in uptake, intracellular transport and basolateral secretion of fat-soluble vitamins and carotenoids by mammalian enterocytes. Prog. Lipid Res. 50 (4), 388–402.

Reboul, E., Soayfane, Z., et al., 2012. Respective contributions of intestinal Niemann-Pick C1-like 1 and scavenger receptor class B type I to cholesterol and tocopherol uptake: in vivo v. in vitro studies. Br. J. Nutr. 107 (09), 1296–1304.

Reiter, E., Jiang, Q., et al., 2007. Anti-inflammatory properties of α- and γ-tocopherol. Mol. Aspects Med. 28 (5-6), 668–691.

Rhee, S.G., Chae, H.Z., et al., 2005. Peroxiredoxins: a historical overview and speculative preview of novel mechanisms and emerging concepts in cell signaling. Free Radic. Biol. Med. 38 (12), 1543–1552.

Rietjens, I.M.C.M., Boersma, M.G., et al., 2002. The pro-oxidant chemistry of the natural antioxidants vitamin C, vitamin E, carotenoids and flavonoids. Environ. Toxicol. Pharmacol. 11 (3–4), 321–333.

Ristow, M., 2014. Unraveling the truth about antioxidants: mitohormesis explains ROS-induced health benefits. Nat. Med. 20 (7), 709–711.

Ristow, M., Zarse, K., et al., 2009. Antioxidants prevent health-promoting effects of physical exercise in humans. Proc. Natl. Acad. Sci. USA 106 (21), 8665–8670.

Rumsey, S.C., Levine, M., 1998. Absorption, transport, and disposition of ascorbic acid in humans. J. Nutr. Biochem. 9 (3), 116–130.

Sagan, L., 1967. On the origin of mitosing cells. J. Theor. Biol. 14 (3), 225–IN226.

Salminen, A., Lehtonen, M., et al., 2008. Terpenoids: natural inhibitors of NF-κB signaling with anti-inflammatory and anticancer potential. Cell. Mol. Life Sci. 65 (19), 2979–2999.

Schofield, C.J., Ratcliffe, P.J., 2004. Oxygen sensing by HIF hydroxylases. Nat. Rev. Mol. Cell. Biol. 5 (5), 343–354.

Schweizer, U., Streckfuß, F., et al., 2005. Hepatically derived selenoprotein P is a key factor for kidney but not for brain selenium supply. Biochem. J. 386 (Pt 2), 221–226.

Sena, L.A., Chandel, N.S., 2012. Physiological roles of mitochondrial reactive oxygen species. Mol. Cell. 48 (2), 158–167.

Sibrian-Vazquez, M., Escobedo, J.O., et al., 2010. Homocystamides promote free-radical and oxidative damage to proteins. Proc. Natl. Acad. Sci. 107 (2), 551–554.

St. Clair, D.K., Oberley, T.D., et al., 1994. Expression of manganese superoxide dismutase promotes cellular differentiation. Free Radic. Biol. Med. 16 (2), 275–282.

Stefanson, A., Bakovic, M., 2014. Dietary regulation of Keap1/Nrf2/ARE pathway: focus on plant-derived compounds and trace minerals. Nutrients 6 (9), 3777–3801.

Stocker, R., Macdonald, P., 2013. The benefit of coenzyme Q10 supplements in the management of chronic heart failure: a long tale of promise in the continued absence of clear evidence. Am. J. Clin. Nutr. 97 (2), 233–234.

Sun, Q., Collins, R., et al., 2009. Structural basis for the inhibition mechanism of human cystathionine γ-lyase, an enzyme responsible for the production of H2S. J. Biol. Chem. 284 (5), 3076–3085.

Suzuki, T., Yamamoto, M., 2015. Molecular basis of the Keap1–Nrf2 system. Free Radic. Biol. Med. 88 (Part B), 93–100.

Taguchi, K., Fujikawa, N., et al., 2012. Keap1 degradation by autophagy for the maintenance of redox homeostasis. Proc. Natl. Acad. Sci. 109 (34), 13561–13566.

Tapia, P.C., 2006. Sublethal mitochondrial stress with an attendant stoichiometric augmentation of reactive oxygen species may precipitate many of the beneficial alterations in cellular physiology produced by caloric restriction, intermittent fasting, exercise and dietary phytonutrients: "Mitohormesis"; for health and vitality. Med. Hypotheses 66 (4), 832–843.

Tapia, E., Soto, V., et al., 2012. Curcumin induces Nrf2 nuclear translocation and prevents glomerular hypertension, hyperfiltration, oxidant stress, and the decrease in antioxidant enzymes in 5/6 nephrectomized rats. Oxid. Med. Cell. Longev. 2012, 14.

Traber, M.G., 2013. Mechanisms for the prevention of vitamin E excess. J. Lipid Res. 54 (9), 2295–2306.

Turunen, M., Olsson, J., et al., 2004. Metabolism and function of coenzyme Q. Biochim. Biophys. Acta (BBA) Biomembr. 1660 (1–2), 171–199.

Ursini, F., Heim, S., et al., 1999. Dual function of the selenoprotein PHGPx during sperm maturation. Science 285 (5432), 1393–1396.

van Haaften, R.I.M., Haenen, G.R.M.M., et al., 2003. Inhibition of various glutathione S-transferase isoenzymes by RRR-α-tocopherol. Toxicol. In Vitro 17 (3), 245–251.

Vrolijk, M.F., Opperhuizen, A., et al., 2015. The shifting perception on antioxidants: the case of vitamin E and β-carotene. Redox Biol. 4, 272–278.

Wade, L., Nadeem, N., et al., 2013. α-tocopherol induces proatherogenic changes to HDL2 & HDL3: an in vitro and ex vivo investigation. Atherosclerosis 226 (2), 392–397.

Wang, R., 2012. Physiological implications of hydrogen sulfide: a whiff exploration that blossomed. Physiol. Rev. 92 (2), 791–896.

Wang, X., Hai, C., 2015. Redox modulation of adipocyte differentiation: hypothesis of "Redox Chain" and novel insights into intervention of adipogenesis and obesity. Free Radic. Biol. Med. 89, 99–125.

Winterbourn, C.C., Hampton, M.B., 2008. Thiol chemistry and specificity in redox signaling. Free Radic. Biol. Med. 45 (5), 549–561.

Wollesen, F., Brattstrom, L., et al., 1999. Plasma total homocysteine and cysteine in relation to glomerular filtration rate in diabetes mellitus. Kidney Int. 55 (3), 1028–1035.

Yang, G., Zhao, K., et al., 2012. Hydrogen sulfide protects against cellular senescence via S-sulfhydration of Keap1 and activation of Nrf2. Antioxid. Redox. Signal. 18 (15), 1906–1919.

Yekollu, S.K., Thomas, R., et al., 2011. Targeting curcusomes to inflammatory dendritic cells inhibits NF-κB and improves insulin resistance in obese mice. Diabetes 60 (11), 2928–2938.

Yin, J., Ren, W., et al., 2016. L-cysteine metabolism and its nutritional implications. Mol. Nutr. Food Res. 60 (1), 134–146.

Yoshida, Y., Saito, Y., et al., 2007. Chemical reactivities and physical effects in comparison between tocopherols and tocotrienols: physiological significance and prospects as antioxidants. J. Biosci. Bioeng. 104 (6), 439–445.

Yun, J., Finkel, T., 2014. Mitohormesis. Cell Metab. 19 (5), 757–766.

Zelko, I.N., Mariani, T.J., et al., 2002. Superoxide dismutase multigene family: a comparison of the CuZn-SOD (SOD1), Mn-SOD (SOD2), and EC-SOD (SOD3) gene structures, evolution, and expression. Free Radic. Biol. Med. 33 (3), 337–349.

Zeng, C., Zhong, P., et al., 2015. Curcumin protects hearts from FFA-induced injury by activating Nrf2 and inactivating NF-κB both in vitro and in vivo. J. Mol. Cell. Cardiol. 79, 1–12.

Zuo, X., Dong, D., et al., 2013. Homocysteine restricts copper availability leading to suppression of cytochrome c oxidase activity in phenylephrine-treated cardiomyocytes. PLoS ONE 8 (6), e67549.

Essentials II: Heavy Metals, Retinoids, and Precursors

Gold is for the mistress, silver for the maid – Copper for the craftsman, cunning at his trade.

"Good!" cried the Baron, sitting in his hall, "But iron, cold iron, is the master of them all." *From Cold Iron by Rudyard Kipling, 1910.*

Swiss longsword 15th or 16th century.

Figure by Rama (own work) [CeCILL (http://www.cecill.info/licences/Licence_CeCILL_V2-en.html) or CC BY-SA 2.0 fr (http://creativecommons.org/licenses/by-sa/2.0/fr/deed.en)], via Wikimedia Commons.

Heavy metals and retinoids have been grouped together because while they are essential for life, they also are potentially very toxic. Each is regulated by a constellation of influx and efflux transporters, chaperones, and storage proteins that prevent the free nutrient from causing tissue damage. While deficiency symptoms of these nutrients have been relatively well studied, the possibility that degenerative diseases in the brain and other organ systems can be caused or exacerbated by dysregulation of nutrient transport, distribution, and/or storage is under intensive investigation (Lutsenko et al., 2010; Craddock et al., 2012; Hare et al., 2013). Of particular interest is the recent link between high intracellular iron, retinol signaling, and the pathogenesis of obesity and its comorbid conditions (Trasino and Gudas, 2015), as discussed below.

EII.1 IRON

Iron is a nutrient essential for all organisms, humans as well as microbes. Plants, bacteria, and fungi take up iron by releasing low-molecular-weight siderophores (from the Greek: iron carrier) that bind ferric iron (Fe^{3+}) by high-affinity ligands (Hider and Kong, 2010). Uptake by the organism is followed by Fe^{3+} reduction to the ferrous form (Fe^{2+}) for metabolic utilization (Ganz and Nemeth, 2015). Pathogens such as *Salmonella enterica* (Boyer et al., 2002) and others including *Escherichia coli*, mycobacteria, *Shigella*, and S*taphylococcus*, and parasites such as *Plasmodia* (Weinberg, 1999) use siderophores and other systems to extract iron in an iron-poor environment to support their proliferation and metabolism (Andrews et al., 2003).

Most of the 3–4 g of elemental iron in the adult human body is taken up by the bone marrow and incorporated into heme for the synthesis of hemoglobin. Iron as heme is also used for myoglobin biosynthesis and for heme and iron–sulfur-containing enzymes in the mitochondrial respiratory chain. Ionic iron is required for a host of enzymes including myeloperoxidase (for oxygen-dependent killing in phagocytic cells) and the peroxisomal enzyme, catalase, that converts H_2O_2 to water.

The human diet contains iron in multiple forms. Iron incorporated into organic molecules such as a porphyrin ring (heme) obtained from animal flesh and similarly, iron bound into a ferritin protein cage obtained from legumes and seeds are both well absorbed by specific intestinal transporters (Theil et al., 2012). Unbound inorganic Fe^{3+} salts and complexes are taken up by the divalent metal transporter 1 (DMT1) that requires proton cotransport (Theil, 2011; Theil et al., 2012). Dietary iron absorption takes place in the proximal duodenum and is modified by complex interactions with zinc, copper, calcium, and ascorbate (Scheers, 2013). Within the enterocyte, iron is released from heme and ferritin, reduced to Fe^{2+}, enters the enterocyte iron pool, and is bound to chaperones for cytoplasmic storage or transport.

Iron is exported from the enterocyte by the basolateral iron transporter, ferroportin (FPN; SLC40A1). FPN is upregulated by the zinc-dependent metal transcription factor 1 (MTF-1) and is expressed on all cells that release iron, including enterocytes, macrophages, hepatocytes, and the placental syncytiotrophoblasts. Iron export requires participation of copper-containing ferroxidases (ceruloplasmin, hephaestin, and others) that oxidize ferrous iron to the ferric form (Fe^{3+}), allowing its uptake by apo-transferrin. It is thought that continuous release of ferric iron from the cell surface maintains a favorable gradient that drives FPN-dependent iron export (Ganz, 2013). Transferrin (Tf) transports iron to target cells where it binds to transferrin receptors (TfR1; TfR2) expressed on the plasma membrane. Iron that enters the plasma in excess of Tf binding capacity can be bound to citric acid or proteins such as serum albumin, thought to be the carrier of nontransferrin-bound iron (NTBI) identified in patients with iron overload.

All cells express TfR to facilitate iron uptake. Posttranscriptional TfR expression is controlled by iron-responsive proteins (IRP) that bind to the iron response element (IRE) on the mRNA and stimulate translation. Following Tf uptake at the target cell by receptor-mediated endocytosis, transferrin-bound iron is exposed to acid in the lysosome, iron is released from Tf, iron is reduced from Fe^{3+} to Fe^{2+} and is utilized by the cell.

The hepatocyte controls iron homeostasis as reviewed by Rochette et al. (2015). Because organisms excrete minimal amounts of iron, iron homeostasis is controlled largely at the absorption step by the synthesis of hepatic bacteriocidal protein (hepcidin) in response to a high iron load. Hepcidin is transported to the enterocyte where it facilitates the degradation of the iron efflux protein, FPN. Deprived of its efflux mechanism, the enterocyte accumulates iron and when the cell is sloughed within 2–5 days, the accumulated iron is excreted. Hepcidin is cleared by urinary excretion and is also known to be bound to α-2 macroglobulin and possibly other proteins and degraded in target cells as illustrated in Fig. EII.1.

The molecular pathways that control iron absorption are complex and still being unraveled (Rochette et al., 2015). Briefly, hepatocytes take up iron on TfR. *HFE* (the gene associated with hereditary hemochromatosis) is involved in monitoring body iron status and directing the hepcidin response as depicted in Fig. EII.2. The HFE gene product interacts with TrF at the Tf binding site. When this site is bound by Tf, ERK and p38 MAP kinase pathways are activated, inducing hepcidin expression. It should be noted that at least 14 different mutations in these pathways have been identified that can dysregulate iron homeostasis.

Autocrine upregulation of hepcidin has been reported in other cells including adipocytes, macrophages, cardiomyocytes, and pancreatic islets (Datz et al., 2013). In the macrophage, hepcidin is upregulated by inflammatory cytokines (especially IL-6) and by infections, as well as lipopolysaccharide (LPS) and microbial products that bind to the toll-like receptors on the macrophage. It has been suggested that macrophage production of hepcidin may increase microbial killing by iron retention and localization to the phagosome. Macrophages also act as "nurse cells," providing iron to developing erythrocytes. Upregulation of macrophage hepcidin reduces iron efflux to precursor erythroblasts and results in a microcytic anemia that resembles iron-deficiency anemia, but is not associated with absolute iron depletion. At the end of their lifespan, erythrocytes are phagocytosed by splenic macrophages and their iron is recycled.

Adipocytes also produce hepcidin in response to inflammatory mediators, especially IL-6 and C-reactive protein. This could explain the hypoferremia and increased iron load reported in obese patients (Vuppalanchi et al., 2014). Adipocytes in obese subjects produce more hepcidin and store iron, suggesting impaired iron homeostasis (Coimbra et al., 2013). Further work is required to understand the relationship of obesity and diet with iron homeostasis. For example do all fat depots produce hepcidin? How does hepcidin production relate to the rapid death of visceral adipocytes and increased infiltrating macrophages observed in a rat model fed a high-fat diet (Strissel et al., 2007)?

Unbound Fe^{2+} is dangerous. It participates in the Fenton reaction and has a high potential for generating oxidative damage. Ferrous iron (Fe^{2+}) is oxidized by hydrogen peroxide to form ferric iron [Fe^{3+}] and the hydroxyl radical. Fe^{3+} is then reduced back to Fe^{2+}, again by hydrogen peroxide, to form a hydroperoxyl radical. In this way hydrogen peroxide and iron create two different oxygen radical species. [1] $Fe^{2+} + H_2O_2 \rightarrow Fe^{3+} + HO\bullet + OH^-$; [2] $Fe^{3+} + H_2O_2 \rightarrow$

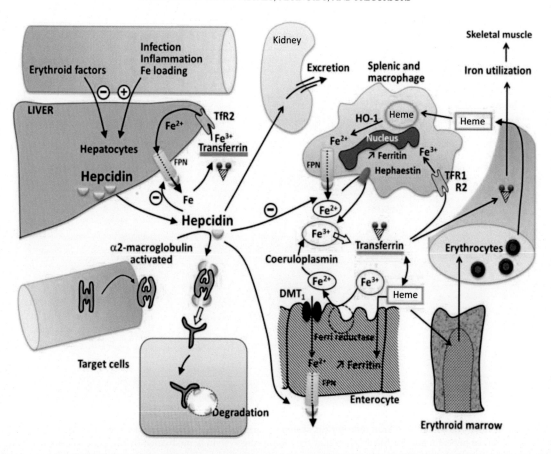

FIGURE EII.1 **Iron homeostasis.** Dietary iron is taken up by the divalent metal transporter 1 (DMT1) on the apical membrane of the enterocyte; heme iron and possibly ferritin are absorbed by specific transporters (not shown). Iron leaves the cell by ferroportin (FPN) and is reduced by copper-containing oxidases such as ceruloplasmin and hephaestin for transfer by transferrin (Tf) and for storage as ferritin. Erythrocyte production requires bone marrow macrophage transfer of iron; macrophages, as most other cells, take up iron from Tf via their TfRs. Senile red cells are phagocytized by splenic macrophages and their heme is degraded by heme oxidase (HO-1) for efflux by FPN or storage as ferritin. Hepcidin clearance is through urinary excretion or binding to α-2 macroglobulin and degradation in the target cell (Rochette et al., 2015). *Source: Reproduced from Rochette, L., Gudjoncik, A., et al., 2015. The iron-regulatory hormone hepcidin: a possible therapeutic target? Pharmacol. Ther. 146, 35–52 with permission from Science Direct*

$Fe^{2+} + HOO\bullet + H^+$. Iron is stored in the cytoplasm as ferritin, a globular protein complex with 24 subunits that allows controlled uptake and release of the potentially toxic metal. Under iron-replete conditions, apoferritin binds to free Fe^{2+} iron and oxidizes it to the insoluble Fe^{3+} form for storage. Although most ferritin is intracellular, a small fraction of ferritin circulates in the serum and is taken as an indicator of total body iron content. It must also be noted that ferritin is posttranscriptionally regulated by iron and also is an acute-phase protein elevated in response to inflammation. It is possible that increased iron retention in the cell due to inflammation-induced hepcidin secretion stimulates ferritin upregulation by the mechanism described below (Gabrielsen et al., 2012). Iron is also found in hemosiderin, a large, intracellular yellow-brown insoluble protein produced by phagocytic digestion of iron. It is believed to represent aggregates of ferritin, but with much higher iron content (Fig. EII.3).

The elegant system for regulation of iron utilization and storage was discovered almost 30 years ago (Casey et al., 1988) and has been recently reviewed (Wilkinson and Pantopoulos, 2014). Briefly, the concentrations of transferrin receptor (utilization) and ferritin (storage) are posttranslationally regulated by iron availability. The mRNAs for these proteins contain highly conserved IRE sequences. The 5′ untranslated region of the mRNA for H and L ferritin contains one IRE, while the 3′ untranslated region of the mRNA for transferrin receptor (TfR1) has five IREs. The IREs are binding sites for two cytoplasmic iron-regulatory proteins, IRP1 and IRP2. In iron deficiency, IRP1 and IRP2 bind the ferritin IRE with high affinity and inhibit its translation. At the same time, they bind to the IREs in TfR1 mRNA and protect it from degradation. In contrast, if iron is too high, IRP1 and IRP2 do not bind, thus the mRNA of ferritin is translated and the mRNA for TfR1 is degraded.

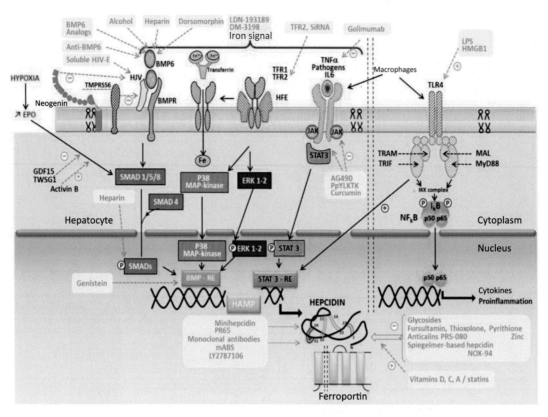

FIGURE EII.2   **Molecular components of iron-monitoring systems.** Current understanding of the systems that monitor iron status and culminate in the synthesis of hepcidin are illustrated. Note that the monitoring system is modulated by dietary components such as genestein, zinc, and vitamins D, C, and A. In addition to iron status, the system is activated by infection, inflammation, and hypoxia (Rochette et al., 2015). *Source: Reproduced from Rochette, L., Gudjoncik, A., et al., 2015. The iron-regulatory hormone hepcidin: a possible therapeutic target? Pharmacol. Ther. 146, 35–52 with permission from Science Direct.*

FIGURE EII.3   **The structure of heme B.** Heme B is the heme found in myoglobin and hemoglobin. Hemoglobin binds oxygen in the lung where the pH is high and CO_2 is low and releases it at the more acidic, CO_2-rich tissues. This property is due to specific amino acids in the vicinity of each of the four hemes on hemoglobin. *Source: Figure by [User:william sam | Yikrazuul] (Template:Defender 007 uss) (Public domain), via Wikimedia Commons.*

EII.1.1 Iron Overload

Because of its reactive nature, excess accumulation of iron has been associated with deleterious effects in the heart, liver, endocrine tissue, and other sites. HFE was the first gene found to be mutated in patients with hemochromatosis but, as indicated above, at least 14 mutations have been identified to date that interfere with the iron-monitoring pathway, as reviewed by Roy (2013). Briefly, in the hepatocyte, the HFE gene product competes with Tf for a binding site on TfR. When Tf displaces HFE from TfR, HFE associates with TfR2 and signals hepcidin production. If HFE is mutated, binding to TrF2 is impaired and hepcidin is not produced and iron absorption is not regulated. Theoretically a mutation in any of the accessory proteins or cofactors in this pathway could also impair iron homeostasis.

Dietary iron requirements are commensurate with iron loss, primarily through bleeding, and the need for growth, as in the growing fetus and child. While the dietary recommended intake (DRI) for both adult men and postmenopausal women is 8 mg/day of elemental iron, premenopausal women require 18 mg/day to compensate for monthly blood loss and pregnant women require 27 mg/day to support fetal growth (Micronutrients, 2001). Lean meat, including fowl and seafood, are rich sources of heme iron. Nonheme sources include nuts, beans, and fortified grains. Many countries, including the United States and Canada, fortify wheat, flours, and cereals with iron and other nutrients. Measurement of iron available from commonly consumed cereals in the amount usually eaten revealed that iron intake may be higher than recommended and contribute to adverse effects (Whittaker et al., 2001) as described elsewhere (iron requirements in specific comorbid conditions are discussed in the appropriate chapters).

EII.2 ZINC

Zinc was recognized as essential for all living things in 1869 (Raulin, 1869), but not demonstrated as essential to humans until 100 years later (Prasad et al., 1961, 1963). As the second most abundant trace element in mammalian physiology, zinc plays multiple roles in intermediary metabolism and is essential for the function of over 300 enzymes (Vallee and Falchuk, 1993). Although it is a divalent cation, Zn^{2+} does not participate in redox activity. It can, however, act as a Lewis acid, accepting a pair of electrons. Zinc is the only metal that is a cofactor for all enzyme types. It plays an essential catalytic cofactor role in >50 enzymes; these enzymes are inactivated upon removal of zinc, and activity is restored with the addition of zinc (Kambe et al., 2015).

Additionally, zinc plays a structural role, for example, in zinc finger proteins. These small protein structural motifs contain one or more zinc ions coordinately bound with cysteine (and sometimes histidine) in order to stabilize the protein into finger-like folds. Zinc fingers act as transcription factors, and regulate expression of a variety of proteins involved in cellular differentiation, proliferation, adhesion, transcription, and other critical functions. Zinc also regulates gene expression in association with multiple transcription factors ($\sim$3000) and with a metal-binding transcription factor (MTF) that binds to a metal response element in the promoter region of the regulated gene. MTF1 regulates diverse activities including lipid peroxidation, apoptosis, immunity, and neuronal function (Vallee and Falchuk, 1993; King, 2011). In addition, Zn^{2+} has been called "the calcium of the 21st century" because of its role in cell signaling (Frederickson et al., 2005).

Dietary zinc is widely distributed in the food supply. It is especially high in animal flesh and seafood; oysters contain the highest concentration of zinc (https://ods.od.nih.gov/factsheets/Zinc-HealthProfessional/). The US DRI for Zn has been set at 8–11 mg/day (Hellwig et al., 2006). However, despite its wide distribution, inadequate zinc intake is common, especially in individuals with low socioeconomic status. A recent study using NHANES 2003–06 data stratified on food insecurity revealed that 12% and 17% of the food insecure population consumed inadequate zinc (Kirkpatrick et al., 2015). One of the reasons for widespread zinc inadequacy is that dietary zinc forms complexes with nonabsorbable organic plant compounds such as phytate (in the seed coat of grains) and dietary fibers. Zinc also competes for absorption with divalent cations such as ferrous iron and calcium in the diet. Thus assessment of zinc intake must be considered in the context of bioavailability inhibitors (Krebs, 2000). For example, zinc in a "meat and potato" meal has few bioavailability inhibitors, whereas the same content of elemental zinc in a vegetarian meal containing plant fibers, inorganic iron, and calcium is less well absorbed. Furthermore, the content of zinc is often lower in vegetarian diets, especially those consumed in developing countries (Foster et al., 2013). Certain populations are especially vulnerable to zinc deficiency, especially those with high requirements such as pregnant and lactating women, infants and growing children, and individuals living on the edge of poverty whose diets provide inadequate quantities of bioavailable zinc (Ruxton et al., 2013).

In the face of widely varying dietary zinc availability, the body can adapt by modifying zinc absorption and excretion. Thus, with a high dietary zinc intake, synthesis of the zinc-binding protein, metallothionein (MT), is induced and binds Zn in the enterocyte until it is utilized or the cell is sloughed. Mammalian MT contains two Zn—cysteine clusters (Zn_4Cys_{11}; Zn_3Cys_9) and binds heavy metals such as Zn, cadmium (Cd), copper (Cu), mercury (Hg), and bismuth (Bi). MT is also induced by hormones (dexamethasone; glucagon; epinephrine; norepinephrine), cytokines (IL-1; IL-6; TNF; INFγ; angiotensin II) and likely by physical and chemical stressers. In contrast, when dietary zinc is low, zinc secreted into the gut with pancreatic enzymes (eg, carboxypeptidase) and other gastrointestinal secretions, comprises a large enteropancreatic/hepatic circulation that provides sufficient endogenous zinc for enterocyte growth and survival.

It must be noted that, as with iron, microbes require zinc. Mammalian hosts have evolved metal-dependent systems for reducing pathogen invasiveness by targeting their requirement for Zn. Members of the S100 family of EF-hand, Ca-binding proteins sequester zinc in the intestinal lumen, thereby starving pathogens and also inducing an inflammatory response and phagocytic chemotaxis to destroy the microbe. Neutrophils (PMN) contain a S100 protein (calprotectin) that binds manganese (Mn) and zinc and enhances sensitivity of bacteria to destruction by superoxide. The proposed mechanism is that by binding Mn, calprotectin inactivates bacterial Mn-superoxide dismutase (MnSOD) (Kehl-Fie et al., 2011).

Zinc distribution is regulated by two gene families of zinc transporters: the zinc transporter (ZnT) and the Zip (Zrt- and Irt-like protein) families. ZnT and Zip proteins appear to have opposite roles in cellular zinc homeostasis: ZnT transporters reduce cytoplasmic zinc concentration by promoting zinc efflux from cells or into intracellular vesicles, while Zip transporters increase cytoplasmic zinc by promoting extracellular and, perhaps, vesicular zinc transport into cytoplasm (Cousins et al., 2006; Lichten and Cousins, 2009). If dietary zinc is limiting, transporters are transcriptionally regulated to distribute zinc to its most needed sites. In states of severe Zn deficit, tissue zinc can also be conserved by reducing insulin-like growth factor 1 (IGF-1), thereby slowing the growth rate (Fukada et al., 2014). Thus, reduced linear growth and failure to thrive in children is a clinical symptom of zinc inadequacy. A recent analysis of the risk/benefit of zinc supplementation to malnourished children reported that the moderate benefits from exogenous Zn outweighed the risk from side effects. On the other hand, they proposed that adequate linear growth would probably have been achieved more effectively and with fewer side effects (nausea, copper malabsorption) with a high-quality, protein- and zinc-replete diet (Mayo-Wilson et al., 2014).

Zinc sequestration in the target cell is accomplished by at least a dozen different MT proteins, each with 60−68 amino acids, approximately one-third of which are cysteines. MT proteins are upregulated by metal-binding transcription factor (MTF-1) to increase cellular zinc-binding capacity in response to increased zinc exposure. MT upregulation is especially high in enterocytes, pancreatic acinar cells, kidney and liver cells. Each MT protein binds seven Zn^{2+} cations into Zn-S clusters; each zinc ion is bound with different affinities, allowing MT to act as a selective zinc acceptor or donor. MT concentrations are upregulated by cytokines, thus zinc sequestered in the hepatocytes can be used for biosynthesis of acute-phase proteins, as recently reviewed by Haase and Rink (2014). Additionally, in trauma, a cytokine-mediated increase in vascular permeability allows Zn-bound albumin to move from the vascular to the extravascular compartment carrying bound Zn. Upregulation of ZIP14 in hepatocytes, white adipose tissue, and muscle is also induced by proinflammatory cytokines, specifically IL-6. While the Zn^{2+} cation does not respond to redox changes, Zn-S clusters respond to the redox state by altering their stability. MT also responds to NO produced in excess by iNOS, releasing metals to the surrounding tissue (Zangger et al., 2001). Thus, MT oxidation leads to Zn^{2+} release, while Zn^{2+} binding is enhanced when MT is reduced.

In mammals, Zn^{2+} absorption is facilitated by ZIP4, localized to the plasma membrane of the enterocyte. Cytoplasmic zinc is bound by MT and other zinc metalloenzymes that regulate zinc content available for transport to the portal blood. Zn not transferred to the blood is lost as the enterocytes slough at the end of their lifespan. Of the ∼2−3 g of zinc in the human body, most (∼60%) is stored in skeletal muscle and in bone (∼30%). Serum zinc comprises only ∼0.1% of total body zinc; 80% of serum zinc is loosely bound to albumin and the remainder is tightly bound to α_2-macroglobulin. ZIP transporters mediate cellular uptake in response to signals; for example, inflammation-induced IL-6 upregulates ZIP14 and increases zinc uptake into human cells (Liuzzi et al., 2005) (Fig. EII.4).

EII.3 COPPER

Copper is a cofactor for enzymes essential for cellular respiration, iron homeostasis, pigment formation, neurotransmitter production, peptide biogenesis, connective tissue biosynthesis, and antioxidant defense. Copper accepts and donates single electrons as it transitions between its two redox states, Cu^+ and Cu^{2+}. When unbound this redox capability allows copper to participate in the Fenton reaction, creating reactive oxygen species (ROS). While initially considered entirely deleterious, Cu-mediated ROS generation has been implicated in the process of macrophage killing and is a major participant in the innate immune response. To control its dangerous oxidative ability, biological organisms tightly regulate copper ion metabolism.

Several trafficking pathways have evolved that bind and compartmentalize copper ions; pathways include influx and efflux transporters. Cu-binding metallochaperones transport the ion to different subcellular sites, for incorporation into cuproenzymes and for storage, largely bound to MT as reviewed by Leary and Winge (2007) and Lutsenko (2010). Copper exists in the body in two valence states: Cu^{1+}, localized to the intracellular compartment comprises ∼95% of total body copper, and Cu^{2+}, present primarily in the extracellular space and comprises the remaining ∼5% (Zhang et al., 2014). Distribution is maintained, in part, by the human copper transporter gene product,

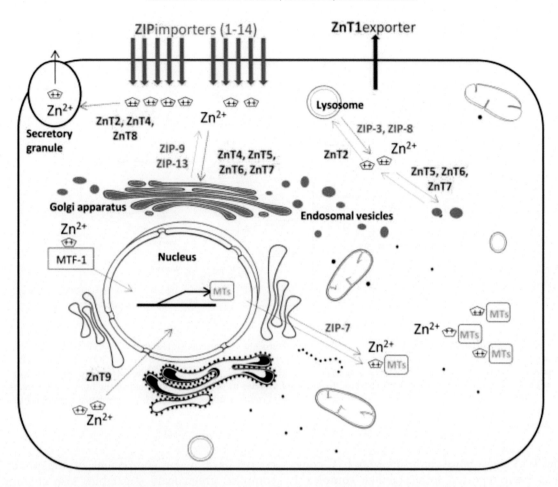

FIGURE EII.4 **Zinc transporter and binding proteins.** *Arrows* indicate the direction of ZIP (green) and ZnT transporters (red). ZIP1, 2, and 4 are induced in zinc-deficient conditions, while ZnT-1 and -2 members are induced by zinc administration. Metallothionein (MT) donates free zinc (Zn^{2+}) to provide Zn for enzymes and other Zn requiring proteins. *Source: Reproduced from Bonaventura, P., Benedetti, G., et al. 2015. Zinc and its role in immunity and inflammation. Autoimmun. Rev. 14 (4), 277—285 (Bonaventura et al., 2015) with permission from Science Direct.*

hCTR1. This small (191 amino acids) protein does not bind Cu^{1+}, but appears to form a transport pore to mediate cellular uptake. Recently, it was demonstrated that hCTR1 is rapidly internalized by the presence of elevated extracellular copper resulting in a dramatic reduction in hCTR1-mediated copper uptake. The CTR1 protein was localized inside the cytoplasm and when extracellular copper was removed, it promptly recycled to the plasma membrane and mediated copper uptake (Molloy and Kaplan, 2009). Thus, CTR1 is an acute reversible mechanism that regulates intracellular copper concentrations to maintain homeostasis (Fig. EII.5).

Copper is an essential cofactor of the electron transfer chain (ETC) and is directly involved in the generation of mitochondrial membrane potential. Cu deficiency induces the formation of ROS, large, misshapen mitochondria, and anemia (Ruiz et al., 2014). Mitochondrial complex IV (cytochrome c oxidase; COX) has two copper centers with a total of three Cu atoms as extensively reviewed by Horn and Barrientos (2008) and discussed elsewhere (refer to Chapter 5: Diabetic Cardiomyopathy and Congestive Heart Failure: Mitochondrial Production and Management of Reactive Oxygen Species). Recently, ^{13}C-metabolic flux analysis of Chinese hamster ovary (CHO) cells was employed to quantitate changes in energy flux with limiting copper availability (Nargund et al., 2015). ^{13}C-MFA quantifies carbon fluxes (rates of reactions) in metabolic networks and provides a system-wide perspective of metabolic restructuring in response to a change in the medium, in this case, Cu availability. The authors reported that Cu depletion increased glycolytic fluxes (25—79% relative to glucose uptake) while lower flux through the TCA (15—23%) and the pentose phosphate pathway (74%) was observed. Reduced pyruvate entry into the TCA cycle compromises fuel oxidation and ATP production. Additionally, 33% of the pyruvate produced by glycolysis was redirected to lactate and malate and not

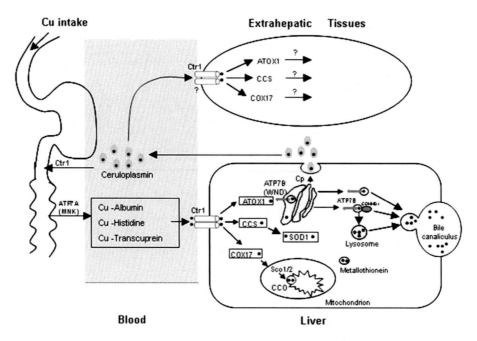

FIGURE EII.5 Copper distribution. Cu is transported from the enterocyte via an energy requiring Cu-ATPase and circulates bound to albumin, histidine, and transcuprin. CTR1 facilitates uptake into the liver where Cu is bound to chaperones (COX17; CCS; ATOX1) for utilization in the mitochondrion, incorporated into Zn-Cu superoxide dismutase (SOD1), or localized to the Golgi network for incorporation into ceruloplasmin. Elevated intracellular Cu can be transferred to secretory vesicles by ATP7B and excreted in the bile. Ceruloplasmin enters the circulation and delivers Cu to extrahepatic tissues, including the intestinal cells for uptake on hCTR1. Cu can be stored bound to metallothionein for later use (Kang, 2011). *Source: Reproduced from Kang, Y.J., 2011. Copper and homocysteine in cardiovascular diseases. Pharmacol. Ther. 129(3): 321–331 with permission from Science Direct.*

through the TCA cycle. Since compromise of ETC complex IV has been reported to destabilize complex I (Suthammarak et al., 2009), these data suggest that Cu deficiency disrupts the ETC, reduces ATP formation, and causes redox imbalance and oxidative stress. Furthermore, the switch to inefficient anaerobic glycolysis is associated with excess lactate production with the potential for reduced pH and metabolic acidosis.

Several critical cuproenzymes ($\sim$12) are known (Prohaska, 2011), including lysyl oxidase (crosslinks collagen and elastin), Zn-Cu superoxide dismutase (converts superoxide to hydrogen peroxide), membrane-bound hephaestin (oxidizes iron for release from enterocytes, macrophages, and hepatocytes). Several cuproenzymes (eg, circulating ceruloplasmin) appear to have overlapping functions as ferroxidases that converting ferrous to ferric iron for loading onto apo-transferrin. Copper also upregulates and activates hypoxia inducible factor 1 (HIF-1), increasing myocardial vascular endothelial growth factor (VEGF) and enhancing angiogenesis (Feng et al., 2009). The regulations and functions of these proteins are discussed in the appropriate chapters.

Rich sources of dietary copper include organ meats, shellfish, legumes, nuts, and seeds, including chocolate. Whole grains, olives, and avocadoes are good sources, and it must be remembered that since

copper is a major component of the photosynthetic complex, green leafy vegetables also contain copper. If dietary copper is inadequate, transport systems are upregulated to increase absorption. Estimations of adequate copper intake (AIs) to maintain health are difficult to calculate, since there are no robust, sensitive and specific biomarkers for copper status. However, the Scientific Committee of the European Food Safety Authority has proposed AIs of 1.6 mg/day for men and 1.3 mg/day for women (EFSA, 2015). Copper deficiency is rarely reported, largely because it is difficult to diagnose since an appropriate biomarker has not been identified (Harvey and McArdle, 2008). Ceruloplasmin, the primary copper-containing serum protein, is also a positive acute-phase protein, thus elevated ceruloplasmin is not diagnostic for adequate copper status (Milne, 1994).

Mammals regulate the total body copper load entirely by absorption and biliary excretion (Fig. EII.5). Copper release from food requires an acid stomach, thus use of proton pump inhibitors has been associated with symptomatic copper deficiency (Plantone et al., 2015). Copper is absorbed in the proximal small intestine via a conserved homotrimeric plasma membrane protein, Ctr1 (Nose et al., 2006); absorption has been estimated at approximately 50% of intake. Copper that is not stored in the enterocyte is transported to the portal circulation by a Cu-ATPase (ATP7A). In the portal circulation, copper bound to histidine, albumin, or transcuprein is transported to the liver and incorporated into ceruloplasmin, the principal (80–95%) copper carrier in the systemic circulation. Copper is also bound to MT and other storage proteins in the liver. When hepatocytes detect elevated copper status, a second transport protein, ATP7B, moves from the Golgi to the lysosomes and transports Cu into the lysosome lumen. Lysosomes interact with dynactin, a multisubunit protein that binds to motor proteins dynein and kinesin 2 and moves lysosomes along microtubules to the canalicular pole where the lysosomes undergo exocytosis and release Cu into the bile. Thus biliary copper concentration is directly proportional to the size of the hepatic pool (Polishchuk et al., 2014).

Interactions with several dietary components can induce copper depletion. A mutual antagonism between copper and zinc absorption has been documented in rats (Oestreicher and Cousins, 1985) and humans provided with zinc supplements (Duncan et al., 2015). This antagonism is thought to be due in part to zinc-induced upregulation of enterocyte MT. Since Cu also binds to MT, excess Zn intake, even with adequate Cu intake can bind up available Cu in the enterocyte, preventing its absorption and promoting its loss when the enterocyte is sloughed. Zinc-induced copper malabsorption can prevent iron oxidation by the cuproenzyme, hephaestin, and lead to iron malabsorption and microcytic anemia as frequently described (Prasad et al., 1978) and recently reviewed (Plum et al., 2010). The ratio of dietary copper to zinc is important in heavy metal homeostasis, thus if zinc must be supplemented, copper intake should be assessed (Sandstead, 1995; Duncan et al., 2015). It should be noted that several cosmetic and personal products can contain zinc that can be absorbed via the dermal or oral route. For example, zinc-containing denture fixative has been frequently associated with hematologic and neurologic symptoms of copper deficiency (Barton et al., 2011).

Fructose feeding also reliably decreases copper absorption in the rat. Ctr1 was upregulated in rats fed a Cu-deficient diet, but addition of moderate fructose intake downregulated Ctr1, compromising Cu transport across the enterocyte (Song et al., 2012). The carbohydrate source also modifies copper bioavailability in humans. Healthy human subjects were fed a copper-deficient diet with fructose or starch as the carbohydrate source for 11 weeks. Although ceruloplasmin and serum copper levels did not change, erythrocyte Cu/Zn superoxide dismutase (SOD) was significantly lower in the fructose group; repletion with 3 mg Cu/day for 3 weeks normalized SOD levels (Reiser et al., 1985). Since fructose metabolism by ketohexokinase (fructokinase) in the enterocyte depletes intracellular ATP and compromises active transcellular

calcium absorption (Tharabenjasin et al., 2014), it is possible that copper absorption, requiring CuATPase may also be compromised. Additional clinical trials are necessary to determine whether the Western diet, replete with intentional and unintentional sucrose and fructose consumption, modifies heavy metal homeostasis.

Obesity and aging may be associated with altered copper distribution. Chronic inflammation associated with these conditions increases the secretion of the acute-phase protein, ceruloplasmin. In the face of reduced copper intake or intracellular availability, inflammation-driven ceruloplasmin biosynthesis may deplete intracellular hepatocyte copper status. Copper concentrations in liver, femur, small intestine, muscle, and testes were lower in genetically obese mice, while serum copper and ceruloplasmin concentrations were elevated compared with age-matched lean mice (Kennedy et al., 1986). Chronic inflammation is known to be associated with both obesity and aging. Although hepatocyte secretion of ceruloplasmin is known, a recent study revealed that ceruloplasmin is also an adipokine secreted from isolated adipocytes; adipocytes from obese subjects secreted more ceruloplasmin than cells from nonobese subjects (Arner et al., 2014).

Things should be made as simple as possible, but not simpler.
Source: Albert Einstein

EII.3.1 Retinoids and Precursors

It has been just over 100 years since McCollum determined the essential nature of a substance contained in extracts of butter or eggs which we now know as "vitamin A" (McCollum and Davis, 1913). The term "retinoid" has been defined to characterize vitamin A (retinol) and its analogs with or without biological activity as well as molecules that are not structurally related to retinol but elicit biological vitamin A or retinoid activity (Goodman, 1984). As would be expected from such a diverse group, retinoids are multifunctional and have been intensively studied (Blomhoff et al., 1991; Blaner and Olson, 1994; Vogel et al., 1999). The molecular function of retinoids in visual excitation was characterized many years ago in work that won a Nobel Prize (Wald, 1968) (Fig. EII.6).

Retinoids play major roles in the maintenance of epithelial cells, immune competence, reproduction, embryonic development (Cunningham and Duester, 2015), mitochondrial bioenergetics (Acin-Perez et al., 2010), and cell differentiation (Mongan and Gudas, 2007) as reviewed by Allenby et al. (1993). Additionally, retinoic acid has been shown to modulate lipid metabolism in a rat model as illustrated in Fig. EII.7 (Bonet et al., 2012).

Retinoic acid (all-*trans*-retinoic acid = ATRA) regulates gene transcription by binding to nuclear receptors known as retinoic acid receptors (RAR: α, β, and γ) and the retinoid X receptors (RXR:α, β, and γ) activated by 9-*cis* retinoic acid (9-cisRA). RAR and RXR must dimerize before they can bind to the DNA. RAR forms a heterodimer with RXR (RAR-RXR), but it does not readily form a homodimer (RAR-RAR). RXR, on the other hand, may form a homodimer (RXR-RXR) and will form heterodimers with many other nuclear receptors as well, including the thyroid hormone receptor (RXR-TR), the vitamin D3 receptor (RXR-VDR), the peroxisome proliferator-activated receptor (RXR-PPAR), and the liver "X" receptor (RXR-LXR). RAR and RXR act as ligand-activated transcription factors, binding to the retinoic acid response element (RARE) in the promoter region of specific genes to regulate transcription.

Preformed vitamin A is found in dairy products, fish, and meat; vitamin A is stored in the liver, thus animal liver and fish liver oils are rich in the nutrient. The most available plant sources are the orange and yellow carotenoid pigments, β carotene; α carotene, and β cryptoxanthin: these provitamin A forms can be converted to vitamin A in the body. Other carotenoids found in food, such as lycopene (red tomatoes), lutein, and zeaxanthin (green leafy vegetables), are not converted into vitamin A but have other bioactive functions (Fig. EII.8).

Current understanding of homeostatic mechanisms for retinoid absorption, transport, and distribution have been comprehensively reviewed (D'Ambrosio et al., 2011). Briefly, upon ingestion, dietary retinoids and proretinoid carotenoids must be removed from the food matrix and emulsified in the presence of dietary fatty acids and bile acids. Vitamin A esters are hydrolyzed primarily by pancreatic triglyceride lipase, enhanced by pancreatic lipase-related protein 2 as reviewed by Reboul and Borel (2011). Dietary carotenoids are transported intact into the enterocyte by scavenger receptor class B, type I (SR-B1). Retinol uptake

FIGURE EII.6 Physiologically active retinoid structures. Retinol is converted to a variety of derivatives as shown. The alcohol, retinol, is converted to retinoic acid by two oxidation steps. Alcohol dehydrogenase 4 (ADH4) is a zinc-dependent enzyme (Brandt et al., 2009) that converts retinol to retinaldehyde; this enzyme also converts alcohol to acetaldehyde. Retinaldehyde is oxidized to retinoic acid by retinaldehyde dehydrogenase. *Source: Reproduced from Rhee, E.-J., Nallamshetty, S., et al., 2012. Retinoid metabolism and its effects on the vasculature. Biochim. Biophys. Acta (BBA) Mol. Cell Biol. Lipids 1821 (1), 230–240 (Rhee et al., 2012) with permission from Science Direct.*

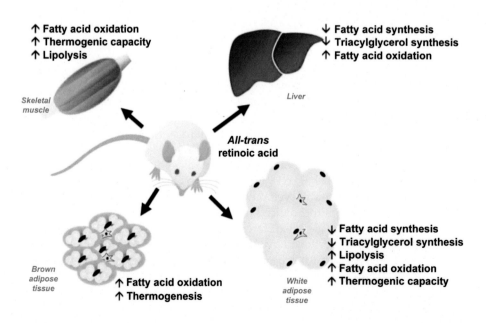

FIGURE EII.7 Observed actions of all-*trans* retinoic acid on lipid metabolism in rodents. While supplemental retinoic acid has been shown to reduce body fat and improve insulin sensitivity in rodents by enhancing systemic fat oxidation and energy utilization, the molecular mechanisms remain incompletely understood (Bonet et al., 2012). *Source: Reproduced from Bonet, M.L., Ribot, J., et al., 2012. Lipid metabolism in mammalian tissues and its control by retinoic acid. Biochim. Biophys. Acta (BBA) Mol. Cell Biol. Lipids 1821 (1), 177–189 with permission from Science Direct.*

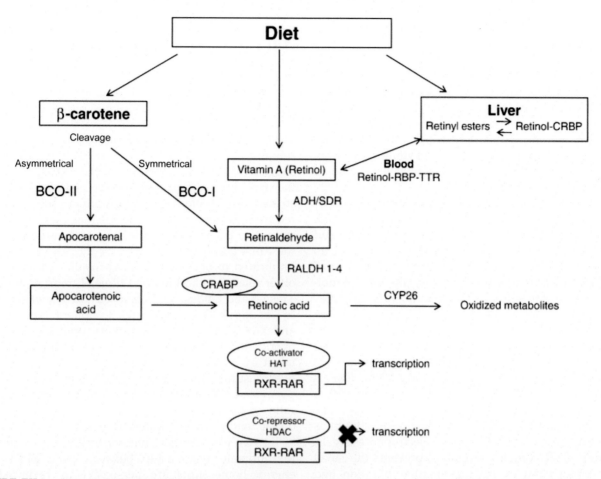

FIGURE EII.8 **Vitamin A bioavailability and conversions.** Dietary sources are hydrolyzed and cleaved in the gut and are transported with chylomicrons in the lymphatics to the liver. Retinal-binding protein (RBP) binds retinol in a complex with transthyretin for transfer to the cells. Retinal uptake by the cell is followed by binding to cellular RBP and conversion to active metabolites, primarily all-*trans* retinoic acid. *Source: Reproduced from Rhee, E.-J., Nallamshetty, S., et al., 2012. Retinoid metabolism and its effects on the vasculature. Biochim. Biophys. Acta (BBA) Mol. Cell Biol. Lipids 1821 (1), 230–240 with permission from Science Direct.*

occurs by a saturable process at low concentrations (<10 μM) and by nonsaturable processes at higher concentrations. Thus, passive absorption could account for the increased risk for toxicity from high-dose supplementation with preformed vitamin A. Bioavailability of vitamin A has been reported as ∼75–100%, while carotenoid bioavailability is less (3–90%). This difference has been attributed to the presence of plant fiber in carotenoid sources. An alternate explanation comes from preliminary evidence that at physiological levels preformed vitamin A is absorbed by a carrier-mediated process, while carotenoids appear to be absorbed by nonspecific lipid transport proteins, thereby reducing their bioavailability.

Once in the enterocyte, proretinoid carotenoids are cleaved to retinoids. Despite early confusion as to symmetrical versus nonsymmetrical cleavage, it has now been established through cDNA cloning that only the beta-carotene 15/15′ monooxygenase1 (Bcmo1) gene product that cleaves proretinoids symmetrically is

significant for retinoid formation from dietary carotenoids. It is of interest that Bcmo1 is a soluble, Fe^{2+}-containing, 63-kDa protein that is not only expressed in the proximal enterocytes, but also the liver, kidney, lungs, skin, testis, the retinal pigment epithelium within the eye, and in a number of embryonic tissues, suggesting that these tissues can also cleave intact carotenoids. In the enterocyte, absorbed retinol is bound to cellular retinol-binding protein II (CRBPII), reported to be solely expressed in the intestinal mucosa and is esterified primarily by lecithin:retinol acyltransferase (LRAT) and to a lesser extent by other esterases.

Retinyl esters are packaged into nascent chylomicrons with triglyceride, cholesterol, carotenoids, other fat-soluble vitamins (D, E, K), and lipid-soluble components in the diet (eg, lycopene, lutein) and secreted into the lymphatics for transport to the heart and systemic circulation (refer to Chapter 6: Atherosclerosis and Arterial Calcification: Lipid Transport and Oxidation). Most (∼66–75%) of the retinyl esters are transported to the liver and stored in lipid droplets in stellate cells (also called fat-storing cells, lipocytes, Ito cells, or perisinusoidal cells). It is now understood that a substantial portion of lipoprotein-transported retinoids are hydrolyzed by lipoprotein lipase and enter the peripheral tissues directly. Retinyl ester hydrolysis appears to facilitate retinol uptake by peripheral tissues such as heart, skeletal muscle, adipose tissue, mammary gland, and perhaps lung. Alternate pathways for uptake and processing of chylomicron retinyl ester by the hepatocyte have been reviewed in detail (D'Ambrosio et al., 2011).

In times of dietary retinoid insufficiency, the retinyl ester stores in the hepatic stellate cells are mobilized. Retinyl ester is hydrolyzed back to retinol and returned to the hepatocyte where it is bound to retinol-binding protein (RBP or RBP4). RBP is a small (21 kDa) protein with a single binding site for one molecule of all-*trans*-retinol. The retinol–RBP complex enters the circulation for transport to peripheral tissues. While RBP secretion is highly regulated by the retinoid status of the peripheral tissues, there is also preliminary evidence that RBP is an adipokine secreted by adipocytes (Tamori et al., 2006). RBP is transported in a 1:1 protein–protein complex with a 55-kDa serum protein, transthyretin (TTR), and one molecule of thyroxine. TTR was previously known as prealbumin due to its behavior on electrophoresis gels. The RBP–TTR complex formation prevents the small RBP–retinol complex from glomerular filtration and urinary loss. Although the uptake mechanism is not entirely clear, a model has been proposed that circulating retinol:RBP binds a membrane-bound protein (STRA6) at the peripheral tissues. STRA6 facilitates retinol import where it is converted to a retinyl ester by LRAT. It has also been suggested that STRA6 transport is bidirectional, and can facilitate efflux of retinol if intracellular concentrations are too high.

EII.4 RETINOID AND IRON INTERACTIONS IN OBESITY AND METABOLIC SYNDROME

As indicated above, elevated iron status is intimately linked to obesity, metabolic syndrome, and Type II diabetes. Vitamin A, as retinol bound to RBP, regulates the flow of pyruvate into the mitochondria. Retinol is an essential component of the mitochondrial signalosome comprised of protein kinase Cδ (PKCδ), the p66Shc adapter protein, and cytochrome c as recently reviewed by Shabrova et al. (2015). This complex enters the mitochondrial intermembranous space and operates upstream of the pyruvate dehydrogenase complex (PDHC). It acts both as a sensor of the reduction potential of cytochrome c (reflecting the ETC workload) and as a signal transducer that regulates PDHC conversion of pyruvate to acetyl CoA. The authors propose that the PKCδ signaling pathway is, in essence, a "thrifty gene." Activation of PKCδ signaling facilitates glucose oxidation and fat storage when food is scarce and downregulates pyruvate oxidation when food is in excess.

The authors also provide evidence in mice that high retinol-RBP (RBP4) levels increase PKCδ signaling and suggest that chronic activation of the PKCδ pathway in the face of overnutrition would generate oxidative stress, force preference for glycolytic substrate and reduce utilization of fatty acids as fuel, promoting their storage as fat. This forms the basis of the argument that the RBP4 adipokine produced by the adipose tissue is implicated in obesity-associated insulin resistance and other related comorbid conditions (Tamori et al., 2006; Trasino and Gudas, 2015) as discussed elsewhere (refer to Chapter 4: Type II Diabetes, Peripheral Neuropathy, and Gout). In short, the authors propose that iron-overloaded adipocytes secrete excess RBP4. RBP4 is taken up by the mitochondria and in the presence of overconsumption, chronic PKCδ stimulation forces pyruvate oxidation by the mitochondria and alters the ratio of macronutrient oxidation to a pattern seen in obesity and its comorbidities (Lopaschuk et al., 2010).

Ex vivo evidence and literature citations supporting the intimate relationship between iron, RBP4, and energy homeostasis are persuasive as reviewed by Fernández-Real et al. (2008) and Trasino and Gudas (2015). In brief, iron deficiency results in inactivation of hepatic retinol palmitate and reduces retinol transport as RBP4 to vitamin A-dependent tissues, possibly because retinyl palmitate hydrolase requires an iron cofactor. Iron therapy even without added vitamin A increases the retinol supply to the tissues. Conversely, intracellular iron excess increases ferritin biosynthesis in experimental models and in humans. Reduction of iron load in adipocyte cultures and in phlebotomized humans reduces serum ferritin. Obesity and type II DM are associated with high intracellular iron load, high serum ferritin concentrations, and high circulating RBP4 levels. Since excess RBP4 has been implicated in dysregulation of mitochondrial function, this area is in need of further investigation.

EII.5 CAROTENOIDS

Carotenoids were originally thought to function only as precursors to vitamin A and as quenchers of free radicals. Over 600 carotenoids have been identified in natural food sources, especially highly colored plant sources such as yellow-orange squash

FIGURE EII.9   **Structure of β carotene.** This structure is a mirror image of retinol and upon ingestion is cleaved by 15-15′ monooxygenase to retinol. Other cleavage sites are known. *Source: Figure by NEUROtiker (own work) (Public domain), via Wikimedia Commons.*

and carrots and dark green leafy vegetables in which the carotenoid pigment is masked by the green chlorophyll. Carotenoids play a major role in the light reaction of photosynthesis. Carotenoids and other complex bioactive components in food are modified by the gut microbiota and by several of the host detoxification enzymes. Thus the chemical forms in which they are absorbed, circulate, and reach the tissues are highly variable. Supplementation with carotenoids or, indeed, any bioactive component extracted from food should be approached with caution. Large randomized clinical trials using β carotene supplements to potentially reduce lung cancer in high-risk subjects demonstrated β carotene supplements actually increased cancer risk (Albanes et al., 1996; Leo and Lieber, 1999) (Fig. EII.9).

Less than 10% of carotenoids act as precursors of vitamin A. Since rodents do not accumulate carotenoids, they may not be the best models of carotenoid actions. Recently, many more functions have been described as reviewed by Bendich and Olson (1989), Rao and Rao (2007), and Bonet et al. (2012). Of particular interest is the action of carotenoids in modulating adipocyte function (Luisa Bonet, 2015). In addition to their actions as vitamin A precursors, carotenoids can enhance the immune response, inhibit mutagenesis and nuclear damage, and protect against photo-induced tissue damage including macular degeneration (Sommerburg et al., 1998). More work is required in these areas.

References

Acin-Perez, R., Hoyos, B., et al., 2010. Control of oxidative phosphorylation by vitamin A illuminates a fundamental role in mitochondrial energy homoeostasis. FASEB. J. 24 (2), 627–636.

Albanes, D., Heinonen, O.P., et al., 1996. α-tocopherol and β-carotene supplements and lung cancer incidence in the alpha-tocopherol, beta-carotene cancer prevention study: effects of base-line characteristics and study compliance. J. Natl. Cancer Inst. 88 (21), 1560–1570.

Allenby, G., Bocquel, M.T., et al., 1993. Retinoic acid receptors and retinoid X receptors: interactions with endogenous retinoic acids. Proc. Natl. Acad. Sci. 90 (1), 30–34.

Andrews, S.C., Robinson, A.K., et al., 2003. Bacterial iron homeostasis. FEMS Microbiol. Rev. 27 (2-3), 215–237.

Arner, E., Forrest, A.R.R., et al., 2014. Ceruloplasmin is a novel adipokine which is overexpressed in adipose tissue of obese subjects and in obesity-associated cancer cells. PLoS ONE 9 (3), e80274.

Barton, A.L., Fisher, R.A., et al., 2011. Zinc poisoning from excessive denture fixative use masquerading as myelopolyneuropathy and hypocupraemia. Ann. Clin. Biochem. 48 (4), 383–385.

Bendich, A., Olson, J.A., 1989. Biological actions of carotenoids. FASEB J. 3 (8), 1927–1932.

Blaner, W.S., Olson, J.A., 1994. Retinol and retinoic acid metabolism. Retinoids Biol. Chem. Med. 2, 229–256.

Blomhoff, R., Green, M.H., et al., 1991. Vitamin A metabolism: new perspectives on absorption, transport, and storage. Physiol. Rev. 71 (4), 951–990.

Bonaventura, P., Benedetti, G., et al., 2015. Zinc and its role in immunity and inflammation. Autoimmun. Rev. 14 (4), 277–285.

Bonet, M.L., Ribot, J., et al., 2012. Lipid metabolism in mammalian tissues and its control by retinoic acid. Biochim. Biophys. Acta (BBA) Mol. Cell Biol. Lipids. 1821 (1), 177–189.

Boyer, E., Bergevin, I., et al., 2002. Acquisition of Mn(II) in addition to Fe(II) is required for full virulence of Salmonella enterica Serovar Typhimurium. Infect. Immun. 70 (11), 6032–6042.

Brandt, E.G., Hellgren, M., et al., 2009. Molecular dynamics study of zinc binding to cysteines in a peptide mimic of the alcohol dehydrogenase structural zinc site. Phys. Chem. Chem. Phys. 11 (6), 975–983.

Casey, J.L., Hentze, M.W., et al., 1988. Iron-responsive elements: regulatory RNA sequences that control mRNA levels and translation. Science 240 (4854), 924–928.

Coimbra, S., Catarino, C., et al., 2013. The role of adipocytes in the modulation of iron metabolism in obesity. Obes. Rev. 14 (10), 771–779.

Cousins, R.J., Liuzzi, J.P., et al., 2006. Mammalian zinc transport, trafficking, and signals. J. Biol. Chem. 281 (34), 24085–24089.

Craddock, T.J.A., Tuszynski, J.A., et al., 2012. The zinc dyshomeostasis hypothesis of Alzheimer's disease. PLoS ONE 7 (3), e33552.

Cunningham, T.J., Duester, G., 2015. Mechanisms of retinoic acid signalling and its roles in organ and limb development. Nat. Rev. Mol. Cell. Biol. 16 (2), 110–123.

D'Ambrosio, D.N., Clugston, R.D., et al., 2011. Vitamin A metabolism: an update. Nutrients 3 (1), 63.

Datz, C., Felder, T.K., et al., 2013. Iron homeostasis in the metabolic syndrome. Eur. J. Clin. Invest. 43 (2), 215–224.

Duncan, A., Yacoubian, C., et al., 2015. The risk of copper deficiency in patients prescribed zinc supplements. J. Clin. Pathol. 68 (9), 723–725.

EFSA, 2015. Scientific Opinion on Dietary Reference Values for copper. Eur. Food Saf. Auth. 13 (10), 4253–4304.

Feng, W., Ye, F., et al., 2009. Copper regulation of hypoxia-inducible factor-1 activity. Molecular Pharmacology. 75 (1), 174–182.

Fernández-Real, J.M., Moreno, J.M., et al., 2008. Circulating retinol-binding protein-4 concentration might reflect insulin resistance–associated iron overload. Diabetes. 57 (7), 1918–1925.

Foster, M., Chu, A., et al., 2013. Effect of vegetarian diets on zinc status: a systematic review and meta-analysis of studies in humans. J. Sci. Food Agric. 93 (10), 2362–2371.

Frederickson, C.J., Koh, J.-Y., et al., 2005. The neurobiology of zinc in health and disease. Nat. Rev. Neurosci. 6 (6), 449–462.

Fukada, T., Hojyo, S., et al., 2014. Zinc signal in growth control and bone diseases. Zinc Signals in Cellular Functions and Disorders. Springer, Tokyo, pp. 249–267.

Gabrielsen, J.S., Gao, Y., et al., 2012. Adipocyte iron regulates adiponectin and insulin sensitivity. J. Clin. Invest. 122 (10), 3529–3540.

Ganz, T., 2013. Systemic iron homeostasis. Physiol. Rev. 93 (4), 1721–1741.

Ganz, T., Nemeth, E., 2015. Iron homeostasis in host defence and inflammation. Nat. Rev. Immunol. 15 (8), 500–510.

Goodman, D.S., 1984. Biosynthesis, absorption, and hepatic metabolism of retinol. Retinoids 2, 2–39.

Haase, H., Rink, L., 2014. Zinc signals and immune function. BioFactors 40 (1), 27–40.

Hare, D.J., Ayton, S., et al., 2013. A delicate balance: iron metabolism and diseases of the brain. Front. Aging Neurosci. 5, 34.

Harvey, L.J., McArdle, H.J., 2008. Biomarkers of copper status: a brief update. Br. J. Nutr. 99 (Supplement S3), S10–S13.

Hellwig, J.P., Otten, J.J., et al., 2006. Dietary Reference Intakes: The Essential Guide to Nutrient Requirements. National Academies Press, Washington, DC.

Hider, R.C., Kong, X., 2010. Chemistry and biology of siderophores. Nat. Prod. Rep. 27 (5), 637–657.

Horn, D., Barrientos, A., 2008. Mitochondrial copper metabolism and delivery to cytochrome c oxidase. IUBMB Life 60 (7), 421–429.

Kambe, T., Tsuji, T., et al., 2015. The physiological, biochemical, and molecular roles of zinc transporters in zinc homeostasis and metabolism. Physiol. Rev. 95 (3), 749–784.

Kang, Y.J., 2011. Copper and homocysteine in cardiovascular diseases. Pharmacol. Ther. 129 (3), 321–331.

Kehl-Fie, T.E., Chitayat, S., et al., 2011. Nutrient metal sequestration by calprotectin inhibits bacterial superoxide defense, enhancing neutrophil killing of Staphylococcus aureus. Cell Host Microbe. 10 (2), 158–164.

Kennedy, M.L., Failla, M.L., et al., 1986. Influence of genetic obesity on tissue concentrations of zinc, copper, manganese and iron in mice. J. Nutr. 116 (8), 1432–1441.

King, J.C., 2011. Zinc: an essential but elusive nutrient. Am. J. Clin. Nutr. 94 (2), 679S–684S.

Kirkpatrick, S.I., Dodd, K.W., et al., 2015. Household food insecurity is a stronger marker of adequacy of nutrient intakes among Canadian compared to American youth and adults. J. Nutr. 145 (7), 1596–1603.

Krebs, N.F., 2000. Overview of zinc absorption and excretion in the human gastrointestinal tract. J. Nutr. 130 (5), 1374S–1377S.

Leary, S.C., Winge, D.R., 2007. The Janus face of copper: its expanding roles in biology and the pathophysiology of disease. Meeting on copper and related metals in biology. EMBO Rep. 8 (3), 224–227.

Leo, M.A., Lieber, C.S., 1999. Alcohol, vitamin A, and ß-carotene: adverse interactions, including hepatotoxicity and carcinogenicity. Am. J. Clin. Nutr. 69 (6), 1071–1085.

Lichten, L.A., Cousins, R.J., 2009. Mammalian zinc transporters: nutritional and physiologic regulation. Annu. Rev. Nutr. 29 (1), 153–176.

Liuzzi, J.P., Lichten, L.A., et al., 2005. Interleukin-6 regulates the zinc transporter Zip14 in liver and contributes to the hypozincemia of the acute-phase response. Proc. Natl. Acad. Sci. USA 102 (19), 6843–6848.

Lopaschuk, G.D., Ussher, J.R., et al., 2010. Myocardial fatty acid metabolism in health and disease. Physiol. Rev. 90 (1), 207–258.

Luisa Bonet, M., Canas, J.A., et al., 2015. Carotenoids and their conversion products in the control of adipocyte function, adiposity and obesity. Arch. Biochem. Biophys. 572, 112–125.

Lutsenko, S., 2010. Human copper homeostasis: a network of interconnected pathways. Curr. Opin. Chem. Biol. 14 (2), 211–217.

Lutsenko, S., Bhattacharjee, A., et al., 2010. Copper handling machinery of the brain. Metallomics. 2 (9), 596–608.

Mayo-Wilson, E., Jean, A.J., et al., 2014. Zinc supplementation for preventing mortality, morbidity, and growth failure in children aged 6 months to 12 years of age. Cochrane Database Syst. Rev. Available from: http://dx.doi.org/10.1002/14651858.CD009384.pub2.

McCollum, E.V. and M. Davis (1913). The necessity of certain lipids in the diet during growth.

Micronutrients, 2001. Dietary Reference Intakes for Vitamin A, Vitamin K, Arsenic, Boron, Chromium, Copper, Iodine, Iron, Manganese, Molybdenum, Nickel, Silicon, Vanadium, and Zinc. I. o. M. Food and Nutrition Board. National Academy Press, Washington, DC.

Milne, D., 1994. Assessment of copper nutritional status. Clin. Chem. 40 (8), 1479–1484.

Molloy, S.A., Kaplan, J.H., 2009. Copper-dependent recycling of hCTR1, the human high affinity copper transporter. J. Biol. Chem. 284 (43), 29704–29713.

Mongan, N.P., Gudas, L.J., 2007. Diverse actions of retinoid receptors in cancer prevention and treatment. Differentiation 75 (9), 853–870.

Nargund, S., Qiu, J., et al., 2015. Elucidating the role of copper in CHO cell energy metabolism using 13C metabolic flux analysis. Biotechnol. Prog. 31 (5), 1179–1186.

Nose, Y., Kim, B.-E., et al., 2006. Ctr1 drives intestinal copper absorption and is essential for growth, iron metabolism, and neonatal cardiac function. Cell Metab. 4 (3), 235–244.

Oestreicher, P., Cousins, R.J., 1985. Copper and zinc absorption in the rat: mechanism of mutual antagonism. J. Nutr. 115 (2), 159–166.

Plantone, D., Renna, R., et al., 2015. PPIs as possible risk factor for copper deficiency myelopathy. J. Neurol. Sci. 349 (1–2), 258–259.

Plum, L.M., Rink, L., et al., 2010. The essential toxin: impact of zinc on human health. Int. J. Environ. Res. Public Health. 7 (4), 1342–1365.

Polishchuk, E.V., Concilli, M., et al., 2014. Wilson disease protein ATP7B utilizes lysosomal exocytosis to maintain copper homeostasis. Dev. Cell. 29 (6), 686–700.

Prasad, A.S., Halsted, J.A., et al., 1961. Syndrome of iron deficiency anemia, hepatosplenomegaly, hypogonadism, dwarfism and geophagia. Am. J. Med. 31, 532–546.

Prasad, A.S., Schulert, A.R., et al., 1963. Zinc and iron deficiencies in male subjects with dwarfism and hypogonadism but without ancylostomiasis, schistosomiasis or severe anemia. Am. J. Clin. Nutr. 12 (6), 437–444.

Prasad, A.S., Brewer, G.J., et al., 1978. Hypocupremia induced by zinc therapy in adults. JAMA 240 (20), 2166–2168.

Prohaska, J.R., 2011. Impact of copper limitation on expression and function of multicopper oxidases (ferroxidases). Adv. Nutr. Int. Rev. J. 2 (2), 89–95.

Rao, A.V., Rao, L.G., 2007. Carotenoids and human health. Pharmacol. Res. 55 (3), 207–216.

Raulin, J., 1869. Etudes clinique sur la vegetation. Ann. Sci. Nat. Bot. 11, 93–299.

Reboul, E., Borel, P., 2011. Proteins involved in uptake, intracellular transport and basolateral secretion of fat-soluble vitamins and carotenoids by mammalian enterocytes. Prog. Lipid Res. 50 (4), 388–402.

Reiser, S., Smith, J.C., et al., 1985. Indices of copper status in humans consuming a typical American diet containing either fructose or starch. Am. J. Clin. Nutr. 42 (2), 242–251.

Rhee, E.-J., Nallamshetty, S., et al., 2012. Retinoid metabolism and its effects on the vasculature. Biochim. Biophys. Acta (BBA) Mol. Cell Biol. Lipids 1821 (1), 230–240.

Rochette, L., Gudjoncik, A., et al., 2015. The iron-regulatory hormone hepcidin: a possible therapeutic target? Pharmacol. Ther. 146, 35–52.

Roy, C.N., 2013. An update on iron homeostasis: make new friends, but keep the old. Am. J. Med. Sci. 346 (5), 413–419.

Ruiz, L.M., Jensen, E.L., et al., 2014. Adaptive responses of mitochondria to mild copper deprivation involve changes in morphology, OXPHOS remodeling and bioenergetics. J. Cell. Physiol. 229 (5), 607–619.

Ruxton, C.H.S., Derbyshire, E., et al., 2013. Micronutrient challenges across the age spectrum: is there a role for red meat? Nutr. Bull. 38 (2), 178–190.

Sandstead, H.H., 1995. Requirements and toxicity of essential trace elements, illustrated by zinc and copper. Am. J. Clin. Nutr. 61 (3), 621S–624S.

Scheers, N., 2013. Regulatory effects of Cu, Zn, and Ca on Fe absorption: the intricate play between nutrient transporters. Nutrients. 5 (3), 957–970.

Shabrova, E., Hoyos, B., et al., 2015. Retinol as a cofactor for PKCδ-mediated impairment of insulin sensitivity in a mouse model of diet-induced obesity. FASEB. J. 30 (3), 1339–1355.

Sommerburg, O., Keunen, J.E.E., et al., 1998. Fruits and vegetables that are sources for lutein and zeaxanthin: the macular pigment in human eyes. Br. J. Ophthalmol. 82 (8), 907–910.

Song, M., Schuschke, D.A., et al., 2012. High fructose feeding induces copper deficiency in Sprague–Dawley rats: a novel mechanism for obesity related fatty liver. J. Hepatol. 56 (2), 433–440.

Strissel, K.J., Stancheva, Z., et al., 2007. Adipocyte death, adipose tissue remodeling, and obesity complications. Diabetes. 56 (12), 2910–2918.

Suthammarak, W., Yang, Y.-Y., et al., 2009. Complex I function is defective in complex IV-deficient Caenorhabditis elegans. J. Biol. Chem. 284 (10), 6425–6435.

Tamori, Y., Sakaue, H., et al., 2006. RBP4, an unexpected adipokine. Nat. Med. 12 (1), 30–31.

Tharabenjasin, P., Douard, V., et al., 2014. Acute interactions between intestinal sugar and calcium transport in vitro. Am. J. Physiol.–Gastrointest. Liver Physiol. 306 (1), G1–G12.

Theil, E.C., 2011. Iron homeostasis and nutritional iron deficiency. J. Nutr. 141 (4), 724S–728S.

Theil, E.C., Chen, H., et al., 2012. Absorption of iron from ferritin is independent of heme iron and ferrous salts in women and rat intestinal segments. J. Nutr. 142 (3), 478–483.

Trasino, S.E., Gudas, L.J., 2015. Vitamin A: a missing link in diabetes? Diabetes Manag. (Lond. England). 5 (5), 359–367.

Vallee, B.L., Falchuk, K.H., 1993. The biochemical basis of zinc physiology. Psysiol. Rev. 73 (1), 79–118.

Vogel, S., Gamble, M., et al., 1999. Retinoid Uptake, Metabolism and Transport. Springer Verlag, Heidelberg, pp. 31–96.

Vuppalanchi, R., Troutt, J.S., et al., 2014. Serum hepcidin levels are associated with obesity but not liver disease. Obesity 22 (3), 836–841.

Wald, G., 1968. The molecular basis of visual excitation. Nature. 219 (5156), 800–807.

Weinberg, E.D., 1999. The role of iron in protozoan and fungal infectious diseases. J. Eukaryot. Microbiol. 46 (3), 231–238.

Whittaker, P., Tufaro, P.R., et al., 2001. Iron and folate in fortified cereals. J. Am. Coll. Nutr. 20 (3), 247–254.

Wilkinson, N., Pantopoulos, K., 2014. The IRP/IRE system in vivo: insights from mouse models. Front. Pharmacol. 5, 176.

Zangger, K., Öz, G., et al., 2001. Nitric oxide selectively releases metals from the amino-terminal domain of metallothioneins: potential role at inflammatory sites. FASEB J. 15 (7), 1303–1305.

Zhang, S., Liu, H., et al., 2014. Diabetic cardiomyopathy is associated with defective myocellular copper regulation and both defects are rectified by divalent copper chelation. Cardiovasc. Diabetol. 13 (1), 1–18.

Essentials III: Nutrients for Bone Structure and Calcification

EIII.1 KLOTHO

The gene and protein of α-Klotho (Klotho) were discovered in 1997 and found in the kidney, brain, and parathyroid glands. The gene was so named because mice deficient in the Klotho exhibit accelerated aging, while overexpression extends lifespan. Klotho deficiency was characterized by severe arteriosclerosis, impaired endothelial function, and reduced angiogenesis as reviewed by Lewin and Olgaard (2015). Klotho is a transmembrane protein that converts fibroblastic growth factor 1 (FGF1) into a receptor specific for FGF23, secreted by the osteocyte (Urakawa et al., 2006). FGF23 binding induces renal phosphate excretion and decreases formation of the vitamin D hormone $1,25(OH)_2D_3$ (1,25D; calcitriol) by downregulating its hydroxylating enzyme, 1-alpha-hydroxylase (1-α-hydroxylase) (Fig. EIII.1).

While Klotho is expressed in the parathyroid gland and animal studies conclusively show that FGF23 inhibits parathyroid hormone (PTH), human studies are indirect and inconclusive (Bhattacharyya et al., 2012). Evidence of FGF23 action in the pituitary is minimal. The authors also cited fragmentary data suggesting that iron may be involved in FGF23 processing or action, but more work is required to determine whether iron status modifies bone mineral metabolism either in patients with chronic kidney disease (CKD) or in normal individuals.

The extracellular domain of Klotho is cleaved (ectodomain shedding) and is secreted into the plasma, urine, and cerebrospinal fluid. Soluble Klotho, an endocrine factor that modulates insulin and insulin-like growth factor action and has been proposed as an antiaging factor (Kurosu et al., 2005) and non-FDF23-mediated actions of soluble Klotho have been examined in reduction of acute kidney injury and fibrosis, uremic cardiomyopathy, vascular calcification, and endothelial dysfunction (Lewin and Olgaard, 2015). For example, soluble Klotho has been shown to play a role in activation of calcium channels (TRPV5,

TRPV6) as well as the potassium channel (ROMK1) and in suppression of the sodium-dependent phosphate cotransporters described below. Klotho has been implicated in cytokine signaling and actions of insulin-like growth factor 1, as well as Wnt, and transforming growth factor β signaling.

EIII.2 INTEGRATED REGULATION OF CALCIUM, PHOSPHORUS, AND VITAMIN D HOMEOSTASIS

The concentrations of calcium (Ca), phosphorus (P), and vitamin D are tightly regulated in the biological organism. The bulk of Ca and P are found in the bone as hydroxyapatite crystals $[Ca_5(PO_4)_3(OH)]$ incorporated into organic bone matrix. Historically, a threshold of ionic Ca × P concentration product ($55 \text{ mg}^2/\text{dL}^2$) was considered a risk factor for ectopic soft tissue calcification in stage 3−5 chronic renal failure patients (Uhlig et al., 2010). These patients are at high risk for metastatic calcification, especially in the vasculature. However, the concept that high serum concentrations of these sparingly soluble ions could spontaneously crystalize in tissues was severely flawed.

The actual physiological chemistry of Ca and P in bone mineral and tissues has been elucidated (O'Neill, 2007). Briefly, bone mineral hydroxyapatite does not form spontaneously. The osteoclast creates a protected microniche, the matrix vesicle, in which supraphysiological concentrations of Ca^{2+} and HPO_4^{2-} form $CaHPO_4$ crystals. Under physiologic mineral concentrations, these crystals dissociate (refer to Chapter 9: Osteoporosis and Fracture Risk: Bone Cells and Normal Remodeling Mechanisms). The second step is conversion of $CaHPO_4$ to the more complex crystal, hydroxyapatite $[Ca_5(PO_4)_3(OH)]$ on the osteoclast-denuded bone matrix. This complex crystal is stable, can undergo intercrystalline exchange with reformation of crystals, and can exchange new ions for calcium and/or adsorb ions onto

FIGURE EIII.1 **The triumph of death, or the three fates.** Clotho, in Greek Mythology, is one of the three fates who spin the thread of life. The gene for Klotho, described below, was named because the protein has been associated with longevity (Lewin and Olgaard, 2015). *Source: From public domain: Photograph of a Flemish tapestry (probably Brussels, ca 1510–1520. Currently in the Victoria and Albert Museum, London.*

the crystal surface. The two-phase crystallization process, the stability of bone crystal at physiologic pH and tightly regulated ion concentrations are the reasons why we do not spontaneously turn to stone and conversely why our bones do not spontaneously dissolve.

In addition to their function in bone, calcium, phosphate, and other ions play major roles in intracellular signaling, metabolic processes, and energy metabolism. Thus, their ionic concentrations in the serum must be tightly controlled through gastrointestinal absorption, bone storage, and renal excretion. Several hormones and signaling systems contribute to maintenance of physiologic calcium and phosphorus concentrations. Current understanding of this complex regulation has been reviewed (Blau and Collins, 2015).

Mechanisms that control the integrated homeostatic system that achieves this delicate balance are as yet unclear, but current research has identified key components. Briefly, osteocytes embedded in the bone matrix detect a rise in P from the diet and secrete fibroblast growth factor 23 (FGF23) to normalize serum P

concentrations. FGF23 binds αKlotho-FGF23 receptors in the kidney (Bhattacharyya et al., 2012) and inhibits the production of active vitamin D (1,25D) by activating mitogen-activated protein kinase to decrease the transcription, translation, and activity of the enzyme, 1-α-hydroxylase (CYP27B1), that hydroxylates circulating 25D in the 1 position. FGF23 also upregulates the renal 24-hydroxylase enzyme (CYP24A1), creating the inactive isomer, 24–25D. Normally, ~80–85% of filtered phosphate is reabsorbed in the kidney. To lower elevated serum phosphate, FGF23 binding downregulates the activity of sodium-dependent phosphate transporters 2a and c (NaPi2a/c) and decreases phosphate reabsorption in the kidney tubules. This action depends on coordination with PTH-related protein receptor (PTHr)-stimulated activity of Na/H exchange factor 1 which appears to stabilize both PTH/PTHr and NaPi2a/c via a protein kinase A pathway. The net result is return of serum phosphate [P_i] concentrations to normal by increasing renal excretion and inhibiting vitamin D-dependent phosphorus absorption. If serum [Ca^{2+}] is too high, calcitonin is released from the thyroid parafollicular cells, and fragmentary evidence suggests that calcitonin may play a role in calcium deposition in the bone, however the actual function of this hormone remains a mystery (Felsenfeld and Levine, 2015).

Thus, while both FGF23 and PTH increase urinary phosphate excretion by reducing proximal tubular expression of the sodium-dependent phosphate cotransporters thereby lowering serum phosphate levels, they differ functionally in that FGF23 decreases serum levels of 1,25D by inhibiting renal mitochondrial CYP27B1 (1-α-hydroxylase) and stimulating CYP24A1 (24-hydroxylase) while PTH increases renal CYP27B1 activity with production of 1,25D, consequently enhancing the absorption of P (and Ca) at the enterocyte (Fig. EIII.2).

EIII.2.1 Parathyroid Hormone

PTH secretion from the parathyroid gland is stimulated by low concentrations of serum-ionized calcium. Changes in extracellular calcium are sensed by G-protein-coupled, calcium-sensing receptors (CaRs) on parathyroid cells (Brown et al., 1994); CaRs have also been identified on diverse tissues including the gastrointestinal tract and skin (Alfadda et al., 2014). PTH secretion is downregulated by calcitriol (1,25D) bound to the vitamin D receptor (VDR) that heterodimerizes with retinoic acid X receptors to bind vitamin D-response elements within the *PTH* gene, decreasing PTH transcription. FGF23 also decreases PTH function by binding the FGFR-Klotho receptor in

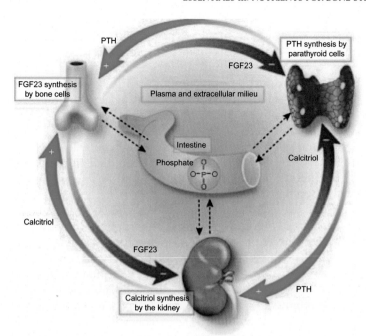

FIGURE EIII.2 **Regulation of the PTH-Vitamin D-FGF23 axis.** Bone osteocytes are stimulated by 1,25D and by high serum phosphate levels to secrete FGF23. FGF23 binds Klotho-FGFr1 in the kidney tubules, inducing phosphaturia; binding suppresses 1-α hydroxylase and downregulates 1,25D (calcitriol) biosynthesis. At the parathyroid gland, FGF23 inhibits parathyroid hormone (PTH) secretion: inhibitory action of FGF23 on PTH secretion has been demonstrated in experimental models, but not in humans (Blau and Collins, 2015). The net result is normalization of serum phosphate levels. Reduction in serum-ionized calcium stimulates PTH secretion. At the kidney, PTH binds the kidney tubule via its receptor (PTH-R), upregulates 1-α hydroxylase, and increases 1,25D biosynthesis. 1,25D acts with PTH to increase osteoclast action, releasing [Ca^{2+}] and [Pi] from the bone matrix. These regulatory loops maintain serum calcium and phosphate levels within normal limits. It is likely that additional sensors monitor electrolyte status and stimulate regulatory response as indicated by the dotted lines (Torres and Brauwere, 2011). *Source: Figure from Torres, P.A.U., Brauwere, D.P.D., 2011. Three feedback loops precisely regulating serum phosphate concentration. Kidney Int. 80 (5), 443−445, with permission from Science Direct.*

the parathyroid gland. Similarly, elevated P decreases PTH secretion, although the sensor is not yet known.

PTH activates 1-α hydroxylase (CYP27B1) in the kidney, thereby increasing 1,25D formation and generating a PTH feedback loop. 1,25D induces the calcium transporter, TRPV6, and other calcium translocation gene products (eg, calbindin) in the enterocyte, raising serum calcium and reducing stimulation for PTH biosynthesis and secretion. PTH acts on the osteoclast by increasing membrane permeability to calcium and activating membrane-bound adenyl cyclase. Osteoclast numbers and content of lytic enzymes and acids are stimulated and bone resorption is enhanced, resulting in increased calcium and phosphorus release.

EIII.2.2 FGF23

The regulation of FGF23 secretion by osteocytes in response to high serum phosphate levels has been described (Kaneko et al., 2015). FGF23 regulation is also subject to control by a complex network of hormones, growth factors, and metabolites, including leptin, cortisol, iron, calcium, and VDRs. While 1,25D bound to VDR induces human FGF23 production in osteocytes by binding multiple VDR elements (VDRE), it represses FGF23 transcription in the adipocyte, suggesting the actions are cell-specific. The 1,25D/VDR complex also represses PTH transcription in the parathyroid gland. Thus, FGF23 not only enhances hyperphosphaturia, but also reduces 25D hydroxylation to 1,25D via repression of CYP27B1 in the kidney, thereby

maintaining a feedback loop to prevent excess 1,25D/VDR action to upregulate FGF23 by the osteocyte (Fig. EIII.3).

EIII.3 MINERAL METABOLISM

EIII.3.1 Calcium

Total body calcium content is approximately 1000−1200 g (1 kg; 25−30 mM), of which 99% exists as bone mineral (hydroxyapatite [Ca$_5$(PO$_4$)$_3$(OH)]). About 1% of bone is labile and serves as a rapidly available source of Ca^{2+}, Mg^{2+}, and P$_i$ ions. (At physiological pH, inorganic phosphorus circulates as a mixture of HPO$_4$$^{2-}$ and H$_2$PO$^-$ ions and is generally designated as P$_i$.) The molar ratio of Ca to P in bone is about 2:1, thus, with bone resorption, Ca and P ions are released in that ratio. In blood, the serum total Ca concentration is 10^{-3} M/L and is distributed as freely ionized (48%), protein bound (46% bound mainly to albumin) with the remainder complexed to citrate and P$_i$ ions. Cytosolic Ca^{2+} is maintained at much lower concentrations (approximately 10^{-6}) by means of transport proteins that extrude calcium from the cell or sequester it within cell organelles. In order to maintain optimal calcium availability in these diverse depots, extracellular Ca^{2+} must be maintained constant in the face of large variations of intake and output. Homeostatic regulatory systems control Ca^{2+} fluxes between compartments and organs and keep plasma [Ca^{2+}] close to 2.5 mM.

Physiologic processes **Pathologic processes**

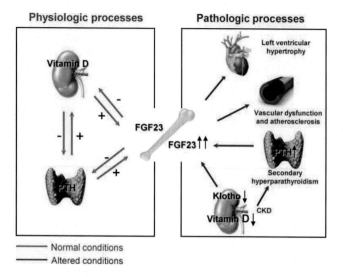

— Normal conditions
— Altered conditions

FIGURE EIII.3 **FGF23 homeostasis and action in disease.** Following absorption, dietary phosphorous raises the serum P concentration and stimulates FGF23 production by osteocytes/osteoblasts in the bone. Phosphate regulation is under three different feedback mechanisms. Active vitamin D activates the FGF23 promoter in the bone; FGF23 inhibits 1,25D production in the kidney. PTH activates vitamin D in the kidney; FGF23 shuts down the PTH promoter in the parathyroid glands (in animals). In renal failure, Klotho is underexpressed and FGF23 does not bind, thus phosphate is retained. Elevated phosphate upregulates FGF23. FGF23 inactivates active vitamin D, decreasing serum calcium and stimulating secondary hyperparathyroidism seen in CKD. FGF23 binds the cardiomyocyte by a non-Klotho FGF receptor 4 and stimulates pathways that lead to left ventricular hypertrophy (Grabner et al., 2015). Vascular dysfunction and calcification are discussed elsewhere (refer to Chapter 5: Diabetic Cardiomyopathy and Congestive Heart Failure and Chapter 6: Atherosclerosis and Arterial Calcification) (Donate-Correa et al., 2012). *Source: Reproduced from Donate-Correa, J., et al., 2012. FGF23/Klotho axis: phosphorus, mineral metabolism and beyond. Cytokine Growth Factor Rev. 23 (1—2), 37—46 with permission from Science Direct.*

Intracellular Ca^{2+} is a major second messenger and cofactor for proteins and enzymes that regulate neurotransmission, hormonal secretion, and other critical cellular processes. Extracellular Ca^{2+}, in addition to its function to support bone mineralization, is a cofactor for adhesion molecules and clotting factors and is involved with other homeostatic systems including volume and blood pressure regulation. Control of the serum Ca^{2+} concentration by PTH and 1,25D has been well described (Jones et al., 1998) and is discussed above.

EIII.3.2 Calcium Sources, Bioavailability, and Absorption

Absorption of dietary calcium has been comprehensively reviewed (Hoenderop et al., 2005; Tharabenjasin et al., 2014). Briefly, dietary calcium is absorbed by two routes, an active transcellular process and a passive paracellular mode. When dietary calcium availability is low, transcellular transport accounts for most (~80%) of the calcium absorbed. Transcellular active Ca^{2+} transport occurs mainly in the proximal intestine in three stages. Ca^{2+} crosses the apical membrane of the absorptive cell by a Ca^{2+}-sensitive channel, the transient receptor potential vanilloid family calcium channel 6 (TRPV6). Upon entry into the cell, Ca^{2+} is bound to calbindin (CaBP9k) and shuttled to the basal membrane where it is extruded from the enterocyte by the energy-requiring transporters, Ca^{2+}-ATPase isoform 1b (PMCA1b) and the Na^+/Ca^{2+} exchanger isoform 1 (NCX1). Energy is supplied directly by adenosine triphosphate (ATP) hydrolysis or indirectly via the Na^+ gradient generated by basolateral Na-K ATPase. Calbindin is thought to facilitate cytosolic Ca^{2+} diffusion from apical to basal enterocyte membrane and also to act as a Ca^{2+} donor to preserve the low intracellular Ca^{2+} concentration required for signaling and other functions (Hoenderop et al., 2005). During physiological stress (growth, pregnancy, or lactation) or nutritional stress (calcium-deficient diet), calbindin D-9k transcription is induced by 1,25D/VDR.

Passive paracellular Ca^{2+} absorption occurs throughout the length of the intestine, as well as in the renal proximal tubule and the thick ascending limb of Henle's loop. Significant regulation of paracellular transport has recently been described (Alexander et al., 2014); the authors examined evidence that paracellular Ca^{2+} movement may also require secondary activation energy. The most important determinants of paracellular Ca^{2+} absorption are Ca^{2+} solubility, permeability of the tight junctions, and the duration of transit; increased retention of luminal contents in more distal intestinal segments increases paracellular transport. The authors provide evidence that claudins, components of the epithelial tight junctions, create cation-permeable paracellular pores, permitting Ca^{2+} flux between intestinal epithelial cells. Claudins are small transmembrane proteins that span the cellular membrane four times and bind intracellular scaffold proteins. Preliminary evidence exists that 1,25D increases claudin expression and augments paracellular Ca^{2+} absorption (Fujita et al., 2008).

The major dietary sources of calcium are dairy products, including milk, cheese, and yogurt. The milk sugar, lactose, has a stimulating effect on calcium absorption. Dark green vegetables such as spinach, kale, broccoli, and seaweed also contain calcium, but some foods (spinach, collard greens, sweet potatoes, and beans) contain oxalates that bind luminal Ca^{2+} and prevent its absorption. Phytic acid, (inositol

FIGURE EIII.4 **Phytate.** Phosphorous in whole grains, legumes, and nuts is in the form of phytic acid (inositol hexakisphosphate; phytate), the principal phosphorus storage form in plants. Phytate is located in the bran and seed coat. It is not digestible by humans and nonruminant animals and is a poor source of phosphorus if intact. The phosphate groups that comprise phytate chelate calcium, iron, and zinc to form insoluble complexes, reducing human absorption not only of phosphate, but the minerals it binds. There is some evidence that rodents can absorb metals bound to phytate (Reddy and Cook, 1991). *Phytase*, an enzyme produced by yeast used to leaven bread, and likely by microbes in the human gut, can degrade phytate and liberate phosphate and minerals for absorption. *Source: From Harbinary (own work) (Public domain), via Wikimedia Commons.*

hexakisphosphate (IP6); inositol polyphosphate; phytate) shown in Fig. EIII.4, found in the seed coats of grains, beans, seeds, nuts, and soy isolates, also binds Ca^{2+}. Phytate is degraded by phytase in yeast. This explains the improved Ca^{2+} bioavailability from leavened bread in which the grains have been exposed to yeast fermentation. Phytase is also produced by several strains of bacteria in the gut (Sharma and Trivedi, 2015); the authors proposed that phytase-producing *Bacillus subtilis* also has promising probiotic features. Integrated actions of enhancers and inhibitors result in dietary calcium absorption in the range of 30—60%. Other factors that modulate calcium absorption include the subject's age, habitual calcium intake, skeletal requirements, vitamin D status, the state of the homeostatic system, and the bioavailability of the calcium in food.

The sugar content of the diet also alters Ca^{2+} bioavailability as recently reviewed by Tharabenjasin et al. (2014). As with calcium, transcellular sugar transport occurs mainly in the proximal intestine. Glucose and galactose are absorbed on a sodium-glucose cotransporter (SGLT1). Glucose can be metabolized in the enterocyte, but usually reaches high concentrations in the cytoplasm and is passively exported across the basolateral border on the facilitative glucose transporter 2 (GLUT2). Fructose, on the other hand, is absorbed by the passive transporter, GLUT5, transported unbound across the cytoplasm and exits the basolateral border on GLUT2. Although fructose is primarily metabolized in the liver, the enterocyte

expresses all fructose-metabolizing enzymes and catabolizes 10—30% of ingested fructose (Patel et al., 2015). Fructose metabolism is initiated by the energy-requiring ketohexokinase (KHK; also called fructokinase) enzyme, producing fructose-1-phosphate. Using transgenic mouse models, the authors determined that the observed upregulation of GLUT5-mediated transport by chronically high dietary fructose requires enterocyte fructose metabolism by KHK as well as GLUT5 trafficking to the apical membrane. Because fructose metabolism by KHK requires energy, it depletes enterocyte ATP availability also required for energy-dependent, basolateral calcium efflux. In short, in the rodent, chronic high fructose intake upregulates GLUT5, decreases enterocyte ATP, and compromises Ca^{2+} absorption.

In the serum $[Ca^{2+}]$ is maintained in a normal range by hormones such as PTH and active vitamin D as described above and recently reviewed (Areco et al., 2015). Low dietary calcium and reduced Ca^{2+} availability stimulates PTH secretion and activates renal 1-α-hydroxylase (CYP27B1) hydroxylating 25D to its active form, 1,25D. Together, PTH and 1,25D mobilize bone Ca^{2+} and increase net renal Ca^{2+} reabsorption. Calcium absorption at the enterocyte is enhanced by VDR-dependent transcription of calcium transporters, TRPV6 and TRPV5, and of CaBP9k. In a mouse model fed a low-calcium diet with glucose or starch as a carbohydrate source, the expected adaption to a low-calcium diet was demonstrated (Douard et al., 2014). When fructose was substituted for glucose, adaptation did not occur. Renal 1-α-hydroxylase (CYP27B1) and 1,25D levels decreased and the expected 1,25D/VDR-dependent increase in calbindin transcription was not present. At the same time, renal 24-hydroxylase (CYP24A1) levels increased, converting 25D to its inactive form, 24,25D. When adult mice were fed fructose with a diet adequate in calcium for 3 months, serum levels of 1,25D were decreased concomitant with decreased renal CYP27B1. FGF23 was increased. The authors concluded that in this rodent model, chronically high fructose intake, even with a normal intake of calcium, compromised vitamin D-mediated calcium absorption. Further studies are required to determine the link between fructose intake and reduction in 1,25D generation. It is also important to test whether these results are generalizable to the human obese population; these individuals have low levels of 25D in their serum and at the same time have inadequate calcium but substantial fructose in their diets.

Fractional absorption of calcium is high during childhood, adolescence, pregnancy, and lactation when skeletal needs are great. If dietary calcium is limited, fractional absorption is increased. With aging, this adaptation declines and calcium absorption is compromised. As of 2011, the Institute of Medicine has set the recommended dietary allowance (RDA) (adequate intake (AI)) for calcium (diet + supplements) at 1000 mg/day for adult women with an increase to 1200 mg/day after age 50 (revised AI (2011) https://www.nlm.nih.gov/medlineplus/magazine/issues/winter11/articles/winter11pg12.html). The tolerable upper limit (UL) is 2000 mg/day. Despite these recommendations, a recent survey (Bailey et al., 2010) revealed that a large fraction of the US population does not achieve the AI for calcium. After the age of 50, less than 50% of males and females in the US achieved AIs for calcium and vitamin D, even with supplementation. Data from the NHANES and other surveys have also shown that dietary calcium intake from foods is lower than the AI (Food and Nutrition Board, 2010), especially for women and for men >30 years.

EIII.3.3 Phosphorus

Phosphate ions (P_i) play a major role in maintenance of nucleic acids and cyclic nucleotides, phospholipids, and intermediates of energy metabolism. Pi are also integral to the formation of hydroxyapatite [$Ca_5(PO_4)_3(OH)$] crystal in bone. Most phosphate ions in the human body are bound up in bone crystal; only ~0.1% are in the extracellular fluid and circulate as inorganic phosphate species (P_i). As with calcium, circulating (P_i) is tightly regulated. In addition to the clinical and pathological sequelae from hypo- and hyperphosphatemia, recent observations suggest that high [Pi], even within the normal range, is linked with increased morbidity and mortality (Donate-Correa et al., 2012) (refer to Chapter 7: Chronic Kidney Disease, Chapter 6: Cardiovascular Disease, Chapter 5: Cardiomyopathy, and Chapter 9: Osteoporosis).

Phosphorus in food is variably absorbed. Organic phosphorus in dairy products, meat, and fish is ~40–60% absorbed, while phosphorus in grains, legumes, and plants is bound to poorly absorbed phytate and fiber, thus only 10–30% is absorbed (Donate-Correa et al., 2012). While dairy products contain substantial phosphorus, they also contain calcium, known to bind phosphate ions and reduce their bioavailability. The rate at which phosphorus is absorbed may be reflected in an immediate spike in serum [P_i] that can stimulate FGF23. Phosphorus from organic sources must be removed from its protein binders in the stomach and proximal small intestine, thus it is slowly absorbed and the rise in serum [P_i] is more gradual and presumably less likely to stimulate robust FGF23 secretion (Calvo et al., 2014).

In contrast to organic phosphorus in foods, inorganic phosphate salts are well absorbed; the bioavailability of phosphate salts has been estimated to approach 100%. Phosphate salts are widely used in food processing to enhance flavor, improve color, and moisture and to extend shelf-life of products. US labeling laws do not require food processors to indicate on the label that the product has been treated with phosphates, thus the actual contribution of added phosphate salts to the total phosphorus load is impossible to determine. Several analytical studies have found that the phosphorus content of foods processed with phosphate salts can be underestimated by 25–70% (Calvo et al., 2014). The unexpected addition of phosphorus salts to foods such as milk, eggs, and vegetables adds to total dietary phosphorus load. For example, French fries, normally not a rich source of phosphorus, can contain modified food starch, sodium acid pyrophosphate, and disodium dihydrogen pyrophosphate, all phosphate salt additives Generally Recognized as Safe (GRAS). (Under sections 201(s) and 409 of the Federal Food, Drug, and Cosmetic Act, GRAS substances can be added to food for human consumption.) More insidious is the practice by meat processors of bathing their products in a phosphate salt bath. This bath both enhances the color of the product and decreases microbial growth in raw meat (Sherman and Mehta, 2009). While factory-wrapped meat may indicate on the label that the meat has been processed with a phosphate bath, the butcher is under no such requirements. Meat in the display case may, or may not, contain phosphate additives, thus the presence and content of phosphate salts may be more difficult to identify.

Many cola-type beverages, calorically sweetened as well as diet cola, contain phosphoric acid. This inorganic phosphate in addition to its deleterious effects on tooth enamel, is also well absorbed (80–100%), especially if the soda is not consumed with food. Well-absorbed phosphate salts consumed alone, without the binders and chelators contained in a meal, have potential to cause a rapid spike in serum P_i (P_i) and trigger subsequent FGF23 secretion, even in normal individuals. High dietary phosphorus (400 mg vs 1200 mg dietary phosphorous) increased serum [P_i] at 2 hours and significantly decreased flow-mediated dilation in healthy human subjects (Shuto et al., 2009). CKD elevates FGF23 and high phosphorus intake exacerbates risk for heart and vascular complications. Public health implications (Calvo and Uribarri, 2013) for normal individuals include ventricular hypertrophy (Grabner et al., 2015) and cardiovascular disease.

Thus, both normal individuals and CKD patients at all stages of the disease should avoid sources of inorganic phosphorus salts. Natural foods that have been processed in any way, cola-type beverages, phosphate-enhanced raw and processed meats, baked goods, instant mixes, cereals, and processed cheeses are all sources of well-absorbed phosphorus. For example, 50 g of unprocessed Brie cheese contains <100 mg P while the same amount of processed American cheese contains >500 mg P. On the other hand, while unprocessed whole grains, nuts, and seeds contain phosphorus as phytates, this P source is poorly absorbed. Patients should be taught to read labels and to avoid foods, supplements, and frequently used over-the-counter medications such as antacids and laxatives and to maintain their dietary phosphorus intake close to the recommended intake of 700 mg/day for adults of both sexes (Food and Nutrition Board, 1997).

EIII.3.4 Magnesium

Dietary magnesium (Mg) is obtained from plants and animals, it is also a component of many drugs (antacids, laxatives), and is available as a dietary supplement. Magnesium is the central atom in chlorophyll, thus it is present in most plant foods, especially green leafy vegetables. Legumes, whole grains, nuts, and chocolate are also rich sources. (The US Department of Agriculture maintains a nutrient database that lists nutrient content of many foods arranged by nutrient content and food name. http://ndb.nal.usda.gov/). In general, foods with high fiber content are good Mg sources, but the processes of milling and refining remove significant amounts of Mg; sometimes as much as 80% of Mg content is lost in processed foods. Meats, fruits, and fish are intermediate Mg sources and dairy products are poor sources. "Hard" water that has percolated through deposits of limestone contains significant Ca and Mg carbonates. Mg is poorly bioavailable with a fractional absorption rate of 35–40%, making milk of magnesia an osmotic laxative. Net Mg absorption increases with increased dietary intake. Mg supplements as oxide, citrate, and chloride are available. Liquid-soluble supplements containing Mg citrate, lactate, chloride, and aspartate are more bioavailable than oxide and sulfate salts (Ranade and Somberg, 2001).

The adult human body contains ∼25 g of Mg, about half of which is found in the bones; less than 1% of Mg is in the blood at concentrations between 0.75 and 1.1 mM. To achieve this level, a dietary recommended intake for Mg of 420 mg for men and 320 mg for women has been established (Food and Nutrition Board, 1997). Magnesium homeostasis is controlled by uptake of dietary Mg at the intestines, release from storage in the bone, and selective excretion and reabsorption by the kidneys. Normally, Mg excretion in the urine is ∼120 mg/day, but this level can fall sharply when total Mg status is low.

Mg is absorbed primarily in the small intestine by paracellular and transcellular processes. Paracellular Mg^{2+} absorption is proportional to the luminal Mg^{2+} within a low, narrow concentration range. At higher luminal concentrations, a reverse paracellular serosal to mucosal Mg^{2+} secretion has been reported (Karbach and Rummel, 1990). The "gatekeeper" for transepithelial Mg transport, transient receptor potential melastatin 6 (TRPM6) is not expressed in the small intestine, explaining in part the poor proximal Mg^{2+} absorption (Groenestege et al., 2006). Mg^{2+} absorption in cecum and colon appears to be primarily transcellular, mediated by expression of TRPM6 and TRPM7 on the luminal side of the enterocyte. TRPM6 expression in the colon is not dependent on 1,25D, thus transcellular Mg^{2+} transport is independent of 1,25D signaling. Molecular characteristics of these and other general and tissue-specific transporters that maintain Mg^{2+} distribution across cells and organs have been reviewed (de Baaij et al., 2015); recently a unique mitochondrial Mg transporter was reported (Trapani and Wolf, 2015).

Early estimates (∼1980) that Mg was required for >300 enzyme reactions have been revised upward to ∼600 enzymes for which Mg is cofactor or activator as recently reviewed by Caspi et al. (2012) and de Baaij et al. (2015). Mg^{2+} is cofactor for MgATP, required to provide energy for numerous enzymatic reactions essential for life. It is required for the structure and activity of DNA and RNA polymerases, enzymes for DNA repair (mismatch repair, topoisomerases, and most others), glycolysis, and enzymes that maintain genome and genetic stability. Mg also plays an essential role in bone development and remodeling and in the activity of the endogenous antioxidant, glutathione. It is a powerful antagonist of Ca^{2+} action, especially in the heart. In concert with calcium and potassium, Mg maintains nerve signaling, muscle contraction, and normal sinus rhythm (Iseri and French, 1984).

In the bone, Mg^{2+} ions bound to the surface of hydroxyapatite crystals modulate crystal size, stability, and rate of formation (Salimi et al., 1985). Surface Mg^{2+} is ∼30% of total bone Mg^{2+} and comprises an exchangeable pool that equilibrates with extracellular fluid. It is likely that this pool supports serum Mg^{2+} and is available to other tissues during periods of dietary deprivation. The remainder of bone Mg^{2+} is embedded in the matrix and released during bone remodeling (Alfrey et al., 1974). Mg^{2+} induces osteoblast proliferation and reduces secretion of proinflammatory cytokines implicated in osteoclastic bone resorption. It is also required to maintain calcium homeostasis by increasing PTH activity and tissue responsiveness to PTH. The impacts of Mg^{2+}

deficiency on processes that lead to osteoporosis are discussed elsewhere (refer to Chapter 9: Osteoporosis and Fracture Risk: Targets for Dietary Intervention in Osteoporosis and Fracture Risk).

Current understanding of kidney regulation of Mg reabsorption has been reviewed (de Baaij et al., 2015). Briefly, Mg is filtered by the glomerulus and $\sim 95\%$ is reabsorbed in the tubules. Paracellular reabsorption in the proximal tubule accounts for $\sim 10-25\%$ of the reabsorbed Mg^{2+}. In the thick ascending limb of Henle's loop, paracellular transport depends on activity of the Na^+-K^+-$2Cl^-$ cotransporter and accounts for $50-70\%$ of Mg^{2+} reabsorption. Fine adjustments of Mg^{2+} reabsorption (10%) occur in the distal convoluted tubule (DCT) with the aid of TRPM6. Epidermal growth factor, estrogen, and insulin regulate Mg^{2+} transport in the DCT. Upon binding to their receptors, these anabolic agents trigger an intracellular signaling cascade that increases TRPM6 expression and channel activity.

EIII.3.5 Causes and Assessment of Magnesium Deficiency

Mg status is difficult to assess, partly because Mg is largely intracellular and only about 1% of total body Mg is in the serum. Approximately one-third of serum Mg is bound to proteins, including albumin and globulins. Only 92% of the remaining serum Mg is ionized (61% of total serum Mg is Mg^{2+}), the remainder is complexed as salts (phosphate, citrates, and others). Although total serum Mg varies with ionized Mg^{2+}, total value is dependent on serum albumin concentration (Elin, 2010) and can be falsely low under conditions of hypoalbuminemia. Normal serum Mg ranges from 1.5–2.0 mM/L. Severe Mg deficiency assessed by total serum Mg levels is <0.75 mM/L; values between 0.75 and 0.85 mM/L are considered adequate. Even if an accurate ionized Mg^{2+} concentration is obtained (many laboratories do not routinely provide this test), total body Mg is uncertain because serum Mg appears to be homeostatically maintained in equilibration with the exchangeable bone Mg pool (Alfrey et al., 1974). A Mg load test has been proposed to assess the Mg status of a patient suspected of being severely Mg-depleted. However, this test is rarely used clinically and is not yet standardized.

Symptoms of magnesium deficiency include loss of appetite, nausea, vomiting, fatigue, and weakness. Severe Mg depletion (<0.4 mM) results in numbness, muscle contractions and cramps, tetany, seizures, arrhythmias, and coronary spasms. Abnormal potassium and calcium handling accompany hypomagnesemia. Increased urinary K^+ excretion results when low intracellular Mg levels release K^+ channel inhibition in the distal tubule and allow K^+ to be excreted. High sodium intake or activation of the renin–angiotensin–aldosterone-mediated sodium reabsorption may exacerbate K^+ loss (Huang and Kuo, 2007). Hypocalcemia occurs because the bone and renal tubules are refractory to PTH in magnesium deficiency.

Dietary surveys (NHANES) have consistently shown that the mean dietary intakes of Mg are below recommendations (dietary Mg of 420 mg for men and 320 mg for women) for all members of the population. Supplements and Mg-containing medications can increase total Mg intake, but even with these added Mg sources, intakes were inadequate in some ethnic groups and with increasing age (Ford and Mokdad, 2003; King et al., 2005). In addition to dietary deficiency, conditioned Mg deficiency has been reported secondary to use of medications such as diuretics, alcohol, and with certain diseases (de Baaij et al., 2015).

Gastrointestinal losses have been reported in patients with gluten-sensitive enteropathy (celiac disease), Crohn's disease, and other malabsorptive states. Mg is lost in the chronic diarrhea associated with these conditions because water reabsorption is required for Mg reabsorption, thus loss of water with diarrhea will also waste Mg. Fat malabsorption increases Mg loss because fatty acids form an insoluble complex (soap) with Mg, increasing Mg losses. Hypomagnesemia following bariatric surgery has been reported, however it is uncertain whether the patients also had depleted status prior to surgery. Again, only serum and not total body Mg was measured (Stein et al., 2014). Mg deficiency reported with chronic alcoholism can be due to a number of factors. Deficient Mg intake due to a poor diet can be exacerbated by the Mg lost due to the diuretic action of alcohol. Gastrointestinal Mg losses in the alcoholic can be due to vomiting, diarrhea, and fat malabsorption. Additionally, renal dysfunction, vitamin D deficiency, alcoholic ketoacidosis, hyperaldosteronism, or liver disease can compromise Mg status in the alcoholic (Rivlin, 1994).

Low serum Mg concentrations have been reported in 13.5–47.7% of patients with type II DM, compared with 2.5–15% of nondiabetic patients, as reviewed by Pham et al. (2007). Although several risk factors such as poor diet, gastrointestinal losses, and concomitant drug use could increase Mg loss in diabetic patients, these patients also have been observed to have increased urinary Mg excretion. A recent study using a rat kidney model has identified attenuated Mg^{2+} reabsorption through TRPM6 in the distal tubules as a presumptive cause of hypermagnesiuria in diabetes (Takayanagi et al., 2015). The authors reported that reduced Mg reabsorption occurred early in diabetic progression, suggesting that Mg loss could exacerbate insulin resistance and reduced insulin production by the β cell (refer to Chapter 4: Type II Diabetes, Peripheral Neuropathy, and Gout).

Diuretics targeting the loop of Henle and the DCT have long been known to reduce absorption of Mg, as well as other electrolytes such as potassium and calcium. Furosemide (Lasix) inhibits the activity of the Na^+, K^+, $2Cl^-$ cotransporter (NKCC2) in the thick ascending limb of Henle, also reducing transepithelial potential for paracellular Mg^{2+} transport. In fact, furosemide increases Mg loss to a greater extent than sodium loss and can result in Mg wasting in patients on high doses of the drug (Quamme, 1981). Some individuals can compensate by increasing the DCT magnesium transporter, TRPM6 (van Angelen et al., 2012). Thiazide diuretics also induce renal Mg^{2+} wasting (Hollifield, 1986) especially in elderly patients and those with chronic heart failure. In a mouse model, thiazide reduced the renal expression of TRPM6 (Nijenhuis et al., 2005). Mg loss is also reported with ethacrynic acid (Edecrin), but not with potassium-sparing diuretics such as amiloride (Midamor) and spironolactone (Aldactone) (Sarafidis et al., 2010). Proton pump inhibitor drugs (esomprazole; Nexium and lansoprazole; Prevacid) can also cause hypomagnesemia when taken over an extended period, usually over 1 year (Begley et al., 2016).

EIII.3.6 Excess Dietary Magnesium

While excessive Mg from food does not cause deleterious effects in individuals with normal kidney function, high Mg intake from supplements or medication can result in diarrhea due to the osmotic activity of unabsorbed salts in the gastrointestinal tract and stimulation of gastric motility. Mg doses >5000 mg/day have been associated with fatalities. Serum levels 1.74–2.61 mM/L can result in symptoms culminating in cardiac arrest; risk for hypermagnesemia is elevated in patients with CKD (Musso, 2009). Tolerable ULs for magnesium have been established (Food and Nutrition Board, 1997); the UL for supplemental Mg intake (in addition to dietary Mg) for adults of both sexes is 350 mg/day.

Mg-rich supplements can decrease the absorption of several medicines:

- Oral bisphosphonates (alendronate; Fosamax) and Mg supplements should be separated by at least 2 hours.
- Mg forms insoluble complexes with tetracyclines (demeclocycline; Declomycin) and doxycycline (Vibramycin) as well as quinolone antibiotics (ciprofloxacin; Cipro and levofloxacin; Levaquin). Antibiotics should be taken at least 2 hours before or 4–6 hours after Mg supplement.

EIII.4 VITAMIN METABOLISM

EIII.4.1 Vitamin D (Cholecalciferol)

Vitamin D is a cholesterol-derived vitamin (cholecalciferol—D3), obtained from a limited number of food sources such as oily salt water fish (salmon, mackerel, sardines), liver, and egg yolks. Vitamin D is formed commercially by exposing a fungus-derived sterol (ergosterol) to ultraviolet radiation to form ergocalciferol (D2). Commercially grown mushrooms are exposed to ultraviolet radiation and are a good source of D2. Preliminary studies have suggested that mushrooms may also interact with gut microbiota, enhancing innate and adaptive immunity as discussed elsewhere (Feeney et al., 2014) (refer to Essentials IV: Diet, Microbial Diversity and Gut Integrity: Diet Exposures that Enhance Antimicrobial Protection). Vitamin D as D2 or D3 is added to milk and other food staples as a supplement. (While originally assumed to be equivalent, recent studies suggest that vitamin D2 is only 20–40% as effective as D3 in raising serum 25D.) Most milk sold in the United States is fortified with 100 international units (IU)/cup, however cheese, ice cream, and other dairy products may not contain supplemental vitamin D. Heat treatment results in variable losses of vitamin D depending on the foods and processes tested. For example, eggs cooked in the oven for ~40 minutes at normal temperatures, as in a quiche, retained ~40% of original vitamin while eggs fried or boiled retained >80% of the vitamin (Jakobsen and Knuthsen, 2014).

Vitamin D content in food is listed as IU; the biological activity of 40 IU is equal to 1 μg vitamin D. The RDA is 800 IU (20 μg) for adults (Food and Nutrition Board, 2010). D3 is hydroxylated in the liver and secreted into the circulation as $25(OH)D_3$ (25D). In this form, the half-life of vitamin D is relatively long (15 days), making it the most likely indicator of vitamin D status (Jones, 2008). It should be noted that serum 25D concentration does not reflect body stores, nor does it indicate the action of vitamin D hormone (1,25D) at target tissues. In contrast to the long half-life of serum 25D, the half-life of 1,25 is much shorter, <15 hours.

The Institute of Medicine has concluded that serum 25D levels <30 nM (<12 ng/mL) reflect risk for vitamin D deficiency. Sufficient levels are >50 nM (>20 ng/mL), while levels >125 nM (>50 ng/mL) have been associated with adverse effects (Food and Nutrition Board, 2010). A 2011 report issued by the National Academy of Sciences used skeletal effects to define vitamin D deficiency (inadequacy) as a serum $25(OH)D_3$ level of <20 ng/mL (Ross et al., 2011). This report has generated significant controversy and many investigators believe the vitamin D threshold is too low. Assay of 25D levels in biological samples varies considerably between antibody-based and chromatography-based assays. These method-related variabilities should be somewhat resolved with the standard reference material for 25D that became available in 2009 (http://www.nist.gov/mml/csd/vitamind_071409.cfm).

EIII.4.2 Vitamin D Action

While early studies focused on the importance of vitamin D in maintaining skeletal integrity, it is now widely appreciated that this vitamin is essential for at least five physiological systems, including the immune system (both the innate and adaptive), pancreas and metabolic homeostasis, heart and cardiovascular systems, muscle and brain, as well as the control of the cell cycle, and thus of neoplastic processes. The most recent Cochrane library analysis concluded that supplemental vitamin D3 appeared to decrease mortality in elderly people living independently or in institutional care, however other forms of vitamin D such as vitamin D2 (ergocalciferol) or as an active form of vitamin D, including the vitamin D hormone, 1α-hydroxyvitamin D (alfacalcidol) or 1,25-dihydroxyvitamin D (calcitriol) had no statistically significant beneficial effects on mortality. Vitamin D3 combined with calcium increased nephrolithiasis and both alfacalcidol and calcitriol increased hypercalcaemia (Bjelakovic et al., 2014).

Current understanding of vitamin D activation and function has been comprehensively reviewed (Berridge, 2015) as shown in Fig. EIII.5. Using data presently available, the author develops the hypothesis that vitamin D functions as a guardian of phenotypic stability by means of its ability to maintain redox and Ca^{2+} signaling systems. Vitamin D controls two systems critical to redox and Ca^{2+} signaling, nuclear factor-erythroid-2-related factor (Nrf2) and Klotho. Vitamin D hormone (1,25D) binds its receptor (VDR), a member of the steroid receptor superfamily present in the cytoplasm of many cells, including peripheral blood mononuclear cells. The 1,25D/VDR complex forms a heterodimer with the retinoid X receptor and regulates gene expression by binding to VDRE in the promoter region of genomic coding sequences. In this way, vitamin D downregulates the gene for PTH and regulates serum calcium concentration as described above.

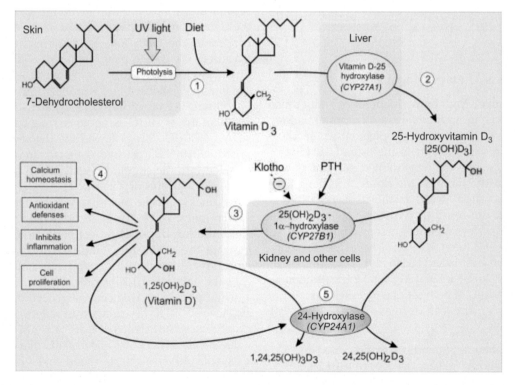

FIGURE EIII.5 **Vitamin D homeostasis.** Solar UV B at wavelengths of 290–320 nanometers photolyse 7-dehydrocholesterol in the human skin to produce cholecalciferol. Vitamin D is hydroxylated in the liver by CYP27A1 and circulates as 25D until taken up by the kidney and other cells where it is again hydroxylated by 1-α-hydroxylase (CYP27B1). 1,25D induces expression of multiple genes that control calcium homeostasis, antioxidant defenses, inflammation, and proliferation. 1,25D is inactivated by 24-hydroxylase (CYP24A1); activity of this enzyme is increased by 1,25D, thereby forming a feedback regulation loop. *Source: Reproduced from Berridge, M.J., 2015. Vitamin D cell signalling in health and disease. Biochem. Biophys. Res. Commun. 460 (1), 53–71 with permission from Science Direct.*

Vitamin D as 1,25D/VDR regulates several genes produced by osteoblasts and osteocytes including RANKL, SPP1 (osteopontin), and the bone-Gla protein (BGP) osteocalcin, involved in bone mineral remodeling. Vitamin D also facilitates calcium absorption by upregulating TRPV6, CaBP9k, and claudin 2. TRPV5, klotho, and Npt2c are also regulated by 1,25D/VDR to control renal calcium and phosphate reabsorption. VDR may also function without 1,25D to regulate genes that affect Wnt signaling (Haussler et al., 2013). In addition, 1/25D/VDR regulates biosynthesis of Nrf2, a transcription factor that binds to the antioxidant response element in the promoter region of a wide spectrum of genes for detoxification and antioxidant systems. As described elsewhere (refer to Essentials I: Life in an Aerobic World: Nutrients, Transcriptional Control of Antioxidant Protection), Nrf2 is bound in the cytoplasm until released by oxidative stress and by several nonnutrient components of foods such as curcumin, flavonoids, and sulforophane (Fig. EIII.6).

EIII.4.3 Vitamin D Deficiency

Vitamin D intake in the United States was assessed using NHANES data 2005–2006. Less than 7% of males and females over the age of 51 years met the adequate intake (AI) for vitamin D through the diet; supplement use increased the percent of subjects that exceeded the AI, but only 37% of the population reported using supplements (Bailey et al., 2010). Vitamin D deficiency, assessed as low serum levels of 25D, appears to be widespread globally, especially in developing countries. A variety of factors reduce the availability and utilization of vitamin D including age (very young and very old), female sex, winter season, dark skin pigmentation, malnutrition, lack of sun exposure, a covered clothing style, and obesity (Arabi et al., 2010; Hayden et al., 2015). Endogenous vitamin D synthesis in the skin is only reliably available year-round at latitudes between 40°N and 40°S. This is important because approximately one-third of the world's citizens (2.3 billion) live between 90°N and 40°N where levels of UVB are low or nonexistent for a significant portion of the year. Since few foods naturally contain vitamin D, individuals whose diets do not contain these foods are at risk. Also at risk are individuals who live in urban areas without regular exposure to the sun and dark-skinned individuals whose melanin production acts as a sunscreen to limit vitamin D biosynthesis in the skin. It is of note that excessive use of sunscreen has also been shown to reduce endogenous vitamin D biosynthesis (Faurschou et al., 2012).

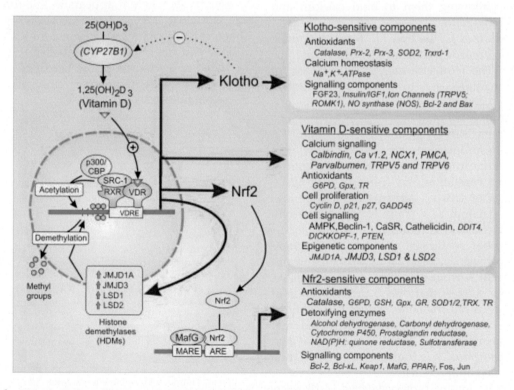

FIGURE EIII.6   **Vitamin D-dependent gene expression.** In addition to genes sensitive to VDR binding, vitamin D upregulates genes controlling calcium homeostasis and redox signaling pathways through Klotho and Nrf2-sensitive pathways. Vitamin D also controls epigenetic activities through histone demethylases. *Source: Reproduced from Berridge, M.J., 2015. Vitamin D cell signalling in health and disease. Biochem. Biophys. Res. Commun. 460 (1), 53–71 with permission from Science Direct.*

Obese patients are at risk for low serum levels of 25D, presumably because fat-soluble vitamin D is sequestered in adipocytes. Elderly patients appear to be at increased risk due to age-associated decline in vitamin D biosynthesis and reduced sun exposure in this population. As use of computers and electronic devices proliferates, vitamin D deficiency is expected to rise. Indeed, analysis of a nationally representative sample (6275 children aged 1−21 years) from NHANES 2001−2004 found vitamin D deficiency in 9% with insufficiency in 61% based on serum 25D levels <29 ng/mL.

EIII.4.4 Vitamin K (Phylloquinone)

Plants and microorganisms synthesize a family of compounds with classical vitamin K activity, that is, the conversion of specific peptide-bound glutamate (Glu) to γ-carboxyglutamate (Gla). The vitamin K family members, phylloquinone (vitamin K1) and the menaquinones (vitamin K2) are under study for their roles in bone metabolism and as potential therapeutic agents for skeletal diseases. The best known of the vitamin K-dependent proteins (Gla proteins) is γ-carboxylase, an integral protein bound to the endoplasmic reticulum membrane. This protein posttranslationally modifies glutamate residues in clotting factors, thereby allowing them to bind calcium and amplify the coagulation cascade. In the process, vitamin K is oxidized to the epoxide form, and must be reduced by epoxide reductase to regenerate the active cofactor. The reductase enzyme is a target for warfarin (Coumadin), used to treat hypercoagulant diseases.

Dietary vitamin K is found mainly as phylloquinone (K_1) in dark green leafy vegetables ($\sim$400−500 μg/$\frac{1}{2}$ cup). Brussels sprouts or broccoli contain less ($\sim$75−150 μg/$\frac{1}{2}$ cup); certain vegetable oils, grains, dairy, and other vegetables and fruits ($\sim$5−50 μg/$\frac{1}{2}$ cup) are rather poor sources. Animal products are also poor sources of vitamin K. Vitamin K can also be synthesized by bacteria and is found in fermented foods. Intestinal anaerobes synthesize a variety of menaquinones ($K_{2,3}$), however evidence for absorption and utilization of vitamin K from this source is inconclusive. The current data suggest that only a minor portion of the vitamin K required for bone integrity is obtained from bacterial biosynthesis. It should be noted that vitamin K is fat-soluble and requires dietary fat and bile acids for absorption. It is transported in circulation by triglyceride-rich lipoproteins.

As a cofactor for γ carboxyglutamyl carboxylase (GGCX), vitamin K catalyzes conversion of specific peptide-bound glutamate residues in specialized proteins to form γ-carboxyglutamate (Gla) proteins (Shearer and Newman, 2014). The best-characterized Gla proteins are factors VII, IX, and X as well as proteins C, S, and Z in the coagulation cascade. Since clotting factor activation is essential to life, the liver sequesters vitamin K for essential coagulation proteins, while other, less essential proteins have limited access to the cofactor. Thus, under conditions of marginal vitamin K status, Gla proteins in other tissues can remain undercarboxylated.

Several Gla proteins have been identified in the organic bone matrix including BGP, matrix-Gla protein (Mgp), protein-S, Gla-rich protein (GRP), periostin, and periostin-like factor (PLF). The postulated roles of these proteins in maintaining bone structure have been reviewed (Rubinacci, 2009). The vitamin also acts as a ligand for the steroid and xenobiotic receptor (SXR) (Azuma and Inoue, 2009). Genes upregulated by this transcription factor may be important in bone matrix integrity and bone strength.

Mgp, although found in the bone, has been studied more extensively for its ability to bind calcium and prevent arterial calcification as discussed elsewhere (refer to Chapter 6: Atherosclerosis and Arterial Calcification: Targets for Atherosclerotic Prevention and Control). Osteocalcin (OCN) is a Gla protein synthesized, carboxylated, and deposited in bone matrix by osteoblasts where it appears to fine-tune bone mineral structure. As described elsewhere (refer to Chapter 9: Osteoporosis and Fracture Risk: Integration of Bone and Intermediary Metabolism), OCN is decarboxylated and released in the bloodstream in the process of osteoclast remodeling and plays a role in glucose homeostasis.

References

Alexander, R.T., Rievaj, J., et al., 2014. Paracellular calcium transport across renal and intestinal epithelia. Biochem. Cell Biol. 92 (6), 467−480.

Alfadda, T.I., Saleh, A.M.A., et al., 2014. Calcium-sensing receptor 20 years later. Am. J. Physiol. Cell Physiol. 307 (3), C221−C231.

Alfrey, A.C., Miller, N.L., et al., 1974. Effect of age and magnesium depletion on bone magnesium pools in rats. J. Clin. Invest. 54 (5), 1074−1081.

Arabi, A., El Rassi, R., et al., 2010. Hypovitaminosis D in developing countries—prevalence, risk factors and outcomes. Nat. Rev. Endocrinol. 6 (10), 550−561.

Areco, V., Rivoira, M.A., et al., 2015. Dietary and pharmacological compounds altering intestinal calcium absorption in humans and animals. Nutr. Res. Rev. 28, 83−99.

Azuma, K., Inoue, S., 2009. Vitamin K function mediated by activation of steroid and xenobiotic receptor. Clin. calcium. 19 (12), 1770−1778.

de Baaij, J.H.F., Hoenderop, J.G.J., et al., 2015. Magnesium in man: implications for health and disease. Physiol. Rev. 95 (1), 1−46.

Bailey, R.L., Dodd, K.W., et al., 2010. Estimation of total usual calcium and vitamin D intakes in the United States. J. Nutr. 140 (4), 817—822.

Begley, J., Smith, T., et al., 2016. Proton pump inhibitor associated hypomagnasaemia—a cause for concern? Br. J. Clin. Pharmacol. 81, 753—758.

Berridge, M.J., 2015. Vitamin D cell signalling in health and disease. Biochem. Biophys. Res. Commun. 460 (1), 53—71.

Bhattacharyya, N., Chong, W.H., et al., 2012. Fibroblast growth factor 23: state of the field and future directions. Trends Endocrinol. Metab. 23 (12), 610—618.

Bjelakovic, G., Gluud Lise, L., et al., 2014. Vitamin D supplementation for prevention of mortality in adults. Cochrane Database Sys. Rev. Available from: http://dx.doi.org/10.1002/14651858. CD007470.pub3.

Blau, J., Collins, M., 2015. The PTH-Vitamin D-FGF23 axis. Rev. Endocr. Metab. Disord. 16 (2), 165—174.

Brown, E.M., Pollak, M., et al., 1994. Cloning and characterization of an extracellular Ca2+ -sensing receptor from parathyroid and kidney: new insights into the physiology and pathophysiology of calcium metabolism. Nephrol. Dial. Transplant. 9 (12), 1703—1706.

Calvo, M.S., Moshfegh, A.J., et al., 2014. Assessing the health impact of phosphorus in the food supply: issues and considerations. Adv. Nutr. Int. Rev. J. 5 (1), 104—113.

Calvo, M.S., Uribarri, J., 2013. Public health impact of dietary phosphorus excess on bone and cardiovascular health in the general population. Am. J. Clin. Nutr. 98 (1), 6—15.

Caspi, R., Altman, T., et al., 2012. The MetaCyc database of metabolic pathways and enzymes and the BioCyc collection of pathway/genome databases. Nucleic Acids Res. 40 (D1), D742—D753.

Donate-Correa, J., Muros-de-Fuentes, M., et al., 2012. FGF23/Klotho axis: phosphorus, mineral metabolism and beyond. Cytokine Growth Factor Rev. 23 (1—2), 37—46.

Douard, V., Patel, C., et al., 2014. Chronic high fructose intake reduces serum 1,25 (OH)(2)D(3) levels in calcium-sufficient rodents. PLoS ONE 9 (4), e93611.

Elin, R.J., 2010. Assessment of magnesium status for diagnosis and therapy. Magnes. Res. 23 (4), 184—198.

Faurschou, A., Beyer, D., et al., 2012. The relation between sunscreen layer thickness and vitamin D production after ultraviolet B exposure: a randomized clinical trial. Br. J. Dermatol. 167 (2), 391—395.

Feeney, M.J., Dwyer, J., et al., 2014. Mushrooms and health summit proceedings. J. Nutr. 144 (7), 1128S—1136S.

Felsenfeld, A.J., Levine, B.S., 2015. Calcitonin, the forgotten hormone: does it deserve to be forgotten? Clin. Kidney J. 8 (2), 180—187.

Food and Nutrition Board, 1997. Dietary Reference Intakes: Calcium, Phosphorus, Magnesium, Vitamin D and Fluoride. National Academy Press, Washington, DC.

Food and Nutrition Board, 2010. Dietary Reference Intakes for Calcium and Vitamin D. National Academy Press, Washington, DC.

Ford, E.S., Mokdad, A.H., 2003. Dietary magnesium intake in a national sample of U.S. adults. J. Nutr. 133 (9), 2879—2882.

Fujita, H., Sugimoto, K., et al., 2008. Tight junction proteins claudin-2 and -12 are critical for vitamin D-dependent Ca2+ absorption between enterocytes. Mol. Biol. Cell. 19 (5), 1912—1921.

Grabner, A., Amaral, A.P., et al., 2015. Activation of cardiac fibroblast growth factor receptor 4 causes left ventricular hypertrophy. Cell Metab. 22 (6), 1020—1032.

Groenestege, W.M.T., Hoenderop, J.G., et al., 2006. The epithelial Mg2+ channel transient receptor potential melastatin 6 is regulated by dietary Mg2+ content and estrogens. J. Am. Soc. Nephrol. 17 (4), 1035—1043.

Haussler, M.R., Whitfield, G.K., et al., 2013. Molecular mechanisms of vitamin D action. Calcif. Tissue Int. 92 (2), 77—98.

Hayden, K.E., Sandle, L.N., et al., 2015. Ethnicity and social deprivation contribute to vitamin D deficiency in an urban UK population. J. Steroid Biochem. Mol. Biol. 148, 253—255.

Hoenderop, J.G.J., Nilius, B., et al., 2005. Calcium absorption across epithelia. Physiol. Rev. 85 (1), 373—422.

Hollifield, J.W., 1986. Thiazide treatment of hypertension: effects of thiazide diuretics on serum potassium, magnesium, and ventricular ectopy. Am. J. Med. 80 (4, Suppl. 1), 8—12.

Huang, C.-L., Kuo, E., 2007. Mechanism of hypokalemia in magnesium deficiency. J. Am. Soc. Nephrol. 18 (10), 2649—2652.

Iseri, L.T., French, J.H., 1984. Magnesium: nature's physiologic calcium blocker. Am. Heart J. 108 (1), 188—193.

Jakobsen, J., Knuthsen, P., 2014. Stability of vitamin D in foodstuffs during cooking. Food Chem. 148, 170—175.

Jones, G., 2008. Pharmacokinetics of vitamin D toxicity. Am. J. Clin. Nutr. 88 (2), 582S—586S.

Jones, G., Strugnell, S.A., et al., 1998. Current understanding of the molecular actions of vitamin D. Physiol. Rev. 78 (4), 1193—1231.

Kaneko, I., Saini, R.K., et al., 2015. FGF23 gene regulation by 1,25-dihydroxyvitamin D: opposing effects in adipocytes and osteocytes. J. Endocrinol. 226 (3), 155—166.

Karbach, U., Rummel, W., 1990. Cellular and paracellular magnesium transport across the terminal ileum of the rat and its interaction with the calcium transport. Gastroenterology 98 (4), 985—992.

King, D.E., Mainous, A.G., et al., 2005. Dietary magnesium and C-reactive protein levels. J. Am. Coll. Nutr. 24 (3), 166—171.

Kurosu, H., Yamamoto, M., et al., 2005. Suppression of aging in mice by the Hormone Klotho. Science 309 (5742), 1829—1833.

Lewin, E., Olgaard, K., 2015. The vascular secret of Klotho. Kidney Int. 87 (6), 1089—1091.

Musso, C.G., 2009. Magnesium metabolism in health and disease. Int. Urol. Nephrol. 41 (2), 357—362.

Nijenhuis, T., Vallon, V., et al., 2005. Enhanced passive Ca(2+) reabsorption and reduced Mg(2+) channel abundance explains thiazide-induced hypocalciuria and hypomagnesemia. J. Clin. Invest. 115 (6), 1651—1658.

O'Neill, W.C., 2007. The fallacy of the calcium-phosphorus product. Kidney Int. 72 (7), 792—796.

Patel, C., Douard, V., et al., 2015. Fructose-induced increases in expression of intestinal fructolytic and gluconeogenic genes are regulated by GLUT5 and KHK. Am. J. Physiol. Regul. Integr. Comp. Physiol. 309 (5), R499—R509.

Pham, P.-C.T., Pham, P.-M.T., et al., 2007. Hypomagnesemia in patients with type 2 diabetes. Clin. J. Am. Soc. Nephrol. 2 (2), 366—373.

Quamme, G.A., 1981. Effect of furosemide on calcium and magnesium transport in the rat nephron. Am. J. Physiol. Renal Physiol. 241 (4), F340—F347.

Ranade, V.V., Somberg, J.C., 2001. Bioavailability and pharmacokinetics of magnesium after administration of magnesium salts to humans. Am. J. Ther. 8 (5), 345—357.

Reddy, M.B., Cook, J.D., 1991. Assessment of dietary determinants of nonheme-iron absorption in humans and rats. Am. J. Clin. Nutr. 54 (4), 723—728.

Rivlin, R.S., 1994. Magnesium deficiency and alcohol intake: mechanisms, clinical significance and possible relation to cancer development (a review). J. Am. Coll. Nutr. 13 (5), 416—423.

Ross, A.C., Manson, J.E., et al., 2011. The 2011 report on dietary reference intakes for calcium and vitamin D from the Institute of Medicine: what clinicians need to know. J. Clin. Endocrinol. Metab. 96 (1), 53—58.

Rubinacci, A., 2009. Expanding the functional spectrum of vitamin K in bone. Focus on: "Vitamin K promotes mineralization, osteoblast to osteocyte transition, and an anti-catabolic phenotype by {gamma}-carboxylation-dependent and -independent mechanisms". Am. J. Physiol. Cell Physiol. 297 (6), C1336—1338.

Salimi, M.H., Heughebaert, J.C., et al., 1985. Crystal growth of calcium phosphates in the presence of magnesium ions. Langmuir. 1 (1), 119–122.

Sarafidis, P.A., Georgianos, P.I., et al., 2010. Diuretics in clinical practice. Part II: electrolyte and acid-base disorders complicating diuretic therapy. Expert Opin. Drug Saf. 9 (2), 259–273.

Sharma, A., Trivedi, S., 2015. Evaluation of in vitro probiotic potential of phytase-producing bacterial strain as a new probiotic candidate. Int. J. Food Sci. Technol. 50 (2), 507–514.

Shearer, M.J., Newman, P., 2014. Recent trends in the metabolism and cell biology of vitamin K with special reference to vitamin K cycling and MK-4 biosynthesis. J. Lipid Res. 55 (3), 345–362.

Sherman, R.A., Mehta, O., 2009. Phosphorus and potassium content of enhanced meat and poultry products: implications for patients who receive dialysis. Clin. J. Am. Soc. Nephrol. 4 (8), 1370–1373.

Shuto, E., Taketani, Y., et al., 2009. Dietary phosphorus acutely impairs endothelial function. J. Am. Soc. Nephrol. 20 (7), 1504–1512.

Stein, J., Stier, C., et al., 2014. Review article: the nutritional and pharmacological consequences of obesity surgery. Aliment. Pharmacol. Ther. 40 (6), 582–609.

Takayanagi, K., Shimizu, T., et al., 2015. Downregulation of transient receptor potential M6 channels as a cause of hypermagnesiuric hypomagnesemia in obese type 2 diabetic rats. Am. J. Physiol. Renal Physiol. 308 (12), F1386–F1397.

Tharabenjasin, P., Douard, V., et al., 2014. Acute interactions between intestinal sugar and calcium transport in vitro. Am. J. Physiol. Gastrointest. Liver Physiol. 306 (1), G1–G12.

Torres, P.A.U, Brauwere, D.P.D, 2011. Three feedback loops precisely regulating serum phosphate concentration. Kidney Int. 80 (5), 443–445.

Trapani, V., Wolf, F.I., 2015. Mitochondrial magnesium to the rescue. Magnes. Res. 28 (2), 79–84.

Uhlig, K., Berns, J.S., et al., 2010. KDOQI US commentary on the 2009 KDIGO clinical practice guideline for the diagnosis, evaluation, and treatment of CKD–mineral and bone disorder (CKD-MBD). Am. J. Kidney Dis. 55 (5), 773–799.

Urakawa, I., Yamazaki, Y., et al., 2006. Klotho converts canonical FGF receptor into a specific receptor for FGF23. Nature. 444 (7120), 770–774.

van Angelen, A.A., van der Kemp, A.W., et al., 2012. Increased expression of renal TRPM6 compensates for Mg2+ wasting during furosemide treatment. Clin. Kidney J. 5 (6), 535–544.

Essentials IV: Diet, Microbial Diversity, and Gut Integrity

EIV.1 MICROBIAL DIVERSITY, FUNCTION, AND DIET

Microbial abundance, as colony-forming units/gram of luminal content, increases distally from the jejunum ($\sim10^4$) to the ileum ($\sim10^7$). Some Gram-negative aerobes and obligate anaerobes predominate in the small intestine while in the colon, with $\sim10^{12}$ colony-forming units/g, anaerobes predominate and amount to approximately 50–60% of the fecal mass (O'Hara and Shanahan, 2006). In general, gut bacteria belong to three major bacterial phyla: the Gram-positive *Firmicutes* and *Actinobacteria* and the Gram-negative *Bacteroidetes* (with over 20 genera). The *Firmicutes* is the largest bacterial phylum, comprising over 200 genera, including *Lactobacillus*, *Mycoplasma*, *Bacillus*, and *Clostridium* species. Microbial patterns appear to be transferred from the mother during the perinatal period and are unique to the individual throughout life. However, microbial patterns can be modulated by environmental factors, especially diet (He et al., 2015) as discussed below. Microbial genomes frequently undergo genetic rearrangements, duplications, and lateral gene transfers, allowing them flexibility to adapt to environmental conditions. Furthermore, while microbial inhabitants play key roles in optimal intestinal and organism-wide functions, the fact remains that these same commensal microbes can be pathogens if they or their toxins translocate the intestinal epithelial cell (IEC) barrier and enter the systemic circulation.

The diversity of gut microbial species is known to be altered in obesity (Manco et al., 2010). Altered diversity not only changes the specificity of nutrient harvest but contributes to microbial translocation, metabolic endotoxemia, chronic inflammation, and comorbidities characteristic of the obese state (Neves et al., 2013; Bischoff et al., 2014). Thus the intact obese patient can have metabolic endotoxemia with two to fourfold more circulating bacterial products such as lipopolysaccharide (LPS) in the bloodstream. LPS is known to increase risk for cardiometabolic complications following injury, however data on the influence of obesity-associated metabolic endotoxemia on the response to injury are scarce. Studies have suggested that recent infection and "leaky gut" are implicated in risk for subsequent ischemic stroke (Grau et al., 1995). These observations have been confirmed in animal models administered LPS prior to transient cerebral artery occlusion. Not only was infarct volume larger, but long-term prognosis was worsened with LPS administration (Doll et al., 2015). While it is possible that preexisting endotoxemia could trigger more severe hyperinflammation with trauma, it is also possible that an organism chronically exposed to metabolic endotoxemia could develop immune tolerance and reduced inflammatory response to injury. Clinical trials are needed to resolve this question. In the interim, a brief review of the selected diet components that reduce bacterial translocation is presented below.

EIV.1.1 Prebiotics and Microbiota Diversity

Prebiotics are defined as dietary components that modify the growth and diversity of existing microbiota, whereas probiotics are living bacterial species given to augment specific microbial species in the gut (Gibson and Roberfroid, 1995). Yogurt and other fermented foods contain living bacteria and have been used without significant adverse effects for hundreds of years. Early studies reported that oligosaccharides such as inulin increase the abundance of beneficial *Bifidobacteria* (Gibson, 1999). More recent work suggests that adults consuming a diet rich in plant-based foods and a low consumption of meat have a highly diverse microbiota with greater abundance of *Prevotella* compared with *Bacteroides*. This ratio is reversed when a diet containing a low proportion of plant-based foods is consumed. Other factors, including consumption of highly milled grains and processed foods, frequent use of antibiotics, and reliance on disinfectants and other drugs can modify the diversity of the microbiome (Greiner et al., 2014). Changes in microbial numbers and diversity have been measured both by metagenomic analysis in humanized gnotobiotic mice (Turnbaugh et al., 2009) and in human fecal samples (David et al., 2014).

EIV.1.2 Biochemical Functions of the Microbiota

The multiple functions of the normal human microbiota have been reviewed (Vaziri et al., 2016). A major function of the microbiota is to sequentially degrade substances that cannot be digested and absorbed in the upper intestine. This includes not only plant-based polysaccharides, but endogenously produced mucin. *Bacteroides* secrete enzymes that degrade glycans, as well as linear or branched homo- or heteropolymers of monosaccharide residues. Other bacteria ferment mono- and oligosaccharides, producing hydrogen, carbon dioxide, ethanol, lactate, and the short-chain fatty acids (SCFA), acetate, propionate, and butyrate; SCFA provide an energy source for colonocytes as well as other barrier cells including regulatory T lymphocytes (refer to Chapter 3: Type I Diabetes and Celiac Disease: Evidence for Targeted Nutrient Modification in Type I DM).

Yet other microbial functions include production of nutrients that are essential for the host. Production of the B vitamins, biotin, riboflavin, niacin, and thiamin (Burkholder and McVeigh, 1942) and other vitamins such as vitamin K (Bentley and Meganathan, 1982) and folate (Sugahara et al., 2015) have been demonstrated.

It has also been revealed that the microbiota has the capacity to produce amino acids from diverse nitrogen sources, utilizing energy generated from dietary and endogenous carbohydrates as reviewed by Metges (2000). Microbial actions on amino acids include deamination, decarboxylation, and release of hydrogen sulfide (H_2S) from sulfur-containing amino acids (methionine, cysteine) (Fig. EIV.1).

While the unique microbiome of an individual begins in early life and is influenced by maternal transfer (Dominguez-Bello et al., 2010) as well as by infant feeding practices (Cabrera-Rubio et al., 2012), prebiotic consumption throughout life can positively modify microbial diversity. The extensive experimental data generated to describe microbial response to environmental factors including specific nutrient and nonnutrient components of the diet and the influence of these factors on immunity throughout life has been reviewed by Lagier et al. (2012) and Jain and Walker (2015).

EIV.1.3 Antibiotics and Antibacterial Agents

Throughout this text the importance of a diverse and metabolically active microbiota, not only in the gut, but in the lungs, skin, and other mucosal surfaces, has been stressed. This is especially true in light of our growing understanding of the role of bacterial metabolites on the maintenance of immune modulation and prevention of

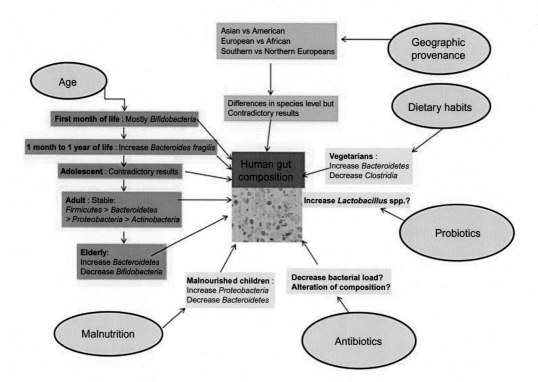

FIGURE EIV.1 **Multiple influences on microbial composition** (Lagier et al., 2012). *From Frontiers in Cellular and Infection Microbiology (open-access article distributed under the terms of the Creative Commons Attribution License).*

autoimmune disease such as type I diabetes as discussed in Chapter 3, Type I Diabetes and Celiac Disease. Antibiotics decimate the microbiota, allowing overgrowth of pathogenic strains and reducing the beneficial effects on the innate and adaptive immune systems.

Similarly, the widespread use of antimicrobial agents such as triclosan in our immediate environment can modify human commensal microbe populations. Triclosan is approved by the Food and Drug Administration (FDA) (http://www.fda.gov/ForConsumers/ConsumerUpdates/ucm205999.htm) and evaluation of its safety is ongoing. It can be found in products such as clothing, kitchenware, furniture, and toys. It also may be added to antibacterial soaps and body washes, toothpastes, and some cosmetics. Its use in pesticides is being evaluated by the FDA and the Environmental Protection Agency. Despite its ubiquitous use, it is known to have potential to cause dysbiosis (altered microbiota) and to impair thyroid function. A recent examination of human samples from the NHANES study indicated that the urinary level of triclosan was positively associated with an increased body-mass index (Lankester et al., 2013). The widespread contamination of the water supply with triclosan and similar chemicals has resulted in detectable levels in the serum of individuals of all ages. The authors review the multiple biological effects of this product and their deleterious implications on the microbiota and other organ systems (Yueh and Tukey, 2016).

EIV.2 INFLUENCE OF DIET ON GUT BARRIER FUNCTION

The intestinal epithelial cell (IEC) barrier consists of a monolayer of cells reinforced by desmosomes attached to keratin filaments in the cytoplasm. The monolayer is linked by tight junctions that limit paracellular transport of luminal content, adherens junctions that provide mechanical attachment and gap junctions that regulate communication between adjacent cells. The molecular structure of the barrier is described in Chapter 2, The Obese Gunshot Patient: Injury and Septic Shock. Several dietary components have been implicated in regulation of the gut barrier as described below.

EIV.2.1 Diet Components That Compromise Gut Barrier Function

EIV.2.1.1 Grasses Containing Gluten, Wheat Germ Agglutinin, and Antigenic Proteins

The major protein in wheat, rye, barley, and other grasses of the genus *Triticum* is gluten, comprised of roughly equal parts of gliadin and glutenin peptides. Gluten can be fractioned into the ethanol-soluble prolamines and ethanol-insoluble glutenins. While soluble prolamines have been more extensively studied, there is evidence that glutenins can also damage the intestine. A common feature of the prolamines of wheat is a high content of glutamine (>30%) and proline (>15%), whereas the nontoxic prolamines of rice and corn have a lower glutamine and proline content (Schuppan, 2000). Prolamines of genetically related rye (secalins) and barley (hordeins) also have glutamine- and proline-rich toxic peptide sequences. In oats, the prolamines (avenins) have a different amino acid profile and appear less toxic.

Upon ingestion, glutamine and proline-rich gliadin peptides are resistant to intestinal digestion and rapidly deaminated by tissue transglutaminase (Sollid, 2002). The resultant glutamic acid-rich fragments are especially efficient as T-cell antigens. Genetically susceptible celiac patients develop a well-studied immune reaction to gluten (refer to Chapter 3: Type I Diabetes and Celiac Disease) in which gliadin peptides bind to the chemokine receptor, CXCR3, on enterocytes (Lammers et al., 2008), leading to myeloid differentiation primary response, 88(MyD88)-dependent zonulin release and increased intestinal permeability. As discussed below, wheat and cereal proteins may also act on tight junctions of nonceliac patients. Gliadin action on tight junctions was demonstrated in vitro in cultured Caco-2 cells and IEC-6 monolayers as well as ex vivo in small intestinal biopsies from both celiac and nonceliac patients (Drago et al., 2006; Hollon et al., 2015). Recently, peripheral blood mononuclear cells from celiac disease (CD) patients responded to a pepsin digest of the gliadin wheat fraction with robust IL-1α and IL-1β production, suggesting that toll-like receptors (TLR2 and TLR4), as well as inflammasome activation, may be involved in the toxic effect of wheat on the CD patient. Non-CD patients were not studied (Palová-Jelínková et al., 2013).

"Wheat," pictured in Fig. EIV.2, actually describes a diverse array of cultivated species and genotypes of the *Triticum* genus. The coding regions for wheat proteins are polymorphic, thus each genotype produces unique proteins as comprehensively reviewed by Kucek et al. (2015). The proteins (glutens, albumins, and globulins) and the indigestible fructose polymers (fructans) in wheat are not efficiently cleaved by digestive enzymes and the intact proteins are more likely to stimulate an immune reaction. Wheat strains differ widely in their immunoreactivity. Selected varieties of ancient grains, including spelt, heritage (varieties produced before 1950) as well as some strains of modern wheat can produce fewer reactive prolamins and fructans, but no wheat varieties are safe for consumption

FIGURE EIV.2 **Wheat and grasses.** Wheat and other grasses contain proteins with potential to increase intestinal permeability. While gluten specifically affects patients with celiac disease, other components of wheat and cereal grains are being studied as agents that trigger symptoms in nonceliacgluten-sensitive individuals (Catassi et al., 2013). *From Bluemoose (own work) (GFDL (http://www.gnu.org/copyleft/fdl.html), CC-BY-SA-3.0 (http://creativecommons.org/licenses/by-sa/3.0/) or CC BY-SA 2.5-2.0-1.0 (http://creativecommons.org/licenses/by-sa/2.5-2.0-1.0)), via Wikimedia Commons.*

by patients diagnosed with CD. Specifically, the D genome in wheat is associated with celiac epitope expression. This genome is widespread because it also codes for the high molecular weight (HMW) proteins that are essential for bread-making quality.

Since different strains of wheat as well as other grains contain diverse proteins and can also be potential antigens, it is likely that insensitivity to grasses and grains is more common than currently described. The concept of nonceliac gluten sensitivity (NCGS) has been extensively reviewed (Fasano, 2011; Fasano et al., 2015). In general, although the popular gluten-free diet (GFD) is currently at "fad" proportions, these authors acknowledge that there is increasing and indisputable evidence that NCGS exists among certain segments of the population. The prevalence of NCGS has been estimated to be at least sixfold higher (~6%) than that of CD. To identify potential crossreactive components in a GFD that could cause patient symptoms, the reactivity of affinity-purified polyclonal and monoclonal α-gliadin 33-mer peptide antibodies against food items in the GFD was tested. Significant reactivity was seen with cow's milk, milk chocolate, milk butyrophilin, whey protein, casein, yeast, oats, corn, millet, instant coffee, and rice as measured using ELISA and dot blot (Vojdani and Tarash, 2013).

These studies suggest that chronic exposure to wheat and/or grain peptides has potential to increase intestinal permeability even in individuals without evidence of CD. The proinflammatory chemokine, CXCL10, decreased the transepithelial resistance of monolayers of normal colonocytes (NCM 460) by diminishing the mRNA expression of cadherin-1 (CDH1) and tight junction protein 2 (TJP2) (Valerii et al., 2015). In an ex vivo study, the authors exposed peripheral blood monocytes (PBM) from patients with NCGS to wheat proteins. CXCL10 secretion from PBM was demonstrated in response to wheat proteins; the extent of secretion depended on the cereal source from which the protein was obtained.

EIV.2.1.2 Lectins Such as Wheat Germ Agglutinin

Lectins are present in a variety of plants, especially in seeds, where they serve as defense mechanisms against other plants and fungi. Lectins are not degraded by heat and tend to have high resistance to digestive proteolysis. They bind to carbohydrate moieties on IECs where they modulate intestinal permeability and metabolic functions (Vasconcelos and Oliveira, 2004). Lectins cross the gut–epithelial barrier both by transcytosis and paracytosis, bind to virtually all cell types, and can cause damage to several organs. Lectin activity is currently being studied in a wide variety of cereal grains and legumes (de Punder and Pruimboom, 2013).

EIV.2.1.3 Chemicals Additives That Compromise the Gut Barrier

The FDA in the United States is charged with assuring that chemicals added to food are safe for consumption. However, a recent review of the FDA searchable database has revealed that of the more than 10,000 chemicals allowed to be added to foods, published feeding studies are available for fewer than 38% of FDA-regulated additives and 93% lack reproductive or developmental toxicity data (Neltner et al., 2013). While a large number of these additives have been classified as GRAS (Generally Regarded as Safe), few studies were conducted to examine the impact of these chemicals on the innate IEC barrier. Recently, several animal studies have called into question the impact of these ubiquitous additives on the innate IEC barrier.

Maltodextrin (MDX), a polysaccharide created from corn and other starches through chemical and enzymatic processing, is commonly added to processed foods, cosmetics, and medications as a filler, thickener, texturizer, or coating agent. Mice were given MDX in water at concentrations (55.5 g/L) similar to its concentration in infant formula. After 2 weeks of feeding, a dramatic breakdown of the antimicrobial mucous layer separating gut bacteria from the intestinal epithelium surface was visualized (Nickerson et al., 2014). Enhanced infiltration of commensal bacteria was seen in colonic sections by fluorescent in situ hybridization using a universal eubacteria probe.

Emulsifiers, polysorbate 80 (P80) and carboxymethylcellulose (CMC) are commonly used to prevent fat separation in foods such as salad dressings, ice cream, and mayonnaise. Concentrations of P80 (1% w/v) and CMC (2% w/v) are permitted in human foods. Wildtype (WT) mice and two engineered murine strains (ES) prone to developing shifts in microbiota composition and inflammation were given emulsifiers at approved human concentrations for 12 weeks. Low-grade obesity with metabolic syndrome and low-grade inflammation were observed in WT mice, while ES mice developed robust intestinal inflammation. Confocal microscopy, using mucus-preserving Carnoy fixation, indicated that in WT mice the closest bacteria resided, on average, about 25 μm from epithelial cells with no bacteria observed within 10 μm. The mucus layer was reduced by more than 50% by emulsifier treatment, and a twofold increase in bacterial contact with the epithelium was observed. Since mucus biosynthesis was not changed, these observations suggest that addition of emulsifiers caused degradation of the mucus barrier (Chassaing et al., 2015).

EIV.2.1.4 Iron

Despite its essential nature, iron exerts toxic effects on biological organisms, in part by generating oxidative damage (refer to Essentials II: Heavy Metals, Retinoids, and Precursors: Iron. Also refer to Chapter 2, The Obese Gunshot Patient: Injury and Septic Shock: Micronutrients: Pathogens and Response to Injury). Iron supplementation increased tight junction permeability in human Caco-2 cell monolayers incubated with ferrous iron salts; monolayer permeability was reversed when iron was removed from the medium. At higher iron concentrations, monolayer apoptosis and necrosis were observed at 24 hours (Ferruzza et al., 2003). Increased pathogen growth associated with epithelial damage (Kortman et al., 2012) as well as decreased transepithelial resistance and increased apical–basolateral mannitol flux (Hansen et al., 2010) were also observed in calves fed high dietary iron.

EIV.2.2 Diet Exposures That Reduce Gut Permeability

EIV.2.2.1 Flavonoids and Bioactive Components in Food

The flavonoid family consists of >4000 natural products containing a 15-carbon skeleton comprised of two phenyl rings and a central heterocyclic ring. Derivatives include reduction of the 2(3) carbon–carbon double bond (flavanones), reduction of the keto group (flavanols), and hydroxylation at various positions. Flavonoids such as isoflavones, flavonolignan, lignans, stilbenoids,

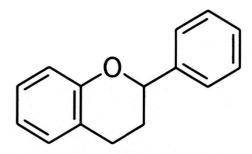

FIGURE EIV.3 **Structure of the flavan backbone (3,4-dihydro-2-phenyl-2H-1-benzopyran).** Flavonoids are categorized according to the saturation level and opening of the central pyran ring as anthocyanins, flavonols, flavanones, flavones, flavan-3-ols, isoflavones, and several other related structures. In nature, flavonoids occur either in free or conjugated forms. In plants they are mainly present as glycosides linked to one or more sugar moieties, but some flavonoids are present as aglycones (Huang et al., 2009). *From Calvero (self-made with ChemDraw) (Public domain), https://commons.wikimedia.org/w/index.php?curid=1891350/, via Wikimedia Commons.*

curcuminoids, and hydrolysable tannins are characterized by a 15-carbon flavan structure in their skeleton which contains both aromatic (A and B Rings) and heterocyclic (C Ring) rings (Fig. EIV.3), and are widely distributed in plant materials, especially fruits, vegetables, and cereals. (The Agricultural Research Service maintains a searchable Ethnobotanical and Phytochemical Database at http://www.ars.usda.gov/services/docs.htm?docid=8875. Information is available on the chemical concentrations in plants, the activities of the chemical, and on medicinal foods.)

Flavonoids undergo significant structural modification following ingestion as demonstrated in a rat in situ perfusion model (Crespy et al., 2003). Flavonoids are variably absorbed by the human enterocyte. Net transfer across the brush border ranged from 78% (kaempferol) to 35% (catechin); transfer rates of bacterial metabolic products of flavonoids have not been well studied. Absorbed intermediate forms can be conjugated, usually in the liver, thus the presence of conjugated forms provided evidence of a robust, but highly varied enterohepatic circulation.

Interactions between flavonoids and the gastrointestinal microbiota can be extensive. However, while diet components were shown to play a major role in human microbial number and diversity (David et al., 2014), other investigators found only minor differences between microbiota of omnivores and vegans with respect to their ability to metabolize dietary isoflavones to plasma equol (Wu et al., 2016). Diverse metabolic pathways as well as endogenous and exogenous factors that may account for the wide variability in the bioavailability of flavonoids have recently been reviewed (Guo and Bruno, 2015). A wide range of biological effects specific to obesity and metabolic syndrome have been demonstrated for polyphenols and other bioactive compounds in the food supply. One of the flavonoids, quercetin, has been studied intensively for its ability to activate mitogen activated protein kinase and adenosine monophosphate-activated kinase signaling pathways (Miles et al., 2014; Nabavi et al., 2015).

Flavonoids are lipophilic and have a tendency to accumulate in biological membranes, especially in lipid rafts where they interact with signal receptors and transducers, modifying their function via lipid-phase actions. Flavonoids form complexes with metals that appear to function as molecular "fasteners" that penetrate into the hydrophobic sites of membranes or proteins and initiate their adhesion and aggregation. These processes may participate in cell–cell interaction and protein cluster formation and may provide protection against infection and disease. By binding transition metals such as iron and copper, flavonoid complexes also have antioxidant potential; their lipophilic nature may facilitate protection of membrane lipids as reviewed by Tarahovsky et al. (2014).

Several studies have examined the impact of flavonoids on gut permeability; most trials were conducted in vitro using Caco-2, T84, and IEC-6 model systems. (The Caco-2 and T84 cell lines are derived from human colon carcinoma. The cells form a polarized monolayer, well-defined brush border on the apical surface and intercellular junctions similar to the normal colonocyte monolayer. IEC6 cells are derived from normal rat small intestine, are able to synthesize fibronectin and collagen, and express IEC surface antigens normally found in vivo.) While cell culture models are widely used to test the integrity of the IEC monolayer, it must be noted that they do not test possible influences of the microbiota, mucus layer, and the immune system. In general, cell culture and animal studies have indicated that dietary flavonoids found in fruits, vegetables, coffee, tea, and other sources exert regulatory and positive effects on expression and surface location of tight junction proteins (Romier et al., 2009; Ulluwishewa et al., 2011; Noda et al., 2012). Bioavailable flavonoids also facilitate progression to the antiinflammatory phase following injury or infection as recently reviewed by Ribeiro et al. (2015). Preliminary evidence suggests that flavonoids reduce activation of complement; thereby reducing inflammatory cell adhesion to the vascular endothelium, modulate the production of inflammatory mediators including eicosanoids, cytokines, and chemokines, and downregulate inflammatory proteins (iNOS) and cascades (NFκB). Although many of these observations have been made in cell culture and experimental models, there is growing interest in using specific flavonoids, metabolites, and/or synthetic analogs as

therapeutic modalities to treat human disease (Biasutto and Zoratti, 2014; Devi et al., 2015).

EIV.2.2.2 Zinc

Zinc deficiency is known to result in epithelial barrier leak in the glycemic index tract. Precise effects of zinc on epithelial tight junctions (TJs) are only beginning to be described and understood. Along with nutritional regimens like methionine restriction and compounds such as berberine, quercetin, indole, glutamine, and rapamycin, zinc has the potential to function as a TJ modifier and selective enhancer of epithelial barrier function. The impact of zinc supplementation on the TJ in Caco-2 cells revealed that 50 and 100 μM of zinc (control 2 μM) increased transepithelial electrical resistance but simultaneously increased paracellular leak to D-mannitol (Carrasco-Pozo et al., 2013). Since this study was done in vitro, the implications of zinc availability on intestinal permeability in the intact organism remain unclear.

EIV.2.2.3 Vitamin A (Retinol)

Vitamin A is fundamental in maintaining the integrity of epithelia, as reviewed by Villamor and Fawzi (2005). Early studies demonstrated that vitamin A deficiency resulted in reduced numbers of mucus-secreting goblet cells in rat intestine (Rojanapo et al., 1980). A later study detected only mild villus atrophy and reduced disaccharidase activities in vitamin A-depleted rats despite serum and liver retinol depletion. Further work is required to characterize the impact of vitamin A on intestinal permeability.

EIV.2.3 Diet Exposures That Enhance Antimicrobial Protection

Intestinal and intestinal-associated immune cells are constantly exposed to microbes and secrete antimicrobial peptides (AMPs) to defend the barrier. AMPs disrupt the integrity of bacterial, viral, and fungal membranes, bind LPSs, and exert chemotactic signals for immune cells. They also modulate the expression of cytokines and chemokines. As recently reviewed (Campbell et al., 2012), the human cathelicidin antimicrobial peptide (CAMP) is secreted in semen, saliva, and sweat, where it also provides barrier protection. The CAMP gene is also expressed in myeloid bone marrow cells, and the proprotein, hCAP18, is packaged in neutrophil-specific granules. The hCAP18/LL37 peptide binds to and neutralizes LPS, thus preventing its interaction with the LPS-binding protein and subsequent activation of TLR-4, NFκB signaling and cytokine release from host cells. LL-37 protects against LPS-induced sepsis in animals. Other AMPs such as defensins have also been characterized.

EIV.2.3.1 Vitamins A, D, and Histone Deacetylase Inhibitors

Active vitamin D (1,25D) induces expression of the human CAMP gene; this induction is specific to humans and nonhuman primates. Activation of the TLR up-regulated expression of the vitamin D receptor and the vitamin D-1-hydroxylase genes in the human macrophages, leading to CAMP induction (Liu et al., 2006). Histone deacetylase inhibitors from dietary intake (sulforaphane) and from bacterial metabolism (butyrate) as well as retinoic acid, regulate AMP production to enhance barrier function. It is likely that the impact of these nutrients on the microbiota can explain, in part, early observations that vitamin A, carotene, and other nutrient supplements enhanced mucosal immunity, especially in malnourished children as reviewed by Villamor and Fawzi (2005) and Hall et al. (2011).

EIV.2.3.2 Mushrooms and Fungi

Preliminary, evidence has been presented that mushrooms modify gut microbiota, enhancing adaptive immunity and improving immune cell function, as reviewed by Feeney et al. (2014). Mushrooms added to a purified diet increased bacterial diversity, increasing *Bacteroidetes* and decreasing *Firmicutes* phyla in mice. Intestinal injury and infection were resolved more rapidly in mushroom-fed mice. The authors proposed that the mechanisms through which mushrooms protect the gut include activating TLR (AMP secretion was not measured); it is also possible that they function through modulating the adaptive immune response. Mushrooms contain polysaccharides which could be metabolized by commensal bacteria and thereby improve integrity of the intestinal enterocyte barrier function. In view of the fact that commercially grown mushrooms are a major source of vitamin D (refer to Essentials III: Nutrients for Bone Structure and Calcification: Vitamin D), well-designed research to more clearly characterize mechanisms is needed.

References

Bentley, R., Meganathan, R., 1982. Biosynthesis of vitamin K (menaquinone) in bacteria. Microbiol. Rev. 46 (3), 241–280.

Biasutto, L., Zoratti, M., 2014. Prodrugs of quercetin and resveratrol: a strategy under development. Curr. Drug Metab. 15 (1), 77–95.

Bischoff, S., Barbara, G., et al., 2014. Intestinal permeability—a new target for disease prevention and therapy. BMC Gastroenterol. 14 (1), 189.

Burkholder, P.R., McVeigh, I., 1942. Synthesis of vitamins by intestinal bacteria. Proc. Natl. Acad. Sci. 28 (7), 285–289.

Cabrera-Rubio, R., Collado, M.C., et al., 2012. The human milk microbiome changes over lactation and is shaped by maternal weight and mode of delivery. Am. J. Clin. Nutr. 96 (3), 544–551.

Campbell, Y., Fantacone, M., et al., 2012. Regulation of antimicrobial peptide gene expression by nutrients and by-products of microbial metabolism. Eur. J. Nutr. 51 (8), 899–907.

Carrasco-Pozo, C., Morales, P., et al., 2013. Polyphenols protect the epithelial barrier function of caco-2 cells exposed to indomethacin through the modulation of occludin and zonula occludens-1 expression. J. Agr. Food Chem. 61 (22), 5291–5297.

Catassi, C., Bai, J.C., et al., 2013. Non-celiac gluten sensitivity: the new frontier of gluten related disorders. Nutrients 5 (10), 3839–3853.

Chassaing, B., Koren, O., et al., 2016. Dietary emulsifiers impact the mouse gut microbiota promoting colitis and metabolic syndrome. Nature. (advance online publication). Available from: http://dx.doi.org/10.1038/nature18000, Published online 04 May 2016.

Crespy, V., Morand, C., et al., 2003. The splanchnic metabolism of flavonoids highly differed according to the nature of the compound. Am. J. Physiol. Gastrointest. Liver Physiol. 284 (5), G980–G988.

David, L.A., Maurice, C.F., et al., 2014. Diet rapidly and reproducibly alters the human gut microbiome. Nature 505 (7484), 559–563.

de Punder, K., Pruimboom, L., 2013. The dietary intake of wheat and other cereal grains and their role in inflammation. Nutrients. 5 (3), 771–787.

Devi, K.P., Malar, D.S., et al., 2015. Kaempferol and inflammation: from chemistry to medicine. Pharmacol. Res. 99, 1–10.

Doll, D.N., Engler-Chiurazzi, E.B., et al., 2015. Lipopolysaccharide exacerbates infarct size and results in worsened post-stroke behavioral outcomes. Behav. Brain Funct. 11 (1), 1–9.

Dominguez-Bello, M.G., Costello, E.K., et al., 2010. Delivery mode shapes the acquisition and structure of the initial microbiota across multiple body habitats in newborns. Proc. Natl. Acad. Sci. 107 (26), 11971–11975.

Drago, S., El Asmar, R., et al., 2006. Gliadin, zonulin and gut permeability: effects on celiac and non-celiac intestinal mucosa and intestinal cell lines. Scand. J. Gastroenterol. 41 (4), 408–419.

Fasano, A., 2011. Zonulin and its regulation of intestinal barrier function: the biological door to inflammation, autoimmunity, and cancer. Physiol. Rev. 91 (1), 151–175.

Fasano, A., Sapone, A., et al., 2015. Nonceliac gluten sensitivity. Gastroenterology 148 (6), 1195–1204.

Feeney, M.J., Dwyer, J., et al., 2014. Mushrooms and health summit proceedings. J. Nutr. 144 (7), 1128S–1136S.

Ferruzza, S., Scarino, M., et al., 2003. Biphasic effect of iron on human intestinal Caco-2 cells: early effect on tight junction permeability with delayed onset of oxidative cytotoxic damage. Cell. Mol. Biol. Paris-Wegmann 49 (1), 89–100.

Gibson, G.R., 1999. Dietary modulation of the human gut microflora using the prebiotics oligofructose and inulin. J. Nutr. 129, 1438S–1441S.

Gibson, G.R., Roberfroid, M.B., 1995. Dietary modulation of the human colonic microbiota: introducing the concept of prebiotics. J. Nutr. 125 (6), 1401–1412.

Grau, A.J., Buggle, F., et al., 1995. Recent infection as a risk factor for cerebrovascular ischemia. Stroke 26 (3), 373–379.

Greiner, A.K., Papineni, R.V.L., et al., 2014. Chemoprevention in gastrointestinal physiology and disease. Natural products and microbiome. Am. J. Physiol. Gastrointest. Liver Physiol. 307 (1), G1–G15.

Guo, Y., Bruno, R.S., 2015. Endogenous and exogenous mediators of quercetin bioavailability. J. Nutr. Biochem. 26 (3), 201–210.

Hall, J.A., Grainger, J.R., et al., 2011. The role of retinoic acid in tolerance and immunity. Immunity 35 (1), 13–22.

Hansen, S., Ashwell, M., et al., 2010. High dietary iron reduces transporters involved in iron and manganese metabolism and increases intestinal permeability in calves. J. Dairy Sci. 93 (2), 656–665.

He, B., Nohara, K., et al., 2015. Transmissible microbial and metabolomic remodeling by soluble dietary fiber improves metabolic homeostasis. Sci. Rep. 5, 10604.

Hollon, J., Puppa, E., et al., 2015. Effect of gliadin on permeability of intestinal biopsy explants from celiac disease patients and patients with non-celiac gluten sensitivity. Nutrients 7 (3), 1565.

Huang, W.-Y., Cai, Y.-Z., et al., 2009. Natural phenolic compounds from medicinal herbs and dietary plants: potential use for cancer prevention. Nutr. Cancer 62 (1), 1–20.

Jain, N., Walker, W.A., 2015. Diet and host-microbial crosstalk in postnatal intestinal immune homeostasis. Nat. Rev. Gastroenterol. Hepatol. 12 (1), 14–25.

Kortman, G.A.M., Boleij, A., et al., 2012. Iron availability increases the pathogenic potential of *Salmonella Typhimurium* and other enteric pathogens at the intestinal epithelial interface. PLoS ONE 7 (1), e29968.

Kucek, L.K., Veenstra, L.D., et al., 2015. A grounded guide to gluten: how modern genotypes and processing impact wheat sensitivity. Comp. Rev. Food Sci. Food Saf. 14 (3), 285–302.

Lagier, J.-C., Million, M., et al., 2012. Human gut microbiota: repertoire and variations. Front. Cell. Inf. Microbiol. 2, 136.

Lammers, K.M., Lu, R., et al., 2008. Gliadin induces an increase in intestinal permeability and zonulin release by binding to the chemokine receptor CXCR3. Gastroenterology. 135 (1), 194–204, e193.

Lankester, J., Patel, C., et al., 2013. Urinary triclosan is associated with elevated body mass index in NHANES. PLoS ONE 8 (11), e80057.

Liu, P.T., Stenger, S., et al., 2006. Toll-like receptor triggering of a vitamin D-mediated human antimicrobial response. Science 311 (5768), 1770–1773.

Manco, M., Putignani, L., et al., 2010. Gut microbiota, lipopolysaccharides, and innate immunity in the pathogenesis of obesity and cardiovascular risk. Endocr. Rev. 31 (6), 817–844.

Metges, C.C., 2000. Contribution of microbial amino acids to amino acid homeostasis of the host. J. Nutr. 130 (7), 1857S–1864S.

Miles, S.L., McFarland, M., et al., 2014. Molecular and physiological actions of quercetin: need for clinical trials to assess its benefits in human disease. Nutr. Rev. 72 (11), 720–734.

Nabavi, S.F., Russo, G.L., et al., 2015. Role of quercetin as an alternative for obesity treatment: you are what you eat! Food Chem. 179 (0), 305–310.

Neltner, T.G., Alger, H.M., et al., 2013. Data gaps in toxicity testing of chemicals allowed in food in the United States. Reproduct. Toxicol. 42 (0), 85–94.

Neves, A.L., Coelho, J., et al., 2013. Metabolic endotoxemia: a molecular link between obesity and cardiovascular risk. J. Mol. Endocrinol. 51 (2), R51–R64.

Nickerson, K.P., Homer, C.R., et al., 2014. The dietary polysaccharide maltodextrin promotes *Salmonella* survival and mucosal colonization in mice. PLoS ONE 9 (7), e101789.

Noda, S., Tanabe, S., et al., 2012. Differential effects of flavonoids on barrier integrity in human intestinal Caco-2 cells. J. Agr. Food Chem. 60 (18), 4628–4633.

O'Hara, A.M., Shanahan, F., 2006. The gut flora as a forgotten organ. EMBO Rep. 7 (7), 688–693.

Palová-Jelínková, L., Dáňová, K., et al., 2013. Pepsin digest of wheat gliadin fraction increases production of IL-1β via TLR4/MyD88/TRIF/MAPK/NF-κB signaling pathway and an NLRP3 inflammasome activation. PLoS ONE 8 (4), e62426.

Ribeiro, D., Freitas, M., et al., 2015. Proinflammatory pathways: the modulation by flavonoids. Med. Res. Rev. 35 (5), 877–936.

Rojanapo, W., Lamb, A.J., et al., 1980. The prevalence, metabolism and migration of goblet cells in rat intestine following the induction of rapid, synchronous vitamin a deficiency. J. Nutr. 110 (1), 178–188.

Romier, B., Schneider, Y.-J., et al., 2009. Dietary polyphenols can modulate the intestinal inflammatory response. Nutr. Rev. 67 (7), 363–378.

Schuppan, D., 2000. Current concepts of celiac disease pathogenesis. Gastroenterology 119 (1), 234–242.

Sollid, L.M., 2002. Coeliac disease: dissecting a complex inflammatory disorder. Nat. Rev. Immunol. 2 (9), 647–655.

Sugahara, H., Odamaki, T., et al., 2015. Differences in folate production by bifidobacteria of different origins. Biosci. Microb. Food Health 34 (4), 87–93.

Tarahovsky, Y.S., Kim, Y.A., et al., 2014. Flavonoid–membrane interactions: involvement of flavonoid–metal complexes in raft signaling. BBA Biomemb. 1838 (5), 1235–1246.

Turnbaugh, P.J., Ridaura, V.K., et al., 2009. The effect of diet on the human gut microbiome: a metagenomic analysis in humanized gnotobiotic mice. Sci. Translat. Med. 1 (6), 6ra14.

Ulluwishewa, D., Anderson, R.C., et al., 2011. Regulation of tight junction permeability by intestinal bacteria and dietary components. J. Nutr. 141 (5), 769–776.

Valerii, M.C., Ricci, C., et al., 2015. Responses of peripheral blood mononucleated cells from non-celiac gluten sensitive patients to various cereal sources. Food Chem. 176 (0), 167–174.

Vasconcelos, I.M., Oliveira, J.T.A., 2004. Antinutritional properties of plant lectins. Toxicon 44 (4), 385–403.

Vaziri, N.D., Zhao, Y.-Y., et al., 2016. Altered intestinal microbial flora and impaired epithelial barrier structure and function in CKD: the nature, mechanisms, consequences and potential treatment. Nephrol. Dial. Transplant. 31, 737–746.

Villamor, E., Fawzi, W.W., 2005. Effects of vitamin A supplementation on immune responses and correlation with clinical outcomes. Clin. Microbiol. Rev. 18 (3), 446–464.

Vojdani, A., Tarash, I., 2013. Cross-reaction between gliadin and different food and tissue antigens. Food Nutr. Sci. 04 (01), 13.

Wu, G.D., Compher, C., et al., 2016. Comparative metabolomics in vegans and omnivores reveal constraints on diet-dependent gut microbiota metabolite production. Gut 65 (1), 63–72.

Yueh, M.-F., Tukey, R.H., 2016. Triclosan: a widespread environmental toxicant with many biological effects. Ann. Rev. Pharmacol. Toxicol. 56 (1), 251–272.

Essentials V: Nutrition Support in Critically Ill Patients[1,2]

The Society of Critical Care Medicine and the American Society for Parenteral and Enteral Nutrition have prepared guidelines that offer basic recommendations for nutrition support. These guidelines are supported by review and analysis of the pertinent available current literature, by other national and international guidelines, and by the blend of expert opinion and clinical practicality. The "intensive care unit" (ICU) or "critically ill" patient is not a homogeneous population. Many of the studies on which the guidelines are based are limited by sample size, patient heterogeneity, variability in definition of disease state and severity of illness, lack of baseline nutrition status, and lack of statistical power for analysis. Whenever possible, these factors are taken into account and the grade of statement will reflect the power of the data. One of the major methodological problems with any guideline is defining the exact population to be included.

The strength of the evidence is indicated by the grade of each recommendation. Grading criteria are

- Grade A   Supported by at least two level I investigations
- Grade B   Supported by one level I investigation
- Grade C   Supported by level II investigations only
- Grade D   Supported by at least two level III investigations
- Grade E   Supported by level IV or level V evidence

Level of evidence

Level I	Large, randomized trials with clear-cut results; low risk of false positives or negatives
Level II	Small, randomized trials with uncertain results; moderate risk of false positives or negatives
Level III	Nonrandomized, contemporaneous controls
Level IV	Nonrandomized, historical controls
Level V	Case series, uncontrolled studies, and expert opinion

EV.1 NUTRITION SUPPORT RECOMMENDATIONS: 2015

Because guidelines evolve based on clinical experience, the most recent points have been summarized below by Rosenthal et al. (2015):

- Enteral nutrition (EN) should be provided to critically ill patients for preservation of gut mucosa and improved immune system function.
  - Early EN should begin immediately following resuscitation;
  - Usually within 24−48 hours of arrival in the ICU.
- EN directly into the small bowel should be attempted in mechanically ventilated patients at risk for aspiration.
- Patient's risk for nosocomial infection should determine type of feeding selected.
  - Patients at high risk for infection should receive an immune-enhancing diet;
  - If low risk, patients should be started on a standard or concentrated polymeric formula;
  - If intolerance occurs, soluble fiber or peptide-based formula should be considered;
  - Supplementation with antioxidants (vitamins A, C, and E) should be considered, as severe depletion is seen in critically ill patients;
  - Adequate phosphate should be given, as hypophosphatemic patients have prolonged ventilator dependency.

Detailed feeding strategies abstracted from SCCM/ASPEN guidelines (McClave et al., 2009).

---

[1]From Guidelines for the Provision and Assessment of Nutrition Support Therapy in the Adult Critically Ill Patient. Canadian Guidelines 2015 (http://www.criticalcarenutrition.com/docs/CPGs%202015/Summary%20CPGs%202015%20vs%202013.pdf).

[2]Note that anthropometric measurements have also not been found to accurately reflect nutritional status or to monitor response to feeding. One obvious confounder is the degree of edema in the tissues measured for skinfold and circumferences.

EV.2 STRATEGIES TO INITIATE ENTERAL FEEDING

1. *Assess patient's nutrition status (grade E):*

   Because traditional assessment tools (albumin, prealbumin, and anthropometry) can be altered by the acute-phase response and other metabolic alterations due to the injury, assessment should include:
   - Evaluation of weight loss and previous nutrient intake (prior to admission);
   - Level of disease severity and comorbid conditions;
   - Function of the glycemic index (GI) tract.

2. *Assess patient's ability to maintain volitional intake (grade C):*

   "Standard therapy (STD)" refers to the food ingested under the patient's own volition. "Nutrition support therapy" (also referred to as "specialized" or "artificial" nutrition therapy) refers to the provision of enteral tube feeding (EN) or parenteral nutrition (PN).
   - Early EN is essential to maintain gut integrity, modulate cellular stress, and maintain systemic immune response, thereby attenuating disease severity.
     - Gut permeability changes are time-dependent. Channels open within hours of injury.
   - EN maintains gut functional integrity by preserving tight junctions between intraepithelial cells, stimulating blood flow, and inducing secretion of trophic endogenous agents (eg, cholecystokinin, gastrin, bombesin, and bile salts).
   - EN also maintains structural integrity by maintaining villous height and supporting the mass of secretory IgA-producing immunocytes which can comprise the gut-associated lymphoid tissue (GALT) and in turn contribute to mucosal-associated lymphoid tissue (MALT) at distant sites such as the lungs, liver, and kidneys.
   - Consequences of increased gut permeability include increased bacterial challenge (engagement of GALT with enteric organisms), risk for systemic infection, and greater likelihood of multiorgan dysfunction syndrome (MODS).
   - As disease severity worsens, increases in gut permeability are amplified. Under these conditions, EN is more likely to favorably impact outcome parameters of infection, organ failure, and hospital length of stay (compared to PN).
   - Further justification for early EN therapy includes delivery of immune-modulating agents and use of enteral formulations as an effective means for stress ulcer prophylaxis.

3. *Assess patient's ability to tolerate EN (grade B):*

   EN is the preferred route of feeding over PN for the critically ill patient who requires nutrition support therapy.
   - The beneficial effects of EN versus PN are well documented in many prospective randomized controlled trials in patients with critical trauma, burns, head injury, major surgery, and acute pancreatitis.
     - The most consistent outcome using EN is a reduction in infectious morbidity (usually central line infections and pneumonia).
     - Further benefits include reductions in hospital length of stay and cost of nutrition therapy. In head injury cases, more complete return of cognitive function has been reported.

4. *Start EN early (grade C): enteral feeding should be started early within the first 24–48 hours following admission. The feedings should be advanced toward goal over the next 48–72 hours (grade E).*
   - Attaining access and initiating EN should be considered as soon as fluid resuscitation is completed and the patient is hemodynamically stable.
     - A "window of opportunity" exists in the first 24–72 hours following admission or the onset of a hypermetabolic insult.
     - Feedings started within this time frame (compared to feedings started after 72 hours) are associated with less gut permeability, diminished activation, and release of inflammatory cytokines (ie, tumor necrosis factor and reduced systemic endotoxemia).

5. *Delay EN in hemodynamically compromised patients (grade E):*

   If patient requires significant hemodynamic support (high-dose catecholamine agents, alone or in combination with large-volume fluid or blood product resuscitation to maintain cellular perfusion), EN should be withheld until the patient is fully resuscitated and/or stable.
   - If EN is provided to patients who are at the height of critical illness and prone to GI dysmotility, sepsis, and hypotension, putting them at increased risk for subclinical ischemia/reperfusion bowel injury.
     - Absorption of intraluminal nutrients increases enterocyte energy demand, putting the intestine at risk for ischemia, particularly in the presence of splanchnic hypoperfusion. It is generally concluded that a significant mismatch of oxygen demand and supply clearly plays a major role in the development of nonocclusive mesenteric ischemia.
     - Ischemic bowel is a rare complication of EN (occurring in <1% of cases).

- EN-related ischemic bowel has been reported most often in the past with use of surgical jejunostomy tubes. However, more recently, this complication has been described with the use of nasojejunal tubes.
- EN intended to be infused into the small bowel should be withheld in patients who are hypotensive (mean arterial blood pressure <60 mmHg), particularly if clinicians are initiating use of catecholamine agents (eg, norepinephrine, phenylephrine, epinephrine, dopamine) or escalating the dose of such agents to maintain hemodynamic stability.
  - EN may be provided with caution to patients into either the stomach or small bowel on stable low doses of pressor agents.
  - Any signs of intolerance (abdominal distention, increasing nasogastric tube output or gastric residual volumes, decreased passage of stool and flatus, hypoactive bowel sounds, increasing metabolic acidosis and/or base deficit) should be closely scrutinized as possible early signs of gut ischemia.

Note: Improved mortality in hemodynamically compromised patients given early enteral feedings has been reported (Khalid et al., 2010). The authors noted that animal studies have demonstrated that although GI blood flow is reduced in critical illness and remains depressed despite fluid replacement and normalization of blood pressure and cardiac output, enteral feeding increases blood flow. These authors suggest that early EN in humans actually improves the oxygen delivery:consumption ratio.

6. *Bowel sounds need not be present (grade B)*: In the ICU patient population, the literature does not support a requirement of bowel sounds or evidence of passage of flatus and stool for the initiation of enteral feeding.
   - GI dysfunction in the ICU setting occurs in 30—70% of patients depending on the diagnosis, premorbid condition, ventilation mode, medications, and metabolic state.
     - Proposed mechanisms for GI dysfunction in the ICU or postoperatively include three general categories: (1) mucosal barrier disruption, (2) altered motility and atrophy of the mucosa, and (3) reduced mass of GALT.
   - Bowel sounds are only indicative of contractility and do not necessarily relate to mucosal integrity, barrier function, or absorptive capacity.
     - Success at attaining nutrition goals within the first 72 hours ranges from 30% to 85%.
     - When ICU enteral feeding protocols are followed, rates of GI tolerance in the range of 70—85% can be achieved.

7. *Either gastric or small bowel feeding is acceptable in the ICU setting (grade C): withholding of enteral feeding for repeated high gastric residual volumes alone may be sufficient reason to switch to small bowel feeding (the definition for high gastric residual volume is likely to vary from one hospital to the next, as determined by individual institutional protocol) (grade: E).* (Multiple studies have evaluated gastric vs jejunal feeding in various medical and surgical ICU settings. Several studies showed reduced aspiration and reflux when the level of infusion was moved from the stomach down through the intestine. No significant difference in mortality was reported between gastric and postpyloric feeding.)
   - An enteral access tube placed in the small bowel reduces risk for aspiration or intolerance to gastric feeding. (The definition for high gastric residual volume varies from one hospital to the next (determined by individual institutional protocol). See guidelines below for recommendations on gastric residual volumes, identifying high-risk patients, and reducing chances for aspiration.)

EV.3 WHEN TO USE PN

1. *In a previously healthy individual, if early EN is not feasible or available the first 7 days following admission to the ICU, no nutrition support therapy (ie, use of STD therapy only) should be provided. (grade C).*
2. *In the patient who was previously healthy prior to critical illness with no evidence of protein-calorie malnutrition, use of PN should be reserved and initiated only after the first 7 days of hospitalization (when EN is not available) (grade E).* (Note that evidence for the first statement (Grade C) is stronger than for the second statement (Grade E))
   - These two recommendations are the most controversial in the guidelines as reviewed by McClave et al. (2009). Studies comparing use of PN with STD therapy in the absence of EN, revealed that use of STD therapy was associated with significantly reduced infectious morbidity.
   - As the duration of illness increased, SDT therapy became less effective, and PN was associated with lower mortality and shorter hospital stays.
     - The length of time on STD therapy before starting PN is not clear. McClave and the Guidelines Committee expressed concern that providing STD therapy beyond 7 days would lead to a deterioration of nutrition status and have adverse effects on the clinical outcome.

3. *If there is evidence of protein-calorie malnutrition on admission and EN is not feasible. For these patients, when EN is not available, there should be little delay in initiating PN after admission to the ICU and adequate resuscitation (grade C).*
   * PCM is defined by a recent weight loss of >10–15% or actual body weight <90% of ideal body weight (IBW).
4. *If a patient is expected to undergo major upper GI surgery and EN is not feasible, PN should be provided under very specific conditions:*
   * *If the patient is malnourished, PN should be initiated 5–7 days preoperatively and continued into the postoperative period (grade B).*
     * PN should not be initiated in the immediate postoperative period but should be delayed for 5–7 days (should EN continue not to be feasible).
   * *PN therapy provided for a duration of <5–7 days would be expected to have no outcome effect and may result in increased risk to the patient. Thus, PN should be initiated only if the duration of therapy is anticipated to be ≥7 days (grade B).*
     * Patients undergoing major upper GI surgery (esophagectomy, gastrectomy, pancreatectomy, or other major reoperative abdominal procedures), especially with preexisting protein-calorie malnutrition and the PN provided under specific conditions showed consistent benefit of PN over STD therapy.
     * Whereas critically ill patients in some analyses had increased mortality using PN versus STD therapy, surgical patients experienced no treatment-induced mortality.
   * *It is imperative to be aware that the beneficial effect of PN is lost if given only postoperatively (grade B).*
     * An aggregate of multiple studies showed a significant 10% increase in complications with PN compared to STD therapy.
     * Because of the adverse outcome effect from PN initiated in the immediate postoperative period, Klein et al. recommended delaying PN for 5–10 days following surgery if EN continues not to be feasible.

EV.4 STRATEGIES FOR ENTERAL FEEDING

1. *The target goal of EN (defined by energy requirements) should be determined and clearly identified at the time of initiation of nutrition support therapy. Energy requirements may be calculated by predictive equations or measured by indirect calorimetry (grade C).*
   * Over 200 predictive equations (including Harris–Benedict, Scholfield, Ireton–Jones, etc.) have been published in the literature.

* *Predictive equations should be used with caution. They provide a less accurate measure of energy requirements than indirect calorimetry in the individual patient (grade E).*
  * In the obese patient, the predictive equations are even more problematic without availability of indirect calorimetry. (See section on Energy requirements for the obese patient, below. The level of the recommendation is grade E.)
* Energy requirements may be calculated either through simplistic formulas (25–30 kcal/kg/d), published predictive equations, or the use of indirect calorimetry.
  * Calories provided via infusion of propofol should be considered when calculating the nutrition regimen. (Propofol (Diprivan) is marketed by AstraZeneca as an emulsion. This product contains 1% propofol, 10% soybean oil, and 1.2% purified egg phospholipid (emulsifier), with 2.25% of glycerol as a tonicity-adjusting agent, and sodium hydroxide to adjust the pH.)
* EN provision that approaches goal nutrient requirements has shown improved outcome.
  * Studies (McClave et al., 2009) in which patients received a greater volume of EN had fewer complications including infections, shorter hospital stays, and a trend to lower mortality than patients receiving inadequate EN volume.
2. *Greater than 50–65% of goal calories are recommended in order to achieve the clinical benefit of EN over the first week of hospitalization (grade C).*
   * Studies cited (McClave et al., 2009) show that >50–60% (versus 37–40%) of goal calories prevented increased intestinal permeability in burn and bone-marrow transplant patients, promoted faster return of cognitive function in head injury patients, and improved outcome from immune-modulating enteral formulations.
   * Evidence suggests that the impact of early EN on patient outcome is dose-dependent. Although small feedings (trickle or trophic feeds defined as 10–30 mL/hour) do appear to prevent mucosal atrophy, these small volumes do not appear to be sufficient to achieve optimal endpoints.
3. *If unable to meet energy requirements (100% of target goal calories) after 7–10 days by the enteral route alone, consider initiating supplemental PN (grade E). Initiating supplemental PN prior to this 7–10-day period in the patient already receiving EN does not*

improve outcome and may be detrimental to the patient (grade: C).

- EN is used early on to maintain gut integrity, reduce oxidative stress, and modulate systemic immunity. If the patient is receiving some EN, adding PN in the first 7–10 days adds cost and does not appear to provide benefit. In one small study, EN supplemented with PN resulted in increased mortality compared with EN alone.

4. *The adequacy of protein provision must be assessed throughout. In patients with body mass index (BMI) <30, protein requirements should be in the range of 1.2–2.0 g/kg actual body weight per day, and may likely be even higher in burn or multitrauma patients (grade E).*

- In the critical care setting, *protein appears to be the most important macronutrient* for healing wounds, supporting immune function, and maintaining lean body mass.
  - For most critically ill patients, protein requirements are proportionately higher than energy requirements and therefore are not met by provision of routine enteral formulations.
  - The decision to add protein modules should be based on an ongoing assessment of adequacy of protein provision.
- Determination of protein requirements is difficult but may be derived (with limitations) from nitrogen balance, simplistic equations (1.2–2.0 g/kg/day) or nonprotein calorie: nitrogen ratio (70:1–100:1).
  - The use of additional modular protein supplements is a common practice, as standard enteral formulations tend to have a high nonprotein calorie:nitrogen ratio.
- *Serum protein markers (albumin, prealbumin, transferrin, C-reactive protein) are not validated for determining adequacy of protein provision and should not be used in the critical care setting in this manner.*

EV.5 RECOMMENDATIONS FOR THE OBESE PATIENT

Severe obesity adversely increases risk of comorbidities (eg, insulin resistance, sepsis, infections, deep venous thrombosis, and organ failure) in the intensive care unit. A moderate degree of weight loss will result in increased insulin sensitivity, improved nursing care, and reduced risk of comorbidities. If the patient is provided with 60–70% of caloric requirements, steady weight loss results. At the same time, protein should be infused at a dose of 2.0–2.5 g/kg IBW/day to approximate protein requirements and neutral nitrogen balance, allowing for adequate wound healing.

- A retrospective study by Choban and Dickerson (2005) indicated that provision of protein at a dose of 2.0 g/kg IBW/day is insufficient for achieving neutral nitrogen balance when the BMI is >40. Use of BMI and IBW is recommended over use of adjusted body weight. (Data indicated that a minimum protein intake of approximately 1.7 g/kg IBW/day would be necessary to achieve nitrogen equilibrium for those with classes I and II obesity, but that a minimum of ~1.8 g/kg IBW/day would be needed for reaching nitrogen equilibrium in morbidly obese (class III) patients. If the patient is critically ill, obese, and hypercatabolic, increased protein may be needed to reduce net protein catabolism and potentially achieve nitrogen equilibrium (~1.9 g/kg IBW/day in patients with class I or II obesity and ~2.5 g/kg IBW/day or more may be necessary to achieve nitrogen equilibrium for those with class III obesity)).

In the critically ill obese patient, permissive underfeeding or hypocaloric feeding with EN is recommended (grade D).

- For all classes of obesity where BMI is >30, the goal of the EN regimen should not exceed 60–70% of target energy requirements or 11–14 kcal/kg actual body weight per day (or 22–25 kcal/kg IBW/day).
  - Determining energy requirements is discussed in guideline C1 (grade D).
- Protein should be provided in a range ≥2.0 g/kg IBW/day for class I and II patients (BMI 30–40), ≥2.5 g/kg IBW/day for class III (BMI ≥ 40).

Choban and Dickerson noted that predictive equations for energy requirements based on normal populations, such as the Harris–Benedict equations, have been shown to underestimate the resting energy expenditure of obese individuals when IBW is used and overestimate energy expenditure when actual body weight is used in the equations. A review of the utility of predictive equations for obese patients has been published (Stucky et al., 2008):

- These authors equations developed for estimating needs (IJEE) that predicted energy expenditure accurately in hospitalized patients including the obese patient:

$$IJEEv = 1784 - 11(A) + 5(W) + 244(S) + 239(T) + 804(B)$$

$$IJEEs = 629 - 11(A) + 25(W) - 609(O)$$

where IJEE, kcal/day; v, ventilator dependent; s, spontaneously breathing; A, age (years); W, wt (kg); S, sex ($m = 1$, $f = 0$) and T, trauma; B, burn; O, obesity (present = 1, absent = 0).

The use of ideal versus actual body weight was examined and it was determined that the use of actual body weight was preferable. They did note, however, the correlation between measured energy expenditure and actual body weight was lost in patients with a BMI >60 kg/m². These data indicate that as the degree of obesity and the severity of concomitant illness increase, it becomes progressively more difficult to estimate energy needs by any method.

EV.6 MONITORING TOLERANCE AND ADEQUACY OF EN

1. *In the ICU setting, evidence of bowel motility (resolution of clinical ileus) is not required in order to initiate EN in the ICU (grade E).*
   - EN is safe before overt evidence of enteric function, including bowel sounds and passage of flatus and stool, is present. EN promotes gut motility, and as long as the patient remains hemodynamically stable, it is safe and appropriate to feed through mild to moderate ileus.
   - Patients should be monitored for tolerance of EN (determined by patient complaints of pain and/or distention, physical exam, passage of flatus and stool, abdominal radiographs) (grade E).
2. *Inappropriate cessation of EN should be avoided (grade E).*
3. *Holding EN for gastric residual volumes < 500 mL in the absence of other signs of intolerance should be avoided (grade B).*
4. *The time period that a patient is made nil per os (NPO) prior to, during, and immediately following the time of diagnostic tests or procedures should be minimized to prevent inadequate delivery of nutrients and prolonged periods of ileus. Ileus may be propagated by NPO status (grade C).*
   - Healthcare providers who prescribe nutrition formulations tend to underorder calories, and patients only receive approximately 80% of what is ordered.
   - This combination of underordering and inadequate delivery results in patients receiving only 50% of target goal calories from 1 day to the next.
   - Cessation of feeding occurs in >85% of patients for an average of 20% of the infusion time (the reasons for which are avoidable in >65% of occasions).
   - Patient intolerance accounts for one-third of cessation time, but only half of this represents true intolerance. Other reasons for cessation include remaining NPO after midnight for

diagnostic tests and procedures in another third of patients, with the rest being accounted for by elevated gastric residual volumes and tube displacement.
   - In one level II study, patients randomized to continue EN during frequent surgical procedures (burn wound debridement under general anesthesia) had significantly fewer infections than those patients for whom EN was stopped for each procedure.
   - Gastric residual volumes do not correlate well to incidence of pneumonia, measures of gastric emptying, or to incidence of regurgitation and aspiration.
     - Four level II studies indicated that raising the cutoff value for gastric residual volume (leading to automatic cessation of EN) from a lower number of 50−150 mL to a higher number of 250−500 mL does not increase risk for regurgitation, aspiration, or pneumonia.
     - Decreasing the cutoff value for gastric residual volume does not protect the patient from these complications, often leads to inappropriate cessation, and may adversely affect outcome through reduced volume of EN infused.
     - Gastric residual volumes in the range of 200−500 mL should raise concern and lead to the implementation of measures to reduce risk of aspiration, but automatic cessation of feeding should not occur for gastric residual volumes <500 mL in the absence of other signs of intolerance.
5. *Use of enteral feeding protocols increases the overall percentage of goal calories provided and should be implemented (grade C).*
   - ICU or nurse-driven protocols which define goal infusion rate, designate more rapid startups, and provide specific orders for handling gastric residual volumes, frequency of flushes, and conditions or problems under which feeding may be adjusted or stopped, have been shown to be successful in increasing the overall percentage of goal calories provided.
6. *Patients placed on EN should be assessed for risk of aspiration (grade E). Steps to reduce risk of aspiration should be employed (grade E).*
   - Aspiration is one of the most feared complications of EN. Factors that have been shown to identify patients at increased risk for aspiration include: use of a nasoenteric tube, an endotracheal tube and mechanical ventilation, age >70 years, reduced level of consciousness, poor nursing care, location in the hospital, patient position, transport out of the ICU, poor oral health, and use of bolus intermittent feedings.

- Pneumonia and bacterial colonization of the upper respiratory tree are more closely associated with aspiration of contaminated oropharyngeal secretions than regurgitation and aspiration of contaminated gastric contents.
- The following measures have been shown to reduce risk of aspiration:
  - *In all intubated ICU patients receiving EN, the head of the bed should be elevated 30–45°. (grade C).* (McClave et al., 2009 cited a paper comparing supine to semirecombent patient position. The paper reported that elevating the head of the bed 30–45° reduced the incidence of pneumonia from 23% to 5% ($P = 0.018$)).
  - *For high-risk patients or those shown to be intolerant to gastric feeding, delivery of EN should be switched to continuous infusion (grade D).*
  - *Agents to promote motility such as prokinetic drugs (metoclopramide and erythromycin) or narcotic antagonists (naloxone and alvimopan) should be initiated where clinically feasible (grade C).* (McClave et al., 2009 showed a table on the effects of adding prokinetic agents such as erythromycin or metoclopramide. The agents improved gastric emptying and tolerance of EN, but had little effect on the clinical outcome for ICU patients. Naloxone infused through the feeding tube (to reverse the effects of opioid narcotics at the level of the gut and improve intestinal motility) was shown in one level II study to significantly increase the volume of EN infused, reduce gastric residual volumes, and decrease the incidence of ventilator-associated pneumonia (compared to placebo)).
  - *Diverting the level of feeding by postpyloric tube placement should be considered (grade C).*
  - *Use of chlorhexidine mouthwash twice a day should be considered to reduce risk of ventilator-associated pneumonia (grade C).* (Chlorhexidine mouthwashes twice daily were shown in two studies to reduce respiratory infection and nosocomial pneumonia in patients undergoing heart surgery. In contrast, studies evaluating use of chlorhexidine in general ICU populations have shown little outcome effect.)
- Several methods may be used to reduce the risk of aspiration. Changing the level of infusion of EN from the stomach to the small bowel has been shown to reduce the incidence of regurgitation and aspiration, although the results from three meta-analyses suggest that any effect in reducing pneumonia is minimal.
- Other steps to decrease aspiration risk would include reducing the level of sedation/analgesia when possible, minimizing transport out of the

ICU for diagnostic tests and procedures, and moving the patient to a unit with a lower patient: nurse ratio.

7. *Blue food coloring and glucose oxidase strips, as surrogate markers for aspiration, should not be used in the critical care setting (grade E).*
   - Blue food coloring is an insensitive marker for aspiration, and was shown to be associated with mitochondrial toxicity and patient death. (The United States Food and Drug Administration through a Health Advisory Bulletin (Sep. 2003) issued a mandate against the use of blue food coloring as a monitor for aspiration in patients on EN.)
   - The basic premise for use of glucose oxidase (that glucose content in tracheal secretions is solely related to aspiration of glucose-containing formulation) has been shown to be invalid, and its use is thwarted by poor sensitivity/specificity characteristics (Singh et al., 2015).
8. *Development of diarrhea associated with enteral tube feedings warrants further evaluation for etiology (grade E).*
   - Diarrhea in the ICU patient receiving EN should prompt an investigation for excessive intake of hyperosmolar medications, such as sorbitol, use of broad-spectrum antibiotics, *Clostridium difficile* pseudomembranous colitis, or other infectious etiologies. Most episodes of nosocomial diarrhea are mild and self-limiting.
   - Assessment should include an abdominal exam, fecal leukocytes, quantification of stool, stool culture for *Clostridium difficile* (and/or toxin assay), serum electrolyte panel (to evaluate for excessive electrolyte losses or dehydration), and review of medications. An attempt should be made to distinguish infectious diarrhea from osmotic diarrhea.

Note that bowel necrosis has been reported at the site of postoperative placement of the jejunal feeding tube. The cause of mesenteric ischemia progressing to septic shock is unknown, but may relate to bacterial overgrowth or damage from hyperosmotic feeding (Melis et al., 2006).

EV.7 SELECTION OF APPROPRIATE ENTERAL FORMULATION

In selecting the appropriate enteral formulation for the critically ill patient, the clinician must first decide if the patient is a candidate for a specialty immune-modulating formulation. Patients most likely to show a favorable outcome (appropriate candidates for use of immune-modulating formulations) include those undergoing major elective GI surgery, trauma (abdominal trauma index scores >20) as described (Tinkoff et al., 2008), burns (total body surface area >30%), head and neck cancer, and critically ill patients on mechanical ventilation (who are not severely septic). (The Organ Injury Scale (OIS) was devised by the committee of the American Association for the Surgery of Trauma in 1987. The OIS was devised to categorize injury severity for individual organs. In 1988, the committee published the first OIS grading systems for spleen, liver, and kidney (OISI-1), followed by similar grading of other organ systems followed.)

1. *Immune-modulating enteral formulations (supplemented with agents such as arginine, glutamine, nucleic acid, ω-3 fatty acids, and antioxidants) should be used for the appropriate patient population (major elective surgery, trauma, burns, head and neck cancer, and critically ill patients on mechanical ventilation), with caution in patients with severe sepsis (for surgical ICU patients, grade A; for medical ICU patients, grade B.).*
   - *ICU patients not meeting criteria for immune-modulating formulations should receive standard enteral formulations (grade B).*

      Mechanisms through which immune-modulating agents act include the discovery of specialized immune (myeloid suppressor) cells, whose role is to regulate the availability of arginine, necessary for normal T lymphocyte function.
   - Myeloid suppressor cells can cause severe arginine deficiency which impact production of nitric oxide and negatively affect microcirculation.
     - Immune-modulating diets containing arginine and ω-3 fatty acids appear to overcome the regulatory effect of myeloid suppressor cells.
     - Agents such as RNA nucleotides increase total lymphocyte count, lymphocyte proliferation, and thymus function.
     - In a dynamic fashion, the ω-3 fatty acids eicosapentaenoic acid (EPA) and docosahexaenoic acid (DHA) displace ω-6 fatty acids from the cell membranes of immune cells. This effect reduces systemic inflammation through the production of alternative biologically less active prostaglandins and leukotrienes.
     - EPA and DHA (fish oils) have also been shown to downregulate expression of nuclear factor-kappa B (NFκB), intracellular adhesion molecule 1 (ICAM-1), and E-selectin, which in effect decreases neutrophil attachment and transepithelial migration to modulate systemic and local inflammation.

- In addition, EPA and DHA help to stabilize the myocardium and lower the incidence of cardiac arrhythmias, decrease incidence of acute respiratory distress syndrome (ARDS), and reduce the likelihood of sepsis.
- Glutamine, considered a conditionally essential amino acid, exerts a myriad of beneficial effects on antioxidant defenses, immune function, production of heat shock proteins, and nitrogen retention.
- Addition of agents such as selenium, ascorbic acid (vitamin C), and vitamin E provides further antioxidant protection.

     Multiple meta-analyses have shown that use of immune-modulating formulations is associated with significant reductions in duration of mechanical ventilation, infectious morbidity, and hospital length of stay compared to use of standard enteral formulations, however immune-modulation had no overall impact on mortality. The beneficial effects of these formulations appear more pronounced in surgical patients than in critically ill patients (Heyland, 2000) (Heyland et al. (cited in McClave et al., 2009) showed that the greatest beneficial effect was seen in surgery patients with significant reductions in infectious morbidity (relative risk; RR = 0.53; 95% CI 0.42−0.68; $P \leq 0.05$) and hospital length of stay (weighted mean difference; WMD = −0.76; 95% CI −1.14 to −0.37; $P < 0.05$). In contrast, aggregating the data from studies in medical ICU patients showed no effect on infections (RR = 0.96; 95% CI 0.77−1.20; P = NS) but a similar reduction in hospital length of stay (WMD = −0.47; 95% CI −0.93 to −0.01; P = 0.047)). These same five meta-analyses showed no overall impact on mortality from use of immune-modulating formulations.
- It is possible that the use of arginine-containing formulations in medical ICU patients who are severely septic could pose a risk.
  - Use of arginine-containing formulations resulted in greater mortality than standard EN and PN formulations.
  - The mechanism proposed for this adverse effect was that in severe sepsis, arginine may be converted to nitric oxide contributing to hemodynamic instability.
  - Upon review of this controversy, the Guidelines Committee felt that immune-modulating formulations containing arginine were safe enough to use in mild to moderate sepsis, but that caution should be employed if utilized in patients with severe sepsis.

Based on the strength and uniformity of the data in surgery patients, the Guidelines Committee felt that a grade A recommendation was warranted for use of these formulations in the surgical ICU. (For any patient who does not meet the criteria mentioned above, there is a decreased likelihood that use of immune-modulating formulations will change outcome. In this situation, the added cost of these specialty formulations cannot be justified and therefore standard enteral formulations should be used.) *The reduced signal strength and heterogeneity of the data in nonoperative critically ill patients in a medical ICU was felt to warrant a grade B recommendation.*

2. *Patients with ARDS and severe acute lung injury (ALI) should be placed on an enteral formulation characterized by an antiinflammatory lipid profile (ie, ω-3 fish oils, borage oil) and antioxidants (grade A).*

- Three level I studies involving patients with ARDS, ALI, and sepsis, and using an enteral formulation fortified with ω-3 fatty acids (in the form of EPA), borage oil (γ-linolenic acid (GLA)), and antioxidants significantly reduced length of stay in the ICU, duration of mechanical ventilation, organ failure, and mortality compared to use of a standard enteral formulation. Controversy remains as to the optimal dosage, makeup of fatty acids, and ratio of individual immune-modulating nutrients which comprise these formulations.

  To receive optimal therapeutic benefit from the immune-modulating formulations, at least 50–65% of goal energy requirements should be delivered (grade C).

- Because the benefit of EN in general, and specifically the added value of immune-modulating agents, appears to be a dose-dependent effect, significant differences in outcome are more likely to be seen between groups randomized to either an immune-modulating or a standard enteral formulation in those patients who receive a "sufficient" volume of feeding. These differences may not be as apparent when all patients who receive *any* volume of feeding are included in the analysis.

  If there is evidence of diarrhea, soluble fiber-containing or small peptide formulations may be utilized (grade E).

- Those patients with persistent diarrhea (in whom hyperosmolar agents and *C. difficile* have been excluded) may benefit from use of a soluble fiber-containing formulation or small peptide semielemental formulation. The laboratory data, theoretical concepts, and expert opinions would support the use of the small peptide enteral formulations but current large prospective trials are not available to make this a strong recommendation.

EV.8 ADJUNCTIVE THERAPY

EV.8.1 Probiotics

Probiotics are defined as microorganisms of human origin, which are safe, stable in the presence of gastric acid and bile salts, and when administered in adequate amounts confer a health benefit to the host. Multiple factors in the ICU induce rapid and persistent changes in the commensal microbiota, including broad-spectrum antibiotics, prophylaxis for stress gastropathy, vasoactive pressor agents, alterations in motility, and decreases in luminal nutrient delivery. Probiotic agents act by competitively inhibiting pathogenic bacterial growth, blocking epithelial attachment of invasive pathogens, eliminating pathogenic toxins, enhancing mucosal barrier, and favorably modulating the host inflammatory response. Unfortunately for the general ICU patient population, there has not been a consistent outcome benefit demonstrated. The most consistent beneficial effect from use of probiotics has been a reduction in infectious morbidity demonstrated in critically ill patients involving transplantation, major abdominal surgery, and trauma.

1. *Administration of probiotic agents has been shown to improve outcome (most consistently by decreasing infection) in specific critically ill patient populations involving transplantation, major abdominal surgery, and severe trauma (grade C).*

   No recommendation can currently be made for use of probiotics in the general ICU population due to a lack of consistent outcome effect. It appears that each species may have different effects and variable impact on patient outcome, making it difficult to make broad categorical recommendations. Similarly, no recommendation can currently be made for use of probiotics in patients with severe acute necrotizing pancreatitis, based on the disparity of evidence in the literature and the heterogeneity of the bacterial strains utilized.

- While some of these studies would warrant a grade B recommendation, the Guidelines Committee felt that the heterogeneity of the ICU populations studied, the difference in bacterial strains, and the variability in dosing necessitated a downgrade to a grade C recommendation.

- As the ease and reliability of taxonomic classification improve, stronger recommendations for use in specific populations of critically ill patients would be expected.

- Probiotics in severe acute pancreatitis are currently under scrutiny due to the results of two level II single center studies showing clinical benefit (significantly reduced infectious morbidity and hospital length of stay), followed by a larger level I multicenter study showing increased mortality in those patients receiving probiotics.

EV.8.2 Antioxidant Vitamins and Minerals

Antioxidant vitamins (including vitamins E and ascorbic acid) and trace minerals (including selenium, zinc, and copper) may improve patient outcome, especially in burns, trauma, and critical illness requiring mechanical ventilation.

2. *A combination of antioxidant vitamins and trace minerals (specifically including selenium) should be provided to all critically ill patients receiving specialized nutrition therapy (grade B).*
   * A meta-analysis aggregating data from studies evaluating various combinations of antioxidant vitamins and trace elements showed a significant reduction in mortality with their use (RR = 0.65; 95% CI 0.44−0.97; P = 0.03).
   * Parenteral selenium, the single antioxidant most likely to improve outcome, has shown a trend toward reducing mortality in patients with sepsis or septic shock (RR = 0.59; 95% CI 0.32−1.08; P = 0.08).
   * Additional studies to delineate compatibility, optimal dosage, route, and optimal combination of antioxidants are needed. Renal function should be considered when supplementing vitamins and trace elements.

EV.8.3 Amino Acids and Fiber

The addition of enteral glutamine to an EN regimen (nonglutamine supplemented) has been shown to reduce hospital and ICU length of stay in burn and mixed ICU patients, and mortality in burn patients alone compared to the same EN regimen without glutamine.

3. *The addition of enteral glutamine to an EN regimen (not already containing supplemental glutamine) should be considered in burn, trauma, and mixed ICU patients (grade B).*
   * The glutamine powder, mixed with water to a consistency which allows infusion through the feeding tube, should be given in two or three divided doses to provide 0.3−0.5 g/kg/day.
   * While glutamine given by the enteral route may not generate a sufficient systemic antioxidant effect, its favorable impact on outcome may be explained by its trophic influence on intestinal epithelium and maintenance of gut integrity.
   * Enteral glutamine should not be added to an immune-modulating formulation already containing supplemental glutamine.
4. *Soluble fiber may be beneficial for the fully resuscitated, hemodynamically stable critically ill patient receiving EN who develops diarrhea.*

5. *Insoluble fiber should be avoided in all critically ill patients. Both soluble and insoluble fiber should be avoided in patients at high risk for bowel ischemia or severe dysmotility (grade C).*
   * Three small level II studies using soluble partially hydrolyzed guar gum demonstrated a significant decrease in the incidence of diarrhea in patients receiving EN.
   * However, no differences in days of mechanical ventilation, ICU, length of stay, or MODS have been reported.
   * Insoluble fiber has not been shown to decrease the incidence of diarrhea in the ICU patient.
   * Cases of bowel obstruction in surgical and trauma patients who were provided enteral formulations containing insoluble fiber have been reported.

EV.9 WHEN INDICATED, MAXIMIZE EFFICACY OF PN

1. *If EN is not available or feasible, the need for PN therapy should be evaluated (see guidelines B1, B2, B3, C3) (grade C). If the patient is deemed to be a candidate for PN, steps to maximize efficacy (regarding dose, content, monitoring, and choice of supplemental additives) should be used (grade C).* (For these patients, a number of steps may be used to maximize the benefit or efficacy of PN while reducing its inherent risk from hyperglycemia, immune suppression, increased oxidative stress, and potential infectious morbidity. The grade of the first recommendation is based on the strength of the literature for guidelines B1−3 and C3, while that of the second is based on the supportive data for guidelines G2−6).
   * The patient is well nourished prior to admission, but after 7 days of hospitalization, EN has not been feasible or target goal calories have not been met consistently by EN alone.
   * On admission, the patient is malnourished and EN is not feasible.
   * A major surgical procedure is planned, the preoperative assessment indicates that EN is not feasible through the perioperative period, and the patient is malnourished.
2. *In all ICU patients receiving PN, mild permissive underfeeding should be considered at least initially. Once energy requirements are determined, 80% of these requirements should serve as the ultimate goal or dose of parenteral feeding (grade C).*
3. *Eventually, as the patient stabilizes, PN may be increased to meet energy requirements (grade E).*

4. *For obese patients (BMI ≥ 30), the dose of PN with regard to protein and caloric provision should follow the same recommendations given for EN guideline (grade D).*

- "Permissive underfeeding" in which the total caloric provision is determined by 80% of energy requirements (calculated from simplistic equations such as 25 kcal/kg actual body weight per day, published predictive equations, or as measured by indirect calorimetry) will optimize efficacy of PN. (In two studies, lower-dose hypocaloric PN was shown to reduce the incidence of hyperglycemia and infections, ICU and hospital length of stay, and duration of mechanical ventilation compared to higher eucaloric doses of PN.)
  - This strategy avoids the potential for insulin resistance, greater infectious morbidity, or prolonged duration of mechanical ventilation and increased hospital length of stay associated with excessive energy intake.

5. *In the first week of hospitalization in the ICU, when PN is required and EN is not feasible, patients should be given a parenteral formulation without soy-based lipids (grade D).*

- This recommendation is controversial and is supported by a single level II study (which was also included in the hypocaloric vs eucaloric dosing in guideline G2 above).
  - The recommendation is supported by animal data, with further support from EN studies, where long-chain fatty acids have been shown to be immunosuppressive.
  - Currently in North America, the choice of parenteral lipid emulsion is severely limited to a soy-based 18-carbon ω-6 fatty acid preparation (which has proinflammatory characteristics in the ICU population). Over the first 7 days, soy-based lipid-free PN has been shown to be associated with a significant reduction in infectious morbidity (pneumonia and catheter-related sepsis), decreased hospital and ICU length of stay, and shorter duration of mechanical ventilation compared to use of lipid-containing PN.

6. *A protocol should be in place to promote moderately strict control of serum glucose when providing nutrition support therapy (grade B). A serum glucose range of 110–150 mg/dL may be most appropriate (grade E).*

- Strict glucose control, keeping serum glucose levels between 80 and 110 mg/dL, has been shown in a large single center trial to be associated with reduced sepsis, reduced ICU length of stay, and lower hospital mortality when compared to conventional insulin therapy (keeping blood glucose levels <200 mg/dL).

The effect was more pronounced in surgical ICU than medical ICU patients.

- However, an as yet unpublished large level I multicenter European study suggested that moderate control (keeping glucose levels between 140 and 180 mg/dL) might avoid problems of hypoglycemia and subsequently reduce the mortality associated with hypoglycemia compared to tighter control.
- With a paucity of data, the Guidelines Committee felt that attempting to control glucose in the range of 110–150 mg/dL was most appropriate at this time.

Note: Evidence for the beneficial effect of strict glucose control in the hospitalized diabetic patient has been discussed (Clement et al., 2004). When indicated, continuous insulin infusion has been recommended:

- Furnary et al. (2003) reported that continuous insulin infusion reduced mortality in patients with diabetes with perioperative hyperglycemia following coronary artery bypass grafting.
- Van den Berghe et al. (2001) reported that intensive intravenous insulin therapy was associated with a 45% reduction in ICU mortality with a mean blood glucose of 103 mg/dL (5.7 mmol/L), as compared with the conventional treatment arm, where mean blood glucose was 153 mg/dL (8.5 mmol/L) in a mixed group of patients with and without diabetes.

More recently, the American College of Physicians published best practice advice recommending a target blood glucose level of 7.8–11.1 mmol/L (140–200 mg/dL) if insulin therapy is used in SICU/MICU patients. They cautioned that targets less than 7.8 mmol/L (<140 mg/dL) should be avoided to reduce risk for hypoglycemia and its deleterious sequelae (Qaseem et al., 2014).

7. *When PN is used in the critical care setting, consideration should be given to supplementation with parenteral glutamine (grade C). (Of note, the dipeptide form of parenteral glutamine upon which most of these data are based is widely used in Europe but not commercially available in North America (referring both to the United States and Canada). All three reports which showed a positive clinical effect were level II studies, warranting the grade C recommendation.)*

- The proposed mechanism of this benefit relates to generation of a systemic antioxidant effect, maintenance of gut integrity, induction of heat shock proteins, and use as a fuel source for rapidly replicating cells.
- The addition of parenteral glutamine (at a dose of 0.5 g/kg/day) to a PN regimen has been shown to reduce infectious complications, ICU length of

stay, and mortality in critically ill patients, compared to the same PN regimen without glutamine. (A meta-analysis by Heyland combining results from nine studies confirmed a trend toward reduced infection (RR = 0.75; 96% CI 0.54–1.04; $P = 0.08$) and a significant reduction in mortality (RR = 0.67; 95% CI 0.48–0.92; $P = 0.01$) in groups receiving PN with parenteral glutamine vs those groups getting PN alone.)

- Use of L-glutamine, the only source of parenteral glutamine available in North America, is severely limited by problems with stability and solubility (100 mL water per 2 g glutamine).

8. *In patients stabilized on PN, periodically repeated efforts should be made to initiate EN. As tolerance improves and the volume of EN calories delivered increases, the amount of PN calories supplied should be reduced. PN should not be terminated until ≥ 60% of target energy requirements are being delivered by the enteral route (grade E).*

- Because of the marked benefits of EN for the critically ill patient, repeated efforts to initiate enteral therapy should be made. To avoid the complications associated with overfeeding, the amount of calories delivered by the parenteral route should be reduced appropriately to compensate for the increase in the number of calories being delivered enterally. Once the provision of enteral feeding exceeds 60% of target energy requirements, PN may be terminated.

References

Choban, P.S., Dickerson, R.N., 2005. Morbid obesity and nutrition support: is bigger different? Nutr. Clin. Pract. 20 (4), 480–487.

Clement, S., Braithwaite, S.S., et al., 2004. Management of diabetes and hyperglycemia in hospitals. Diabetes Care. 27 (2), 553–591.

Furnary, A.P., Gao, G., et al., 2003. Continuous insulin infusion reduces mortality in patients with diabetes undergoing coronary artery bypass grafting. J. Thoracic Cardiovas. Surg. 125 (5), 1007–1021.

Heyland, D.K., 2000. Parenteral nutrition in the critically-ill patient: more harm than good?. Proc. Nutr. Soc. 59 (03), 457–466.

Khalid, I., Doshi, P., et al., 2010. Early enteral nutrition and outcomes of critically ill patients treated with vasopressors and mechanical ventilation. Am. J. Crit. Care. 19 (3), 261–268.

McClave, S.A., Martindale, R.G., et al., 2009. Guidelines for the provision and assessment of nutrition support therapy in the adult critically ill patient. J. Parenter. Enteral Nutr. 33 (3), 277–316.

Melis, M., Fichera, A., et al., 2006. Bowel necrosis associated with early jejunal tube feeding: a complication of postoperative enteral nutrition. Arch. Surg. 141 (7), 701–704.

Qaseem, A., Chou, R., et al., 2014. Inpatient glycemic control: best practice advice from the clinical guidelines committee of the American college of physicians. Am. J. Med. Qual. 29 (2), 95–98.

Rosenthal, M.D., Vanzant, E.L., et al., 2015. Evolving paradigms in the nutritional support of critically ill surgical patients. Curr. Probl. Surg. 52 (4), 147–182.

Singh, S., Suresh, S., et al., 2015. Treating every needle in the haystack: hyperammonemic encephalopathy and severe malnutrition after bariatric surgery—a case report and review of the literature. J. Parente. Enteral Nutr. 39 (8), 977–985.

Stucky, C.-C.H., Moncure, M., et al., 2008. How accurate are resting energy expenditure prediction equations in obese trauma and burn patients? J. Parente. Enteral Nutr. 32 (4), 420–426.

Tinkoff, G., Esposito, T.J., et al., 2008. American association for the surgery of trauma organ injury scale i: spleen, liver, and kidney, validation based on the National Trauma Data Bank. J. Am. Coll. Surg. 207 (5), 646–655.

Van den Berghe, G., Wouters, P., et al., 2001. Intensive insulin therapy in critically ill patients. N. Engl. J. Med. 345 (19), 1359–1367.

Index

Note: Page numbers followed by "*f*" and "*t*" refer to figures and tables, respectively.

Printed in the United States
By Bookmasters